LEGAL ASPECTS OF IMPLEMENTING THE KYOTO PROTOCOL MECHANISMS

Legal Aspects of Implementing the Kyoto Protocol Mechanisms: Making Kyoto Work

Edited by

DAVID FREESTONE

and

CHARLOTTE STRECK

OXFORD
UNIVERSITY PRESS

OXFORD
UNIVERSITY PRESS

Great Clarendon Street, Oxford OX2 6DP

Oxford University Press is a department of the University of Oxford.
It furthers the University's objective of excellence in research, scholarship,
and education by publishing worldwide in

Oxford New York

Auckland Cape Town Dar es Salaam Hong Kong Karachi
Kuala Lumpur Madrid Melbourne Mexico City Nairobi
New Delhi Shanghai Taipei Toronto

With offices in

Argentina Austria Brazil Chile Czech Republic France Greece
Guatemala Hungary Italy Japan South Korea Poland Portugal
Singapore Switzerland Thailand Turkey Ukraine Vietnam

Oxford is a registered trade mark of Oxford University Press
in the UK and in certain other countries

Published in the United States
by Oxford University Press Inc., New York

First published 2005

British Library Cataloguing in Publication Data

Library of Congress Cataloging in Publication Data

Legal aspects of implementing the Kyoto Protocol mechanism : making Kyoto work /
edited by David Freestone and Charlotte Streck.
p. cm.
Includes bibliographical references and index.
ISBN 0–19–927961–6 (hard cover : alk. paper)
1. United Nations Framework Convention on Climate Change (1992). Protocols, etc.,
1997 Dec. 11. 3. Greenhouse gases—Law and legislation. 3. Global warming—Law
and legislation. 4. Climatic changes. I. Freestone, David. II. Streck, Charlotte.
K3593.A41992Z4 2005
344.04′6342—dc22 2004023871

ISBN 0–19–927961–6

3 5 7 9 10 8 6 4 2

Typeset by Newgen Imaging Systems (P) Ltd., Chennai, India
Printed in Great Britain
on acid-free paper by
Biddles Ltd., King's Lynn, Norfolk

Foreword

Climate change is one of the most significant challenges that mankind faces in the twenty-first century. We still do not fully understand the implications of the significant changes which might take place in the delicate climate system upon which we depend. Recent studies suggest that the impact on biodiversity of minor increases in global temperatures have been grossly underestimated. Similarly, it has been suggested that relatively modest increases in ocean temperature may threaten drastic changes in major ocean currents such as the Gulf Stream. Two things however are certain: first, any changes in global climate patterns will have the most severe impacts on the poorest and most vulnerable sections of society, and second that mitigating climate change will eventually demand drastic adaptations to our existing carbon dependent economy.

While the Kyoto Protocol is a step in the right direction, the targets it sets can only be seen as a modest beginning; much more significant international efforts will be necessary even to stabilize greenhouse gas emissions. This is why the mechanisms defined under the Kyoto Protocol are such important legal instruments. The innovative market mechanisms which it introduces provide opportunities for businesses in industrialized countries to offset the reduction of greenhouse gases which they finance elsewhere in the world against their own targets. They also provide developing countries and those with economies in transition with a new income stream to finance climate friendly technology which contributes to their sustainable development. It is very exciting that the World Bank has played such a pioneering role in the emerging carbon market and has mapped a path through the technical, legal and institutional challenges which all investors in greenhouse gas reduction projects face.

This volume is a testament to the convening power of the Bank to bring together some of the foremost thinkers and practitioners in this complex field to provide an analysis of the key issues faced by market entrants. The sharing of the knowledge and understandings in this book will, I am sure, have a major significance for the implementation of greenhouse gas reduction projects all around the globe. Despite the uncertainties surrounding the coming into force of the Kyoto Protocol, the launch of national and regional trading schemes such as the European Emissions Trading Scheme means that the carbon market now has an existence independent of the vicissitudes of Kyoto. The further growth of this market promises more resources for investment in sustainable development projects and a growing impact on global emissions of greenhouse gases. I am delighted to have been invited to write the foreword for this book at this critical juncture. I congratulate the editors and contributors, and commend it to a wide readership.

Roberto Dañino
Senior Vice-President and General Counsel
The World Bank

Washington D.C.
July 2004

Preface

In the hectic final forty-eight hour session of the Protocol negotiations in Kyoto in December 1997, when the Clean Development Mechanism was born, few understood the potential power of the instrument they were creating to contribute to both sustainable development and the mitigation of climate change. Seven years on, with a wealth of experience and insights behind us, the Carbon Finance Business of the World Bank and its many and various partners see this potential clearly and feel driven to ensure that it can reach its full potential. In a nutshell, the Clean Development Mechanism (CDM) of the Kyoto Protocol is an extraordinary first step towards the creation of a market for global environmental goods. It facilitates trade in greenhouse gas emissions reductions ('carbon' in the lingua franca of the marketplace); this makes it the first global environmental externality to be subjected to a global regulatory and incentive system. As the market develops, it should lead to the costs of climate change being internalized within the price of carbon-intensive goods and services in the industrialized counties, laying a foundation of legitimacy and practice for this policy globally over the longer term.

For the developing countries the CDM is an unprecedented opportunity to attract private capital, privately held clean technology and know-how, in exchange for greenhouse gas emissions reductions that can be used by industrialized counties to reduce the costs of their compliance with obligations to meet their emissions reduction obligations under the Kyoto Protocol, and potentially also under some other regime. The fundamentals necessary to establish a market are satisfied because of the large differences in the marginal cost of abatement of greenhouse gases in industrialized and developing countries. Many developing countries have long functioned under energy pricing regimes that barely reflected resource costs, or for other reasons have been unable to attract investment in modern energy efficient and environmentally clean technologies. Hence, in these economies, the marginal cost of abating greenhouse gas emissions is measured in dollars per tonne of carbon abated, not in tens or hundreds of dollars per tonne, as in the most energy efficient of the OECD economies.

Despite the fact that the correct fundamentals for the establishment of a market exist and that the CDM is a mechanism for trade and compliance that is beguilingly simple, it has proven to be anything but easy to implement. First, the most powerful of the G77, and those with the greatest opportunity to benefit from carbon trade, have long persisted with a vision of the CDM more akin to the transfer of grant resources represented by the Global Environment Facility, than to a market mechanism. Indeed, there is no single widely shared vision of the CDM. Understandably then, the transition from the concept of a market mechanism to its implementation has been plagued by lack of shared vision and of objectives lodged within a governance structure based on compromise and caution rather than decisive guidance.

The agreement of the detailed "rules of the game" for the implementation of the CDM in Marrakesh in 2001 has irrevocably shifted its focus to the market and trade philosophy of CDM and away from north–south transfers—based on historical equity considerations—that have driven overseas development assistance. Nevertheless tensions exist between stakeholder groups over the scale and scope of CDM which can limit realization of its potential. Many NGOs and European decision-makers still wish to be assured that the CDM will not be so successful in creating low-cost high-volume carbon assets that it will remove pressure from governments and companies to take tough and expensive decisions to limit greenhouse gas emissions reductions at home. Nor is there a shared view on the activities and technologies that the CDM should support. There are stakeholders that seek to limit the types of transactions eligible under the CDM. Others want the CDM to be as flexible as possible in lowering the costs of compliance with the emission reduction obligations that the overwhelming majority of industrialized countries have voluntarily assumed so that the costs of meeting these targets will not be an intolerable drag on the competitiveness of those economies in tough economic times. Simply following the rules of the game is also no guarantee that the CDM will invariably stand both for sustainable development locally as well as climate change mitigation, as Article 12 of the Kyoto Protocol would appear to mandate.

From my standpoint, the success of the CDM is crucial to effective long-term climate change mitigation. I see the CDM as the cornerstone in an essential "global compact" between the industrialized and developing countries. Under this compact the industrialized countries will provide the incentives for the transfer of largely private, capital, technology, and know-how to developing countries to enable them to pursue 'low-carbon' development paths. In return, the developing countries will provide industrialized countries with access to lower cost emissions reductions to help them reduce the cost of compliance with increasingly tough quantitative limitations on carbon emissions, while also remaining open to a dialogue about some form of carbon-intensity targets under sustained economic and energy use growth. At their current low-energy intensity of GDP, it is not credible to talk of commitments by developing countries to low-carbon economies without assuring them of the flow of capital and technology that would support this economic transformation. And it is not realistic to imagine this happening on the required scale through any other means than market-driven trade.

Here lies the contradiction between a vision of the CDM as a facility limited to support small-scale renewable energy and efficiency measures and the need to accommodate investment on a scale that will transform the infrastructure for growth in developing countries toward prosperous low-carbon economies over the next twenty to thirty years. The CDM must support large projects that are meaningful to this economic transformation, including fuels shifts and clean energy technology between and within fossil fuels. On the other hand, carbon trade must benefit smaller poorer countries and the rural poor through small projects in riskier business environments in order to build ownership and ensure inclusion in fostering truly global stewardship of the "global commons".

A vision of the CDM as a major vehicle for carbon trade should in no way compromise its role as a catalyst for sustainable development. In line with this thinking carbon trade should demonstrably contribute to sustainable development locally. The CDM constitutes just one mechanism in a family of project-based mechanisms to serve this purpose. In addition to Joint Implementation, project-based Assigned Amount Units (AAUs) trade, or "greening of AAUs", can be used to maximize the contribution of Kyoto's flexible mechanisms to sustainable development.

In seeking to support a flourishing trade in carbon as a global environmental commodity on a scale commensurate with long-term climate change mitigation, standardization of process and documentation is essential. This book pays particular attention to the documentation that underpins the development of a "carbon asset" and enables it to be commoditized for secondary market trade. The definition of emission reductions as private rights and the development and standardization of the Emissions Reductions Purchase Agreements (ERPAs) has been an important achievement of the first years of the Bank's Carbon Finance Business. Without the standardization of terms and definitions, and contracts, it would be impossible to readily assess risk and uncertainty at the project level, and to build a foundation for private sector investment in carbon for compliance purposes. Equally important, it would be impossible for financiers to effectively value and monetize ERPAs and to use them to leverage financing that will bring projects that will achieve emissions reductions to financial closure and into production. This challenge has largely been met and this book is a milestone in the development of the carbon market.

Likewise, contracts for the purchase of emissions reductions must embody the concept of equitable benefit and risk sharing that is embraced by the Protocol. This is not a philanthropic twist on commercial contracts but simply good business practice in the face of enormous uncertainty about the future value and price of carbon assets which are being purchased now at a fraction of the marginal cost of abatement in the OECD. The work reported here puts in concrete terms these commercial and equity considerations in carbon purchase agreements.

Right now, the missing ingredient in market creation for carbon as a commodity is a substantial body of case methodologies that inform market participants, and in particular, project sponsors and developers on the volume and compliance status of different types of CDM projects. Although the start of the European Emissions Trading Scheme is imminent, and the Kyoto Protocol commitment period starts in 2012, the long lead times necessary for the development of projects which can take advantage of the Kyoto Protocol mechanisms means that time is running out for sponsors to develop projects to create CDM projects which support local sustainable development. Nevertheless, the regulatory body for the CDM, the Executive Board, has yet to agree on the body of rules and reference cases that will remove or substantially reduce regulatory uncertainty and hence risk for CDM investors.

If I am correct that the CDM and the trade of emission reductions generated by greenhouse gas reducing projects is of fundamental importance to the long-term mitigation of climate change, then it is beholden on us all to support and strengthen the regulatory bodies of the UNFCCC to complete the foundations for the development of a vibrant market in a commodity which can lead to both climate change mitigation and sustainable development. I see this volume as an important contribution to this process.

Ken Newcombe
Senior Manager
Carbon Finance Business
The World Bank

Washington D.C.
July 2004

Acknowledgements

First and foremost the editors wish to express their thanks to the authors who have contributed to this volume. Many hours of dedicated work by busy professionals are captured in this volume and we are honoured to share this with a wide public.

We are also grateful to Robert O'Sullivan for his help in researching a number of important aspects of the book and for reviewing a number of chapters. Richard Tarasofsky provided valuable comments in the early stages of the book. However our special thanks go to Shéhan de Sayrah who has reviewed and copy-edited all the contributions, we would not have been able to finish the book without him.

Furthermore it is a pleasure to acknowledge with gratitude financial support from the PCF*plus* programme and the Legal Vice Presidency of the World Bank. The Bank Carbon Finance Business led by Ken Newcombe has provided the catalyst and the incentive for us to develop our understanding of the emerging carbon market and its legal ramifications.

We are also grateful to Martinus Nijhoff Publishers for permission to draw on sections from David Freestone's paper previously published in the *Liber Amicorum for Ibrahim S.I. Shihata.*

Finally, we would like to express our special thanks to our respective spouses, Jenny Freestone and Michael Streck, for their support and tolerance during the many late nights and weekends spent working on this volume.

David Freestone
Charlotte Streck

July 2004

Contents—Summary

IV. ARTICLE 12 OF THE KYOTO PROTOCOL: CLEAN DEVELOPMENT MECHANISM

V. CARBON SEQUESTRATION

VI. JI AND CDM: CARBON CONTRACTS

VII. ARTICLE 17 OF THE KYOTO PROTOCOL: EMISSIONS TRADING

VIII. CONCLUSION

APPENDICES

Detailed Contents

III. ARTICLE 6 OF THE KYOTO PROTOCOL: JOINT IMPLEMENTATION

VI. JI AND CDM: CARBON CONTRACTS

List of Contributors

Kai-Uwe Barani Schmidt is a programme officer at the UNFCCC Secretariat. He holds a Masters Degree in Macroeconomics and Finance from the University of Geneva. After working for an international NGO dedicated to supporting advanced research and researchers from developing countries he joined the UNFCCC Secretariat in January 1995. He supported negotiations on AIJ and cooperative mechanisms. Presently his main focus is on supporting the work of the CDM Executive Board, inter alia, managing the accreditation work under the EB and providing support to the registration of CDM project activities.

Benoît Bosquet is a Senior Natural Resources Management Specialist in the World Bank Carbon Finance Unit. He coordinates the BioCarbon Fund, which started operations in May 2004, and manages four energy and forestry projects for the Prototype Carbon Fund in Bulgaria, Moldova, and Romania. He holds degrees in applied linguistics and business administration from Belgium, an MA in International Relations from the Johns Hopkins University School of Advanced International Studies, and a PhD in Policy Studies from the University of Maryland.

James Cameron is a practising barrister and counsel to the law firm of Baker & McKenzie where he heads the Global Climate Change, Clean Energy, and Emissions Trading Practice group. He was a founding director of FIELD—the Foundation for International Environmental Law and Development and taught at the Universities of Cambridge and London. He advises leading companies as well as governments and the European Commission on climate change issues.

Josh Carmody is a senior associate and senior member of the Baker & McKenzie Global Clean Energy & Climate Change team in Melbourne, Australia. He advises a number of multilateral and private carbon funds, project sponsors and financiers, governments, and other participants in international carbon and renewable energy markets.

Robert Casamento is a senior manager in Deloitte's UK Energy Markets Team and is a qualified chartered accountant and trained as an ISO 14001 environmental management system auditor. He specializes in the risk management and financial consequences of emerging environmental regulations and sustainable development issues and has published numerous discussion papers and articles on emissions trading, risk management, and assurance issues. He is currently Project Director of the World Economic Forum's Global GHG Register Initiative and in 2003 authored a report for the European Commission assessing the capacity of Central and Eastern European countries to host and maintain Joint Implementation projects.

Bradnee Chambers is a senior legal research fellow at the Centre for International Sustainable Development Law in Canada and is also Senior Programme Coordinator of the United Nations University's Institute for Advanced Studies (UNU/IAS) in Tokyo, and Coordinator of their Multilateralism and Sustainable Development Programme. He previously worked at the UNCTAD Transnational Corporation Division on National Legislation and the Regulations Relating to Transnational Corporations. He has published widely in these fields and regularly participates in the meetings of the Conference of the Parties and Subsidiary Bodies of the UNFCCC and CBD, as well as WTO Ministerials, and CSD. He is a convening lead author in the Millennium Ecosystem Assessment.

Charles Cormier is a Senior Environment Specialist in the Environmental and Socially Sustainable Development division of the World Bank Institute, Washington, DC, where his task is to promote capacity building on climate change issues in developing countries. Prior to joining the Bank, he worked for Environment Canada and was involved in international negotiations of multilateral environmental agreements, including the Basel Convention and the United Nations Framework Convention on Climate Change. Mr Cormier has had first hand experience in drafting and implementing regulations to ensure that Canada meets its international obligations.

Chris Dodwell heads the multi-disciplinary team responsible for the UK implementation of the EU Emissions Trading Scheme, including the preparation of the UK National Allocation Plan in the Department for Environment, Food and Rural Affairs. He has advised on a number of UK government initiatives including the UK Emissions Trading Scheme, the Climate Change Levy Agreements and, most recently, the promotion of combined heat and power. He has been a member of the UK team negotiating the EU Emissions Trading Directive since its proposal in October 2001. He qualified as a solicitor with Freshfields Bruckhaus Deringer, and has an LLM in Environmental Law from University College London.

Robert Dornau was the Conference Director of the 2004 Carbon Expo, the first trade fair for the carbon market, working as a consultant with the World Bank and the International Emissions Trading Association (IETA). From 2001 to 2004 he was manager of IETA responsible for coordinating working groups on 'Contracts', 'CDM' and 'GHG Registries', and the development of the IETA Emission Trading Master Agreement. From 2001 to 2004 he was an official observer of the Executive Board (EB) of the Kyoto Protocol CDM. Before joining IETA he was project manager for Deutsche Boerse Group and a research fellow at the Center for European Economic Research. Robert holds a Masters Degree in Economics from the University of Konstanz, Germany.

Nathalie Eddy has been working on climate change issues since 1998, with the UN Climate Change Secretariat in Bonn, Germany (pre-Kyoto), the Center for International Environmental Law (CIEL), US Climate Action Network (USCAN), and Greenpeace. During her last three years with USCAN and Greenpeace, Nathalie focused her climate work on the public participation provisions of the Clean Development Mechanism of the Kyoto Protocol. She authored and presented several briefing papers on public participation in CDM at the international climate change negotiations and supported the creation of the International Forum of Indigenous Peoples and Local Communities on Climate Change.

David Freestone is Deputy General Counsel, Legal Advisory Services at the World Bank, Washington, DC. Prior to moving to the World Bank, he was Professor of International Law at the University of Hull, England. Professor Freestone has published widely in the field of international environmental law, is on the Board of Editors of the *British Yearbook of International Law* (BybIL) and *Yearbook of International Environmental Law* (YbIEL) (OUP) and is founder and Editor-in-Chief of the *International Journal of Marine and*

Coastal Law. His recent books include *International Law and Sustainable Development* (edited with Alan Boyle, 1999, OUP) and *Legislating for Sustainable Fisheries* (with Willian Edeson and Elly Gudmundsdottir, 2001, World Bank).

Markus Gehring is Lead Counsel at the Centre for International Sustainable Development Law. He is also a PhD candidate conducting legal research on a WTO/GATT law project at the Institute for International Affairs, University of Hamburg, Germany. His research involves methods of implementing the precautionary principle in WTO law. He also teaches German constitutional and administrative law in Hamburg. He has worked for the Ministry of Environment of Hamburg and has participated in various environment and international law conferences, including the recent COP6 of the Framework Convention on Climate Change in The Hague, Netherlands.

Mina Guli is currently Counsel with the Project Finance and Co-Financing Group of the World Bank's Legal Department, which provides legal advice and services to the Bank Carbon Finance Unit. She is on a staff exchange from Baker & McKenzie's Global Climate Change Practice Group. Prior to joining Baker & McKenzie, she worked with the Sydney Futures Exchange in the development of contracts for trade in carbon sequestration. Mina has a Bachelors Degree of Science and Law from Monash University, and a Masters in Law from the University of Melbourne.

Peter Hawkes is a member of Baker & McKenzie's Global Clean Energy and Climate Change Practice Group, and Environmental Law Group in London, UK. He specializes in climate change issues, including renewable energy and emissions trading projects, and issues relating to environmental and health and safety law, in particular in contaminated land, waste and water issues, and matters relating to the energy industry. He is actively involved in the International Emissions Trading Association, UK Emissions Trading Group, and the UK Environmental Law Association's Climate Change Working Group.

Anthony Hobley is a senior associate with Baker & McKenzie's Global Climate Change and Clean Energy Practice Group in London, UK. Anthony has a first class Honours Degree in Chemistry with Physics and recently completed a Masters Degree in Environmental Law. He is an expert in environmental law and climate change/clean energy law. His clients include the EU Commission, World Bank, BP, DuPont, and BHP Billiton. Anthony has lectured and written widely on all subjects of environmental law, particularly in relation to contaminated land, renewable energy, and emissions trading under the Kyoto Protocol.

Saleemul Huq is Director of the Climate Change Programme at the International Institute for Environment and Development in London and was previously an Academic Visitor at the Huxley School of Environment at Imperial College, London from where he holds a PhD in Plant Physiology. Dr Huq founded the Bangladesh Centre for Advanced Studies and was awarded the Robert McNamara Fellowship from the World Bank (1986–7) and Duggan Fellowship from the Natural Resources Defence Fund, US (1989). He is an Adviser to UNDP Capacity 21 Programme where he has helped develop the Agenda 21 Programmes in Mongolia, Bhutan, Gambia, and Kazakhstan.

Franck Lecocq is an economist with the Development Economics Research Group of the World Bank. He has worked on many aspects of the economics of global environmental problems, including selection of mitigation paths, choice of instrument under uncertainty, post-Kyoto mitigation regimes, and project-based mechanisms. He also advises the World Bank Carbon Finance Business operations, in particular with regard to the state and trends of the carbon market. Mr Lecocq graduated from Ecole Polytechnique (France) and holds a PhD in Economics from Ecole Nationale du Genie Rural, des Eaux et des Forets, France.

Jürgen Lefevere is administrator at the Climate Change and Energy Unit of the European Commission Environment Directorate General where he works on the EU emissions allowance-trading scheme. Previously as Director of the FIELD Climate Change and Energy Programme he was involved in preparatory work for the Emissions Trading Green Paper as well as the Emissions Trading Directive and the GHG monitoring and reporting guidelines. He has acted as legal adviser to the Alliance of Small Island States (AOSIS) and as representative of Samoa in the UNFCCC. He is currently a guest lecturer at the University of Leuven and University College London.

Marcela Main S. has worked with the Chilean Comisión Nacional del Medio Ambiente (CONAMA), which is in charge of Energy and Climate issues, since 1998. She is responsible for climate change within the Department of International Affairs within CONAMA, and is currently drafting the Tradable Emissions Permit Framework Law for Chile.

Simon Marr is currently seconded to the German Ministry for Environment where he is working on the implementation of an emissions trading system in Germany. Dr Marr is a Senior Fellow at Ecologic—Institute for International and European Environmental Policy in Berlin, Germany where he provides legal advice on aspects of climate change, as well as other fields of environmental protection.

Ernestine Meijer is a senior researcher at the Center on Environmental and Land Use Law of New York University School of Law. She studied international law in Amsterdam (the Netherlands) and Grenoble (France) and has also acted as counsel to a number of environmental NGOs in several cases on issues of national, European, and international environmental law. She was a member of the Taskforce on Traditional Knowledge and Access and Benefit Sharing of the United Nations Development Programme and has worked at the Foundation for International and Environmental Law and Development, and at the Dutch Ministry of Environment.

Maria Netto is a programme officer at the UNFCCC Secretariat. She holds a Masters degree in International Relations with focus on development and economics from the Institute of International Studies in Geneva. After working with a UNDP project in Central America and in an international organization for south–south cooperation she joined the UNFCCC Secretariat in February 1997. She supported negotiations on the financial mechanism of the Climate Change Convention, national communications from non-Annex I Parties, and cooperative mechanisms. Presently her main focus is on supporting the work of the CDM Executive Board, inter alia, supporting the methodological work and registration of CDM project activities.

Ian Noble is an adviser to the Carbon Finance Unit of the Environment Department at the World Bank, Washington, DC. He was previously Professor of Global Climate Change Research in the Institute of Advanced Studies at the Australian National University and Chief Executive Officer of the Cooperative Research Centre for Greenhouse Accounting at the Research School of Biological Sciences, Australian National University. He was also co-chair of the Working Group on the Millennium Assessment of the Intergovernmental Panel on Climate Change.

Robert O'Sullivan is a Junior Professional Associate in the Cofinancing and Project Finance Practice Group in the World Bank's Legal Department, which provides legal advice and services to the Carbon Finance Unit of the Bank. He holds an LLM in International Law, and undergraduate degrees in law, biology, and philosophy.

Anikó Pogány is a senior adviser at the Hungarian Ministry of Environment, where she advises on the implementation of the EU Emissions Trading Scheme in Hungary. Prior to joining the Ministry, she was the Director of Strategy at Pannonpower in Pecs, Hungary, where she managed and concluded the sale of JI credits for the Pecs biomass project to the Prototype Carbon Fund. Between 1996 and 2000 she worked for Ernst & Young's Budapest office in the area of corporate finance. As a senior consultant and later manager, she was expert in financial modelling, business plan preparation and review, valuation and due diligence. She is also a member of the Hungarian Association of Environmental Education.

Dane Ratliff is Assistant Legal Counsel at the Permanent Court of Arbitration at The Hague, and was educated in political science, philosophy, and law. He has represented the Permanent Court of Arbitration at COPs of the UNFCCC, Convention on Biological Diversity, and other meetings of environmental conventions. Mr Ratliff is also a lecturer in environmental and human rights law at the Universities of Wurzburg and Rotterdam.

Hannah Reid is a research associate working primarily with the Climate Change Programme at the International Institute for Environment and Development. Dr Reid has a broad background of working on development, community-based biodiversity management, and climate change-related issues.

Sebastian M Scholz is a PhD research fellow at the Center for Development Research in Bonn, Germany. His doctoral thesis examines forestry projects in the CDM as a means of carbon mitigation and poverty alleviation. Previously he worked for the Berlin-based Institute for Applied Ecology on synergies between the UN Framework Convention on Climate Change and the Convention on Biological Diversity. He gained international work experience in forests and climate change as a research fellow at the Food and Agriculture Organization (FAO) in Rome, Italy and with the World Bank BioCarbon Fund team.

Jaya Singhania is a senior counsel with the Infrastructure Development Finance Company Limited (IDFC), Mumbai, India. She advises the IDFC on issues pertaining to corporate and project finance law as well as on CDM carbon contracts. Her professional experience includes project finance and project development work in India. She holds a Masters Degree in International Law.

Charlotte Streck is counsel in the Cofinancing and Project Finance Practice Group in the World Bank's Legal Department, which provides legal advice and services to the Carbon Finance Unit of the Bank. Dr Streck was educated in both law and biology, and is an adjunct lecturer in environmental law at the University of Potsdam, Germany. Before joining the World Bank in 2000 Dr Streck was a participant in the 'Global Public Policy Networks Project', which provided strategic advice for the Secretary-General of the UN. She has authored and co-authored four books and published several articles on environmental law and policy, and she is a board member of the Global Public Policy Institute.

Jari Väyrynen is an environmental specialist and project manager in the World Bank Carbon Finance Business, managing CDM and JI projects in various countries, including Brazil, Honduras, the Czech Republic, and Poland. He is also managing activities under the PCFplus capacity building, research and training programme. Before joining the World Bank he worked for the Finnish Ministry of the Environment and the European Commission.

Daniel van der Weerd has been legal counsel at the Senter agency of the Netherlands Ministry of Economic Affairs since February 2001. At Senter he works as counsel for Carboncredits.nl, which includes the ERUPT and CERUPT programmes. He is counsel for several bilateral cooperation programmes and for the export credits programmes and is also involved in developing new Joint Implementation Facilities for the Ministry of Economic Affairs of the Netherlands.

Matthieu Wemaere is an associate attorney at Huglo Lepage in Brussels. He advises European clients including government departments on all aspects of environment law. Between 1999 and 2003 he worked with the European Commission's Climate Change Group.

Jake Werksman is Senior Adviser to the Global Inclusion Program of the Rockefeller Foundation. Previously, he was Environmental Institutions and Governance Adviser, United Nations Development Programme, and held various posts at the Foundation for International Environmental Law and Development, including Managing Director. He is currently an adjunct Professor of Law at New York University, and was a lecturer in international economic law at the University of London. He has held visiting academic posts at the United Nations University and the Universities of Edinburgh, Kent, and Connecticut Law School.

Martijn Wilder is a partner at Baker & McKenzie where he is responsible for the global climate change and emissions trading practice and advises governments, major corporations, and multilateral organizations including the World Bank Carbon Finance Unit, the European Commission, UNEP, and leading financial institutions. Mr Wilder has published widely on environmental and international law, is an adjunct lecturer at the University of NSW, on the NSW Greenhouse Office Advisory Panel, and is the President of TRAFFIC (Oceania). He is on the Board of the Australian Branch of the International Law Association and is its representative on the ILA Committee on the Trans-National Enforcement of Environmental Law.

Monique Willis is an associate and senior member of the Baker & McKenzie Global Clean Energy & Climate Change team in Sydney, Australia. She is an adviser to the World Bank's Carbon Finance Unit and other investment institutions participating in the international carbon and renewable energy markets. She advises governments around the world on emissions trading and climate change policy and regulatory issues.

Rutger de Witt Wijnen is a partner at De Brauw Blackstone Westbroek, Amsterdam, where he developed their international emissions trading practice, advising on all aspects of emissions trading and carbon credit creation projects, including issues relating to the EU Emissions Trading Directive and the Kyoto Protocol. He has drafted and negotiated

Emission Reductions Purchase Agreements for CDM and JI projects in Latin America, Asia, and Eastern Europe. He has advised the Dutch Government, acting through the Ministries of the Environment and Economic Affairs, on the Emission Reduction Purchase Facilities with various international financial institutions, such as the IBRD, IFC, CAF, EBRD, and Rabobank International.

Tables of Treaties and Legislations

INTERNATIONAL TREATIES, CONVENTIONS, AND RESOLUTIONS

EUROPEAN UNION TREATIES, REGULATIONS, DIRECTIVES, AND DECISIONS

Regulations

Directives

Decisions

COUNTRY LEGISLATION

Australia

List of Abbreviations

AAU	Assigned Amount Unit
AB	Appellate Body
AC	Abatement Certificate
ACBE	Advisory Committee on Business and the Environment (UK)
AGBM	Ad Hoc Group on the Berlin Mandate
AGO	Australian Greenhouse Office
AICPA	American Institute of Certified Public Accountants
AIE	Accredited Independent Entity
AIJ	Activities Implemented Jointly
AOSIS	Alliance of Small Island States
AP	Accreditation Panel
A&R	Afforestation and Reforestation
BINGO	Business and Industry Non-Governmental Organization
BIT	Bilateral Investment Treaty
BOT	Build Operate Transfer
CAN	Climate Action Network
CBD	Convention on Biological Diversity
CCA	Climate Change Agreement
CDCF	Community Development Carbon Fund
CDM	Clean Development Mechanism
CER	Certified Emission Reductions
CERUPT	Certified Emission Reduction Unit Procurement Tender
CICA	Canadian Institute of Chartered Accountants
CITL	Community Independent Transaction Log
CONAMA	Chilean Environmental Agency (Comisión Nacional del Medio Ambiente) (Chile)
COP	Conference of the Parties
COP/MOP	Conference of the Parties serving as the Meeting of the Parties
CPC	United Nations Provisional Central Product Classification System
CPV	Common Procurement Vocabulary
CSD	Commission on Sustainable Development
CSR	Corporate Social Responsibility
CTE	Committee on Trade and Environment (WTO)

DEFRA	Department of Environment, Food and Rural Affairs (UK)
DETR	Department of the Environment, Transport and the Regions (UK)
DNA	Designated National Authority
DOE	Designated Operational Entity
DoE	Department of Energy (US)
DSB	Dispute Settlement Body
DSU	Dispute Settlement Understanding
DTI	Department of Trade and Industry (UK)
EB	CDM Executive Board
EBRD	European Bank for Reconstruction and Development
ECCP	European Climate Change Programme
ECJ	European Court of Justice
EIA	Environmental Impact Assessment
EIT	Economy in Transition
ENGO	Environmental Non-Governmental Organization
EPA	Environmental Protection Agency (US)
ER	Emission Reduction
ERPA	Emissions Reductions Purchase Agreement
ERU	Emission Reduction Unit
ETS	Emissions Trading Scheme
EU25	European Union 25 member countries
EU ETS	European Union Emissions Trading Scheme
ERUPT	Emission Reduction Unit Procurement Tender
FASB	Financial Accounting Standards Board
FERC	Federal Energy Regulatory Commission
FDI	Foreign Direct Investment
FMU	Fund Management Unit
GATT	General Agreement on Tariffs and Trade
GATS	General Agreement on Trade in Services
GDP	Gross Domestic Product
GEC	Gas Electricity Certificates (Australia)
GEF	Global Environment Facility
GERT	Gas Emission Reduction Trading (Canada)
GHG	Greenhouse Gas
GIS	Green Investment Scheme

GMO	Genetically Modified Organisms
GST	Goods and Services Tax
HCA	Host Country Umbrella Agreement
IASB	International Accounting Standards Board
IBRD	International Bank for Reconstruction and Development
ICC	International Chamber of Commerce
ICJ	International Court of Justice
ICSID	International Centre for the Settlement of Investment Disputes
IDA	International Development Association
IET	International Emissions Trading
IETA	International Emissions Trading Association
IFC	International Finance Corporation
IFI	International Finance Institution
IFRIC	International Financial Reporting Interpretations Committee
IFRS	International Financial Reporting Standards
IIFB	International Indigenous Forum on Biodiversity
IISD	International Institute for Sustainable Development
IPCC	Intergovernmental Panel on Climate Change
IPO	Indigenous Peoples Organization
IREDA	Indian Renewable Energy Development Agency Limited (India)
IRR	Internal Rate of Return
IRS	International Revenue Service (US)
ITL	Independent Transaction Log
JI	Joint Implementation
JI-SC	Joint Implementation Supervisory Committee
lCER	long-term CER
LCIA	London Court of International Arbitration
LGMA	Local Government and Municipal Authorities
LoA	Letter of Approval
LULUCF	Land Use, Land-Use Change or Forestry
MEA	Multilateral Environmental Agreements
MDB	Multilateral Development Bank
MIGA	Multilateral Investment Guarantee Agency
MIT	Multilateral Investment Treaty

MNES	Ministry of Non-Conventional Energy Sources (India)
MoEF	Ministry of Environment and Forests (India)
MOU	Memorandum of Understanding
MRET	Mandatory Renewable Energy Target (Australia)
NAO	National Audit Office (UK)
NAP	National Allocation Plan
NGO	Non-Governmental Organization
ODA	Official Development Assistance
OECD	Organization for Economic Co-operation and Development
OPG	Ontario Power Generation (Canada)
ORER	Office of the Renewable Energy Regulator (Australia)
OTC	Ozone Transport Commission (Canada)
PCA	Permanent Court of Arbitration
PCF	Prototype Carbon Fund
PDD	Project Design Document
PDDA	Chilean Decontamination and Prevention Plan (Plans de Prevención y Descontaminación Atmosférica)
PERT	Pilot Emission Reduction Trading (Canada)
PIN	Project Information Note
PSO	Programme for Co-operation with Countries in Central and Eastern Europe (Programma Samenwerking Oost-Europa) (Netherlands)
QELRC	Quantified Emission Limitation and Reduction Commitments
REC	Renewable Energy Certificates (Australia)
RECLAIM	REgional CLean Air Incentives Market (US)
RINGO	Research-Oriented and Independent Organizations
RMU	Removal Unit
SBI	Subsidiary Body for Implementation
SBSTA	Subsidiary Body for Scientific and Technological Advice
SCM	Subsidies and Countervailing Measures
SEDA	Sustainable Energy Development Authority (Australia)
SM	Standard Methods
STO	Specific Trade Obligation

TBT	Technical Barriers for Trade
tCER	temporary CER
TRIMs	Trade-related Investment Measures
UN/ECE	United Nations Economic Commission for Europe
UNCED	United Nations Conference on Environmental and Development
UNCITRAL	United Nations Commission on International Trade Law
UNDP	United Nations Development Programme
UNEP	United Nations Environment Programme
UNFCCC	United Nations Framework Convention on Climate Change 1992
UNGA	United Nations General Assembly
VROM	Ministry of Housing, Spatial Planning and the Environment (Het ministerie van Volkshuisvesting, Ruimtelijke Ordering en Milienbeheer) (Netherlands)
WG	Working Group of the IPCC
WMO	World Meteorological Organization
WRI	World Resources Institute
WSSD	World Summit on Sustainable Development
WWF	World Wildlife Fund
WTO	World Trade Organization

PART I

INTRODUCTION

1

The UN Framework Convention on Climate Change, the Kyoto Protocol, and the Kyoto Mechanisms

*David Freestone**

Climate change has been described as a defining issue for the twenty-first century.[1] On the one hand it poses an unprecedented and as yet still not fully understood threat to the global climate system upon which mankind depends; on the other hand remedial action requires an extensive, and expensive, overhaul of northern industrial economies which are still heavily dependent on non-renewable carbon-based fossil fuels. For decades scientists have understood the chemical processes by which emissions of carbon dioxide and other gases might warm the planet through the so-called 'greenhouse effect', nevertheless it was not until the 1980s that international concern about anthropogenic impacts on the atmosphere through such emissions came to a head, and it was only in the last two decades of the twentieth century that the United Nations took the first, somewhat faltering, steps towards recognizing and addressing the issue. These first steps led to the conclusion of the 1992 Framework Convention on Climate Change, followed by its innovative 1997 Kyoto Protocol[2] that is designed, *inter alia*, to utilize 'market mechanisms' to assist with the massive reductions of greenhouse gas emissions necessary to arrest the processes of climate change.

The aim of this chapter is to provide an overview of the regime established by the Convention and the process leading up to and after the adoption of the Kyoto Protocol. It introduces some of the key processes and describes some of the key

* David Freestone is Deputy General Counsel, Advisory Services at the World Bank. The views expressed in this chapter are the views of the author and do not necessarily represent the views of the World Bank. He is grateful to Martinus Nijhoff Publishers for permission to draw heavily upon his paper 'The World Bank's Prototype Carbon Fund: Mobilising New Resources for Sustainable Development' in Schlemmer-Schulte, *et al.* (eds), *Liber Amicorum for Ibrahim S.I. Shihata* (The Hague, 2001) 265–341.

[1] D Hunter, J Salzman, and D Zaelke, *International Environmental Law and Policy* (2nd edn, New York, 2002) 589.

[2] FCCC/CP/1997/L.7/Add.1, Decision 1/CP.3 Adoption of the Kyoto Protocol to United Nations Framework Convention on Climate Change, Annex, reprinted in (1998) 37 International Legal Materials (ILM) 22. *Not yet in force.*

concepts that will be further developed in later chapters. It will also discuss the rationale for the involvement of the World Bank in this area, which includes a desire to see mitigation of changes which are likely to impact the poorest countries most severely, and a role in capacity building so that potential project host countries can take advantage of this important financing opportunity for sustainable development while facilitating the development of a global carbon market.

1 The Road to Rio and the UN Framework Convention on Climate Change

The publicity surrounding the discovery and mapping of the growing 'ozone hole' over the Southern Polar region during the early 1980s, together with the swift and successful negotiation of the 1985 Vienna Convention for the Protection of the Ozone Layer[3] followed by its 1987 Montreal Protocol on Substances that Deplete the Ozone Layer[4] focused attention on the much wider and more complex issue of the impacts of greenhouse gases on the global atmospheric system. By the late 1980s climate change was already firmly on the international agenda, and in 1988 the UN General Assembly (UNGA) passed Resolution 43/53 on the Protection of the Global Climate for Present and Future Generations of Mankind.[5] It declared the issue to be one of 'common concern to mankind' and endorsed the action of the UN Environment Programme (UNEP) and the World Meteorological Organization (WMO) jointly establishing an Intergovernmental Panel on Climate Change (IPCC) to 'provide internationally co-ordinated scientific assessments of the magnitude, timing and potential environmental and socio-economic impacts of climate change and realistic response strategies'.[6] Two years later, in 1990, UNGA Resolution 45/212 established an Intergovernmental Negotiating Committee to develop a legal instrument on climate change. The outcome of a further two years of strenuous negotiations was the 1992 UN Framework Convention on Climate Change (UNFCCC).[7]

The UNFCCC was concluded in New York on 9 May 1992.[8] It was opened for signature in June 1992 as a part of the UN Conference on Environment and Development (UNCED) in Rio de Janeiro, Brazil. It now has 193 Parties. The basic objective of the UNFCCC, which is not to reverse greenhouse gas emissions, but rather to stabilize them 'at a level that would prevent dangerous anthropogenic interference with the climate system' is set out in Article 2.[9] Indeed, that article

[3] (1985) 26 ILM 1527. [4] (1987) 26 ILM 1550.

[5] UN General Assembly, *Protection of Global Climate for Present and Future Generations of Mankind* (New York, United Nations) (A/RES/43/53), 70th plenary meeting, 6 December 1988.

[6] The IPCC has three Working Groups (WGs): WGI deals with the science of Climate Change; WGII with impacts and responses; and WGIII with the economic and social dimensions.

[7] 1771 UNTS 107. [8] UN Doc Distr General A/AC.237/18 (Part II)/Add.1, 15 May 1992.

[9] This important limitation of the UNFCCC is pointed out by P Birnie and A Boyle, *International Law and the Environment* (2nd edn, Oxford, 2002) 527.

goes on to require that such stabilization should be achieved 'within a time frame sufficient to allow eco-systems to adapt naturally to climate change, to ensure that food production is not threatened and to enable economic development to proceed in a sustainable manner.'[10]

Article 3 then goes on to enumerate the principles by which the Parties should be guided in their actions to achieve this objective. These include a number of the innovative principles set out in the 1992 Rio Declaration, notably the Precautionary Principle,[11] the principle of inter-generational equity[12] and the principle of 'common but differentiated responsibility.'[13] However the UNFCCC also recognizes a number of other principles such as the special needs of developing country Parties and of those 'that would have to bear a disproportionate or abnormal burden under the Convention',[14] as well as the right of all Parties to promote sustainable development[15] and the need to promote 'a supportive and open international economic system.'[16]

The commitments of the Parties are set out in Article 4(1). Although these are generally hortatory, the full significance of the common but differentiated obligations emerges in Article 4(2) where those developed countries and countries with economies in transition (EITs) listed in Annex I undertake to adopt national policies and measures which will 'demonstrate that developed countries are taking the lead in modifying longer-term trends in anthropogenic emissions consistent with the objective of the Convention.'[17] While a number of particular factors are listed, the article does recognize 'the return by the end of the present decade to earlier levels of anthropogenic emissions . . . would contribute to such

[10] Art 2 UNFCCC.

[11] Art 3(3) UNFCCC. See also Principle 15 of the Rio Declaration (UN Doc Distr General A/Conf.151/5/Rev.1, 13 June 1992). There is an extensive literature on the Precautionary Principle, see for example, D Freestone, 'The Precautionary Principle' in R Churchill and D Freestone (eds), *International Law and Global Climate Change* (London/Dordrecht, 1991) 21. See also T O'Riordon and J Cameron (eds), *Interpreting the Precautionary Principle* (London, 1994) and D Freestone and E Hey (eds), *The Precautionary Principle and International Law: The Challenge of Implementation* (The Hague/London, 1996). Most recently see the comprehensive study by A Trouwborst, *Evolution and Status of the Precautionary Principle in International Law* (The Hague, 2002).

[12] Art 3(1) UNFCCC which provides that 'The Parties should protect the climate system for the benefit of present and future generations of humankind, on the basis of equity and in accordance with their common but differentiated responsibilities and respective capabilities.' See also Principle 3 of the Rio Declaration, n 11 above. On the principle of intergenerational equity, see E Brown Weiss, *In Fairness to Future Generations: International Law, Common Patrimony and Intergenerational Equity* (Transitional Publishers, Inc, for the United Nations University, Tokyo, 1989); C Redgwell, 'Intergenerational Equity and Global Warming' in Churchill and Freestone (eds) n 11 above, 41–56.

[13] Art 3(1) UNFCCC, ibid. The text of the article continues: 'Accordingly the developed country Parties should take the lead in combating climate change and the adverse effects thereof.' See also Principle 7 of the Rio Declaration, n 11 above. For a wider discussion of the principle, see CD Stone, 'Common but Differentiated Responsibilities in International Law' (2004) 98 *American Journal of International Law* 276–301.

[14] Art 3(3) UNFCCC.

[15] Art 3(4) UNFCCC.

[16] Art 3(5) UNFCCC. For a discussion of the legal status of the 'guidance' under Art 3, see Birnie and Boyle, n 9 above, 525.

[17] Art 4(2) UNFCCC; see also other articles of the UNFCCC, for example, Art 12.

modification'. The UNFCCC does not however provide any modality by which this return to 'earlier' levels can be made; nor indeed does it prescribe which particular earlier levels should be used as the benchmark.

However, Article 4(3) does impose an unequivocal obligation on developed countries listed in Annex II[18] to provide 'new and additional financial resources' to meet the agreed costs incurred by developing countries in meeting their communication obligations under Article 12 and to meet the 'agreed full incremental costs' in implementing the measures envisaged in Article 4(1) and which are agreed between the developing country Party and the 'entity or entities referred to in Article 11' (that is, the financial mechanism).[19] The Annex II countries also undertake to assist countries particularly vulnerable to the impacts of climate change in meeting the costs of adapting to those adverse effects[20] and to take all practicable steps to promote, facilitate and finance as appropriate, the transfer of, or access to, environmentally sound technologies and know-how to other parties, particularly developing countries, to enable them to implement the provisions of the UNFCCC.[21]

Articles 7 to 11 follow the pattern of the more recent multilateral environment agreements by establishing a fairly sophisticated institutional framework. The Conference of the Parties, as the 'supreme body' of the UNFCCC, is charged with keeping the implementation of the UNFCCC under regular review and is specifically granted the power to make decisions necessary to promote the effective implementation of the UNFCCC.[22] In addition, the UNFCCC establishes a Permanent Secretariat, and two 'Subsidiary Bodies'. The Subsidiary Body for Scientific and Technological Advice (SBSTA) is established, and its role defined, by Article 9 and the Subsidiary Body for Implementation (SBI) by Article 10. The Subsidiary Bodies meet in conjunction with the Conference of the Parties and also inter-sessionally.[23] Article 11 defines the 'Financial Mechanism' which is to be the chief modality for the mobilization of the new financial resources 'on a grant or concessional basis' required by the UNFCCC.

Article 11 envisages that the financial mechanism is to function under the guidance of, and be accountable to, the Conference of the Parties, but that its operation shall be entrusted to one or more existing international entities. The UNFCCC specifically provides that the financial mechanism shall have an equitable and balanced representation of all Parties within a transparent system of governance. To this end the interim arrangements set out in Article 21(3) specifically designate the Global Environment Facility (GEF) as the international entity

[18] Annex II does not contain EIT countries.

[19] Under Art 21(3) UNFCCC, the Global Environment Facility (GEF), see below, shall be 'the international entity entrusted with the operation of the financial mechanism referred to in Article 11 on an interim basis.'

[20] Art 4(4) UNFCCC.

[21] Art 4(5) UNFCCC.

[22] Art 7 UNFCCC, provided of course that these are 'within its mandate'.

[23] The Subsidiary Bodies commonly meet in the week prior to the COP and inter-sessionally—usually in June at the seat of the Secretariat in Bonn.

entrusted with the operation of the financial mechanism but only 'on an interim basis' and require that it be 'appropriately restructured and its membership made universal to enable it to fulfil the requirements of Article 11.'[24]

2 The Berlin Mandate

The UNFCCC came into force on 21 March 1994. The first session of the Conference of the Parties (COP1) was held in Berlin in April 1995. At the COP the major issue was the need for the Parties to address the vagueness of the commitments in the UNFCCC itself and to develop more concrete obligations and a more precise time frame for the reduction of emissions of greenhouse gases (GHGs) by the industrialized countries in Annex I. The Conference produced a negotiated paper setting both a process and a time line for the development of another instrument to achieve this objective.[25] This became known as the Berlin Mandate.

The Conference also endorsed the concept of a pilot phase for joint implementation activities. This pilot phase was to be called 'activities implemented jointly [AIJ] under the pilot phase'.[26] The objective was to allow investments by Annex I countries in a range of GHG mitigation projects outside their own borders to allow them to begin to understand more clearly the implications of developing such 'off shore' projects. The decision did not envisage Annex I countries being able to claim credit against future reduction targets for investments made under the Activities Implemented Jointly (AIJ) pilot phase.

Building on the mandate of COP1 an ad hoc group was established to advance the concepts of the adequacy of the commitments under Article 4 and the development of a new instrument to crystallize the inspirational wording of the UNFCCC. The Ad Hoc Group on the Berlin Mandate (AGBM) met regularly at

[24] Art 11 requires the Conference of the Parties at its first session to review the interim arrangements in Art 21(3) and to decide whether to maintain these arrangements. Within four years the financial mechanism shall be reviewed and 'appropriate measures' taken. These very cautious provisions indicate that the negotiators were not entirely happy with passing sole responsibility to the newly established and yet unproven GEF and it is equally clear that some fundamental changes in its governance structure would be necessary before the Parties would be willing to confirm its status as a [or the] financial mechanism for the UNFCCC. Note the similar provisions in the Convention on the Protection of Biological Diversity, Arts 21 and 39 and also Agenda 21, ch 33.16 (a)(iii).

[25] At the final Plenary session of the Conference, the President Angela Merkel introduced the review of the adequacy of commitments in Art 4(2)(a) and (b). (FCCC/CP/1995/L.14, review of the adequacy of commitments in Art 4(2)(a) and (b).) She said that the industrialized countries' agreement to specific measures and the developing countries' agreement to reaffirm and advance existing commitments meant that the process would advance. India, for the G77, and China said the decision to launch a process to strengthen the commitments made COP1 a success, noting that consultations were not easy. The EU stated its understanding that the wording 'developed countries/other Parties' in section II(2)(a) must be interpreted as 'developed countries and/or other Parties' and means that this sub-para applies to Annex I Parties within the European Community, individually or jointly among themselves, in accordance with Art 4(2)(a) and (b) UNFCCC.

[26] FCCC/CP/1995/L.13 AIJ pilot phase.

COP2 and at the intervening Subsidiary Bodies meetings. The culmination of this process, which involved a series of submissions by the Parties of observations and of draft language, was the COP3 meeting in Kyoto during which the Kyoto Protocol was negotiated.

The Kyoto Protocol sets out the emission reduction commitments of Annex I countries in its Annex B, and establishes three market mechanisms which are available to Annex I countries to help them meet these commitments. The Kyoto Protocol is discussed more fully in section 5 below, and the market mechanisms it contains are discussed in sections 6 to 8.

3 The Buenos Aires Plan of Action

Immediately after the finalization of the Kyoto Protocol, the Parties to the UNFCCC began negotiating the ways in which its mechanisms would operate. Understanding between the Parties to the UNFCCC on how to proceed was advanced at the fourth session of the Conference of the Parties to the UNFCCC (COP4), held in Buenos Aires in November 1998. COP4 resulted in a number of decisions including Decision I/CP.4, the so-called 'Buenos Aires Plan of Action'. The Plan of Action declared the determination of the Parties to the UNFCCC to strengthen the implementation of the UNFCCC and to prepare for the future entry into force of the Kyoto Protocol. It included a resolution to demonstrate, *inter alia*, substantial progress on the Kyoto mechanisms. More significantly, Decision 7/CP established that the deadline for reaching decisions on all three of the market mechanisms would be COP6—the sixth session of the Conference of the Parties which had been planned to be held in November 2000 in The Hague.[27]

4 From The Hague to Marrakesh

COP6 at The Hague was a not a success. Despite the active intervention of the President—Dutch Environment Minister Jan Pronck—the negotiations over the modalities for operating the Kyoto mechanisms encountered a number of major roadblocks—including serious problems with the role that carbon sequestration (or sink) projects would play in helping countries to meet their Annex B commitments under Kyoto. Unable to reach agreement, the COP was adjourned and an extra session (named variously COP6(*bis*) or COP6.5) was convened in July 2001 in Bonn. This by no means solved all the problems but it did enable the parties to reach accommodations on a number of key issues so that by the time that the previously scheduled seventh session of the Conference (COP7) took place in Marrakesh in November of that year the parties were in a position to agree a major

[27] It also outlined the detailed work programme for developing the mechanisms.

package of measures to enable the Kyoto mechanisms to start operations. This bundle of some 262 pages and 39 decisions was called the Marrakesh Accords. As will be seen in the following pages of this book, the Accords provide a much needed framework of 'guidelines, modalities and rules' for moving forward with implementation of the Protocol, however a number of key issues still remained to be settled either through practice or by further COP decisions.

In 2000, however, there were other setbacks to the Kyoto process. During the Clinton administration the US signed the Protocol, but after the November 2000 Presidential election, the incoming Bush administration announced its intention to withdraw from Kyoto. To enter into force the Kyoto Protocol requires ratification by fifty-five Parties to the UNFCCC Convention including ratification by Annex I Parties 'which accounted in total for at least 55 per cent of the total carbon dioxide emissions for 1990 of the Parties included in Annex I.'[28] The result of the decision of the US not to ratify the Protocol is therefore that virtually all the other Annex I parties—including the Russian Federation—will now have to ratify the Protocol in order to bring it into force.[29] Paradoxically the decision of the world's largest GHG emitter to withdraw from the process galvanized the remaining UNFCCC Parties to move forward with the finalization of the Marrakesh Accords and to proceed with the establishment of the bodies—such as the CDM Executive Board—and of the procedures to enable an immediate start of the Kyoto Mechanisms once the Protocol does enter into force.

5 The Kyoto Protocol

The Kyoto negotiations were not easy. On 11 December 1997, in the early morning of the day after that scheduled for the completion of COP3, after more than two days of continuous negotiation, the Parties to the 1992 UNFCCC concluded the long awaited protocol. The 1997 Kyoto Protocol strengthens the commitments of the UNFCCC—particularly Article 4(2)(a) and (b)— by setting out a firm schedule for reductions of GHG emissions by Annex I countries and firm targets to be met within an agreed commitment period. In adopting the Protocol the COP recalled that it had decided in Berlin in 1995 at its first Conference to start a process by which it could take 'appropriate action' for the period beyond 2000 by adopting a protocol or another legal instrument at its third session; and recalled that one aim of that process was to strengthen the commitments in Article 4(2)(a) and (b) for developed country parties and other parties included in Annex I of the UNFCCC, both to require them to elaborate policies and measures and to set quantified limitation and reduction commitments

[28] Art 25(1) Kyoto Protocol.

[29] As of 13 June 2004, the Protocol had 122 parties representing 44.2 per cent of the required emissions.

(QELRCs) within specified time frames for their anthropogenic emission by source, and removal by sinks, of greenhouse gases.[30]

The COP decision also recalled that according to the Berlin Mandate, the process would not introduce any new commitments for developing countries but would reaffirm existing commitments in Article 4(1) and continue to advance the implementation of these commitments in order to achieve sustainable development, taking into account Article 4(3), (5), and (7).

The specific commitments of Annex I Parties are set out in Annex B to the Protocol. These envisage that during the five years of the 'commitment period' from 2008 to 2012, those Parties will have reduced their emissions by an average of 5.2 per cent from 1990 levels. The specific targets (or assigned amounts) however range from Iceland and Australia, which were able to increase their emissions from the 1990 base levels (by 10 and 8 per cent respectively) to the countries of the European Union which accepted an 8 per cent reduction from 1990 levels.

Since 1990 many of the Annex I countries have substantially further increased their GHG emissions through growth in their economies. The commitments to reduce emissions below 1990 levels are therefore commensurably even more rigorous, and the potential economic impacts of these obligations are unparalleled in international environmental law.[31] However, the most innovative aspects of the Protocol may in fact not be these strict commitments but the introduction of so-called market mechanisms, initially called flexibility mechanisms (or flex-mechs) but now known as the Kyoto mechanisms, into the process by which Annex I countries can meet their obligations. Some degree of cooperation had been envisaged by the UNFCCC itself which in Article 4(2)(b) talks of the aims of 'returning [GHG emissions] individually *or jointly*, to their 1990 levels . . .'[32] The wording of the UNFCCC is echoed in Article 3(1) of the Protocol which reaffirms that these commitments may be made individually or jointly, and has also been developed further in Article 4 in response to requests by the EU to allow it to be done within the EU group of nations as a whole. This mechanism, the so-called EU 'Bubble', requires that Parties in Annex I which reach an agreement

[30] The UNFCCC and its Kyoto Protocol both purport to regulate greenhouse gases 'not controlled by the Montreal Protocol.' This form of words is used as some gases that contribute to global climate change also deplete the ozone layer and are already subject to the legal regulatory regime of the Montreal Protocol, n 31 below. To avoid undue repetition this important qualification is not used in the text of this paper but should be borne in mind.

[31] The potential impacts of the economic changes necessary to meet the commitments are different in kind from the very significant commitments made by developed countries under the 1987 Montreal Protocol (as amended and adjusted) to the 1985 Vienna Convention for the Protection of the Ozone Layer (1987) 26 ILM 1527. The text of the Montreal Protocol on Substances that Deplete the Ozone Layer; as changed by the London adjustments and the amendments of 1990, 1992, 1995, and 1997 can be found at <htto://www.unep.org/ozone>.

[32] (Emphasis added.) The idea that the parties might seek to do this jointly was the basis for the AIJ pilot phase; see n 26 above, in which the modalities of international cooperation would be explored.

to fulfil their commitments under Article 3 jointly shall be deemed to have met those commitments if their total combined aggregate emissions do not exceed their assigned amounts calculated in accordance with Annex B. Such an inter-State agreement must be formally concluded and notified to the Parties through the UNFCCC Secretariat and in the event of failure of the group of States to meet the required reductions, then each party then becomes again individually liable for its own levels of emissions.[33]

6 The Kyoto Mechanisms

More radical than the joint responsibility agreements envisaged under UNFCCC Article 4 however are the provisions of Articles 6 and 12 of the Kyoto Protocol—each of which envisages that in various ways emission reductions financed in other countries might be set off against the financiers' GHG reduction targets. The theory behind this approach is that the 'marginal abatement cost', that is, the cost of financing an emission reduction, will usually be far higher in a relatively fuel efficient industrialized country than in a country such as an EIT or a developing country which may have less efficient fuel-use technology. As the global climate system benefits from reductions wherever they are made, then making reductions in an EIT or a developing country as part of a national strategy (which also of course includes the introduction of domestic 'policies and measures' to reduce emissions at home), will make the cost of reaching these reduction targets cheaper and increase the chances that they will actually be reached. The consequential advantages for the project host countries are new resource and technology transfers, including access to cleaner technology and contributions to sustainable development. In addition, Article 17 establishes a system of emission rights trading whereby one Annex I country might directly purchase from another Annex I country some of its rights to emit GHG, known as Assigned Amounts, or some of the emission reductions generated under Articles 6 or 12.

7 Article 6

Article 6 builds directly on the wording of Article 4(2)(b) of the UNFCCC,[34] which envisages that countries listed in Annex I may act individually or jointly to meet their emission reduction objectives. Article 6 of the Kyoto Protocol allows any Annex I country to transfer to, or acquire from, another Annex I country, reductions of GHG emissions, described as Emission Reduction Units (ERUs),

[33] See also Art 4(6) which applies the same principles to cooperation within a 'regional economic integration organization' (the standard treaty language for bodies such as the EU) which is itself a Party to the Protocol.

[34] See n 32 above.

achieved by project activities.[35] These ERUs may be generated by any projects that reduce anthropogenic emissions of designated greenhouse gases or which enhance the anthropogenic removal by sinks[36] of such gases. The key feature of this mechanism is that all the emission reductions need to be brought about, and verified, by investments in specific projects. It is also subject to a number of further requirements. First, these projects require the express approval of both the State Parties acting as transferor and transferee. The Protocol uses the term Parties, hence both States must be parties to the Protocol. Secondly, the reduction of GHGs that take place as a result of the project must also be 'additional' to any that would otherwise have occurred.[37] Thirdly, no State Party can acquire ERUs if it is not itself in compliance with a number of other obligations under the Protocol.[38] Once a minimum number of conditions have been met a host country may participate in an Article 6 Joint Implementation (JI) project either under close review by an independent agent to verify the ERUs (Track Two JI), or if more extensive monitoring and inventory requirements have been met, the host country may verify ERUs nationally (Track One JI). Finally, the acquisition of such ERUs cannot be a substitute for domestic action to reduce GHG emissions, for all such acquisitions must be 'supplemental to domestic actions'.[39]

In order to make it clear that the full risk of any non-compliance with other provisions of the Protocol rests entirely with the acquirer of the ERUs, the Protocol goes on to provide that where an issue of non-compliance is raised, transfer and acquisition transactions may continue but entirely at the risk of the acquirer. As the usual modality for these transfers will be by investment or purchase, this is known as 'buyer risk'.[40]

The final issue of importance is that although the Protocol talks clearly about Party-to-Party transactions it does envisage others such as the private sector (or even international organizations, such as the World Bank) participating in the financing and organizing of such projects. Article 6(3) permits any party to authorize a legal entity to participate in 'action leading to the generation, transfer or acquisition' of ERUs, provided however that that authorization is under the responsibility of the authorizing State Party.

[35] Once verified and reported in accordance with guidelines to be elaborated by the UNFCCC COP/MOP, Art 6(2).

[36] Such removal could include the sequestering of carbon by land use, land-use change or forestry (LULUCF).

[37] The requirement of being additional (or the additionality requirement) can be understood generally as a 'but for' test: 'But for the project' a particular benefit would not have happened. However, defining the scope and weight given to the additionality requirement has been a contentious issue.

[38] See Arts 5 and 7 Kyoto Protocol.

[39] Discussions continue of the meaning of supplemental in this context.

[40] Art 6(4) reads: 'If a question of implementation by a Party included in Annex I of the requirements referred to in this Article is identified in accordance with the relevant provisions of Article 8, transfers and acquisitions of emission reduction units may continue to be made after the question has been identified, *provided that any such units may not be used by a Party to meet its commitments under Article 3 until any issue of compliance is resolved*' (emphasis added).

8 Article 12

This article was the great surprise of the Kyoto Meeting. The Brazilian delegation had proposed a form of Green Development Fund at an earlier meeting but the forms and modalities of the Clean Development Mechanism (CDM) defined in Article 12 were truly the results of the negotiations at Kyoto.[41] Perhaps as a result of the fact that the negotiation of Article 12 took place in a different drafting group from that of Article 6, or perhaps deliberately, each article uses rather different language to describe concepts that are essentially very similar. The most important aspect of Article 12 is the fact it would be used by Annex I countries to finance emission reductions in countries which have not made commitments under the Protocol to meet GHG emission reduction targets, that is, developing countries, which are not listed in Annex I. Article 12(2) sets out clearly the objective which is 'to assist Parties not included in Annex I in achieving sustainable development and in contributing to the ultimate objective of the Convention, and to assist Parties included in Annex I in achieving compliance with their . . . commitments under Article 3.'

The particular appeal as well as the tension of the CDM however lies in the fact that States with commitments would be financing projects in countries without commitments. This should mean of course that developing countries share the benefits of project investments in clean technology within their economies, but also that a far wider pool of possible reductions is possible than would be available through reductions in Annex I countries alone. Despite some initial concerns that this might be seen as the 'thin end of the wedge' in the introduction of commitments to developing countries, in fact the basic assumptions on which the CDM is based are sound. Even if the majority of the responsibility for the historical emissions leading to climate change can be laid at the doors of the developed world, a reduction of emissions anywhere in the world has an equally beneficial impact on the global climate system, and it is in the developing world—notably in India and China—where much of the increase in future emissions of GHG seems likely to take place in the decades to come. Therefore, a mechanism that encourages investments in climate friendly projects in developing countries should also have a beneficial impact on the global climate system, as well as produce a net beneficial impact on sustainable development. There is however also a global community interest in ensuring that Annex I countries or entities only be permitted to claim credit for those projects in developing countries which demonstrably reduce GHG emissions. Hence the Protocol envisaged a supervisory structure—the CDM Executive Board—including an emission reduction verification and certification system. CDM projects will need to be validated and registered as a CDM project by 'operational entities' designated by the COP/MOP, who are also responsible for

[41] J Werksman, 'Unwrapping the Kyoto Surprise' (1998) 7/147 Review of European Community and International Environmental Law 151.

verifying and certifying that emission reductions actually occurred. Validation of the project through the operational entity represents the certification of the project as contemplated by Article 12(5) of the Kyoto Protocol. Operational Entities also certify emission reductions accrued with the project. An emission reduction so certified is therefore termed a Certified Emission Reductions (CER)—to distinguish it from the product of Article 6 projects, which is termed an Emission Reduction Unit (ERU).

The administrative expenses of servicing these bodies will be covered, *inter alia*, by a levy of an administrative fee from certified project activities.[42] An additional 'share of proceeds' will be levied in the form of CERs, which will be used to 'assist developing countries that are particularly vulnerable to the adverse effects of climate change to meet the cost of adaptation.'[43] In a number of other respects CDM projects will resemble Article 6 projects. Participation in CDM projects must be voluntary; projects must manifest real, measurable and long-term benefits relating to mitigation of climate change; and a project activity generating CERs must be 'additional' to that which would have occurred in its absence.[44] Participation in the CDM is also open to the involvement of private and/or public entities,[45] subject to the guidance of the Executive Board. The COP/MOP has overarching responsibility to provide 'authority and guidance' and is responsible for elaborating 'modalities and procedures'[46] for the operation of the CDM. At COP7 it was decided that 'afforestation and reforestation' projects were CDM eligible,[47] and the modalities for such projects were finalized at COP9 in Milan in December 2003,[48] but the COP has yet to reach final decisions on whether other sink projects will be CDM eligible.[49]

[42] Now fixed at 2 per cent. FCCC/CP/2001/13/Add.2, Decision 17/CP.7 Modalities and procedures for the clean development mechanism as defined in Article 12 of the Kyoto Protocol, para 15(a).

[43] Although the modalities have still to be finalized, the GEF Council approved preliminary arrangements for the establishment of such a fund (GEF/C.19/6) and invited the World Bank to be the Trustee. *Summary of Chairs, GEF Council*, 15–17 May 2002.

[44] See Art 12(5)(c).

[45] Compare, on the one hand Art 6(3) which says legal entities, authorized by Annex I Parties, may participate in 'actions *leading to* the generation, transfer or acquisition . . . of emission reduction units' (emphasis added). This could be interpreted as saying that they may not be transferred or acquired (and therefore perhaps not owned?) by such entities. On the other hand, by contrast, Art 12 envisages 'participation . . . in the acquisition of certified emission reductions' suggesting that private/public entities might own these CERs.

[46] Art 12(7) empowers the COP/MOP at its first meeting, to 'elaborate modalities and procedures with the objective of ensuring transparency, efficiency, and accountability through independent auditing and verification of project activities'.

[47] FCCC/CP/2001/13/Add.2, Decision 17/CP.7, para 7.

[48] Decision -/CP.9 Modalities and procedures for afforestation and reforestation project activities under the clean development mechanism in the first commitment period of the Kyoto Protocol.

[49] The treatment of land use, land-use change and forestry project activities under the CDM in future commitment periods will be decided as part of the negotiations on the second commitment period; FCCC/CP/2001/13/Add.2, Decision 17/CP.7, para 7.

9 Article 17

The third mechanism envisaged by the Kyoto Protocol allows the trading of parts of Assigned Amounts, CERs, ERU and removal units (RMUs)[50] among Annex I countries (Article 17 of the Protocol). 'Assigned Amounts' refers to the quantity of GHGs a Party to the Kyoto Protocol is allowed to release in the global atmosphere as calculated on a yearly basis in Annex B of the Protocol (Article 3 of the Protocol). This mechanism is also referred to as 'emission trading'.

10 The Role of the World Bank

Climate change is a matter of serious concern to the World Bank as an international development bank. Analysis by the IPCC estimates that, should the level of CO_2 concentration in the world be doubled, developing countries would suffer economic costs of 5–9 per cent of GDP, several times higher than industrialized countries, and that the poor in the Bank's borrowing countries would be at the greatest disadvantage.[51]

Others have charted the way in which the Bank began, in the last two decades of the twentieth century, to examine more rigorously the environmental impacts of its lending and to become more actively involved in pursuing an environmental agenda, including the financing of environmental projects.[52] In the run-up to the 1992 Rio Earth Summit—UN Conference on Environment and Development—the issues of climate change and biodiversity loss began to assume greater political significance and a number of Western donor countries began to re-examine the way in which they were supporting 'global environmental public goods'. The Brundtland Commission in its seminal 1987 Report, *Our Common Future*, had proposed the 'development of a special international banking programme or facility linked to the World Bank'[53] and in response to a French and

[50] RMUs are generated by land use, land-use change, and forestry activities such as afforestation or reforestation that sequestrate CO_2.

[51] Intergovernmental Panel on Climate Change (IPCC), *Climate Change 1995: The Economic and Social Dimensions of Climate Change* (Cambridge, 1996).

[52] See K Piddington, 'The Role of the World Bank' in A Hurrell and B Kingsley (eds), *The International Politics of the Environment* (Oxford, 1992) 212–27, 215. R Wade, 'Greening the Bank: The Struggle over the Environment' in D Kapur, JP Lewis, and R Webb (eds), *The World Bank: Its First Half Century* (Washington, DC, 1997) 611–734. Also D Freestone, 'Incorporating Sustainable Development Concerns into the Development and Investment Process: The World Bank Experience' in M Fitzmaurice and M Szuniewicz (eds), *Exploitation and Management of Natural Resources in the 21st Century: The Challenge of Sustainable Development* (The Hague, 2003) 91–112.

[53] *Our Common Future* (Oxford, 1987) 338. This idea was developed by the Washington, DC-based World Resources Institute (WRI) as a proposal for a global International Environmental Facility (IEF). See F van Bolhuis (ed.), *Natural Endowments: Financing Resource Conservation for Development* (Washington, DC, 1989) 14.

German proposal at the 1989 World Bank Annual Meetings, a donors meeting was held in Paris in November 1990 to finalize a proposal for a Global Environment Facility (GEF) with a commitment of some US$1 billion over a three year pilot phase. A proposal for the establishment of a Global Environment Trust Fund was then approved by the Bank's Board of Executive Directors by Resolution No 91-5 on 14 March 1991. The GEF itself was established by tripartite procedural resolutions dated 28 October 1991 by the United Nations Development Programme (UNDP), the UN Environment Programme (UNEP), and the World Bank. By the October resolutions the three institutions agreed to cooperate in the implementation of the pilot phase of the facility. The Bank was to continue to act as Trustee of the Fund as well as (with UNDP and UNEP) one of the three implementing agencies.

As discussed above, the text of the UNFCC itself (as well as that of the Convention on Biological Diversity[54] and Agenda 21) indicates some concerns at the governance of the pilot phase GEF. Agenda 21 recommended that the GEF should be 'restructured' so as to, *inter alia*, 'ensure a governance that is transparent and democratic in nature, including in terms of decision-making and operations, by guaranteeing a balanced and equitable representation of the interests of developing countries, as well as giving due weight to the funding efforts of donor countries.'[55] The negotiation of the Restructured GEF, involving some 73 states, was completed in Geneva in March 1994 and a replenishment of some US$2 billion was agreed.[56]

The innovative governance and decision-making structure that was agreed in the restructuring process is set out in the Instrument establishing the Restructured Global Environment Facility.[57] Actually, the Instrument is innovative in a number of ways. The legal status of the Instrument itself may be unique and the collaborative mechanism between the UN family and a Bretton Woods institution which it embodies is also highly novel. The Instrument is not an international treaty. After the text had been negotiated it was approved by three separate resolutions of the three Implementing Agencies—the Executive Directors of

[54] Arts 21 and 39 Convention on the Protection of Biological Diversity.

[55] Ch 33.16(a)(iii). A similar provision is expressly included in the text of Art 21(3) UNFCCC, n 24 above, the final sentence of which reads: '. . . The Global Environment Facility should be appropriately restructured and its membership made universal to enable it to fulfil the requirements of Art 11.'

[56] In March 1998 at a meeting in Paris the negotiations for a second replenishment (GEF-2) were completed with commitments totalling US$2.75 billion for the four-year period to 2002, and in August 2002 a third replenishment was agreed to 2006 of more than US$2.9 billion.

[57] For text of the Instrument for the Establishment of the Restructured Global Environment Facility, Report of the GEF Participants Meeting, Geneva 14–16 March 1994, see (1994) 33 ILM 1273. See further IFI Shihata, 'The World Bank's Contribution to the Development of International Environmental Law' in G Haffner, G Loibl, A Rest, L Sucharipa-Bermann, and K Zemanek (eds), *Liber Amicorum Professor Seidl-Hohenveldern—in honour of his 80th birthday* (Dordrecht, 1998) 631–57.

the World Bank,[58] the Governing Council of UNEP,[59] and the Executive Board of UNDP.[60]

Largely as a result of its role as one of the three Implementing Agencies of the GEF, the World Bank has the largest investment portfolio of climate friendly energy projects in the world. In its Review of the first decade of the GEF, the Second Overall Performance Evaluation Team noted that in this period over US$1 billion had been committed to 270 climate change projects in 120 countries—with an extra US$5 billion in co-financing.[61] The World Bank share of this portfolio is approximately 40 per cent. In addition, since 1995 the Bank has—with the support of the Government of Norway—invested in a number of 'Activities Implemented Jointly' projects, under the AIJ pilot phase established by the Berlin UNFCCC COP1.[62]

11 The Evolution of the World Bank Carbon Finance Portfolio

In February 1997 the World Bank Environment Department established a Global Carbon Initiative to attempt to respond to a broader development agenda and to explore how market based mechanisms—contemplated even in 1997 by the Parties to the UNFCCC—could help reduce the global concentrations of GHGs and contribute to the sustainable development of the Bank's borrowing member countries. Its first product was envisaged as a demonstration multipartite carbon fund. The basic concept of the fund was remarkably simple. It would collect money from both public and private sector entities; use the pooled funds to convert normal energy projects into a portfolio of climate-friendly projects in return for the emission reductions, which they generated. The emission reductions so generated would then be distributed pro rata to those entities that had contributed to the fund. The portfolio approach would spread the risks which result

[58] The Executive Directors of the World Bank adopted the instrument at a meeting in Washington, DC, on 24 May 1994 (Resolution No 94-2, 'Global Environment Facility Trust Fund: Restructuring and First Replenishment of the Global Environment Facility'), and the Board of Governors of the Bank adopted a resolution on 7 July 1994, approving cooperation by the Bank with other international organizations as appropriate to achieve the purposes of the GEF (Resolution No 487, 'Protection of the Global Environment').

[59] The Governing Council of the UNEP adopted the instrument at a special meeting of its Governing Council in Nairobi on 18 June 1994 (Resolution ss.iv.1, 'Adoption of the Instrument for the Establishment of the Restructured Global Environment Facility').

[60] The Executive Board of the UNDP and of the UN Population Fund adopted the instrument at a meeting in New York on 13 May 1994 (Document DP/1994/9, 'Report on the Second Regular Session').

[61] GEF, *The First Decade of the GEF: Second Overall Performance Evaluation* (Washington, DC, 25 January 2002) 15ff.

[62] See n 25 above and also D Freestone, 'World Bank Initiatives Towards AIJ and its Potential Role in the Global Carbon Investment Business', Proceedings of the International Workshop on Joint Implementation and Activities Implemented Jointly (International Energy Agency, 1997).

from total project failure (which some AIJ/JI projects had faced) and also allow economies of scale and the replication of basic transactions. By entering into the field at an early stage of the development of the market for emission reductions the Bank also felt that it could provide comfort to the private sector and provide important practical experiences for others to follow. The costs of developing this demonstration fund were met by the 'New Products Fund' within the Bank that was established by the Strategic Compact—itself approved by the Bank's Board of Executive Directors in March 1997. In June 1997, Mr James Wolfensohn, the Bank's President, announced at the United Nations General Assembly Special Session that the Bank would launch a carbon fund should the UNFCCC Parties be favourably disposed towards such an initiative.

After an extensive consultative process, involving potential contributors to the fund from both the public and private sectors, the World Bank Board of Executive Directors, and a broad spread of potential project host countries as well as other stakeholders including the UNFCC Secretariat, NGOs, and key negotiators, the proposal to establish a Prototype Carbon Fund (PCF) was approved by the Executive Directors of the Bank on 20 July 1999.[63] The PCF built on the World Bank's significant experience in investing in climate friendly technology projects in many parts of the world. It was called a prototype, because its primary intent was that it would be able to act as a 'trail blazer' for other similar funds that might be launched by the other International Financial Institutions (IFIs) or the private sector. Because of its ongoing portfolio and its pipeline of projects, including renewable energy and energy efficiency projects, in both developing countries and EITs, it was envisaged that the Bank would be able to provide baseline, or underlying projects, onto which the additional resources and environmental benefits required by the Kyoto Protocol could be relatively easily grafted.[64] Similarly, after nearly ten years of experience in working at a project level with the 'incremental costs' requirements of GEF funding, many of the basic concepts of baseline projects from which additionality (as required by Kyoto) must be shown, were relatively well understood by its operational staff.

In order to mobilize the sorts of investments necessary to bring about the global changes in energy consumption patterns necessary to meet the Kyoto commitments within the commitment period it has been estimated that resource flows in the order of US$30 billion a year may need to be available.[65] As should be apparent from the discussion above, the negotiations of the details of the Kyoto Protocol regulatory framework were proceeding extremely slowly and in the face of regulatory uncertainty the private sector was clearly reticent to invest in Kyoto projects

[63] For further details of the consultative processes and the PCF itself, see D Freestone, 'The World Bank's Prototype Carbon Fund: Mobilising New Resources for Sustainable Development' in Schlemmer-Schulte, *et al.* (eds), *Liber Amicorum for Ibrahim S.I. Shihata* (The Hague, 2001) 265–341, 282–3.

[64] For additionality, see n 37 above.

[65] AD Ellerman, HD Jacoby, and A Decaux, *The Effect of Developing Countries on the Kyoto Protocol and CO_2 Emissions Trading* (IBRD Research Paper, September 1998).

until the risk elements were reduced. There was also some evidence of mistrust between the Annex I countries and those counties without emission reduction commitments, the majority of which are members of the G77. Bank management thought that a fund, which would be able to make investments in a participative and open way, could assist both host and financing countries to appreciate the benefits to be derived from the Kyoto mechanisms. The development of a new commodity—an emission reduction—as an asset with a saleable value would increase the rate of return of many currently financially non-viable renewable energy and other sustainable development projects and hence make them 'bankable'. In the early stages of the development of these transactions there were, and remain, many lessons to be learned and good practices to be developed. The objective of the prototype fund was to show how these transactions could be concluded in a manner that would assist both the industrialized world and also those host countries interested in developing their capacity to take full advantage of the opportunities offered by the Kyoto mechanisms.

12 Key Features of the PCF

Since its launch in 2000, a great deal has been written about the PCF.[66] It was intended as a prototype for similar funds by other IFIs and by the private sector. However the ongoing regulatory uncertainty resulting from first the delay in the finalization of the rules and modalities for the Kyoto mechanisms and subsequently the withdrawal of the US from Kyoto as well as the delays in its entry into force have put something of a chill over private sector investments in the field. In the early debates on the proposed fund fears were expressed that a World Bank fund would 'crowd out' the private sector from the small emerging market for emission reductions. In fact, in light of the slow development of the market, it can be said that the success of the PCF has been to 'crowd in' the private sector—in the sense that until 2003 most of the private sector investments in projects, which hoped to meet the requirements of the Kyoto Protocol, have been by or in association with the PCF. The PCF has also been the model for further 'boutique' funds which the Bank has launched which are aimed at mobilizing resources for projects

[66] See for example, Freestone, n 63 above; <http://www.carbonfinance.org>, S Huq, *Applying Sustainable Development Criteria to CDM Projects: PCF Experience*, PCF*plus* Report 10 (Washington, DC, 2002); HC de Coninck, and NH van der Linden, *An Overview of Carbon Transactions: General Characteristics and Specific Peculiarities* (ECN Beleidsstudies, March 2003); C Sutter, *Sustainability Check-Up for CDM Projects: How to Assess the Sustainability of International Projects under the Kyoto Protocol* (Berlin, 2003); 'Pioneering Transactions, Catalyzing Markets and Building Capacity: the Prototype Carbon Fund Contributions to Climate Policies' [2003] American Journal of Agricultural Economics 85, Proceedings of the 2003 ASSA Meetings, 703–7; C Streck in F Biermann, R Brohm, and K Dingwerth (eds), *The Clean Development Mechanism—A Playing Field For New Partnerships* (Potsdam 2002). Global Environmental Change and the Nation State; 'Proceedings of the 2001 Berlin Conference on the Human Dimensions of Global Environmental Changes' (Potsdam, Potsdam Institute for Climate Impact Research) 266–73.

likely to be unattractive to the private sector but which have definite sustainable development benefits. In addition the many practical lessons that it has learned have been made available through its website and have also been used by the Bank itself in developing a series of funds managed for particular countries.

The basic design concept is explained above and is set out in the instrument establishing the fund. It was envisaged as a multipartite fund with contributors from both the public and the private sectors. Contributions were fixed at US$5 million for private sector participants and US$10 million for public sector participants. The contributions are supported by promissory notes but drawdowns to the fund are dictated by fund management needs and agreed at the annual Participants Meeting. The PCF itself is a World Bank Trust Fund—a model followed by the subsequent funds.[67] The administration of the fund is in the hands of a Fund Management Unit headed by the Fund Manager and overseen by a Fund Management Committee of senior Bank officials. In addition to an Annual Participants Meeting there is an elected Participants Committee, representing all the Participants, which has met regularly, actually and virtually, between the annual meetings. As well as offering guidance on the implementation of the Fund the Participants Committee reviews all the project concept notes that the Fund Management Unit (FMU) puts forward for support from the Fund and has a one-off opportunity to veto possible projects at this early stage.[68] Input into the policy and management of the Fund is also provided by the Host Country Committee, which is comprised of representatives of those countries which are hosting PCF projects or which have signed MoUs with the Bank expressing an interest in being involved in the process. The Instrument includes project section criteria (designed to ensure that projects meet UNFCCC, Kyoto Protocol, World Bank, and other important project criteria) as well as project portfolio criteria (designed to ensure project diversity in both technology and geographical region). As a prototype the fund was capped at US$180 million[69] and it is envisaged that it will be wound up in 2013 at the end of the Kyoto Protocol Commitment Period.

In the course of implementation of the fund a number of important legal and practical lessons were learned—many of which are discussed in detail in the chapters of this book. From a practical point of view it became clear that it would

[67] In 1978 Sir Joseph Gold, former General Counsel of the International Monetary Fund, identified the proliferation of arrangements bearing the name 'trust' within the practice of international organizations and analysed their key characteristics. He argued that '. . . the establishment of trust funds for administration by existing international organisations under implied or express powers avoids the need to create new organisations with limited purposes and makes use of the experience and demonstrated skills of organisations already in the field.' see J Gold, 'Trust Funds in International Law: The Contribution of the International Monetary Fund to a Code of Principles' (1978) 72 American Journal of International Law 856–66.

[68] Although this has been rarely used, the veto power was intended to provide important Participant input without compromising Fund Management discretion.

[69] As it reached US$145 million in its two closings, the Participants agreed that the fund cap be reached only by 'voluntary supplementary contributions' from existing Participants. This was done in the winter of 2001.

not be feasible for the fund to confine its support—or 'investments' as the Instrument originally called them—to ten to fifteen projects most of which would be drawn from the World Bank portfolio. Opening its website to project proposals from outside the Bank resulted in an avalanche of proposals from the private sector and civil society. Most of these were relatively small projects yielding payments for carbon in the order of US$2–3 million rather than US$10–15 million. To deal with these the FMU established a project proposal format—a Project Information Note (PIN)—which was then subjected to extensive internal review before funding would be released for the development of a PCN. This process has meant that the bulk of projects supported by the PCF are not associated with underlying Bank projects, but rather are from the private sector.

In the course of negotiating the very first project—the Liepaja solid waste project in Latvia—it also became clear that the form of contribution envisaged by the Fund Instrument modelled on the modus operandi of the Global Environment Facility—namely investment in specific project components—was not an effective way of developing a market in 'emission reductions'. At an advanced stage in the negotiations with Latvia the agreement was transformed from a traditional project finance agreement into a totally new instrument—an Emissions Reduction Purchase Agreement (ERPA). The basic structure was that, in return for the project generating Emission Reductions (ERs) verified by an independent certifier, the PCF would pay agreed amounts.[70]

The Liepaja Agreement was with the Republic of Latvia, which underwrote the delivery of the ERs with a pledge of Assigned Amount Units (AAUs) in the event of project failure. Agreements with host countries were initially envisaged as an essential part of the modality for recording the 'consent of the Parties'—required by both Articles 6 and 12.[71] However two issues intervened. First, a significant number of developing countries began to ratify the Protocol and to implement domestic measures designating national authorities (DNAs) and national procedures for CDM project approval. This rendered the conclusion of a formal host country agreement between the Bank and the CDM host—which was always a time-consuming process—not only superfluous but also arguably deleterious to the development of autonomous national processes. Host country agreements with CDM hosts, which had ratified the Kyoto Protocol, were therefore discontinued. Second uncertainties continued with EIT countries as to whether it would be possible for projects in such countries to generate ERUs under the Kyoto regime prior to 2008. The PCF decided therefore to seek reassurance from such countries that assigned amount units corresponding to the amounts of verified emission reductions generated by PCF-financed projects would be available prior to 2008 on a tonne-for-tonne basis. These agreements took a long time to develop, and

[70] This required an amendment to the Instrument, which was agreed unanimously by the Participants and then approved by the Board of the Bank in June 2001. The text of the Latvia ERPA is published in (2003) 1 *World Bank Legal Review* 433–524.

[71] See Arts 6(1)(a) and 12(5)(a).

indeed in some Eastern European countries proved impossible (for example, Hungary). The result has been a much smaller portfolio of PCF projects in Eastern Europe than had originally been envisaged.[72]

13 The Expansion of the Carbon Finance Portfolio

Despite initial suspicions among some of the G77 negotiators over the concept of the PCF, the fund quickly gained popularity in developing countries anxious to take advantage of the inward investments that the CDM promised. A number of Bank programmes financed by donors were launched to build capacity for developing countries to respond to the opportunities that the Kyoto mechanisms might present.[73] The PCF itself held capacity building workshops in conjunction with a number of its early projects to which other countries in the regions were invited. These were so successful that it soon became clear that PCF resources would not be sufficient to finance projects in all the countries that had come forward to sign MoUs with the Bank in order to join the PCF Host Country Committee. It also became very clear from the first projects that the PCF undertook that, despite the number of capacity building programmes being developed by a number of agencies, the ultimate capacity building experience was actually to host a project, which aimed to meet all the Kyoto eligibility requirements. Such an experience would require, for example, the designation of a national authority (or focal point) for Kyoto Protocol activities, the identification of the responsible ministry, the determination of the correct legal authority to sign an ERPA or grant authorization for the transfer of CERs/ERUs from project entities to the PCF Participants. Given the demand for such experiences Bank management determined that it would be desirable to explore alternative sources for financing emission reduction purchase agreements. The Bank was therefore receptive to the approach of the Netherlands Ministry of Housing, Spatial Planning and the Environment (VROM) when it proposed establishing a Netherlands Clean Development Facility within the Bank to purchase up to 16 million tonnes of CO_2 equivalent emission reductions on behalf of the Netherlands. Similar approaches have now been received from the Government of Italy. This has further enhanced the ability of the Bank to widen the number of countries in which carbon purchase transactions can be made and to provide countries and project sponsors through parallel purchase arrangements with the opportunity to leverage further resources.

Given the slow start of the CDM and the reluctance of investors to finance projects designed to be eligible for Kyoto Protocol emission reduction credits, it also became apparent that those countries which had most welcomed the CDM,

[72] The original portfolio criteria envisaged a 'broad balance' in the number of projects between Arts 6 and 12 projects.

[73] The National Strategic Studies Programme financed by the Swiss, Australian, and Canadian Governments.

particularly the poorer countries of Africa, were in fact the countries least likely to see the benefits of such projects. Larger economies with ongoing demand for increased energy production quickly became seen as the most attractive sites for investment and the opportunities for using carbon finance for small sustainable development projects in the poorest countries or the poorest areas of developing countries seemed likely to be missed. Hence the development of two other specialist funds by the Bank, which sought to fill a niche unlikely to be addressed by the private sector. In developing these funds the intent was to provide a distinctive role for the Bank in the growing carbon market—leaving the major attractive investments to the private sector and concentrating on projects with demonstrable sustainable development benefits which could show the way in which carbon finance could assist the Bank's poorest clients. These two funds are the Community Development Carbon Fund[74] and the BioCarbon Fund.[75] The CDCF was developed in association with the UNFCCC Secretariat and the International Emission Trading Association—a not-for-profit body set up to assist the development of carbon trading.[76] The CDCF aims to finance small projects with certifiable community development benefits in the poorest regions of the world—likely to be ignored by the core carbon business. The BioCarbon Fund seeks to demonstrate the benefits of carbon finance in the sequestration sector. This is a controversial area but COP9 in Milan in December 2003 saw the eventual agreement of the UNFCCC Parties to rules relating to reforestation and forestation projects under the CDM. One window of the BCF will finance such projects with the aim that they should generate emission reductions that could be verified as meeting Kyoto Protocol requirements. The second more exploratory window of the fund would finance sustainable agriculture projects and other sequestration projects—currently not eligible under the Kyoto Protocol rules—with the aim of demonstrating how practical solutions can be found to many of the problems surrounding the generation of emission reductions by such activities.

14 Conclusions

This chapter has sought to provide an introduction to the UNFCCC system and to the processes leading up to, and from, the adoption of the Kyoto Protocol. It has aimed to introduce some of the key processes and describe some of the key concepts that will be further developed in later chapters. It has also attempted to map out the rationale for the initial involvement of the World Bank in the area of climate change mitigation and to explain how thinking has evolved during the

[74] The CDCF was announced at the 2002 Johannesburg Summit for Sustainable Development and started operations in June 2003.

[75] The BioCarbon Fund was announced in November 2003 in Japan and started operations in June 2004.

[76] See further <http://www.carbonfinance.org> accessed 13 June 2004.

process of implementing the innovative and groundbreaking Prototype Carbon Fund. The scientific evidence is now clear that it is the poorest countries that are likely to feel the impacts of climate change the most severely and it is also clear that if market forces are left to the private sector, the countries and communities which need the important financing opportunity for sustainable development which carbon finance provides, are the ones least likely to benefit. Here the Bank sees itself as having a major catalytic role in using the emerging carbon market to mobilize additional resources for these countries and communities. The remaining chapters of the book aim to look in much more detail at the legal and policy issues that are an essential aspect of the development of a global carbon market.

2

Climate Change in Business

*James Cameron**

There is no doubt that the threat of climate change is real and dramatic in scale. It looms large and full of foreboding. It can make one feel overawed, nihilistic, or depressed into inactivity. Those feelings are of little use if one also wants to do something about 'anthropogenic' climate change. In so far as there is a global culture, it is one where business enterprise presents itself at centre stage. Over two decades now, what we simplistically describe as 'business' has played a central role in the societal response to climate change. For much of that period, that role was, in the hands of very powerful energy and natural resource interests, malevolent. That was changed in part through the public pronouncement of business leaders like Lord Browne of BP. However, more is to come if business comes to see climate as a story of opportunity. If ventures that reduce dependency on cheap but destructive fossil fuels look profitable, the politics of climate will alter dramatically.

It is perfectly possible to feel confident about changing the way we orient our economic life. The Framework Convention on Climate Change (UNFCCC or the Convention),[1] signed in Rio in 1992, shows the way. The international legal obligation contained in Article 2 of the Convention, can guide and direct thousands of positive policy responses. Business interest in private gain can align with the public good. As policy frameworks become clearer and a new cycle of risk-taking nearer, huge amounts of capital can flow to climate-change solutions, both technological and behavioural. Trustees of pension funds designed to mature, and which keep maturing, decades hence have fiduciary duties already closely aligned with investment to abate the risks of climate change. The moment that long-term capital decision-makers realize this, very substantial amounts of resources will be free to flow into beneficial alternatives, or to change the ways we use fossil fuels to deliver our economic wants and needs.

* James Cameron is a barrister at law, co-founder of Climate Charge Capital, Counsel to Baker & McKenzie in London. His expertise is in the Climate Change, Environment, WTO-International Trade Regulation & Customs, EU law and litigation practices of the firm. The views expressed in this chapter are the views of the author and do not necessarily represent the views of Baker & McKenzie.

[1] United Nations Framework Convention on Climate Change (UNFCCC), 1771 UNTS 107, reprinted in (1992) 31 ILM 849.

Change need not be absolute. Relative shifts, offsets, and compromises offer huge prizes, both economic and environmental. All the possibilities, including biofuels, blended bio-diesel, gas, clean coal technologies, and sequestration, geological, chemical, biological, and nuclear power (perhaps to make hydrogen), can help to stabilize concentrations of greenhouse gases (GHGs) in the atmosphere at below 450 ppm CO_2 concentrations, generally regarded as 'safe'. The real prize today is delivering energy and power needs with renewable resources and/or in distributed systems. All this is perfectly achievable in the next half century, provided the next decade sees sufficient investment in both clean technology and carbon abatement.

When politicians feel the optimism in the business community, when they sense that the lower-carbon economy is achievable, they will be bolder about increasing the cost of carbon, and thereby accelerating the take-up of low-carbon technologies. But the scale of change required is so great that governments need to play a part in reshaping markets, to provide incentives for rapid innovation. Clearly we need cooperative governmental action; we also need what good companies have—that crucial combination of instinctive entrepreneurial risk-taking, and careful and rational management of their own resources.

Working on the climate change issue for any number of years can be a depressing experience. We are used to describing the problem in terms of catastrophic threat, and the governmental response in terms of burden, risk, and a sort of collective denial of wants. Those of us in the OECD world created the problem; as punishment for our collective sins, many of the things that we associate with successful economic life are suddenly considered bad. Think of cars, air-conditioning, central heating, abundant energy, and power at low prices and astonishing physical freedoms created by the use of technology and materials made possible by access to energy—not to mention plastics and lubricants and the myriad refined products from carbon-based fuels. Fossil fuels are now the focus of the debate about business and climate change. Within a lifetime a once-glorious, adventurous enterprise has become one in which intelligent, educated and ambitious people are beginning to feel guilt even for being associated with an oil company. This feeling is not universal, but it is increasingly apparent in Europe amongst the young, and is the way our societies talk about the oil and minerals businesses.

Unlike the US, Europe has never had an active business lobby against constructive action to deal with climate change—even robust governmental intervention in markets in which business is affected. But neither side of the Atlantic has yet seen a powerful movement in the business community for a major shift towards alternative energy and power systems. The voice of the 'booster' may well be heard more strongly in the US, where enthusiasm for new technological solutions and a strong desire to loosen dependency on imported oil may create business leadership to move us away from the oil economy to the hydrogen economy. Certainly the current administration of George W Bush, the Governor of California, one Arnold Schwarzenegger, and the car manufacturers are communicating a strong interest in the potential of hydrogen. I am not able to judge whether any one of the many visions for a hydrogen economy is the right one to follow. Nevertheless, even though concerted business opposition to grappling with the climate change issue emerged in the US and not in Europe, one should not assume that business leadership toward solutions should necessarily reside in Europe rather than the US, or indeed elsewhere in the world—China, India, Brazil perhaps. However, policy-makers in the European Union are clearly convinced that policy structures to be put in place will create stronger incentives for technological advancement in Europe, simply because businesses will learn to value carbon.

For the time being, the debate about business and climate change is locked onto the discussion about the price of carbon. Within that structure, we can see the full range of responses to a policy-driven market, from threat to opportunity and back again. Many years of negotiating international instruments on climate change has produced practical results. The Kyoto Protocol has already succeeded in inspiring domestic action even as it awaits ratification. You may now find yourself with a chief executive officer of a large energy user company, discussing the way accounting rules will treat allowances granted by the European Commission within the context of the Emissions Trading Scheme. In the course of this year and the next, the decision-making machinery of corporations throughout Europe will be grappling with these issues for the first time. Businesses are trying to understand how much asset and liability they now have. An intergovernmental policy-making process to deal with climate change has produced rules that can be enforced, and come with a price-tag attached. Models are being made, to show what the price for carbon does to the options available to a large power company. When does the price argue for a switch from coal to gas? When and where is installing combined-cycle gas turbines economic? When, perhaps with other incentives, is substantial investment in renewables sensible?

If setting a price for carbon is essential, so are a whole suite of policies that deal indirectly with climate change. Increasingly, climate policy influences energy policy. In the UK, for instance, the White Paper setting the policy framework for the future attaches environmental priorities to long-run energy policy.[2] Dealing with security and diversity of supply also contributes to tackling the problem of human-induced climate change, by focusing on reducing or eliminating reliance on fossil fuels to drive our economic life. Such measures include promoting renewable energy; encouraging technological development to reduce CO_2 emissions; regulations for more efficient combustion engines, and for fuels lower in CO_2 and other pollutants; building regulations to encourage energy efficiency in homes, offices, factories, and other commercial buildings; and transport policy to reduce congestion on roads and increase the use of public transport, by charging to use road networks. Consider too the costs of nuclear power; the political risk in securing

[2] DTI, UK, White Paper, 'Our Energy Future—Creating a Low Carbon Economy' <http://www.dti.uk/energy/whitepaper/ourenergyfuture.pdf> accessed 13 June 2004.

ever-increasing amounts of fossil fuels from dangerous parts of the world; and even the ironic fact that a warming climate caused by burning fossil fuels complicates extracting these same fossil fuels from sodden permafrost in the Arctic Circle. At a certain level of abstraction these are all business issues. Whether they affect a company's immediate returns on investment or reporting requirements or shareholder value arguments or make or break a highflying career, will depend on two factors—cost and compliance. Each of these factors is an expression of value, in which the law is an expression of the contemporary political morality that governs the life of the corporation. Remember that the corporation is the product of law, a fiction designed for a purpose; and the law can change.

What of the physical consequences of climate change? Here we are looking at risk in a different way. Where do business assets lie? How easily can a business adapt to climate change, and who ultimately pays the cost? The reinsurance industry knows only too well the increase in cost of weather-related damage. Ten per cent year on year increases in cases from natural disasters. Jeremy Leggett in *The Carbon War* cites the 2 trillion dollars of uninsured assets at the coastal margin of Florida—and therefore exposed to sea-level rise and increased cyclonic activity when the industry as a whole keeps less than half a billion a year in reserve to cover all catastrophe losses everywhere in the world in any one year.[3] In 1990, when my colleagues and I were assisting the Alliance of Small Island States (AOSIS) in their negotiations on the climate convention, we produced a paper on insurance and climate change.[4] It demonstrated the difficulty of getting insurance in the small island states for any costal development, whether hotel, airport, road, or marina. The insurance industry had already determined, back then, that sea-level rise was an uninsurable risk, so certain was its arrival. Think of the implications for business assets at the coastal margin around the world whether in cities or agricultural land or in expensive residential property.

At the ninth session of the Conference of Parties (COP9) to the UNFCCC, held in Milan in December 2003, the Hadley Centre, Britain's leading climate research facility, reconfirming the announcement of the International Panel of Climate Change (IPCC) on the level of human contribution to climate change, produced more detailed modelling than ever before on the regional variations of the phenomena.[5] For effect they chose to focus on the region around northern Italy, where the meeting was taking place. They ran nine models, and then a probability analysis. The models produced a very high level of agreement that the next half-century would see an average summer temperature in the region warmer by more than seven degrees Celsius. To say that this will have business consequences seems too trivial. Indeed much of our discourse about climate change and business

[3] J Leggett, *The Carbon War: Dispatches from the End of the Oil Century* (Allen Lane, 1999).

[4] Insurance against Climate Change? The AOSIS proposal <http://unfccc.int/resource/ccsites/senegal/fs238.htm> accessed 13 June 2004.

[5] Climate change observations and predictions: Recent research on climate change science from the Hadley Centre, December 2003.

trivializes the cost of the consequences of climate change. Cost-benefit analysis is such an unimportant exercise here. We are so focused on the now, the next week, the next quarterly report that we seem unable to communicate to our business community that these risks are just not worth running. The scientific community is solidly agreed on the essential details of the climate-change problem, and indeed on the modelling of its likely effects, providing levels of predictability that most businesses would love in the year ahead, let alone in ten to fifty years. Looking at these risks, valuing the cost of inaction puts the costs of intervention now in proper perspective. The Region of Milan cannot afford this level of warming this fast at any price. Adaptation costs are inevitable but adaptation without strenuous efforts at mitigation is reckless.

For small island states, and indeed for developing countries with much land at or below sea level, the situation is even more serious. Their physical size, small economies, and limited options for adaptation expose them to the full range of climate risk. Their best hope now, as it was in 1990, is in the power of law to change behaviour and realign markets. Enterprises such as the Carbon Disclosure Project,[6] the Institutional Investors Group on Climate Change,[7] and Investor Network on Climate Risk[8] might finally penetrate the consciousness of mainstream investors—those who still channel trillions of dollars towards businesses whose success increases risk for long-term pension holders, as well as the coming generations. Climate-related litigation, particularly in the US, could combine with pressure from insurers on the terms of directors' and officers' insurance policies. High-profile shareholder actions could achieve some success, and cause concern in the boardrooms.

But this is all too negative. The real story about business in climate change ought to be one of optimism and opportunity—a passionate desire to find a better way to satisfy both a deeply sceptical post-modern industrial society and the essential needs of emerging economies. We need a new narrative. We need a story to tell, to inspire and guide—a coherent sequence of ideas, organized, directed and delivered through enterprise, both public and private. Private enterprises can concentrate the creativity, energy and desires of individuals. Together they can form a common purpose, and entrepreneurs can feel free enough to extend themselves fully, in pursuit of advantage for themselves, for their families and for their societies. The freedom to do so without depending on the state adds energy. The new story needs a framework. To understand a novel, you have to appreciate what a novel is, and to have some sense of its structure and form. In the same way, our new story will tell how we can transform the way our societies deliver the services we need—illumination, comfort, motive power, mobility, information, security—within a clear framework of policy, including all regulatory, fiscal, and other measures. The story will be about elegant industrial design, making and building

[6] The Carbon Disclosure Project <http://www.cdproject.net/> accessed 13 June 2004.

[7] Institutional Investors Group on Climate Change <http://www.iigcc.org/index.aspx> accessed 13 June 2004.

[8] The Investor Network on Climate Risk <http://incr.com/> accessed 13 June 2004.

things in harmony with natural systems. Think, for instance, of intelligent electricity networks linking many generators and users, with electricity a process, not a commodity, and users paying not for kilowatt-hours but for reliable illumination and other benefits. Such technological developments will be disruptive. To transform our human systems toward sustainable energy we shall need first of all to transform the way we think. The first and best response is imaginative. After that it is all about determination, money, and power.

Who knows where the tipping point truly lies, and where the last most significant impulse for change comes from? Look at the current situation in the US. The political circle around the White House (and most business elite) do not have a vocabulary to discuss climate change, either as urgent for policy makers or as opportune for business investment. But they do have a highly developed language for security risk. How might that play out? The Pentagon is now exploring the relationship between security risk and rapid climatic change.[9] This kind of scenario planning may have a poor track record; and those who have been working on climate issues for many years may feel no need to amplify the risk they already identify from the solid and reliable consensus of the scientific community. Nevertheless, well established anxieties about energy security, terrorism, politically unstable states, and violent conflict over scarce resources may combine to accelerate innovation of new technologies, and deployment of low-carbon technologies we already know about. If this helps us avoid catastrophe no one will complain. The trouble is, we are currently being led into further conflict over oil. Climate change and oil scarcity are linked in a death embrace. Do we have to fight over supplying our dependency on oil before putting the resources into alternatives? Good sense argues for higher levels of human and capital investment into clean fuels (even if fossil fuel based) and clean power.

What, moreover, of the spectacular opportunities in China, India, and Brazil, for example? In their various ways their dependency on imported fuels inhibits them, even though they possess substantial natural resources of their own. Why should they not lead the industrial revolution in solar technologies, in manufacturing wind turbines, and in designing and building intelligent electricity networks where no networks at all yet exist? All the signs are there today that they could. The single most important policy process to observe is China's long-term energy and power planning In May 2004 Pan Yue, Deputy Director of China's State Environmental Protection administration said, 'We should give full play to the governments green guidance, smash the monopoly interests of some powerful ministries and enterprises and invest large-scale government funds on the development of new energy sources . . .' This at the same time as the Government set ambitious renewable energy targets—10 per cent by 20 per cent.[10] Why should

[9] P Schwartz and D Randall, *An Abrupt Climate Change Scenario and Its Implications for United States National Security* (October 2003) <http://www.ems.org/climate/pentagon_cliamte_change.html> accessed 13 June 2004.

[10] J Kyngne, 'China's Growing Pains Call for Birth of Green Reduction' *Financial Times*, 25 May 2004.

the developing world not establish low-carbon businesses, to compete against those in the North and West that have had decades to capitalize on their investments in oil exploration, refining, and distribution through large multinational corporations of their own. Building global markets in these technological solutions will make obligations on states to reduce their emissions more easy to meet. It will guarantee that tougher obligations can be taken on; and it can mean real alignment of interest between the North and South.

In Europe we are witnessing the birth of a new market, a carbon market. It is a policy-driven market, a unique combination of trading in commodities, government securities and taxes. If the EU Emissions Trading Scheme extends as it should and could, within the next eight years we will create something equivalent to the bullion market attached to the gold standard for the future sustainable development of the global economy.

With the right price signal things could move fast. As they did with oil. Personally, I think the development of the oil economy is inspiring. Jeremy Rifkin tells us that the first gas/petrol station was put up in Detroit in 1911.[11] Five years later the US had more than three million cars, by 1925 twenty-three million.

When Winston Churchill was taking advantage of his new ministerial position at the outrageously young age of 26, he had the temerity to suggest that the British Navy should be fuelled by oil rather than coal. He said, 'To build any large additional number of oil-burning ships meant basing our naval supremacy upon oil. But oil was not found in appreciable quantities in our islands. If we required it we must carry it by sea in peace or war from distant countries. We had, on the other hand, the finest supply of the best steam coal in the world, safe in our mines under our land. To commit the Navy irrevocably to oil was indeed "to take arms against a sea of troubles" .' This was Churchill's chance to put into effect a plan he had that was dismissed by his predecessor Lord Selbourne seven years earlier. Lord Selbourne had said, 'the substitution of oil for coal is impossible because oil does not exist in this world in sufficient quantities'. This sounds like a contemporary argument against renewable energy. Churchill's response was, in effect, if I give the order it will be satisfied. He was acting strategically in times of crises and his courage and clear-sightedness worked. The Anglo Persian Oil Company, later British Petroleum/BP, was the direct beneficiary of the Royal Navy order, setting a pattern for governmental support of this vital industry. Sometimes you need regulatory intervention, other times fiscal or other market instruments, and sometimes you just need a long-term contract at scale. At all times leadership matters.

We know we are not going to stop using fossil fuels for some time yet. We know that huge coal and gas reserves will be used in order to power growing economies until alternatives to fossil fuels can meet massive demand with confidence. Those fossil fuels will have to be burned in a way that heavily restricts, or preferably removes, the threat they will cause to our atmosphere by releasing GHGs—hence

[11] Rifkin, *The Hydrogen Economy*: The Creation of the Worldwide Energy Web and the Redistribution of Power on Earth. (Tarcher/Putnam, New York, 2002) viii, 294.

the need for technologies and processes like justification and sequestration of CO_2. We must investigate many options, and invest in some. Should we opt for chemical sequestration, to fix CO_2 within rocks as calcium carbonate? Should we pump the CO_2 into old oil and gas fields, or aquifers? What are the risks, and who is going to bear and pay for them? If we take too long phasing out fossil fuels for power and mobility it will spoil the chances of the alternatives.

What of nuclear power? Can we risk that technology again, because it can generate substantial amounts of power without CO_2 emissions? I share Professor James Lovelock's view that climate change is so great a risk to the planet that ruling out the use of nuclear power would be foolish.[12] Old nuclear technologies produced damaging waste; and the industry was discredited by its excessive secrecy, the risks it ran with nuclear proliferation and the natural consequence of its allegiance with the military. That does not mean, however, that we should then rule out small-scale nuclear power. By comparison to today's giants, tiny pebble bed reactors might provide distributed generation, or produce hydrogen for the hydrogen economy of the future. A regulatory and market-based system to reduce GHGs depends in large part upon creating a meaningful price for carbon; it cannot deny opportunity to a low-carbon technology simply because in the past the technology has squandered public money and incurred public wrath. New business leaders will have to emerge, who can let go of the past and present an attractive future for consumers, policy-makers, and the risk-averse general public. To set boundaries for those new leaders, laws will be essential.

This great new enterprise requires a vision for a better way of life. Access to energy and the services it provides is central to our existence as a society. Huge responsibility lies in the change we must experience. Our story must be about building resilient systems that people can rely on, that do not trade-off our current needs against those of future generations. The first step is to stop feeling that the mission is impossible. I often think of the boosters of Chicago, the amazing confidence required to build that city, to take swampy land in the middle of a continent and create an agricultural hinterland to supply it—astonishing courage of conviction not dependent on careful cost-benefit analysis. Think, too, of the great monument builders, the lasting offerings to the divine in churches, mosques and temples—all required huge organizational effort, attention to detail, analytical skills, careful husbandry, and motivation beyond the ordinary accumulation of daily work tasks. In the same way, great public works—bridges, tunnels, aqueducts, railways, highways—are built because they make stories real, stories that promise a better life.

The challenge of climate change will bring out the best in our engineers, designers and inventors. It will add meaning and purpose to the provision of professional advice. It will inspire the money-lenders and the merchants to profit by delivering our salvation.

[12] J Lovelock, 'Nuclear Power is the only green solution' *The Independent*, 24 May 2004, <http://argument.independent.co.uk/comentators/story.jsp?story=524230> accessed 13 June 2004.

PART II

THE KYOTO FLEXIBLE MECHANISMS: GENERAL ISSUES

3

Legal Ownership and Nature of Kyoto Units and EU Allowances

*Matthieu Wemaere and Charlotte Streck**

1 Introduction

The Kyoto Protocol is based on the principle of achieving quantitative reductions in greenhouse gas (GHG) emissions by establishing binding emission targets on individual parties to the Protocol. While the overall goal is the reduction of GHG emissions, the Protocol allows for certain flexibility as to where the emission reductions can be generated. The flexibility is created through a mechanism for trading emission rights (Article 17), and two mechanisms for creating project-based emission credits, which are measured and verified against baseline emission levels (Articles 6 and 12).[1] The base units of the Kyoto Protocol are Assigned Amount Units (AAUs) where each unit is equivalent to one metric tonne of carbon dioxide emissions equivalents assigned to an industrialized (Annex I) country pursuant to Article 3 of the Kyoto Protocol. AAUs can be acquired and transferred in accordance with Article 17 of the Kyoto Protocol. Emission rights associated with Article 12 of the Kyoto Protocol and the Clean Development Mechanism (CDM) are Certified Emission Reductions (CERs). CERs are issued by the Executive Board of the CDM on the basis of an independent report that verifies and certifies their existence. Joint Implementation (JI) projects pursuant to Article 6 of the Kyoto Protocol generate Emission Reduction Units (ERUs) which come into existence through a conversion of an AAU or Removal Unit (RMU) from the accounts of the host country. RMUs result from eligible sink activities in accordance with Article 3(3) and (4) of the Kyoto Protocol.[2] The latest

* Matthieu Wemaere is an associate attorney at Huglo Lepage. Charlotte Streck is a legal counsel at the World Bank. The views expressed in this chapter are the views of the authors and do not necessarily represent the views of their employers or partners.

[1] Decision 15/CP.7 sets out the principles, nature and scope of all three mechanisms. Decisions 16/CP.7, 17/CP.7, and 18/CP.7 contain operational rules for JI, the CDM, and emissions trading respectively.

[2] For the rules governing the issuance of RMUs, start at UNFCCC/CP/2001/13/ Add.2, Decision 19/CP.7, Draft Decision -/CMP.1 Modalities for the accounting of assigned amounts under Article 7(4) of the Kyoto Protocol, Annex, para 25.

in addition to the family of Kyoto Units are tCERs and lCERs, defined by the ninth session of the Conference of the Parties to the UN Framework Convention on Climate Change (UNFCCC). Temporary CERs (tCERs) or long-term CERs (lCERs)[3] are associated with land-use change and forestry activities, which create removals in GHG.

This plethora of units, which determine the currencies of the international carbon market, have been recently complemented by national GHG emission rights. Most prominently, the 'Directive establishing a scheme for greenhouse gas emission allowance trading within the European Community' (EU Emissions Trading Directive) defined the tradable item as a transferable allowance to emit one tonne of carbon dioxide equivalent during a specified period.[4] The EU Emissions Allowances Trading Scheme (EU ETS) is intended to help the European Community meet its Kyoto target of an 8 per cent reduction in emissions relative to 1990 levels, during the Kyoto Protocol's first commitment period of 2008–12.[5] The EU Emissions Trading Directive is currently being transposed into twenty-five emission trading systems of EU Member States, whereby tradable allowances need to be allocated to covered installations before the start of the EU ETS in 2005. In each case the concept of the 'Allowance' needs to be integrated into the domestic legal system.

All these units, AAUs, ERUs, CERs, RMUs, Allowances, and verified Emission Reductions, are equal to one metric tonne of carbon dioxide equivalent and will have their own unique serial number. They are first and foremost accounting units, which are tracked and recorded through national registries to be established and maintained by Annex I Parties. But they are also tradable instruments, which share two important determining features:

1. They represent an entitlement to release a certain quantity of GHG emissions into the atmosphere; and
2. They are transferable under certain established conditions.

The creation of transferable rights in the context of emissions trading schemes also represents the main difference between these schemes and instruments rooted in the traditional command and control approach of environmental policy. Whereas in command and control instruments the entitlement to emit is granted on an individual basis as a condition to operate a well-defined installation, this entitlement is not transferable.

[3] COP9 Decision -/CP.9 'Modalities and procedures for afforestation and reforestation project activities under the clean development mechanism in the first commitment period of the Kyoto Protocol', Draft Decision -/CMP.1 'Modalities and procedures for afforestation and reforestation project activities under the clean development mechanism in the first commitment period of the Kyoto Protocol', Annex, para 1 (g), (h).

[4] Art 3 of the Directive 2003/87/EC of the European Parliament and of the Council establishing a scheme for greenhouse gas emission allowance trading within the Community and amending Council Directive 96/61/EC.

[5] The European Community has ratified the Kyoto Protocol on 31 May 2002.

However, despite the general similarity of the listed units, they also differ in significant aspects:

1. They are either created or allocated through a regulating body (international/domestic), and are subject to different legal regimes (mandatory/voluntary); and
2. They differ in the scope of their use, ranging from compliance with international or national legally binding reduction obligations to the fulfilment of voluntary commitments.

The following chapter aims at analysing the legal nature and characteristics of the different units. Clear decisions on the nature and the treatment of these units are important to give legal security and certainty to both governments and private (or public) entities and provide security and confidence to any emission trading system and improve market liquidity. Where legislations define tradable units differently, but where the trading systems are linked, such as in the EU ETS, difference in treatment can make one sub-market of the overall system more attractive than others. Different treatment can thus lead to, intended and—worse—unintended, distortions of the market.

The analysis in this chapter is ambitious in its scope as the different units pertain to markedly different legal frameworks including public international law, both civil and common law systems, and rights based on private contracts. They are also embedded in different legal traditions: civil law and common law systems traditionally define property rights differently—however, applying the same basic criteria based on the maxim '*usus, abusus, fructus*'[6] to characterize property. We will see that we are dealing with a new set of rights, which carries a number of features, which justify the classification of emission rights as instruments '*sui generis*'. These rights may not be assimilated easily by any existing legal system, requiring the nature and ownership to be newly determined by domestic law.

2 The Cornerstones of Emissions Trading

2.1 The Ethical Dimension

Emissions trading is based on the creation of tradable entitlements to emit a certain quantity of a pollutant. This concept, which relies on the privatization of formerly publicly held rights, has drawn significant criticism.[7] The idea of cutting the atmosphere in little pieces and distributing it to polluting industries has been

[6] This maxim refers to the right of an owner to use (*usus*), discharge or transfer (*abusus*), and benefit from (*fructus*) a thing or right he/she possesses.

[7] PS Menell, 'Institutional Fantasylands: From Scientific Management to Free Market Environmentalism' (1992) 15 Harv. J.L. & Pub. Pol'y, 489; MC Blumm, 'The Fallacies of Free Market Environmentalism', ibid 371; G Torres, 'Seventh Annual Lloyd K Garrison Lecture On Environmental Law: Who Owns the Sky?' (2001) 18 Pace Envtl. L. Rev., 227.

interpreted as 'privatization of the atmosphere' which would result in an abuse in the global commons that is the responsibility of the state to protect, not to give away.[8] It has been argued that by looking at the sky as though it were a common resource free of any substantial public interest other than the protection of its quality, governments have taken too narrow a view and neglected their capacity as trustee towards public resources by transferring a significant public resource to private hands, especially if this happens free of charge.[9] These critical voices are complemented by those of environmentalists, who feel insulted by the very notion of 'rights' to pollute, let alone a profitable market based on these rights.[10]

Responding to these concerns requires us to take a step back and to reconsider the nature of the right, which has been created for trade among private entities. The question is two-fold: first, has the government the necessary power to establish an emissions trading system? Secondly, what type of right does the government bestow in private entities that participate in such scheme?

With respect to the first question, it is important to keep in mind the management of global commons is largely based on property rights. This holds true for command-and-control as well as market-based approaches. In the process of formulating and implementing environmental policy the State has a choice of instruments, which represent different types of management systems. The difference between the instruments often lies only in the nature and extent of property rights imposed. The State is entrusted to manage, maintain, and conserve public goods for the benefit of its people. Effective as well as cost-efficient protection of the good should be the determining factor in the choice of the policy instrument, which will eventually depend on the type of good to be protected and the transaction costs related to a certain policy instrument. Regardless of the instrument chosen, however, the State's environmental protection efforts are inevitably property-based in that all solutions to the 'tragedy of the commons'[11] involve the imposition of property rights (private/individual, common, or state/public) on formerly unowned (non-property or open-access) resources.[12] When turning to command-and-control regulations the State asserts public property rights. When creating transferable emission rights, the State imposes a combination of public and private property rights on the environmental goods: it first converts non-property into public property before it converts the same property into private property.[13] The most

[8] See Torres, n 7 above. [9] See ibid, 286.

[10] RW Hahn and GL Hester, 'Where Did All the Markets Go? An Analysis of EPA's Emissions Trading Program' (1989) 6 Yale J. on Reg., 109, 142–3.

[11] G Hardin, 'The Tragedy of the Commons' (1968) 162 Sci., 1243. The 'tragedy of the commons' relates to the problem that free access to the commons leads to its overuse and ruination.

[12] DH Cole, 'From Local to Global Property: Privatizing The Global Environment?: Clearing The Air: Four Propositions About Property Rights And Environmental Protection' (1999) 10 Duke Env L & Pol'y F, 103.

[13] This sequence, however, does not give the answer on how modern property rights evolve. In this context Robert Nelson argues that, indeed, it is the transformation of private property to public property rights is how these rights evolve. The benefits that flow from regulation are seen as being no different than the stream of benefits that comes from any other form of property. See RH Nelson, 'Private Rights to Government Actions: How Modern Property Rights Evolve' [1986] U. Ill. L. Rev., 361, 364.

prominent distinction between the two approaches is that the creation of non-tradable permits determines the extent of pollution-reduction at the installation level, while the creation of tradable permits enables the operators of covered installations to allocate the costs of achieving pollution reductions most efficiently via the market.[14]

But what is this right the State creates? Generally, the assignment of allowances (or AAUs) does not lead to the 'privatization of the atmosphere' itself. It is not the 'air' or the 'atmosphere', not even the 'emissions' which are subject to distribution. Instead the assignment of quantified emission reduction targets or the allocation of allowances for a given period of time aims at regulating the use of the atmosphere (not the access to or conditions to share the 'resource'), but it does not create any right or entitlement for the benefit of allowance holders in the atmosphere itself. The State does not transfer ownership in a parcel of air, but creates the right to release a certain amount of GHGs in the atmosphere. The right that is created through this action is of regulatory nature, the market a 'permit' market.[15] Other than public property, for which government agents manage the resource in question, or private property for which private parties allocate, manage, and transfer rights in the benign light of government, regulatory property is characterized by government agents allocating and managing inalienable rights.[16]

The question where exactly the State's sovereign right to the atmosphere finds its borders therefore does not need to be answered in this context. The State, often mandated through a constitutional obligation, has the right and the obligation to protect and regulate the sustainable use of the atmosphere. Would the legal nature of the atmosphere as *res communis* or common good be recognized, it would not limit the State in fulfilling its mandate in regulating use and access to the resource. When creating international law to ensure the global access and management of public goods, States submit themselves to regimes of shared management and preservation in the understanding that common goods such as oceans[17] or the atmosphere[18] form part of the common property of mankind. The assignment of AAUs in the context of the Kyoto Protocol does not contradict with this general principle. It neither leads to a regulation of the climatic system for the benefit of a limited number of countries (with targets), nor does the allocation of AAUs create any right or entitlement to modify the climatic system. Instead it sets an obligation to preserve the beneficial effects of the climate for future generations.

[14] See Cole, n 12 above.

[15] Most of the markets that exist today do not trade in private property, rather they trade in regulatory property. See B Yandle, 'From Local to Global Commons: Private Property, Common Property, and Hybrid Property Regimes: Grasping For The Heavens: 3-D Property Rights And The Global Commons' (1999) 10 Duke Env L & Pol'y F, 13, 1.

[16] See Yandle, ibid 6.

[17] United Nations Convention on the Law of the Sea, UN Doc A/Conf 62/122, reprinted in (1982) 21 ILM 1261, available at <http://www.un.org/Depts/los/unclos/closindx.htm>.

[18] Because of the opposition of some Parties to Part XI of the 1982 Montego Bay Convention, it was preferred to refer to the concept of 'common concern of mankind' instead of 'common property of mankind' in order to define legally the climatic system in the Preamble of the UNFCCC, further to the UN General Assembly Resolution 43/53 of 6 December 1988.

2.2 The Basic Concepts

Before engaging in a more detailed analysis of the units defined under the Kyoto Protocol, it is advised to briefly describe the main underlying concepts which determine the features of emissions trading schemes: (i) emissions; (ii) allowances; (iii) allocation of authorizations to emit in the context of cap and trade schemes; and (iv) emission credits generated in baseline and credit systems.[19]

2.2.1 Anthropogenic GHG emissions

Anthropogenic GHG emissions are the human-induced release of substances into the atmosphere that contribute to the phenomenon commonly referred to as 'global warming'. While the natural cycle of carbon dioxide (CO_2) and other gases also results in the release of GHG into the atmosphere, it is the increase in the release of such gases as a result of human activities which causes the problem of global warming. Since our control over the natural cycle of GHGs is limited, efforts to curb GHG emissions focus on anthropogenic emissions.

GHG emissions result from a wide range of activities and often represent the unintended by-product of an activity. They are therefore residues of combustion and industrial processes. An interesting parallel can be drawn between waste, which like GHG emissions, is not produced voluntarily. An object, a material, or a substance becomes waste whenever the holder decides so or is obliged to discard it. By analogy, GHG emissions could be regarded as gaseous wastes at the point of time when they are released into the atmosphere. Discharges of hazardous substances into the soil or watercourses represent, under certain circumstances, a disposal operation for liquid wastes, which may be prohibited, or subject to an administrative authorization. However, air pollutants or emissions are not usually defined as wastes because they are not materialized. Wastes may have an economic value, which can be either positive or negative for the holder, who may also be the owner. The economic value of waste is a direct consequence of the evolution of legislation requiring a reduction of wastes (prevention) or an ecologically sound management or disposal of waste, which comes at a certain cost. Again a similar reasoning could be made for GHG emissions, which by themselves have no economic value (gases are usually not reused, with few exceptions such as the use of furnace gases for energy production or liquefaction of CO_2). Any economic value attached to GHG emissions arises from the right to transfer defined units, which represent the authorization to emit a certain quantity of GHGs. But taking in an extensive manner recourse again to the analogy to waste could support the case that the operator of an emitting installation would also be the owner of the emissions.

[19] The present discussion is focusing on the legal ownership and nature of GHG emissions, however, interesting lessons can be drawn from other emission trading programmes covering other air pollutants, such as sulphur dioxides and nitrogen oxides (schemes are in place in the US, Canada, and the Netherlands, among others).

2.2.2 Initial allocation

There are two basic forms of emissions trading systems: cap-and-trade systems which are based on an initial allocation of emission rights through a government or other regulatory body, and baseline-and-credit systems in which emission credits are generated through the implementation of an activity which reduces emissions against a baseline. In theory, both the cap-and-trade approach and baseline-and-credit systems may be subject to absolute or relative emissions caps. Any emissions trading system may also combine features and credits from both approaches.

A key element in cap-and-trade systems is the procedure in which allowances are allocated. Governments establishing such an emissions trading scheme are required to set the total 'fringe' amount of the resource that may be safely and sustainably used before, in a second step, they divide this amount up and allocate that usable total in the form of individual tradable allowances.[20] Under Article 3(1) of the Kyoto Protocol, allowances are assigned to governments by virtue of international law. Governments on the other hand can translate their targets and limitations through the allocation of allowances to private and public entities under national emissions trading schemes. In both instances, the allocation fixes the overall quantity of emissions that can be emitted in a given jurisdiction and therefore puts a cap on emissions. It is the starting point before trading of allowances can happen but it is also the pre-determination of the overall environmental benefits that can be expected from the system. Only voluntary trading schemes operate without a cap, relying on non-binding initiatives to generate emission reductions and participate in the trade. Frameworks based on mandatory schemes regulate through the allocation process the use of the atmosphere to discharge certain substances from a quantitative point of view. The key issue here is to note that the allocation of emission rights in mandatory cap-and-trade systems is of governmental origin but that once allowances are allocated they can be traded without any restriction other than those foreseen by the applicable regime.

2.2.3 Allowances

Allowances generally represent the authorization or entitlement to emit a certain amount of GHG emissions. It is the process of allocation in which a regulatory authority decides how many allowances an emitter receives. When the Kyoto Protocol allocates Assigned Amounts to Parties it distributes allowances, as does a national emissions trading scheme when it develops an allocation plan. The entitlement to emit a certain amount of specified substances is generally not different

[20] BA Ackerman and RB Steward, 'Reforming Environmental Law: The Democratic Case for Market Incentives' (1988) 13, Colum. J. Envtl. L., 171, 184. Ackerman and Steward add another four bureaucratic measures that are required to ensure a functioning trading scheme: A system for monitoring and record-keeping; and the enforcement of the limits both externally (to exclude interlopers) and internally (to constrain insiders' overuse).

from the traditional command-and-control approach under which an operator of an installation needs a permit to release up to a certain quantity of emissions for a specified period of time. However, the transferability, and when different schemes are linked, the fungibility among units, makes the holding of allowances fundamentally different from the holding of a traditional command-and-control permit. An allowance does not only carry the authorization to emit a certain substance, it also embodies a right to be sold and transferred. While the allowance does not create a right in the emissions themselves, it creates a transferable entitlement to emit a certain substance. This right is different depending upon under which regime the right has been created:

1. Article 17 of the Kyoto Protocol establishes the right of Annex I Parties to trade Kyoto Units under the framework of the Protocol. The right can be exercised subject to certain conditions and limitations. Such limitations include the necessity to fulfil a set of eligibility criteria[21] and to maintain a certain minimum amount on the registries of the Party (commitment period reserve).[22] The right to participate in the trade with AAUs may be delegated to private and public legal entities by virtue of government authorization. The AAUs allocated through the Kyoto Protocol could thus be regarded as a mixture of a sovereign right (to do something, for instance to fulfil an international obligation) and a public property right of an Annex I Government (to abuse of assigned units by selling them to another Annex I Party).

2. Allowances can also create property rights or quasi property rights with private entities holding allowances allocated under a domestic scheme. The exact legal nature of the allowances depends very much on the legal system under which the allowance is held and may include an authorization or permit to emit, an administrative or public right, a private property right, a security or financial instrument, a good or a commodity. In most cases the allowance will show features of a number of these categories and have a mixed nature.

Generally, allowances are designed for simplicity, transparency, and security. The strengths of the concept of an allowance derive from their impersonality, their simplicity and explicitness, their usefulness for diffuse and wide-ranging resources, their openness to new entry, and their adaptability to changes in human demand.[23] In this respect they differ from other environmental management systems, which

[21] See ch 22 below.

[22] The Marrakesh Accords require Annex I Parties to hold a minimum level of AAUs, CERs, ERUs, and/or RMUs in a 'commitment period reserve' that cannot be traded. This is calculated as 90 per cent of the assigned amount or the amount of emissions reported in the Party's most recent emission inventory, whichever is the lower figure. Decision 18/CP.7, Annex (6).

[23] CM Rose, 'From Local to Global Commons: Private Property, Common Property, and Hybrid Property Regimes: Expanding The Choices For The Global Commons: Comparing Newfangled Tradable Allowance Schemes to Old-Fashioned Common Property Regimes' (1999) 10, Duke Env L & Pol'y F, 45, 70.

rely on the distribution of responsibilities with respect to a particular source or problem by a difficult set of (negotiated) rights between different affected parties. An interesting hybrid between such more complex and multidimensional rights are verified Emission Credits.

2.2.4 Emission credits

In contrast to allowances, emission credits are generated privately. Governments may recognize these credits under an emissions trading scheme, if these credits and the process of generating the credits fulfil certain defined criteria. However, governments do not allocate these rights. Emission credits represent a reduction of GHG emissions resulting from a well-defined project activity, calculated on the basis of the comparison between the level of verified actual emissions and a counterfactual scenario (defined as baseline scenario). Of private 'origin', these credits are generated by entities as a direct consequence of their investment project-activities. In order to ensure that the emission reductions are real and measurable, most regimes, where defining emission credits, require an independent verification. Under the Kyoto Protocol verified emission reductions constitute the basis for both JI and CDM specific credits. Depending on the risk allocation in the CDM or JI contract, the buyer of emission rights may agree to purchase verified 'emission reductions' (ERs) instead of CERs or ERUs.[24] Verified ERs are contractual rights which confer the right to the benefits accruing with respect to a certain reduction or removal of GHG emissions and are established on the basis of a contract between the parties. Beyond carbon credits' market prices, the value of the right depends in its entirety on the underlying project. Carbon contracts which govern the creation and transfer of verified ERs therefore need to be significantly more complex than allowance trade contracts. The resulting right represents an individually negotiated multidimensional right, which can either be used to fulfil voluntary commitments or can be transformed, and hence 'standardized', into CERs or ERUs. Interesting is the link between baseline-and-credit schemes and cap-and-trade systems which result in an increased demand for emission credits provided that they can be generated at costs below the market price for allowances.

CERs come into existence through issuance by the Executive Board of the CDM, in a number, which equals the verified ERs stated in an independent verification and confirmed in a certification report. The Joint Implementation ERUs represent a hybrid between an allowance and an emissions credit. Based on verified ERs, they are converted and re-labelled AAUs or RMUs included in the cap of the host country and backed by the guarantee of a host country emission reduction activity. To avoid double-counting of emissions reductions within the Annex I cap of AAUs, the host country is required for each ERU it issues to cancel a corresponding amount of RMUs or AAUs from its assigned amount. The host country is

[24] See ch 20 below.

ultimately liable for achieving the target assumed under the Kyoto Protocol and thus also for emission reductions generated by a JI project implemented in its territory.

3 Nature and Legal Ownership of Emission Rights

Environmental markets depend on rights that are well defined, enforced, and transferable. Carrying these features, an allowance represents an asset or (a quasi) property right for the owner. The limits of the property may be defined by the State. A definition of the right, however, is essential to establish the limits to the right. Additionally, careful monitoring and enforcement is important to avoid conflicts and make trades possible. Finally, the rights need to be transferable. Although well-defined and enforced rights allow the owner to enjoy the benefits of using his property, legal restrictions on the sale of that property preclude the potential for gains from trade.[25]

Allowances to emit a certain amount of a pollutant are established by law and the scope of the authorization is generally defined in the legal act establishing the system. They represent a hybrid between a purely public and a purely private right, which has been described as a 'regulatory' right. As such, they find themselves between an administrative grant and private property. While embodying this dualism, the distinction between the public law status and the private law status becomes important when it comes to the treatment of allowances under taxation of accounting schemes.

3.1 Public International Law

3.1.1 The Kyoto Protocol

All tradable units defined by the Kyoto Protocol follow the same definition: they are defined as a unit of 'one metric tonne of carbon dioxide equivalent, calculated using global warming potentials defined by decision 2/CP.3 or as subsequently revised in accordance with Article 5.'[26] All these units are created by an act of international law, which puts obligations upon its member Parties. It does not oblige any private or public entity, other than the signing and ratifying State. Therefore all emission rights under the Kyoto Protocol are created, owned, and held by governments.

These rights need to be reconciled with the statement in the Marrakesh Accords that 'the Kyoto Protocol has not created or bestowed any right, title or entitlement to emissions of any kind on Parties included in Annex I.'[27] The

[25] See Yandle, n 15 above, at n 3.

[26] Decisions 16/CP.7, 17/CP.7, and 18/CP.7 of the Marrakesh Accords.

[27] FCCC/CP/2001/13/Add.2, Decision 15/CP.7 Principles, nature and scope of the mechanisms pursuant to Arts 6, 12 and 17 of the Kyoto Protocol, preamble; draft Decision -/CMP.1 (Mechanisms), preamble.

meaning of this sentence needs to be understood in the political context of the Kyoto Protocol:

1. As elaborated above, the Kyoto Protocol does not create any rights to emissions or the atmosphere, it only creates the right to a limited pollution in a defined time frame.
2. The allocation of AAUs does not grandfather any future authorizations to emit a certain quantity of pollutants. It does not bestow Annex I countries with any entitlements beyond the first commitment period (2008–12) of the Kyoto Protocol.

Another concern is that the Kyoto Protocol also clearly envisages the participation of entities other than States in the Kyoto mechanisms.[28] The treaty itself, however, does not transfer any direct rights or obligations to entities other than its Parties. Since the Kyoto Protocol does not bestow any rights and obligations to private sector entities, such rights need to be created either through implementation legislation or through single government acts. Such as an individual authorization. In order to allow private entities to hold, own, and trade units defined under the Kyoto Protocol, they need to be authorized to do so.[29] In some instances letters in which countries approve CDM and JI projects authorize the project participants to participate in the project, and authorize the participants to own and hold CERs or ERUs. The following sentence is an example:

> As authorized representative of the Designated National Authority for [Host Country] under the Kyoto Protocol, I further acknowledge the Project Participants' right, title and interest to all of the greenhouse gas emission reductions generated by the Project (and any CERs which are created out of the Project).

Independently, whether created through general act (law) or through governmental decision, with the authorization to participate in a project and to hold and transfer CERs or ERUs, a government recognizes the right of an entity to an ERU as a CER without prejudice to the obligation to remain responsible to fulfil its Kyoto commitments.[30]

3.1.2 Treatment of Kyoto Units under the WTO

The interface between the Kyoto Protocol and the WTO is part of a broader debate on relationship between WTO and multilateral environmental agreements

[28] Article 6(3) foresees the authorization of private sector entities to participate in JI. Article 12(9) of the CDM refers to the involvement of private or/public sector entities in the mechanism.

[29] Decision 16/CP.7 Guidelines for the implementation of Article 6 of the Kyoto Protocol, draft Decision -/CMP.1 (Article 6), Annex, para 29; Decision 17/CP.7, draft Decision -/CMP.1 (Article 12) Modalities and Procedures for a Clean Development Mechanism Annex, para 33; Decision 18/CP.7 Modalities, rules and guidelines for emission trading under Article 17 of the Kyoto Protocol, draft Decision -/CMP (Article 17) Annex, para 5.

[30] See in particular para 33 of Decision 17/CP. 7.

(MEAs), such as the UNFCCC or the Kyoto Protocol, which should be mutually supportive.[31] MEAs and international trade law, however, can come into conflict, for instance when MEAs' trade measures are used for compliance or enforcement purposes. A preliminary question is to know whether the relationship between the Kyoto Protocol and the WTO is covered by the Doha Agenda, and this depends on specific trade obligations (STOs) that MEAs may contain. In that respect, it is clear that the Kyoto Protocol does not contain provisions that would affect trade in a restrictive manner. The trade-related provisions contained in the Kyoto Protocol are not specific enough to be considered as part of the Doha mandate. These provisions do not conflict directly with the WTO regime. Actually, some interactions of the Kyoto Protocol and the WTO regime could lead to some win-win situations (reduction of trade barriers concerning environmental goods and services related to climate change mitigation, reduction of fuel subsidies as part of a discussion on mainstreaming environment and development).

If there is no conflict between the Kyoto Protocol itself and international trade law, still some tension could arise concerning measures adopted by Parties for the implementation of the Kyoto Protocol. Therefore, implementing measures for climate mitigation that are technical regulations may be notified to the Technical Barriers for Trade (TBT) Committee as foreseen in the TBT Agreement[32] (for example, the forthcoming EU measures on labelling and certification of energy efficiency).

As far as international emissions trading under Article 17 of the Kyoto Protocol is concerned, a key question is to know whether Kyoto Units are 'goods' under the GATT (including Trade-related Investment Measures [TRIMs] and Subsidies and Countervailing Measures [SCM]) or 'services' under the General Agreement on Trade in Services (GATS) (including its fifth protocol on financial services).

As a preliminary remark, it has to be noted that government to government trading of Kyoto Units does not create a market in goods or services, rather it corresponds to a sovereign to sovereign exchange of commitments: the sovereign exchange of assigned amounts could be viewed as the reallocation of the overall assigned amount established by the 'cap' rather than the creation of a market.[33] In that case, there is no interaction with WTO rules.

Things become more complex, however, if one looks at trade in emission allowances between private entities. Due to their ambiguous legal nature, it is not evident which part of the international trade law would be applicable for the trade in emission rights. Allowances do not seem to fall under the definition of 'goods' or 'products' as defined by the General Agreement on Tariffs and Trade (GATT 1994) and thus the transfer and mutual recognition of allowances are not covered by the WTO disciplines.[34] This is confirmed by WTO case law according to which goods

[31] See in particular Art 2(3) of the Kyoto Protocol and Art 3(5) of the UNFCCC stating clearly that the adverse effect on international trade should be minimized.

[32] <http://www.wto.org/english/docs_e/legal_e/17-tbt_e.htm> accessed 16 June 2004.

[33] J Werksmann, 'Greenhouse Gas Emissions Trading And the WTO' (1999) 8(3) RECIEL 1, 3.

[34] See ibid, 1, 2.

are defined as material things with an intrinsic value. In addition, the Harmonized Commodity Description and Coding System elaborated by the World Custom Organization do not cover the category of certificates, if one can accept that allowances are before all certificates whatever their legal nature. Furthermore, it can hardly be argued that Kyoto Units would be a 'service' mainly because they do not represent activities with an economic value. The GATS does not provide a definition of what a service is, and services are defined by contrast with products. Certificates such as the Kyoto Units or emission allowances are not listed among services in the WTO's Services Sectoral Classification List[35] and the United Nations' Provisional Central Product Classification (CPC) system.[36] However, trading of Kyoto Units will eventually involve services, such as exchange services by brokers or trustee services by banks and financial managers, which may well fall under the Annex on Financial Services of the GATS.[37] Therefore, remaining concerns with respect to the applicability of WTO agreements in fact relate more to the domestic implementation of measures related to the Kyoto Protocol and secondary markets on derivatives (financial services) than on the transfer of Kyoto Units and Allowances itself.

It is commonly accepted that a cap-and-trade regime is 'inherently more compatible with rule-based international trading systems than other environmental policy instruments are.'[38] However, one has to keep in mind that the allocation process is likely to affect the competitive relationship between products and services that are governed by international trade agreements. States can be seen as subsidizing industries when they waive a revenue otherwise due. They could also be seen to be directly subsidizing renewable or other 'clean' energies. If these preferences undermined the competitive relationship between 'like' products in similar or domestic markets, they may be regarded as countervailing subsidies as defined in the SCM and subject to a challenge or direct countervailing measures.[39] However, a closer look at the SCM's provisions does not support this assessment: the allocation of emission rights can neither be regarded as a 'financial contribution' nor as an 'income or price support'. It is also questionable whether the transfer of an allowance to private entities does confer a 'benefit' in the sense of Article 1 of the SCM Agreement, since the allocation of an allowance to private entities is accompanied by an obligation limiting the emission of GHG to the number allowances held.[40]

35 <http://trade-info.cec.eu.int/doclib/docs/2003/november/tradoc_114182.pdf> accessed 16 June 2004.

36 <http://unstats.un.org/unsd/cr/family2.asp?Cl=9>accessed 16 June 2004.

37 Para 5 (definitions) GATS Agreement states that negotiable instruments are included within the notion of financial services.

38 A Petsonk, 'Environmental Defence, Presentation at the Transatlantic Dialogue on Climate Change' (Washington, DC, 11 June 2003).

39 This question is particularly interesting taking into account the recent OECD decision to subtract the corresponding financial value of CERs from ODA reported by Annex I countries: ODA Eligibility Issues for Expenditures under the Clean Development Mechanism (CDM), 16 March 2004, DCD/DAC(2004)17.

40 However, allowances have been classified as subsidies by a number of authors. For example see, C Koenig, J-D Braun, and R Fromm, 'Beihilferechtliche Probleme des EG-Emissionsrechtehandels' (2003) 2 ZweR 153.

3.1.3 Domestic legal systems

3.1.3.1 The EU ETS

The EU ETS Directive (Directive 2003/87/EC), Article 3(a), defines an 'allowance' as '. . . an allowance to emit one tonne of carbon dioxide equivalent during a specified period, which shall be valid only for the purposes of meeting the requirements of this Directive and shall be transferable in accordance with the provisions of this Directive . . . '. It also prescribes the main rights embodied in an allowance:

(i) Any person may hold allowances (Article 19(1));
(ii) It can be transferred between persons in the EU and between persons in third countries where allowances are recognized (Article 12(1));
(iii) It can be used for compliance with the permit (Article 12(3));
(iv) It can be 'destroyed' through cancellation (*abusus*) (Article 12(4)).

However, it does not qualify the legal nature and ownership of allowance beyond the main characteristics of a tradable instrument. It is also silent on the legal nature of the allowances, especially whether the allowance is to be regulated as security or financial instrument under the Investment Services or Market Abuse Directives[41] or whether it ought to be treated as a good or, in other words, as a commodity. A very first draft prepared by the European Commission proposed to define allowances as administrative authorization, thus making a clear analogy with SO_2 allowances referred to in § 403(f) of the 1990 amendments to the US Clean Air Act (for further detail, see below). But the Legal Service of the Commission opposed this inclusion into the Commission proposal before it was tabled before the Council and the Parliament, arguing the need to preserve the subsidiarity principle as well as Member States' national legal systems particuliarities. Therefore, the answer to these questions lies with the transposition of the Directive in the legislation of the EU Member States depending on their national legal systems and traditions.

The EU ETS has been adopted in the form of a Directive legally based on Article 175(1) of the EC Treaty (environmental protection) which prescribes a defined objective (to promote GHG emission reductions in an economically efficient manner) but leaves to Member States the choice of the means and form to achieve it. In accordance with the principle of subsidiarity, the intervention of the Community in setting a European emissions trading system has been mainly justified with the

[41] Directive 2004/39/EC of the European Parliament and of the Council of 21 April 2004 on markets in financial instruments amending Council Directives 85/611/EEC and 93/6/EEC and Directive 2000/12/EC of the European Parliament and of the Council and repealing Council Directive 93/22/EEC (Investment Services Directive) [2004] OJ L145/1; Commission Directive 2004/72/EC of 29 April 2004 implementing Directive 2003/6/EC of the European Parliament and of the Council as regards accepted market practices, the definition of inside information in relation to derivatives on commodities, the drawing up of lists of insiders, the notification of managers' transactions and the notification of suspicious transactions (Market Abuse Directive) [2004] OJ L162/70.

necessity to ensure the integrity of the internal market and the need to avoid distortion of competition.[42] However, using the instrument of a Directive, the Community law only regulates and defines the EU ETS to the extent necessary to ensure the smooth operation of the EU ETS in the territory of the EU. The further refinement of the definition and the treatment of allowances under contractual and tax law remains the responsibility of the Member States. It is interesting to note that the private sector has expressed its discomfort with the position of the European Commission to rely on the legal instrument of the Directive.[43] The potential differences in legal treatment (as a permit, good, right, or financial instrument) and under domestic tax and accounting systems are perceived as creating major uncertainties and thus as barriers to the emergence of a liquid market. The less defined, the less secure the allowances are, the less likely potential buyers will be to invest in them. An interesting question is to know whether Member States would have accepted that the Community adopts the EU ETS through a Regulation of direct and immediate application in Member States' national legal systems. Such a Regulation would have had to clearly define the legal nature of emission allowances, but also set the total quantity allocated to the sectors covered by the scheme and no doubt that Member State would have opposed for this latter reason.

3.1.3.2 Under EU Domestic Schemes

EU Member States are currently in the process of transposing the EU ETS Directive into national law. At the time of writing, few of the countries has adopted and promulgated a law implementing the EU ETS. However, the publication of the Directive has kicked off a lively debate on the different features, the strength, and the weaknesses of the Directive. The main discussions hereby centre around the problem of whether the system is legally permitted under constitutional law of the individual Member States,[44] how it relates to traditional instruments of European and domestic law,[45] whether an allowance constitutes property,[46] a subsidy,[47] or a financial instrument, and finally how allowances or the transfer

42 Preamble 7 of the EU ETS.

43 This position has been expressed by many representatives of private banks and trading houses most recently at the occasion of a workshop held in London and co-sponsored by the Foundation of International Environmental Law and Baker & McKenzie (London, King's College, 24–25 May 2004).

44 A Reuter, 'Grund- und Grundrechtsmängel des Emissionhandels in der EU' (2003) 10–11 RdE 262.

45 W Frenz, 'Emissionszertifikathandel und Immissionsschutzrecht' (2003) 2 RdE 32.

46 For many: M Pâques, 'Le système d'échange de quotas d'émission de gaz à effet de serre dans la Communauté européenne, in Aménagement-Environnement', livraison spéciale 2003, L'énergie, 29; M. Moliner, 'Pollution atmosphérique: analyse du droit d'émission négociable sous l'angle du droit privé', Gaz. Pal., 12–14 October 2003, 18; CM Rose, 'Expanding the choices for the global commons: comparing newfangled tradable allowance schemes to old-fashioned common property regimes' (1999) 10, Duke Env L & Pol'y F, 45, 52; cf S Agbayissah and L Segalen, 'Aspects juridiques et comptables des quotas d'émission de gaz à effet de serre' (PriceWaterhouseCoopers and Landwell, April 2003) 31; J Palmisano, 'Are Emission Reductions Property?' Evolution Market Executive Brief, 23 May 2001, <http://www.evomarkets.com/assets/evobriefs/nw_995473605.pdf>.

47 C Koenig, J-D Braun, and R Fromm, 'Beihilferechtliche Probleme des EG-Emissionsrechtehandels' (2003) 2 ZweR, 153.

should be taxed and accounted for.[48] A number of the debates seem to be triggered by issues relating to the definitions of the EU Emissions Trading Directive and their translation into the different Member State languages.[49]

A number of countries have presented draft legislation. Generally, it can be observed that the EU Allowance does not fit easily in any legal system of the EU Members. It is very likely to be considered as a right '*sui generis*' in many jurisdictions, carrying features of a transferable permit (an administrative public right) a non-material good or a commodity (a private tangible asset), and a security (a financial instrument). As an example, the French Ordinance of 15 April 2003 transposing Directive 2003/87/EC defines allowances as 'an immaterial good' which the holder will own as of the date allowances are issued in his account in the national registry, eg on 28 February 2005. Actually, according to French Law, it is interesting to note that what constitutes the 'immaterial good' is not the CO_2 that can be emitted but rather 'the right to transfer', in other words the possibility for the holder of allowances to present a right to emit to another person the allowance. Such a legal qualification results mainly from the fact that allowances, once allocated, are transferable without restrictions other than those provided by national law (basically, it will be the possibility to block transfers of allowances if the verification of actual emissions is not judged satisfactory). It will also depend on the legislative systems how the property rights (and obligations) to the allowance will be defined. The legal treatment of an allowance has an impact on the following questions:

(i) Will the trading be supervised by a regulatory authority, as is the case for financial instruments? This would make the trade for appraised entities easier, however, it would exclude many operators of covered installations since they do not meet the capital retaining requirements of banks or authorized traders.

(ii) Can the allowance be revoked and cancelled (with or without compensation)?

(iii) Can the allowance be encumbered and used as credit security?

(iv) What is the treatment of allowances in case of insolvency and bankruptcy?

(v) Will the use and trade of allowances attract value added taxes (VAT)? If so and in case of cross-border trades, will it accrue within the acquiring or transferring country?

[48] See ch 4 below; IETA, UK Emissions Trading Group and Deloitte & Touche, 'Report on Concepts and Approaches to Implementing Corporate Accounting for Greenhouse Gas Emissions Allowances'; G Wagner, 'Handel mit Emissionsrechten' (2003) 15 Zeitschrift für Bankrecht und Bankwirtschaft 409–24; Global Industrial and Social Progress Research Institute (GISPRI), June 2002 <http://www.gispri.or.jp/english/pdf/GHG2.pdf> 'Accounting for Carbon under the UK Emissions Trading Scheme' <http://www.ieta.org/Documents/Discussion_Papers/IETA_UKETG_accounting_ June2002.pdf>.

[49] Especially the French translation of 'quota' for 'allowance' and 'autorisation d'émettre' for 'permit', see M Pâques, *Nature juridique du quota d'emission du gaz á effet de serre créé par la Directive 2003/87/CE, Quelques Observations* (Bruylant Publishers, Brussels, 2004) forthcoming.

3.1.3.3 SO_x/NO_x Trading in the US

While the European systems are still in the process of being formulated, it may be worth taking a look at the experiences gained with the implementation of an emissions trading scheme on the other side of the Atlantic.

The 'acid deposition control' program established in Title IV of the 1990 Clean Air Act Amendments[50] sought to halve sulphur dioxide (SO_2) emissions by the year 2000. A new transferable emissions 'allowance' scheme was established to accomplish this goal. The regulating authority has set emission quotas for the covered installations. The Act does not specify how sources are to meet their emissions reduction targets, and the law does not even require sources to reduce emissions to the levels set by Congress, but only to possess allowances equal to their actual emissions.[51]

The authorization to emit then gains a quality, which confers a quasi property right upon the holder of the unit. The holder is authorized to hold, sell, or use the unit. However, whereas these rights are enforceable against third parties, they enjoy only limited enforceability against governments. The most elaborate restriction with respect to the use of allocated emission rights can be found in § 403(f) of the 1990 amendments to the Clean Air Act which carefully defines the legal nature of emission allowances:[52]

(f) Nature of allowances
An allowance allocated under this subchapter is a limited authorization to emit sulfur dioxide in accordance with the provisions of this subchapter. Such allowance does not constitute a property right. Nothing in this subchapter or in any other provision of law shall be construed to limit the authority of the United States to terminate or limit such authorization. Nothing in this section relating to allowances shall be construed as affecting the application of, or compliance with, any other provision of this chapter to an affected unit or source, including the provisions related to applicable National Ambient Air Quality Standards and State implementation plans. Nothing in this section shall be construed as requiring a change of any kind in any State law regulating electric utility rates and charges or affecting any State law regarding such State regulation or as limiting State regulation (including any prudency review) under such a State law.

Nothing in this section shall be construed as modifying the Federal Power Act (16 U.S.C. 791a et seq.) or as affecting the authority of the Federal Energy Regulatory Commission under that Act. Nothing in this subchapter shall be construed to interfere with or impair any program for competitive bidding for power supply in a State in which such program is established. Allowances, once allocated to a person by the Administrator, may be received, held, and temporarily or permanently transferred in accordance with this subchapter and the regulations of the Administrator without regard to whether or not a permit is in effect under subchapter V of this chapter or section 7651g of this title with respect to the unit for

[50] Pub. L. No 101-549, 104 Stat. 2399 (codified as amended at 42 U.S.C. 7401–671 (1994)).

[51] See Cole, n 12 above.

[52] 42 USC 7651b (Title 42: The Public Health and Welfare, Chapter 85: Air Pollution Prevention and Control, Subchapter IV-A: Acid Deposition Control, sect. 7651b: Sulfur dioxide allowance program for existing and new units).

which such allowance was originally allocated and recorded. Each permit under this subchapter and each permit issued under subchapter V of this chapter for any affected unit shall provide that the affected unit may not emit an annual tonnage of sulfur dioxide in excess of the allowances held for that unit.

Although the Act clearly states that an allowance does not create a property right, it seems that, between the contracting parties, all normal property rights (*usus, fructus,* and *abusus*) are available.[53] However, the US legislator did not want to create property, in order to avoid being held liable for compensation should the allowance be devalued or withdrawn from the market. This has led some to describe § 403(f) as 'premised on the confusion between property rights in something and the thing itself'.[54] Indeed, it does expressly define an allowance as not creating a property right, yet allows free transfer and thus recognizes the property rights *in* the emission allowance, because utilities can receive, hold (that is, possess), and transfer (that is, alienate) allowances. Utilities and all other allowance holders (such as private NGOs) can exclude all others, besides the government, from interfering with their possession, use, and disposition of allowances.[55]

This means that, although not being a property right itself, allocated emission rights usually carry characteristics of 'property rights' such as transferability or the protection of allowances against arbitrary confiscation. While the State protects its flexibility to create or cancel units in order to manage a cap-and-trade system, or to withdraw or suspend allowances found to have been improperly allocated, it allocates to the private sector 'quasi' property rights which allow free use and transfer of these units. The State creates what has been described as a regulatory right.

4 Conclusion

The discussion of the different features of emission allowances has made it obvious that, although it is in the best interest of a liquid and functioning market, that

[53] Restatement of the Law, Property 1936, The American Law Institute, § 5(e): Complete Property. The totality of these rights, privileges, powers, and immunities which it is legally possible for a person to have with regard to a given piece of land, or with regard to a thing other than land, that are other than those which all other members of society have as such, constitutes complete property in such land or thing other than land. This totality varies from time to time, and from place to place, either because of changes in the common law, or because of alterations by statute. Thus if the law should come to be that no person could build a five-storey building on his land, the totality of privileges that every person has who owns land would be correspondingly diminished. So if a zoning ordinance were passed, the totality of interests would be affected, to the extent of the ordinance, for persons owning land within the district to which the ordinance applied. At any one time and place, however, there is a maximum combination of rights, privileges, powers and immunities in the land that is legally possible, and which constitutes complete property in the land, or thing other than land.

[54] See Cole, n 12 above.

[55] M Gehring, and C Streck, 'Emissions Trading: Lessons from SO_2 and NO_x Emissions Allowance and credit Systems, Legal Nature, Title, Transfer, and Taxation of Emission Allowances and Credits' (World Bank Working Paper, 2004). Available from the authors.

States are reluctant to attach a property right per se to an emission allowance. Instead of clearly defining the boundaries of the rights they create, they often leave it up to the market and the private sector to explore the limits of the rights they hold in 'learning by doing'. Until today, literature on tax or accounting treatment of SO_x/NO_x rights in the US is hard to find. The Kyoto Protocol as well as the EU ETS establish the allowance as an 'authorization' to emit, without giving further detail on the legal nature of these rights.

The Kyoto Protocol creates sovereign rights recognized for Parties trading these Kyoto Units. Kyoto Parties can authorize private entities to participate in trades under the international instrument. Therefore, however, rights to the units have to be recognized on the domestic level and they have to relate to the right to transfer something of economic value, which can be used for compliance or other purposes.

When creating emission rights under national law, from the government's view the main advantage of an ambiguous right is the cautionary approach it represents. The creation of such narrowly and concretely defined rights will pose significant difficulties for regulators who need the flexibility of creating or cancelling such rights. However, at the same time, there is also keen awareness that certainty, for example that emission allowances should be protected against arbitrary confiscation, is vital for the creation of an active trading market. It is the hybrid nature of emission rights which does make it difficult to clearly categorize such rights as public or private. They are a hybrid which can best be described as regulatory right. The right is transferable and enforceable against other private parties, however, it enjoys limited enforceability against the State, which wishes to retain the overall control over the allowances it has created.

As a result, holders of tradable emission rights generally have the exclusive right to hold, transfer, and use these rights and the administrative agency should refrain from arbitrary confiscation or discounting of these rights.[56]

[56] See also J Lin, 'The Legal Aspects of Emissions Trading: A Resource Paper' (NYU, 2004) (not yet published, reviewed and kept by the authors).

4

Accounting for and Taxation of Emission Allowances and Credits

*Robert Casamento**

1 Introduction

The accounting for, and taxation of, emission allowances[1] and credits[2] is becoming an area of increasing interest for a number of shareholders and boards of directors. As governments begin to develop and, in some cases, implement greenhouse gas emission trading schemes, companies and their shareholders will want to know the impact this may have on their cash flows, balance sheets, and income statements.

This chapter looks at the current developments in accounting for, and taxation of, emission allowances and credits and explores the potential impact that different treatments may have on the nature of carbon transactions and corporate management of greenhouse gas (GHG) emissions.

2 Accounting for Emission Allowances

Emissions trading schemes based on the principle of 'cap-and-trade' are not new. The US, for example, have had acid rain (sulphur dioxide and nitrogen oxides) trading schemes for a number of years. The US Environmental Protection Agency (EPA) establishes the targeted level of total emissions and the corresponding

* Robert Casamento is an employee of Deloitte & Touche LLP (UK). He is based in London and works in the Energy Markets team. The views expressed in this chapter are the views of the author and do not necessarily represent the view of Deloitte and should not be relied on for technical accounting or taxation advice. Interested parties should seek specific advice from a qualified individual prior to making any decisions.

[1] For the purposes of this chapter, 'emission allowance' is used to describe an instrument that enables the holder to emit a specific volume of gaseous emissions during a specified period for the purposes of meeting regulatory requirements. The allowance is assumed to be transferable in accordance with the provisions of a defined GHG emissions trading scheme.

[2] 'Emission credit' is used to describe an instrument that asserts that a specific volume of gaseous emissions has been reduced compared to an established baseline.

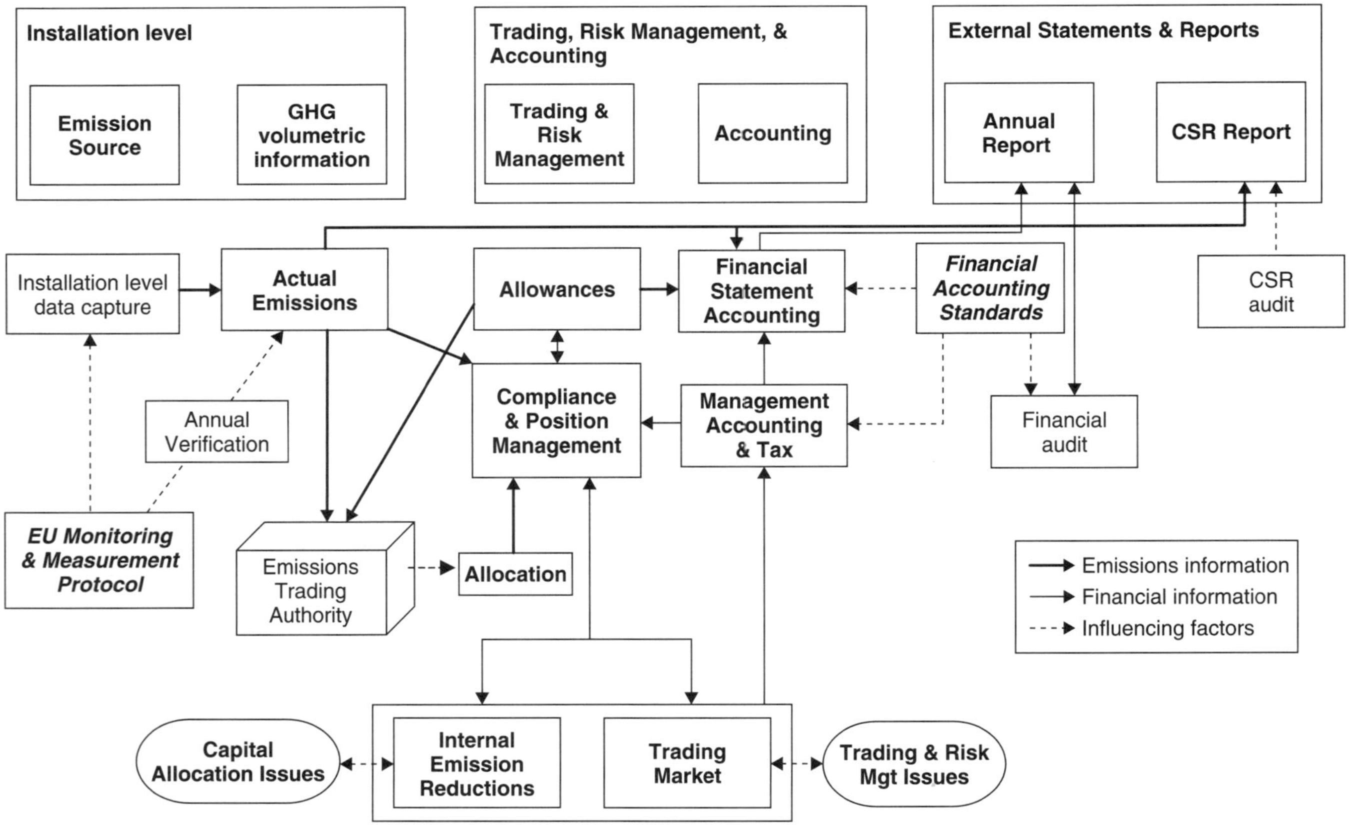

Figure 4.1: Putting emission trading into context

Source: ©2004 Deloitte & Touche LLP.

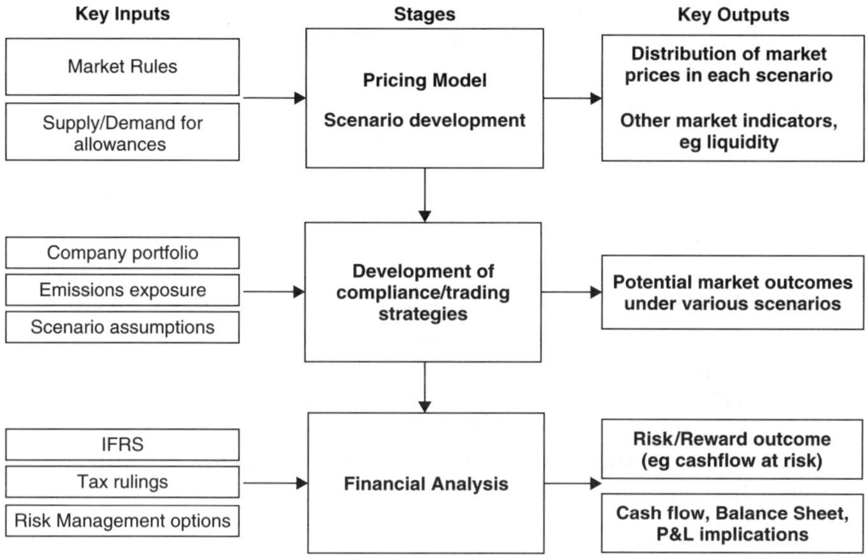

Figure 4.2: Financial Impacts of Emissions Trading Companies will need a structured process to quantify the financial impact of the ET under various scenarios
Source: ©2004 Deloitte & Touche LLP.

allowances. The EPA allocates a portion of the available allowances to participants. The EPA may make the unallocated allowances available to participants through government auction or approved projects (for example, renewable energy). Participants and others may buy and sell allowances. If participants do not have enough allowances at the end of the compliance period, they are required to pay fines.

In terms of accounting, the staff of the US Financial Accounting Standards Board (FASB) believe that most companies account for emissions allowances in a manner similar to that required under the Federal Energy Regulatory Commission's (FERC) Uniform System of Accounts (USofA).[3] FERC requires public utilities and other regulated energy entities to keep books and records in accordance with the USofA. In 1993, FERC issued guidance on the accounting for acid rain emissions allowances. FERC requires companies to recognize emission allowances on a historical cost basis. Allowances obtained from the EPA at no cost are recorded at a zero basis. Allowances that are purchased through government auctions and from third parties are recognized at their purchase price. FERC requires companies to recognize the cost of the allowances in earnings based on a weighted-average cost over the compliance

[3] FASB Emerging Issues Task Force (EITF) Issue No 03-14, *Participants' Accounting for Emissions Allowances under a 'Cap and Trade' Program* (October, 2003).

period. If a company emits more gases than it has allowance for, it is required to accrue the estimated cost of obtaining the needed allowances. Companies typically recognize the gains and losses that result from the sales of allowances in earnings, subject to recognition of any regulatory asset or liability under Financial Accounting Standard (FAS) 71, *Accounting for the Effects of Certain Types of Regulations.*[4] With the emergence of GHG emissions trading schemes many people began to ask whether the accounting for GHG emissions is the same as for acid rain schemes or whether different issues and therefore different treatments apply.

The UK Government introduced a GHG emissions trading scheme, based on the Kyoto Protocol, in 2002. Soon requests from a number of stakeholder groups, including UK companies and government representatives, were being received to develop guidance on accounting for emission allowances in an attempt to understand the potential impact that participating in a GHG emissions trading scheme could have on a company's financial performance. In response, the International Emissions Trading Association, the UK Emissions Trading Group and Deloitte & Touche UK published a discussion paper in May 2002 entitled *Accounting for Carbon under the UK Emissions Trading Scheme.*[5] This paper set out to identify and provide guidance on some of the key issues associated with the accounting for emission allowances. The paper helped crystallize many of the ideas and concepts developed on the topic.

Working groups from the UK Accounting Standards Board, the Canadian Institute of Chartered Accountants (CICA), the American Institute of Certified Public Accountants (AICPA), FASB Emerging Issues Task Force and interested industry organizations closely monitored the UK developments and during the course of 2002 and 2003, numerous meetings were held and discussion papers issued in various jurisdictions.

The International Accounting Standards Board (IASB), which is responsible for establishing international accounting standards, also identified accounting for emission trading schemes as an important emerging issue and added it to its agenda. The decision to look into this issue was supported by the fact that commencing from 2005, many companies in the European Union will be required to report their financial performance according to International Financial Reporting Standards (IFRS) issued by the IASB (and not individual countries' accounting standards as is the current practice). The year 2005 also represents the start date of the European Union (EU) GHG emissions trading scheme, a scheme that is expected to cover some 12,000 installations and 46 per cent of estimated EU carbon dioxide emissions.

In May 2003 the IASB, through the International Financial Reporting Interpretations Committee (IFRIC), which is part of the IASB, issued a draft

[4] ibid.

[5] See <http://www.ieta.org/Documents/Discussion_Papers/IETA_UKETG_accounting_June 2002.pdf>.

interpretation (D1 Emission Rights) for public comment,[6] which sets out to explain how IFRS should be applied in this context.[7] The draft interpretation attempted to answer the key issues common to most GHG emission trading schemes, clustered around the following issues:[8]

- Does an emissions allowance scheme give rise to (i) a net asset or liability or (ii) an asset (for allowances held) and a liability, deferred income and/or income?
- If a separate asset is recognized, what is the nature of that asset?
- If a separate liability, deferred income and/or income is recognized, what is the nature of that item and how is it measured?
- When should a potential penalty, which will be incurred if a participant fails to deliver sufficient allowances to cover its actual emissions, be recognized, and how should it be measured?

After considering a number of alternative treatments the IFRIC proposed that, depending on the circumstances, emissions trading schemes give rise to:

1. **An asset for allowances held.** Emission allowances, whether allocated by government or purchased in the market, are intangible assets and should be accounted for in accordance with International Accounting Standard (IAS) 38, *Intangible Assets*. Allowances that are allocated for less than fair value shall be measured initially at their fair value. Allowances shall not be amortized, but are to be tested for impairment under IAS 36, *Impairment of Assets*.[9]

2. **A government grant.** When allowances are allocated by government for less than fair value,[10] the difference between the amount paid and fair value is a government grant that shall be accounted for in accordance with IAS 20, *Government Grants and Disclosure of Government Assistance*. Accordingly, the grant is initially recognized as deferred income in the balance sheet and subsequently recognized as income on a systematic basis over the compliance period for which the allowances were allocated.[11]

[6] Interpretations are developed by IFRIC, exposed for public comment (called draft interpretations), approved by IFRIC, and then sent to the IASB Board for review and approval as Final Interpretations. Final Interpretations are part of the IASB's authoritative literature. Therefore, financial statements may not be described as complying with the IFRS unless they comply with all the requirements of each applicable Standard and each applicable Interpretation. On a national level, despite some differences in interpretation, nationally based accounting standards are, for the most part, fundamentally consistent with the IASB conceptual framework.

[7] The UK Urgent Issues Task Force (part of the UK Accounting Standards Board) issued an Abstract on Emission Rights around the same time. The Abstract was based on the IFRIC draft interpretation with a small number of changes to reflect UK Financial Reporting Standards.

[8] In preparing its draft interpretation, the IFRIC took into consideration the fact that the features that are present in some of the emerging emissions trading schemes are not present in others.

[9] See para 6, IFRIC [Draft] Interpretation, Emission Rights, May 2003.

[10] Fair value of an asset is defined as 'the amount for which the asset could be exchanged between knowledgeable willing parties in an arm's length transaction' IAS 38, *Intangible Assets*.

[11] See para 7, IFRIC [Draft] Interpretation, Emission Rights, May 2003.

3. **A liability for the obligation to deliver allowances equal to emissions that have been made.** As emissions are made, a liability is recognized for the obligation to deliver allowances to cover those emissions (or to pay a penalty). This liability is a provision that falls under IAS 37, *Provisions, Contingent Liabilities and Contingent Assets*. The liability shall be measured at the best estimate of the expenditure required to settle the present obligation at the balance sheet date. This will normally be the present market price of the number of allowances required to cover emissions made up to the balance sheet date. However, if the participant's best estimate is that some of or the entire obligation will be settled by incurring a cash penalty, it (that is, the participant) shall measure that part of its obligations at the cost of the penalty rather than at the market price of the relevant number of allowances.[12]

This approach treats assets (that is, allowances) independently to the liabilities (that is, obligations). Accordingly, netting off or offsetting of the asset and liability is not permitted.

3 Feedback

Over forty comment letters were received from various sources in reaction to the IFRIC draft interpretation. These included national accounting standard-setting bodies, professional accounting organizations, corporations, and industry associations. Respondents welcomed IFRIC's attempts to provide guidance on this issue and that IFRIC's conclusions were consistent with existing IFRS. However, only two respondents fully supported the proposed interpretation. Many disagreed with various aspects of the model, taking the view that the end result of such accounting treatment would not reflect the business reality or economic substance of the schemes being put in place.

3.1 Key Concerns

The key concern was the lack of symmetry in measuring and reporting the changes in the three elements in the scheme (the allowances, the liability for emissions to date, and the government grant). Respondents were concerned that the lack of symmetry resulted in an entity reporting artificial volatility in its income statement. This arises because the current IFRS contain both a mixed measurement model (whereby some items are measured at cost and others at fair value) and a mixed presentation model (whereby some gains and losses are reported in the income statement and others in equity). In short, the proposed accounting treatment would:

- record a gain in the value of allowances to equity, but the loss related to revaluing the liability would be recorded in profit or loss (P&L);

[12] See ibid, para 8.

- record a loss in the value of allowances against previous gains recognized in equity, but the gain related to revaluing the liability would be recorded in P&L.

Some respondents suggested a solution could be to amend IAS 38 so that gains and losses on allowances are recognized in income rather than equity.

Respondents commented that such treatment creates an inappropriate accounting result, especially when the market price increases. In these circumstances the accounting treatment (as opposed to the market) introduces a volatility in earnings that (as claimed by some respondents) is not intended by the regulators in the design of emission trading schemes. In addition, earnings visibility is critical for some industries, in particular utilities, since both equity and debt investors usually buy utilities for their predictable income and dividend payments. The proposed treatment could introduce significant fluctuations in reported earnings that are not representative of the actual cash flow position or the year-end compliance position. Depending on the materiality[13] of the transactions involved, this could have significant implications on the ability to compare earnings between quarters and between years. It could also have a significant impact on some of the common financial statement ratios that analysts use (for example, the quick and current ratios) when assessing company performance. From a trading perspective, such movements could lead to the creation of artificial trading-based solutions in order to maximize potential revenue opportunities created by timing differences in both gains and losses in the P&L account. Such volatility has the potential to reduce both the understandability and predictive value of financial statements.

Based on these concerns, some respondents believed that allowances are best measured at fair value with changes in fair value being reported in the income statement. Unfortunately, such a treatment is not possible under IAS 38, which only permits fair value as an allowed alternative, and if fair value is used, changes in fair value are reported in equity.

The other key concern for many respondents is whether it is appropriate for the IFRIC to issue an Interpretation at this time, given the situation that emission trading schemes are still in their infancy. As such, the IFRIC should wait until additional schemes are developed before providing authoritative guidance.

4 Other Issues

In addition, respondents also highlighted a number of other matters of concern centred on the fact that many thought the proposals failed to reflect the economic substance of emission trading schemes.

[13] Materiality in this context is used to describe information that is likely to be of significant interest to the users of financial statements.

4.1 Classification of an Emission Allowance as a Financial Instrument or Inventory

D1 Emission Rights proposes that emission allowances should be classified as intangible assets and fall within the scope of IAS 38. IAS 38.7 defines an intangible asset as 'an identifiable non-monetary asset without physical substance held for use in the production or supply of goods or services, for rental to others, or for administrative purposes'. While many respondents agreed with this interpretation, there were alternative views. Some believed that, in a market with adequate liquidity, emission allowances have characteristics more similar to financial instruments or inventory.

Some believe that active markets will develop and participants will actively trade allowances for purposes ranging from compliance to speculation. As these markets develop, the emissions allowances will become readily convertible to cash and, therefore, they will meet the definition of a financial asset because they will convey a right to cash. Proponents of this view believe that a financial assets model provides the best framework for the accounting for emissions allowances and their markets. IFRIC accepted that some features of allowances could more commonly be found in financial assets than intangible assets; however it concluded that emission allowances do not meet the definition of a financial asset as defined in IAS 39, *Financial Instruments: Recognition and Measurement* because they are neither equity instruments nor contracts for cash or other financial assets. The fact that they are tradable, assuming the existence of an active market, does not make them financial assets any more than, say, a readily tradable commodity, which is not considered to be a financial asset.

Despite not meeting the definition of a financial asset, some respondents put forward the view that, as the majority of emission allowances would be used for compliance purposes, the allowances themselves could be considered to be held as a hedge against future obligations. They are therefore, for all intents and purposes, hedging instruments and they strongly recommended that the IASB consider broadening the hedging base for non-derivative financial assets and liabilities under IAS 39 to potentially include allowances.

The other form of asset discussed by many respondents was inventory. Inventory is defined in IAS 2 as 'assets held for sale in the ordinary course of business (finished goods), assets in the production process for sale in the ordinary course of business (work in process), and materials and supplies that are consumed in production (raw materials).'

This definition becomes important as some respondents believe that emissions (and the associated allowances) are an integral part of a production process and are in fact consumed, not used. Accordingly, they should be classified as inventory and not as an intangible asset.

Others pointed to the need to make a clear distinction between emission allowances held primarily to settle obligations with a regulator and those held primarily for trading purposes. These respondents argued that while allowances held to settle obligations meet the definition of an intangible asset (that is, an asset for use in the production or supply of goods or services, for rental to others, or for administrative purposes), allowances held for trading could be seen to meet the definition of an item of inventory as they are held for sale in the ordinary course of business. Accordingly, it was suggested that depending on the purpose for which an allowance is held, some allowances could be classified as intangible assets and others as inventory.

This is further complicated by IAS 38.2(a), which allows for another IAS to be applied to an intangible asset if that other standard deals with a specified type of intangible asset. This includes intangible assets held by an enterprise for sale in the ordinary course of business. Taking into consideration IAS 38.2(a), there is an argument that emission allowances held for trading purposes meet the definition of an intangible asset as they are held by an enterprise for sale in the ordinary course of business. To achieve a classification that is consistent with all aspects of IFRS, some respondents proposed that where emission allowances are held for operational purposes, they should be classified as intangible assets, and where the allowances are held for trading purposes they should be classified as inventory. In other words, the nature of the asset depends on management's intent. Proponents of this approach believe this also makes more sense if you consider the accounting approaches that could be taken by brokers and other position-taking institutions. However, this approach raises a range of other issues, such as what happens if the motivation for holding allowances changes over time? And, more practically, how would a company distinguish between the different allowances held?

4.2 Separate Recognition of the Asset and Liability

Another concern for some was the issue of separate recognition of assets and liabilities. The justification the IFRIC gives to the separate recognition of assets and liabilities is that allowances and obligations have independent existences. While an entity may intend to use the allowances it holds to settle its obligation, it cannot be compelled to do so. Instead, it may choose to sell allowances and reduce emissions, buy allowances at a future date, or pay a penalty. Thus, there is no contractual link between the asset and the liability, even though many entities will hold the allowances solely for the purpose of settling their obligations. The majority of respondents agreed with the IFRIC conclusion because under the existing standards there is no conceptual or practical basis for offsetting the assets or liabilities.

Despite this, some respondents thought that another approach should be applied. They view emission allowances and the obligations as linked transactions and, accordingly, would like to have only a net position reflected. In fact, many argue that where an emission right is being used as a compliance mechanism, a liability does not always arise. The theory supporting this is that the delivery of used emission allowances does not represent an outflow of economic benefits; instead, the economic benefits have been consumed, and the delivery of an allowance acts as a mechanism by which the entity demonstrates or provides evidence to the sponsor (the government) that it has either been allocated or has purchased the necessary allowances to cover its emissions. In so doing, the entity will avoid the requirement to make a cash payment for any shortfall. Only in the situation where an entity emits in excess of the allowances granted would a liability result. If the accounting model were to consider the asset and liability as 'linked' then an entity would only recognize a liability when it pollutes more than allowed under its cap or its allowances. Proponents of this approach assert that it is simpler and easier to apply than the IFRIC model and would better reflect the economic reality and substance of the underlying transactions. As previously noted, however, there is no support for this approach in the existing IFRS.

4.3 Potential Withdrawal of IAS 20 on Government Grants and Disclosure of Government Assistance

Another concern for some respondents is that in January 2003, the IASB took the tentative decision to withdraw and replace IAS 20. Many respondents believed that given this situation, the issuance of formal guidance should be delayed as IAS 20 plays a significant role in the accounting treatment outlined in the draft interpretation.

4.4 Issues not covered in the Draft Interpretation

While the draft interpretation is intentionally limited in scope, many respondents also requested guidance on a range of other issues. These include (but are not limited to):

- Accounting by brokers and other position-taking institutions not subject to an emission limit or cap;
- Guidance on when allowances should be recognized by an entity involved in an emissions trading scheme;
- How to treat penalties; in particular, guidance on the estimation of liability if the penalties take the form of reductions in allowances for subsequent periods or restrictions in operations;
- Disclosure in both financial statements and notes to the accounts. For example, whether allowances allocated or issued for compliance for one year or less should be classified as a current asset in the balance sheet (unless held for use on a continuing basis, or issued several years in advance) or whether information should be given on the conditions of the allowances, on the fair value, on the emission level allowed and on the estimation of the penalty to be incurred; and
- Accounting for multiple year allocations.

5 IFRIC's Response

The IFRIC discussed the main points raised in the comment letters and considered the alternative accounting models for emission trading schemes proposed by respondents. After considering the comments received (including suggested alternative interpretations), the IFRIC confirmed that the proposals set out in the draft interpretation are the most appropriate interpretation of current IFRS.

They did note, however, the key concern regarding the lack of symmetry in the accounting and decided that they would explore whether it would be possible to address this mismatch by amending IAS 38 to define a narrow sub-set of intangible assets, one that considered the unique characteristics of an allowance.

At its meeting in December 2003, the IFRIC considered the characteristics of an allowance that distinguish it from other intangible assets referred to in IAS 38. They concluded that the most faithful way to report an allowance is akin to monetary currency and as such should be accounted for in the same way. As such the IFRIC decided to ask the Board to amend IAS 38 so that any intangible asset:

- that is like a currency, because it has value only because it is used to settle an obligation; and
- whose fair value is determinable by reference to an active market (as defined in IAS 38),

should be measured at fair value with changes in value recognized in profit or loss.

The reasoning behind this approach is based on the following considerations:

- Economic benefits of such an intangible asset are not 'consumed' over the period that an entity holds that asset. Therefore amortization is incompatible with the inherent nature of the asset, as amortization is a method of allocating the cost of the asset on some systematic basis over the period that the asset is held, so as to reflect its consumption;
- fair value most faithfully reflects the resources controlled by the entity; and
- the nature of the asset as akin to currency is such that it should be accounted for in the same way as monetary currency.

The IASB agreed with IFRIC's conclusions and proposed to issue an exposure draft[14] to outline a limited amendment to IAS 38 that will enable emission allowances and liabilities to be measured at fair value, with changes in value recognized in profit and loss.

The IASB also noted that an important reference for the IFRIC's draft interpretation is IAS 20. In light of the IFRIC's work on emission rights, and also as a result of concerns with the tentative decision to withdraw IAS 20 expressed by some of

[14] Exposure drafts are published by accounting standard-setting bodies. They request feedback from interested parties on proposed or amended accounting standards.

the IASB's partner standard-setters, the IASB decided to accelerate work on replacing IAS 20[15] and to consider the implications this may have when accounting for emission allowances.

In addition to these two central issues, the IFRIC also has made comment on a number of other issues raised by respondents:

- **Are allowances inventory?** The IFRIC concluded that the emission rights are not inventory. The IFRIC also considered whether the illustrative example in the final interpretation should clarify that, for an entity that emits in the course of producing inventory, the cost of emissions (that is, the debit entry on recognizing the obligation to deliver allowances) should form part of the cost of inventory and that this cost should be net of the related government grant (that is, net of the credit entry on amortizing the government grant). IFRIC agreed that the final Interpretation should not address whether emission right payments should be a component of the cost of inventory. However, there was general agreement that emission payments should not be considered a penalty (and therefore prohibited from inclusion in the cost of inventory).
- **Should the government grant be recognized in income immediately rather than recognized as a deferred credit?** The IFRIC did not support the suggestion that the government grant should be treated as immediate income.
- **Should the Interpretation specify disclosure requirements?** The IFRIC did not believe additional disclosure requirements were required. Where relevant, participants would be required to provide the disclosures as required by existing standards.
- **Should the Interpretation include guidance for those cases where allowances are not traded in an active market?** The IFRIC did not support providing additional guidance in this area. In reaching this conclusion, the IFRIC observed that trading in allowances for the EU Emissions Trading Scheme (due to start in 2005) had already begun, although it acknowledged that this market was currently relatively illiquid.

Based on these discussions the IASB suggested that the IFRIC re-expose its draft interpretation at the same time as the IASB exposes its intention to withdraw IAS 20 and to amend IAS 38. Documents outlining these changes were expected by the end of 2004. However in October 2004 the IFRIC noted it was unlikely that the Board would issue a final amended IAS 20 for at least another year (ie, end of 2005) and hence that the IFRIC would be unable to finalize the draft interpretation until that time.

Given that the EU Emissions Trading Scheme starts at the begininning of 2005, and given the potential for diversity of accounting for that scheme, the IFRIC reconsidered whether it should finalize its original proposals in the draft interpretation. In

[15] In particular looking at whether IAS 20 could be replaced by extending the requirements for government grants at present contained in IAS 41, *Agriculture*.

doing so the IFRIC realized that this would mean that if allowances were subsequently measures at fair value, changes in value above cost would be recognized in equity rather than in profit or loss. Nonetheless, most of the members present noted that the disadvantage of this treatment would be outweighed by the benefits of providing timely accounting guidance that promotes consistent application of IFRSs. As a result, the draft interpretation is expected to be issued in its final form in November 2004, with an effective date of 1 March 2004

6 Accounting for Emission Credits

The other market for GHG emissions is one in which tradable 'credits' are generated from emission reduction projects. Unfortunately, unlike accounting for emission allowances, no substantive work has been undertaken to account for emission credits under existing IFRS or other national accounting standards. Despite this, project developers and credit purchasers who are engaged in this market are beginning to question when and how credits can be recognized in financial statements.

Typically, the first question asked by project developers and credit purchasers is whether an asset exists or not. The IAS Framework defines assets as resources and entity controls as a result of past events and from which future economic benefits are expected to flow. Project developers or purchasers will need to show that the emission credit is a resource that is controlled by the enterprise as a result of past events (for example, purchase or self-creation) and from which future economic benefits (inflows of cash or other assets) are expected. In practice this usually requires evidence such as a firm commitment (that is, contract) with external parties for the purchase/sale of an emission credit or a market mechanism that gives legal form to, and recognizes the existence of, emission reduction units.

Assuming that an asset exists, the next question is how to recognize it. Upon delivery of qualifying credits a purchaser could, in some circumstances, record them on the balance sheet as an asset at the end of the reporting period. Alternatively, purchase agreements may contain provisions to pay for credits prior to their delivery. Such payments could be considered as prepayments and recorded as such in the balance sheet. When qualifying credits are actually delivered the amounts accumulated in the prepayment account could be reclassified to an emission credit asset account. Another accounting treatment may arise when purchase agreements contain provisions that provide options to purchase credits. In these situations derivative accounting may need to be considered. For example, in certain circumstances, such options could be considered derivative instruments and therefore accounted for under IAS 39. Classification under IAS 39 could have interesting tax consequences particularly if the holder of an option may have to pay tax every time there is an increase in price (even if the option is not exercised for a number of years).

The final key question asked by project developers and credit purchasers is how to value such assets. This can be especially difficult in an illiquid and immature market.

There have been private trades in emissions credits, and a private market developing. Studies by organizations such as Natsource, Evolution Markets, the International Emissions Trading Association, and the World Bank's Carbon Finance Business unit all indicate an ever increasing volume of transactions. The World Bank's third annual *State of the Carbon Market 2003* report[16] estimated that the volume of GHG emission-related trading instrument transactions more than doubled in 2003 to approximately US$70 million annually. This is expected to increase as markets, such as the Chicago Climate Exchange and EU Emissions Trading Scheme (which is considering allowing the conversion of emission credits to emission allowances) begin to operate. Prices for each transaction, however, vary significantly. Nominal prices for emission credit trades have ranged anywhere from 60 cents to US$6.00 per tonne. This range is mainly due to the differences in the terms of the transactions rather than the shift in supply and demand and on the relative differences in the quality of projects. Despite the increasing volume of transactions, the difficulty for practitioners is that the market is still relatively illiquid and most trades are one-off transactions that cannot be deemed sufficiently representative to provide adequate price signals required for accurate value estimation.

Where does this leave us? Effectively, the answer on how to account for emission credits is most likely to come by looking (via analogy) to the work being undertaken by IFRIC on emission allowances. Given that this work is not finalized, it may be some time before authoritative literature is issued. In the interim, accountants will need to look at the specific facts and circumstances of each emission reduction project and employ a consistent approach to classifying and valuing any emission credits in companies' financial statements.

The lack of guidance and the relatively immature state of the emission credit market should not deter companies from recognizing emission credits in their accounts, however, if management is to do so, it should endeavour to report any assumptions they have made for the benefit of the users of the financial statements. Management must also be prepared to accept the potential impact on financial performance if the accounting treatment or circumstances supporting the transaction or valuation change.

7 Tax Consequences of Emission Allowances and Credits

As with any new business activity, the tax consequences of emission allowances and credit transactions need to be considered. Companies looking to participate in transactions associated with emission allowances or credits need to know the possible tax consequences of issues, such as how to treat expenditures for, or the

[16] F Lecocq, and K Capoor, *State and Trends of the Carbon Market* (The World Bank, Washington, December 2003), <http://www.carbonfinance.org> accessed 8 June 2004.

proceeds from sales of, emissions allowances or credits, the impact of cross-border transactions, hedging treatment, commodity tax implications, and the implications of investing in alternative emission reduction activities, such as in plant and equipment, asset acquisitions, and disposals.

Given the infancy of emission trading markets and the fact that much of the tax analysis depends upon specific facts and circumstances particular to each jurisdiction and individual transaction, no authoritative literature on the topic has been developed. Many jurisdictions are adopting a 'wait and see' approach to answering these questions. As a result, there is a significant amount of uncertainty as to the suitable tax treatment. Indeed, at the time of this chapter and to our knowledge, no legislators other than those in the UK had received requests for a ruling on the taxation consequences of emission-related transactions.

In the UK, preliminary tax discussions have focused on how to treat expenditures for, or the proceeds from sales of, emissions allowances, as well as the applicability of transaction taxes (for example, stamp duty, VAT) to sales of permits and credits.

Discussions with UK Inland Revenue Department have confirmed some general principles, but no specific legislation. In general, the taxation treatment will be driven by the accounting treatment adopted. This will typically influence the timing that the income and expenses are recognized for tax purposes. The basic position in the UK is that the trade of emission allowances will be treated as an account of income and as such will receive a full deduction for the costs, while any profits on the sale will be treated as trading profits.

The indirect taxation consequences of emission allowance or credit transactions also need to be considered. Many countries in the world have an indirect tax regime such as Value Added Tax ('VAT'), Goods and Services Tax ('GST'), or sales tax. The standard rate of indirect tax applied to transactions tends to be relatively high; for example, the standard rate of VAT in the European Union ranges from 15 to 25 per cent. With relatively high rates of VAT, participants in emissions trading markets need to identify their compliance obligations and ensure that, as far as possible, indirect tax does not become an additional cost to the business.

It seems likely that the issue of emission allowances by government or regulatory bodies will not be subject to indirect tax, although few tax authorities have confirmed this to date. Indirect tax is of more relevance to transfers and trades of allowances between market participants. In most jurisdictions, the liability to charge VAT is likely to depend on the specifics of each transaction, for example, whether it is domestic or cross-border.

Discussions with the tax authorities in the UK have resulted in HM Customs and Excise confirming that transfers and trades of allowances between market participants are potentially liable to VAT. Although there may be some exceptions, it seems that they expect UK VAT of 17.5 per cent to be charged on most emissions transactions. Many participants will be able to recover this VAT through their UK VAT returns and should not be out of pocket. However, there could be a VAT cost for companies not ordinarily entitled to recover all the VAT they incur, as well as for participants based outside of the UK. UK traders and brokers are currently in discussions with HM Customs and Excise in the UK to clarify the rules and also to ensure that the application of UK VAT is minimized. Undoubtedly, similar negotiations will have to commence in other jurisdictions as trading activity grows.

But what of other jurisdictions? A recent paper has been prepared on taxation issues for the Canadian market, which provides a good discussion on the topic for that market. The paper is entitled, *Taxes and the Environment: Tax Consequences of Emission Reduction Regulations.*[17]

Finally, as with the initial discussions in accounting, participants in emissions trading markets often seek guidance from the US acid rain market. The Internal Revenue Service (IRS) has developed some guidelines on the federal income tax consequences of the acid rain scheme. However, since many of the attributes of emissions allowances or credits may be different from the attributes of acid rain instruments, making reference to the US tax rulings on the acid rain scheme may not be a good indicator of the ultimate tax treatment. Ultimately, an understanding of particular tax issues for emission-related transactions should be sought from taxation specialists applicable to each jurisdiction.

8 Conclusion

If, as some experts predict, the market for GHG emissions allowances and credits develops into a multi-million or even multi-billion dollar market, the potential implications for company financial reporting is significant.

Europe, for example, is already seeing an increase in the interest of shareholders and boards of directors towards the management of climate change risks and opportunities. The development of appropriate accounting guidance is fundamental to such activities.

While the accounting and taxation matters involved are challenging, the accounting profession is taking a very proactive approach to ensure that sufficient guidance is available to enable companies to measure, assess and report the impact of emissions trading schemes in their financial statements.

[17] Published by the Canadian Tax Foundation in Report of Proceedings of the Fifty-Fourth Tax Conference, 2002 Conference Report (Toronto: Canadian Tax Foundation). Copies of *Taxes and the Environment: Tax Consequences of Emission Reduction Regulations* can be obtained from Jay Pon of Deloitte & Touche LLP, Canada (jpon@deloitte.ca).

5

Public Participation in CDM and JI Projects

*Nathalie Eddy**

1 Introduction: Global Recognition of the Benefits of Public Participation

Public access to, and engagement in decision-making that affects the environment enhances the success, sustainability, and accountability of policies and projects.[1] Governments worldwide have increasingly recognized the important role meaningful public participation can play in ensuring the realization of the environmental and development goals of a project or policy.[2] The acknowledgement of public participation as a key element to successful environmental decision-making and project development has been illustrated through several international declarations and bindings treaties. Most recently, 172 countries used the occasion of the 2002 World Summit on Sustainable Development (WSSD) in Johannesburg, South Africa to reiterate national commitments to public participation in environmental and development decision-making. Accordingly, Parties agreed to:

> Ensure access, at the national level, to environmental information and judicial and administrative proceedings in environmental matters, as well as public participation in decision-making, so as to further principle 10 of the Rio Declaration . . . [3]

Ten years prior, countries formally acknowledged, for the first time, the three primary elements of public participation and good environmental governance as access to information, access to meaningful public involvement, and access to

* Nathalie Eddy is an associate for the Access Initiative and the Partnership for Principle 10 with the World Resources Institute (WRI). The views expressed in this chapter are the views of the author and do not necessarily represent the views of WRI.

[1] World Resources Institute, *The World Resources Report 2002–2004: Decisions for the Earth* (Washington, DC, 2003) 47.

[2] IFI Shihata, *The World Bank Inspection Panel: In Practice* (Washington, DC, 2000) 265.

[3] World Summit on Sustainable Development (WSSD), *Plan of Implementation*, para 119*ter* (September 2002).

justice. Principle 10 of the Rio Declaration recognizes that:

> Environmental issues are best handled with the participation of all concerned citizens, at the relevant level. At the national level, each individual shall have appropriate access to information concerning the environment . . . the opportunity to participate in decision-making processes. . . . Effective access to judicial and administrative proceedings, including redress and remedy . . . .[4]

In addition to these broadly supported international declarations, the three elements of public participation also form the basis of a binding multilateral agreement under the United Nations Economic Commission for Europe (UN/ECE). The Aarhus Convention on Access to Information, Public Participation in Decision-Making and Access to Justice in Environmental Matters (Aarhus Convention) requires Parties to comply with fundamental environmental governance norms which are founded on the principle elements of access to information, participation and justice.[5]

While national performance standards vary greatly in the implementation of the three key access principles, certain general characteristics are commonly recognized.[6] Access to information includes public access to documents and timely, culturally appropriate notification of information availability and policy developments. Public participation in decision-making builds on adequate information to ensure civil society an opportunity to provide input, which will be directly considered by decision-makers. Access to justice allows civil society the possibility of requesting a review of a project or policy, and seeking remedial measures for any harms suffered.

As one of the agreements created at the 1992 Rio Earth Summit along with the Rio Declaration, the Parties to the United Nations Framework Convention on Climate Change (UNFCCC) and the subsequent Kyoto Protocol face an opportunity to put the public participation elements of Principle 10 into action in the context of a binding international treaty (the Kyoto Protocol). This chapter provides an introduction to the public participation provisions in the UNFCCC and Kyoto Protocol, and a detailed inventory of the public participation rules under the Clean Development Mechanism (CDM) and Joint Implementation (JI) elaborated in the context of the seventh session of the Conference of the Parties to the UNFCCC in the so-called Marrakesh Accords.[7] Section V offers a comparative

[4] UN Conference on Environment and Development, *Rio Declaration on Environment and Development*, (1992), principle 10.

[5] UN Economic Commission for Europe, *Convention on Access to Information, Public Participation in Decision-Making and Access to Justice in Environmental Matters*, done at Aarhus, Denmark, at Art 1 (25 June 1998) (hereinafter 'Aarhus Convention'). The Aarhus Convention entered into force on 30 October 2001 and currently includes twenty-seven Parties from Eastern and Western European countries.

[6] E Petkova, C Maurer, N Henninger, and F Irwin with J Coyle and G Hoff, *Closing the Gap: Information, Participation and Justice in Decision-Making for the Environment* (World Resources Institute, Washington DC, 2002) 121.

[7] FCCC/CP/2001/13/Add.2. The decision on Implementation Guidelines for JI Projects is contained in the draft Decision 16/CP.7. Decision 17/CP.7 on 'Modalities and procedures for a clean development mechanism as defined in Article 12 of the Kyoto Protocol' and its Annex on 'Modalities and procedures for a clean development mechanism as defined in Article 12 of the Kyoto Protocol.'

analysis of the treatment of access to information, participation, and justice in other multilateral environmental agreements (MEAs), by the World Bank and the World Trade Organization (WTO).

The CDM and JI are the first mechanisms of their kind to enable State Parties to meet their international treaty commitments in part through mitigation projects implemented in other countries. Once implemented, CDM and JI projects stand to set significant precedent for other treaties and institutions while also impacting a wide range of stakeholders. Given the spotlight on these innovative mechanisms, any perceived failure to adequately respond to stakeholders' input and concerns risks not only undermining the integrity of the mechanisms, but also the strength of the Kyoto Protocol as an effective approach to combat dangerous climate change.

2 Public Participation Provisions in the UNFCCC and Kyoto Protocol Negotiating Process

The role and rights of civil society in the overall negotiations of the UNFCCC and the Kyoto Protocol provide a necessary context within which to consider the more specific public participation provisions of the CDM and JI. The UNFCCC includes a broad public participation provision, which calls on all Parties to '[p]romote and cooperate in education, training and public awareness related to climate change and encourage the widest participation this process, including that of non-governmental organizations . . . '[8] UNFCCC and Kyoto Protocol negotiations provide for both formal and informal procedures by which non-governmental organizations (NGOs) can obtain information and contribute directly to the negotiating sessions.

Prior to further consideration of the general UNFCCC rules and the specific CDM and JI provisions regarding public participation, it is essential to recognize the range of interests and entities which fall under the UNFCCC's definition of an 'NGO'. The UNFCCC Secretariat offers five categories from which an NGO applying for UNFCCC observer accreditation may select:

(1) business and industry non-governmental organization (BINGO);
(2) environmental non-governmental organization (ENGO);
(3) indigenous peoples organization (IPO);
(4) local government and municipal authorities (LGMA); and
(5) research-oriented and independent organizations (RINGO).[9]

While the UNFCCC-maintained website boasts that 'The operational rules of the mechanisms are based on openness and transparency'[10] it also recognizes that

[8] UN Framework Convention on Climate Change (UNFCCC), 9 May 1992, entered into force 21 March 1994, 31 ILM 849, Art 4(1)(i); see also Art 6.

[9] UNFCCC, 'Standard Admission Process', <http://unfccc.int/resource/ngo/adm_proc.pdf> accessed 28 March 2004.

[10] UNFCCC, 'Issues' <http://unfccc.int/issues/mechanisms.html> accessed 10 April 2004.

the access is not necessarily provided directly to the stakeholder. The overview text points to the provisions in the Marrakesh Accords which 'provide for businesses, non-governmental organizations and other "legal entities" to participate in the three mechanisms, *albeit* under the responsibility of their governments.'[11]

Most CDM and JI public participation provisions provide for access to English-only documents via the internet. Currently, limited steps have been taken to adapt the rules to provide for culturally appropriate access to information, participation, and justice for indigenous peoples and local communities most directly impacted by the projects. The following sections do not attempt to measure the adequacy of the CDM and JI public participation provisions for all stakeholders. Even in the absence of such a differentiation of the application of the CDM and JI rules, the imbalance of provisions, which allow for engagement of some NGOs, but likely exclude indigenous peoples and local communities, is apparent. The conclusion highlights initiatives such as CDM Watch, the CDM Gold Standard and the International Forum of Indigenous Peoples and Local Communities on Climate, which seek to strategically make use of publicly available CDM information and expand the existing CDM and JI provisions.

2.1 Formal NGO Access to UNFCCC Negotiations

The UNFCCC recognizes the right of NGOs to observe negotiations, barring objections from governmental Parties.[12] In addition, the draft Rules of Procedure (applied in practice but never officially adopted by the Conference of the Parties (COP) to the UNFCCC) provide NGO access to smaller, more technical meetings convened under the UNFCCC.[13]

The text of the UNFCCC allows for NGO observer participation in the negotiations of both the UNFCCC and the Kyoto Protocol providing that:

> Any body or agency, whether national or international, governmental or non-governmental, which is qualified in matters covered by the Convention, and which has informed the secretariat of its wish to be represented at a session of the Conference of the Parties as an observer, may be so admitted unless at least one third of the Parties present object. The admission and participation of observers shall be subject to the rules of procedure adopted by the Conference of the Parties.[14]

The draft Rules of Procedure not only reiterate the UNFCCC's provision for NGO participation, but also add the possibility for observers to actually contribute directly to the negotiating sessions if invited to do so by the COP President.[15] While NGO observers are not accorded a right to vote under this rule, the opportunity to address the full plenary session of the governmental Parties and

[11] UNFCCC, 'Issues' <http://unfccc.int/issues/mechanisms.html> accessed 10 April 2004.

[12] Art 7(6), UNFCCC.

[13] Draft Rules of Procedure, 7(2), FCCC/CP/1996/2 (22 May 1996) <http://unfccc.int/resource/docs/cop2/02.pdf> accessed 20 January 2004.

[14] See n 12 above.

[15] See n 13 above.

benefit from the advantage of instantaneous translation into the six UN languages offers NGOs a meaningful forum in which to contribute to the negotiations.[16] Both the right to attend the UNFCCC negotiations, as well as the opportunity to participate in plenary sessions, may be denied if at least one-third of the Parties present object.[17] While the draft Rules of Procedure leave room for ambiguity, the right to attend the negotiations as an observer and the option to address the plenary are by default available for NGOs to exercise, 'unless the Conference of the Parties decides otherwise.'[18]

In 1998, the UNFCCC Subsidiary Body on Implementation (SBI) further considered the role of NGOs in the negotiations.[19] Significantly, the SBI recognized the importance of information to enable NGOs to be 'effectively involved in the implementation of the Convention' and sought to further strengthen the 'availability of documentation and information to non-governmental organizations . . . '[20]

The COP at its fourth session clarified the participation of NGOs in contact groups, specifying that NGOs may participate *as observers* in contact groups, at the invitation of the presiding officers.[21] At their own discretion, the presiding officers may retract their invitation for NGOs to observe at any point during the contract group discussions.[22] Establishing meetings as open by default, but closed at the discretion of the official body, is common in international negotiations. For example, the provisional Rules of Procedure for the United Nations Security Council provide, 'Unless it decides otherwise, the Security Council shall meet in public.'[23]

Most recently, in March 2003, the UNFCCC secretariat released a more detailed set of guidelines for NGO participation in the climate change negotiations.[24] Based on the draft Rules of Procedure, the guidelines elaborate a general NGO code of conduct with regard to access to the venue of the climate negotiations, etiquette and safety, participation in open and closed meetings and contact groups, and materials distributed by NGOs.[25]

2.2 Informal NGO Access to UNFCCC Negotiations

There are also a number of informal procedures that have developed over the course of the UNFCCC and Kyoto Protocol negotiations, but which are not

[16] See n 13 above. [17] ibid. [18] ibid, at r 30.

[19] UNFCCC, FCCC/SBI/1998/6, <http://unfccc.int/resource/docs/1998/sbi/06.pdf> accessed 15 January 2004. [20] ibid, at para 81(c).

[21] UNFCCC, FCCC/CP/1998/16/Add.1, Decision 18/CP.4 'Attendance of intergovernmental and non-governmental organizations at contact groups at paragraph 1' (20 January 1999). <http://unfccc.int/resource/docs/cop4/16a01.pdf> accessed 17 January 2004. [22] ibid.

[23] Rule 48, UN Doc. S/96/Rev.7 (1982).

[24] UNFCCC, Guidelines for the Participation of Representatives of Non-governmental Organizations at Meeting of the Bodies of the United Nations Framework Convention on Climate Change, March 2003, <http://unfccc.int/resource/ngo/coc_guide.pdf> accessed 28 March 2004.

[25] ibid.

officially recognized in the draft Rules of Procedure or subsequent COP decisions. For example, in addition to taking the plenary floor at the COP President's invitation, several opportunities have arisen in which NGO submissions are requested on a particular issue.[26] Submissions received by NGOs and Parties are then typically compiled into a 'Misc'. document and made available to all Parties and observers at the negotiations and posted on the UNFCCC website. Such invitations offer NGOs the opportunity to contribute views and analysis directly into the negotiating process.

3 CDM And JI Rules

Set within the context of the information and access provided under the UNFCCC and Kyoto Protocol negotiations, the project-based 'flexibility' mechanisms further elaborate more specific rules for the role and rights of civil society. The public participation rules of the CDM and JI address the overall governance of each mechanism, while also establishing minimum performance standards for individual projects. Similar to the trends identified at the national level, the CDM and JI tend to be strongest in providing access to information, weaker in ensuring access to participation, and particularly lacking on allowing access to justice.[27] Both CDM and JI provide for varying degrees of publicly available information and public comment periods. Neither mechanism allows for any stakeholder-triggered project or general policy review. However, the CDM recently elaborated a process which enables stakeholders to submit comments regarding the accreditation of a designated operational entity (DOE).[28]

While the Marrakesh Accords of 2001 established a framework for the CDM rules, the Parties have since further elaborated more detailed provisions. In particular, with regard to the CDM, significant public participation rules have been created or expanded through the work of the CDM Executive Board (EB). The following sections highlight the public participation provisions for both the project phase specific and general governance for the CDM and JI.

3.1 CDM Rules: Summary of Provisions for Access to Information, Access to Participation, and Access to Justice

The CDM rules include public participation provisions for specific stages of CDM project development, as well as for general CDM policy. Table 5.1 provides a

[26] See UNFCCC, FCCC/SBSTA/2001/2, para 27(c) (18 September 2001). <http://unfccc.int/resource/docs/2001/sbsta/02.pdf> accessed 28 January 2004.

[27] Petkova, n 6 above, at 3.

[28] CDM EB, 'Report of the Seventh Meeting', 4 February 2003, Annex 2, paras 17 and 23 (hereinafter 'EB7 Report, Annex 2').

Table 5.1: Summary: CDM public participation provisions*

Access to Information	**Responsible Entity**
CDM Validation Phase	
• Project Design Document (PDD), para 40(b); *DOEs are required to either post a link to the PDD on the UNFCCC website, or post the PDD in its entirety on the UNFCCC website, EB Elaborated Procedures, November 2003; to date, six PDDs have been posted to the CDM website, see <http://cdm.unfccc.int/Validation/publicPDD>*	Validating DOE
• Public comments, para 40(c); *DOEs must make public comments received available on the internet until the DOE requests registration of the project (or withdraws the project from further consideration), EB Elaborated Procedures, November 2003.*	Validating DOE
• Validation report (including how DOE took into account public comments received), para 40(f) and (g); *to date, no validation reports have been posted to the CDM website; for updates, see <http://cdm.unfccc.int/EB/ProjectsUnderReview>*	Validating DOE
CDM Registration Phase	
Final eight weeks after registration request made (unless review triggered)	
• EB decision and its reasons on a review of the registration of a project, para 41(b); *to date, no requests for registration have been received by the EB; for updates, see <http://cdm.unfccc.int/EB/ProjectsForRegistration>*	EB
CDM Verification and Certification Phases	
• Monitoring report, para 62—the information included in the monitoring report is subject to the confidentiality provisions of para 27(h) and is based on the monitoring measures conducted by project participants.	Verifying DOE
• Verification report, para 62(h)—the verifying DOE must provide the verification report to the project participants, the Parties involved and the EB; following the verifying DOE's submission, the EB will in turn make the verification report publicly available.	EB
• Written certification decision/certification report, para 63.	Verifying DOE
Issuance of CERs	
Final fifteen days after issuance request made (unless review requested)	
• EB decision and its reasons on a review of the issuance of CERs, para 65(c).	EB
General Project Life	
UNFCCC secretariat has expanded and improved the CDM website, made CD-ROM copies of website available, and translated key CDM documents into the six UN languages, EB10 Report, Annex 6, para 54.	
• Repository of approved rules, procedures, methodologies and standards, para 5(k).	EB
• CDM registry, para 5(l) and Appendix D, para 9—tracks the issuance, holding, transfer and acquisition of CERs by non-Annex I Parties; *EB10 decided not to establish an interim CDM registry in 2003; the EB has placed a public call for input on the development of a CDM registry, EB10 Report, Annex, para 47.*	EB
• Database of CDM projects (include info on PDD, comments received, verification reports and information on all CERs issued), para 5(m).	EB
• Full text of all EB decisions in six UN languages, para 17; *UNFCCC secretariat has extended translation to include CDM documents and forms beyond EB decisions, EB10 Report, Annex 6, para 54.*	EB / UNFCCC Secretariat

(Continued)

Table 5.1: (*Continued*)

Access to Information	Responsible Entity
• List of all DOEs, para 20(c); *to date, the EB has not approved any designations of Operations Entities—twenty-one applicant entities are currently under review, see <http://cdm.unfccc.int/DOE>*	EB / UNFCCC Secretariat
• New baseline and monitoring methodologies, *after* EB review and approval, para 38; *four new CDM methodologies have been approved and are posted to the CDM website:<http://cdm.unfccc.int/methodologies/approved>, twenty-three proposed methodologies await approval <http://cdm.unfccc.int/methodologies/process>*	EB / UNFCCC Secretariat
• List of CDM projects for which DOE has carried out validation, verification or certification, para 27(f).	Each DOE
• Information from CDM project participants, para 27(h).	Each DOE
• List of NAI Parties which are KP Parties, para 34(a); *to date, this list has not yet been made available.*	UNFCCC Secretariat
• List of AI Parties that do not meet the CDM participation requirements or have been suspended, para 34(b); *to date, this list has not yet been made available.*	UNFCCC Secretariat
• EB's recommendation and COP/MOP decision (based on EB's recommendation, review and hearing) to suspend/withdraw OE designation, para 21.	UNFCCC Secretariat
Public Participation	
Preparation of the PDD	
• Comments by local stakeholders **must be invited** and documented in the PDD para 37(b)/App B, para 2(g).	Project Participant
Validation	
• Following PDD release, Parties, stakeholders and UNFCCC accredited NGOs have 30 days to provide comments on the validation requirements, para 40(c).	Validating DOE
Verification	
• Verifying DOEs **may** interview local stakeholders while conducting on-site inspections, para 62(b).	Verifying DOE
DOE Accreditation	
• *Fifteen-day stakeholder 'call for input' on applicant entities announced on the CDM website, EB7 Report, Annex 2, paras 17 and 23; to date 21 applicant entities have been announced on the CDM website and comments invited, see see <http://cdm.unfccc.int/DOE>*	EB
Technical Reports	
• Any technical reports commissioned by the EB must be open to public comment for eight weeks, para 5(j).	EB
General project life	
• *At the invitation of the EB, observers can make presentations at EB meetings, Rule 27, EB Rules of Procedure.*	EB
Access to Jusrice	
Registration	
• **A Party involved in the CDM project or at least three EB members** may request a review of the registration of a CDM project; the review is **limited** to issues relating to the validation requirements, para 41.	EB performs the review

Issuance of CERs	
• **A Party involved in the CDM project or at least three EB members** may request a review of the proposed issuance of CERs; the review is **limited** to the review issues of fraud, malfeasance or incompetence of the DOEs, para. 65.	EB performs the review
DOE Accreditation	
• EB reviews each DOE and decides whether or not to re-accredit a DOE every three years, para 20(d); *EB may conduct spot-checking which can be triggered by a 'written substantiated complaint' by stakeholders, EB7 Report, Annex 2, para 53.3.*	Stakeholder-triggered review of DOE accreditation

* Unless otherwise noted, all citations refer to the CDM M&P; **text in italic** indicates developments since the agreement of the CDM M&P in Marrakesh in November 2001.

detailed inventory of all of the CDM provisions addressing either access to information, access to participation or access to justice. Table 5.2 offers a stage-by-stage summary of the CDM public participation provision. The following sections offer a narrative to accompany Tables 5.1 and 5.2 to highlight several recent developments in the CDM public participation rules.

3.2 Access to Information in the CDM

The CDM rules require that nineteen separate types of CDM-related information be made 'publicly available'. Beginning with the draft negotiating text, which resulted in the Marrakesh Accords 2001, NGOs continually called upon the Parties to specify the manner and time frame for making CDM documents 'publicly available'.[29] However, while the rules offer limited specific guidelines for the process of making documents available, the UNFCCC secretariat and the EB have gone to great lengths to make extensive CDM-related information available to the public. At the tenth meeting of the CDM Executive Board (EB10) in July 2003, 'the Board welcomed the announcement by the secretariat that it has started to undertake work on categorizing CDM related documentation with a view to improving accessibility and user friendly searches.'[30] In addition, the UNFCCC secretariat has taken several significant steps further strengthening the transparency and accessibility of information for observers. For example, the UNFCCC

[29] See K Baumert and E Petkova, *How Will the Clean Development Mechanism Ensure Transparency, Public Engagement, and Accountability?*, WRI Climate Notes, November 2000; See also CAN Non-paper, *Key Public Participation Points for Consideration at EB4* (9 June 2002) (distributed to CDM Executive Board members and parties attending the sixteenth Meeting of the Subsidiary Bodies of the UNFCCC in Bonn, Germany), available at <http://www.climnet.org/sbsta16/CDMpp.pdf> accessed 28 January 2004.

[30] CDM EB, 'Report of the Tenth Meeting', Annex 6, 29 July 2003, para 43 (hereinafter 'EB10 Report, Annex 6').

Table 5.2: Stage-by-stage summary of CDM public participation provisions

Stages of the CDM	**Publicly Accessible Documents**	**Public Comment Periods**	**CDM Project Review**
Project Participant Proposes initial CDM Project	*None specified*	Project participants **must** invite **local stakeholders** to provide input on the proposed CDM project	*None*
Validation	• Project Design Document (PDD) • Public comments received during thirty-day comment period • Validation Report	Thirty-day comment period open to stakeholders and UNFCCC accredited NGOs	*None*
Registration	• EB decision and its reasons on a review of project registration	*None*	*None for stakeholders, however,* involved Parties or three EB members can request a review of registration
Verification and certification	• Monitoring report • Verification report • Written certification decision/certification report	Verifying designated operational entities ***may*** interview **local stakeholders** while conducting on-site inspections of projects	*None*
Issuance of CDM Certified Emissions Reductions (CERs)	• EB decision and its reasons on a review of the issuance of CERs for a project	*None*	*None for stakeholders, however,* involved Parties or three EB members can request a review of the issuance of CERS
Accreditation of DOEs	• EB's recommendation and COP/MOP decision (based on EB's rec., review and hearing) to suspend/withdraw OE designation	Stakeholders may trigger a spot-check of a DOE with the submission of a 'written substantiated complaint'	Stakeholders may trigger a spot-check of a DOE with the submission of a 'written substantiated complaint'
EB Technical Reports	• Draft technical report	Eight-week public comment period on draft technical reports	*None*
General Project Life		At the invitation of the EB, observers can make presentations at EB meetings	*None*

secretariat has developed an extensive CDM website, made CD-ROM copies of the website publicly available and translated key CDM documents and forms into the six UN languages.[31] The scope and depth of the CDM website continues to expand.[32] The CDM website also hosts the CDM new facility which includes approximately 1,000 subscribers as of April 2004.[33] In addition, the EB has also arranged for regular, informal briefings with interested observers.[34] Not surprisingly, the EB reported in July 2003 that: 'a significant portion of the costs for operating the CDM continues to be related to its governance and the maintaining of information systems and channels, essential for broadening the scope of involvement in the CDM . . .'[35]

While much of the publicly available CDM information is closely interlinked, the following subsections consider each of the six CDM project phases, and then provide a summary of the general project life documents, which are to be made publicly available. The broad definition of 'stakeholders' as 'the public, including individuals, groups or communities affected, or likely to be affected, by the proposed clean development mechanism project activity' contributes to the spectrum of individuals and entities guaranteed information and participation rights under the CDM.[36]

3.2.1 CDM validation phase

The CDM validation stage requires that the validating Designated Operational Entity (DOE) makes three key project documents publicly available. The CDM rules define the validation as 'the process of independent evaluation of a project activity by a designated operational entity against the requirements of the CDM . . .'[37] The project design document (PDD) is the first piece of information regarding a specific CDM project to be made publicly available.[38] Appendix B to the CDM rules outlines the required information to be included in each project design document.[39] The full PDD should be made available, subject to the general confidentiality limitations provided in the text.[40] The EB recently elaborated procedures which require the DOE to either create a website

[31] EB10 Report, Annex 6, n 30 above, para 54.

[32] See UNFCCC-maintained CDM website, <http://cdm.unfccc.int/> accessed 10 April 2004.

[33] ibid. [34] EB10 Report, Annex 6, n 30 above, para 60. [35] ibid, para 78.

[36] Report of the Conference of the Parties on its seventh session, FCCC/CP/2001/13/Add.2 (2001), Decision 17/CP.7, para 1(e) (hereinafter 'CDM Rules'). The CDM definition of 'stakeholders' is more narrow than the definition provided in the Aarhus Convention, which reads ' "The public concerned" means the public affected or likely to be affected by, or having an interest in, the environmental decision-making; for the purposes of this definition, non-governmental organizations promoting environmental protection and meeting any requirements under national law shall be deemed to have an interest.' Aarhus Convention, n 5 above, Art 2(5).

[37] CDM Rules, n 36 above, para 35. [38] ibid, para 40(b). [39] ibid, Appendix B, para 2.

[40] ibid, para 27(h).

for the publication of the CDM-PDD in its entirety and to post a link to the publication on the CDM website, or publish the PDD in its entirety on the CDM website.[41]

The validating DOE then has thirty days to both receive and make publicly available all comments from Parties, stakeholders, and UNFCCC accredited NGOs. To date, ten PDDs have been posted to the CDM website for public comment. The PDDs and all the comments received have remained on the website after the review period closed.[42] Following the close of the validation public comment period, the DOE must submit a validation report to the EB and at the same time make this report publicly available.[43] No CDM project validation reports have been posted to the CDM website.[44] In addition to the PDD, the validation report includes written proof of host party approval of the project and an explanation of how the DOE took due account of the public comments received.[45]

The PDD also provides for two options available to project participants when addressing environmental impacts. At the very least, the PDD must include '[d]ocumentation on the analysis of the environmental impacts, including transboundary impacts.'[46] A complete environmental impact assessment (EIA) is only required when the 'impacts are considered significant by the project participants or the host Party . . . '[47] Even if the impacts are considered 'significant' (thus necessitating an EIA), the process followed, and the extent to which the public is consulted, will depend on the 'procedures as required by the host Party'.[48] To date, the Parties have not elaborated guidelines for determining when an environmental impact should be considered 'significant'. It is therefore difficult to determine if EIAs in CDM projects will serve to provide an additional opportunity for public comments.

3.2.2 CDM registration phase

Eight weeks after the DOE has submitted the validation report to the EB, the project is considered to be a registered CDM project.[49] Should one of the project participants, or at least three members of the EB request a review, the EB will make its decision and its justification thereof publicly available.[50] As no validation reports have been posted to the CDM website, no requests for registration have been received by the CDM EB.[51]

[41] CDM EB, 'Procedures on Public Availability of the CDM Project Design Document and for Receiving Comments as Referred to in Paragraphs 40(b) and (c) of the CDM Modalities and Procedures,' Ver. 3, November 2003, <http://cdm.unfccc.int/Reference/Procedures/public%20availPDD_ver03.pdf> (hereinafter 'PDD Procedures').

[42] See CDM website, <http://cdm.unfccc.int/Validation/publicPDD> accessed 10 April 2004.

[43] CDM Rules, n 36 above, para 40(g).

[44] See <http://cdm.unfccc.int/EB/ProjectsUnderReview> accessed 10 April 2004.

[45] CDM Rules, n 36 above, para 40(f).

[46] ibid, Appendix B, para 2(e)(i).

[47] ibid.

[48] ibid.

[49] ibid, para 41.

[50] ibid, para 41(b).

[51] See <http://cdm.unfccc.int/EB/ProjectsForRegistration> accessed 10 April 2004.

3.2.3 CDM verification and certification phases

Over the course of the verification and certification stages, the verifying DOE must make both the monitoring and certification reports publicly available.[52] Additionally, the EB must make the verification report available upon receipt from the verifying DOE.[53] The issuance of Certified Emission Reductions (CERs) is considered final, fifteen days after the request, unless a party involved in the project, or at least three members of the EB request a review.[54] Similar to the provisions following a review of the registration of a CDM project, the CDM rules again require the EB to make its decision and reasoning regarding the review of the issuance of CERs publicly available.[55]

In addition to the eight types of information which DOEs and the EB must make available to the public at specific stages of a CDM project development, there are also eleven other sources of more general CDM-related information which the EB, DOEs and the UNFCCC secretariat are required to make publicly available.

The CDM rules include provisions which both address specific stages of CDM project development, as well as general governance rules, which apply throughout the entire lifetime of a project. As noted above, no specific projects have been proposed for CDM validation and registration. As a result, there is currently a lack of practical experience in implementing the CDM's public participation provisions. However, while the project stage specific provisions have not yet been applied, the EB and parties have both applied and elaborated the more general rules governing the broader CDM decision-making processes and information management structure. Based on the experience demonstrated in applying the more general rules, certain lessons may be drawn for the project-specific CDM rules.

3.2.4 CDM general project life

The EB is responsible for maintaining and making publicly available the central administrative, procedural, and methodological documents for the CDM. The EB has mostly fulfilled the following responsibilities by posting information to the CDM website maintained by the UNFCCC Secretariat and announcing new information and public comment periods through CDM-related listservs. For each CDM provision addressed below, a description of how, and to what extent, the rules have been applied will be provided.

The CDM rules mandate the EB to make CDM 'rules, procedures, methodologies and standards' publicly available.[56] Accordingly, the CDM website maintains updated information on the CDM rules.[57] The EB is also tasked with creating an interactive on-line registry detailing how many CERs are held,

52 CDM Rules, n 36 above, paras 62 and 63. 53 ibid, para 62(h). 54 ibid, para 65(c).
55 ibid, para 65(c). 56 ibid, para 5(k).
57 See <http://cdm.unfccc.int> accessed 9 February 2004.

issued, and transferred by each account holder.[58] At its tenth meeting in July 2003, the EB decided against establishing an interim CDM registry in 2003.[59] Instead, the EB issued a public call 'to Parties and organizations' to provide input and suggestions regarding the future development of a CDM registry.[60] In addition to the CDM registry of CERs, the EB is also responsible for creating and maintaining a database of detailed project-specific information.[61] As no CDM projects have been validated by a DOE, the EB has not yet established the CDM database.

Not only must all EB decisions be made publicly available, but they must also be translated into the six official languages of the United Nations.[62] While the EB has met fifteen times since its creation at Marrakesh, and issued public EB Reports following each meeting, the number of official EB decisions remains much more limited. In contrast to EB decisions, EB Reports are made publicly available in English only.[63] However, the EB and UNFCCC secretariat have translated key CDM documents, beyond the EB decisions, into the six UN languages.[64]

As the primary contact point for reviewing the accreditation and re-accreditation of DOEs, the EB is also required to maintain an up-to-date and publicly available list of all of the accredited DOEs.[65] To date, the EB has approved two designations of operational entities.[66] However, at the time of writing, nineteen applicant entities are currently under review.[67] In addition to maintaining the list of accredited DOEs, the EB is also responsible for rendering new baseline and monitoring methodologies available to the public pursuant to EB review and approval.[68] Thus far, four new methodologies have been approved by the EB and thirty-three methodologies await approval.[69] The proposed CDM methodologies and all public comments received are available on the CDM website.[70]

In addition to the six documents, which the EB is responsible for making available to the public, each DOE also plays an important role in publicly sharing information regarding their individual CDM projects.[71] Once a DOE is accredited by the EB, it is required to maintain a publicly accessible list of all the projects for which it has served as a validating, verifying or certifying DOE.[72] DOEs are currently in the process of requesting accreditation from the EB, and have therefore not yet begun to list projects or information from participants. The CDM rules also require that all information which a DOE receives from a project participant, must be made publicly available unless the project participant marks the information as proprietary or confidential.[73]

[58] CDM Rules, n 36 above, para 5(l) and Appendix D, para 9.
[59] EB10 Report, Annex 6, n 30 above, para 47. [60] ibid, para 47(a).
[61] CDM Rules, n 36 above, para 5(m). [62] ibid, para 17.
[63] EB Reports are available at <http://cdm.unfccc.int>.
[64] EB10 Report, Annex 6, n 30 above, para 54. [65] CDM Rules, n 36 above, para 20(c).
[66] See < http://cdm.unfccc.int/DOE> accessed 10 April 2004. [67] ibid.
[68] CDM Rules, n 36 above, para 38.
[69] See < http://cdm.unfccc.int/methodologies/approved> accessed 10 April 2004. [70] ibid.
[71] CDM Rules, n 36 above, para 27. [72] ibid, para 27(f). [73] ibid, para 27(h).

The determination of the confidentiality of information will be at the discretion of the project participants, subject to the minimal parameters provided by the CDM rules.[74] CDM project participants are prohibited from marking as confidential any information which serves to measure project additionality, explain the baseline methodology, or contribute to an environmental impact assessment.[75]

The extent to which the information included in the database of CDM projects and the list of DOEs maintained by the EB, and the lists of projects and information from project participants compiled by the DOEs is duplicative remains to be seen. Ideally, the EB and DOEs will be able to create links between their respective sources of information to enable interested stakeholders to easily access the full range of publicly available information about a particular project or DOE.

Finally, the UNFCCC secretariat is required to make three types of CDM information relating to CDM eligibility and accreditation of DOEs publicly available. Regarding eligibility for a party to participate in the CDM, the CDM rules call on the UNFCCC secretariat to publicly maintain both (1) a list of non-Annex I Parties which are Parties to the Kyoto Protocol, and (2) a list of Annex I Parties which either fail to meet the participation requirements of the CDM (for example, national emissions reporting and registry requirements, pursuant to Articles 5 and 7 of the Kyoto Protocol) or have been suspended.[76] As the Kyoto Protocol has not yet entered into force, neither of these lists was available on the CDM website at the time of writing. The secretariat is also responsible for allowing public access to the EB's recommendation and the final decision of the Conference of the Parties to the UNFCCC serving as Meeting of the Parties to the Kyoto Protocol (COP/MOP) in response to an EB-initiated review of the accreditation of a DOE.[77] The sections, which follow, will elaborate the public opportunities for comment and review regarding the accreditation of DOEs.

The trends at both the national and international levels demonstrate that of the three access principles articulated in Principle 10 of the Rio Declaration, access to information tends to be more robust than public participation and access to redress and remedy.[78] Similarly, the extent and scope of CDM information available to the public exceeds the frequency and breadth of the available CDM public comment periods. Whereas the CDM rules require nineteen explicit types of information be made publicly available for interested stakeholders, the rules only provide for a total of six potential public comment periods.

3.3 Public Participation in the CDM

The first and most significant required public comment period occurs during the initial preparation of the PDD.[79] In order for a DOE to validate a proposed CDM

[74] CDM Rules, n 36 above para 27(h). [75] ibid. [76] ibid, para 34 (a) and (b).

[77] ibid, para 21.

[78] See Petkova, n 6 above, at 3; see also C Maurer, 'Aligning Commitments: Public Participation, International Decision-Making, and the Environment', WRI Issue Brief, May 2003, 18.

[79] CDM Rules, n 36 above, para 37(b) and Appendix B, para 2(g).

activity, the DOE must confirm that the PDD indicates that stakeholders were invited to comment on the project.[80] In addition to demonstrating that stakeholders were invited to comment on the proposed project, the PDD must also summarize the comments received and how due account of the public comments was taken.[81] The CDM rules do not offer any guidance as to the time frame or the appropriate channels which should be used in order to communicate effectively with stakeholders.

The validation public comment period is the second call for public input into a CDM project, and is the most clearly defined of the limited CDM public comment windows. After the validating DOE has confirmed that the enumerated elements of a PDD have been met, the DOE will make the PDD publicly available.[82] Once the PDD becomes publicly available, the validating DOE must provide for a thirty-day public comment period, open to input from Parties, stakeholders, and UNFCCC accredited NGOs.[83] As noted above, the EB has elaborated guidance which DOEs must follow while making the PDD available for public comment.[84] While the participation for this validation comment period is broad, the scope of the actual comments is more limited. The CDM rules specify that comments must pertain to the actual validation requirements.[85]

The verification phase of a CDM project offers a potential final public comment period during a specific CDM project phase. According to the CDM rules, the verifying DOE may opt to conduct an on-site inspection of a CDM project.[86] In the process of this on-site inspection, a verifying DOE may elect to interview local stakeholders.[87] The CDM rules do not offer additional guidance for conducting on-site interviews over the course of a verification inspection, nor do they provide criteria for determining when such inspections and interview would be appropriate.

Following the conclusion of the Marrakesh Accord negotiations, the EB developed a process through which stakeholders may provide comments on applicant DOEs announced on the CDM website.[88] At the time of writing, two entities had been accredited and nineteen applicant entities had been announced on the CDM website.[89] Given the critical role DOEs are likely to play in ensuring the environmental integrity of CDM projects, this recently created public comment period is significant.

An additional prospective public comment period relates to decision-making of the EB in the context of EB-commissioned technical reports.[90] The CDM rules requires that all technical reports regarding 'draft methodologies and

[80] CDM Rules, para 37(b) and Appendix B, para 2(g). Note that while para 37(b) specifies 'local' stakeholders, Appendix B refers only to 'stakeholders'. The EB has not yet reported on any clarification of the requirement under these provisions.

[81] ibid.

[82] ibid, para 40(b).

[83] ibid, para 40(c).

[84] See PDD Procedures, n 41 above.

[85] CDM Rules, n 36 above, para 40(c).

[86] ibid, para 62(b).

[87] ibid.

[88] EB7 Report, Annex 2, n 28 above, paras 17 and 23.

[89] See <http://cdm.unfccc.int/DOE>.

[90] CDM Rules, n 36 above, para 5(j).

guidance' commissioned by the EB must be made publicly available for review and comment for a period of eight weeks.[91] Following the close of the eight-week comment period, the technical paper will be put forward to the COP/MOP for its consideration.[92]

Only EB members and alternates are empowered to directly contribute to the CDM EB meetings.[93] For each EB meeting, the UNFCCC secretariat provides a space from which non-EB members may observe the EB meeting.[94] The EB reports that an average of eighteen observers have attended the EB meetings at each session. The secretariat ensures that the venue provided for the observers can accommodate up to fifty individuals.[95] The proceedings of the EB meetings are broadcast on closed-circuit television which enables observers to follow the negotiations either on-site via a large screen in a separate room, or via streaming video on the internet. Building on the existing provisions for observers at EB meetings, the EB Rules of Procedure further opened the doors by allowing observers, 'at the invitation of the EB', to make presentations at the EB meetings.[96] Members of Climate Action Network (CAN) have requested the opportunity to address the EB directly regarding specific CDM issues under EB consideration. At the time of writing, no observer had been invited by the EB to make a presentation. However, in each EB meeting report, the EB reiterates that it 'took note, with appreciation, of information forwarded and correspondence received from the public' since its last meeting.[97] In contrast to the expanding public comment periods, the opportunity for a stakeholder to seek redress or request a review of a CDM project or policy is almost non-existent.

3.4 Access to Justice in the CDM

The CDM rules do not allow for any form of a CDM project review to be triggered by a stakeholder. In order for a review to be performed following either the registration for a CDM project or the issuance of CERs (the only two available windows for a review provided for under the CDM rules) a party involved in the CDM project or at least three members of the EB must make a request.[98] As previously mentioned above, the EB may also conduct a review of the

[91] CDM Rules, n 36 above, para 5(j). [92] ibid. [93] ibid, para 16. [94] ibid.

[95] EB10 Report, Annex 6, n 30 above, para 63. [96] CDM EB, Rules of Procedure, r 27.

[97] See CDM EB, EB12 Meeting Report, 2 December 2003, para 46. This same language is repeated in previous EB reports.

[98] CDM Rules, n 36 above, paras 41 and 65. Project participants may be governmental or private entities who develop the initial project proposal and oversee its implementation. See also N Eddy and G Wiser, 'Public Participation in the Clean Development Mechanism of the Kyoto Protocol', in *The New 'Public'—The Globalization of Public Participation* (ed. Carl Bruch Environmental Law Institute, 2002).

designation of a DOE.[99] Through this review, the EB may conduct spot-checking of a DOE, which can be triggered by 'written substantial complaints' by stakeholders.[100]

4 JI Rules: Summary of Provisions for Access to Information, Access to Participation, and Access to Justice

The JI public participation provisions are much less extensive than those under the CDM rules (see Table 5.3). This may be a result of the different participation requirements under each of the mechanisms. Whereas the CDM allow participation by all non-Annex I and Annex I Parties to the Kyoto Protocol (provided that the Annex I Parties are in compliance with their national reporting requirements under Articles 5 and 7),[101] JI limits participation and eligibility to acquire and/or transfer Emission Reduction Units (ERUs) to Annex I Parties to the Kyoto Protocol.[102] The JI rules seem to presume that JI-eligible parties would have adequate guidelines and procedures for considering public comments.[103] While it is significant that all Parties interested in participating in JI are required to explain their national process for public comments, it remains uncertain if this catch-all provision, combined with the limited additional guidance, will guarantee appropriate access to information, participation and justice in JI project development and implementation.

Unlike the clarifications already achieved through the elaboration of the CDM rules, the provisions for the JI rules do not yet extend beyond the agreements reached in Marrakesh. The combination of the mechanism-general public participation expectation, in addition to the holding pattern in which the JI rules remain, results in limited potential analysis of the JI public participation provisions.

As decided in Marrakesh, Parties interested in hosting JI projects may do so pursuant to two different 'tracks'. Track One allows host Parties which satisfy the methodological and reporting requirements under Articles 5 and 7 of the Kyoto Protocol to implement JI projects based on their established national procedures. Track Two of JI requires a more active role on the part of the JI Supervisory Committee (JI-SC) and results in a process more similar to that required under the CDM rules. The following overview of the JI public participation provisions focuses primarily on the procedures required for Track Two JI projects.

[99] COM Rules, n 36 above, para 20(d).
[100] EB7 Report, Annex 2, n 28 above, para 53.
[101] CDM Rules, n 36 above, para 34.
[102] Report of the Conference of the Parties on its seventh session, FCCC/CP/2001/13/Add.2 (2001), Decision 16/CP.7, para 21 (hereinafter 'JI Rules').
[103] JI Rules, para 20.

Table 5.3: Stage-by-stage summary of JI public participation provisions

Stages of JI	**Access to Information**	**Access to Participation**	**Access to Justice**
Project life—General—JI Track I and Track II JI Projects			
	• Full text of all SC decisions shall be made publicly available in the six official UN languages, para 16 • List of Parties eligible to pursue Track I JI projects or who have been suspended due to an eligibility violation, para 27 (secr) • JI project information according to JI reporting guidelines, para 28 (host)	All JI Parties must 'inform the secretariat of . . . its national guidelines and procedures for approving [JI] projects, including the consideration of stakeholders' comments', para 20	
JI Track II Projects			
Verification	• PDD, para 32 (AIE) • PDD verification determination, summary of comments and how due account was taken, para 34 (AIE) • SC's decision regarding a review of a PDD verification (within six months), SC decision is final, para 35	Thirty-day public comment period open to Parties, stakeholders and UNFCCC accredited observers on PDD and any supporting documentation, para 32	*None for stakeholders*, however involved Parties or three members of the SC can trigger a review up to forty-five days after the PDD verification is made public, para 35
Reporting Emissions Reduced or Sinks Enhanced	• Report from project participants on emissions reduced or sinks enhanced, para 36 • AIE's determination as to whether the reductions/ enhancements were appropriately monitored and calculated, para 38 • SC's decision and reasoning, (if a review is granted), para 39(c)		*None for stakeholders*, however involved Parties or three or more SC members may request a review within fifteen days, para 39
Accreditation of Independent Entities	• SC's decision to suspend or withdraw an IE, para 42 • IE's procedures for handling complaints, para 1(e), Appendix A • IE's procedures for complaints, appeals and disputes, para 1(g)(vi), Appendix A		

Briefly, there are two key provisions, which contribute to public participation in Track One JI projects. First, as mentioned above, the JI rules provide for a general requirement that all Parties involved in JI projects must report their 'national guidelines and procedures for approving Article 6 projects, including the consideration of stakeholders' comments' to the UNFCCC secretariat.[104] The JI rules do not specify if such national information, which Parties communicate to the UNFCCC Secretariat, will be made publicly available. Secondly, the secretariat is required to maintain a list of Parties eligible to host JI projects pursuant to the Track One procedures, and of Parties who have been suspended due to an eligibility violation.[105]

Track Two projects require public access to information at three project stages. The initial JI PDD verification stage requires the accredited independent entity (AIE) to make both its verification determination, including a summary of comments and how due account was taken, as well as the PDD itself publicly available.[106] If the JI-SC conducts a review of the PDD verification, the JI-SC's decision must be made publicly available.[107]

The JI Track Two provisions provide for one public comment period during the PDD verification stage. Accordingly, once an AIE makes a JI PDD publicly available, the public is allocated thirty days to provide input on the PDD and any supporting documentation.[108] There are no opportunities for a stakeholder to trigger a review of any kind in a JI project. However, following the verification of a JI PDD, an involved Party, or at least three JI-SC members may request a PDD review up to forty-five days after the determination.[109] Similarly, an involved Party or at least three JI-SC members may request a review of reported emissions reduced or sinks enhanced within fifteen days of the report.[110]

5 Comparative Analysis: Other International Approaches to Public Participation

In order to consider the CDM and JI public participation provisions in light of precedents set in other comparable arenas, the following section highlights the relative strengths of specific multilateral environmental agreements (MEAs), a multilateral development bank (MDB), and an international trade agreement. While none of the entities surveyed stands out as a leader in all three of the key elements of access outlined under Principle 10, each offers significant lessons to be drawn regarding access to information, public participation or access to justice, with particular attention paid to the treatment and inclusion of indigenous peoples and local communities. Specifically, the public participation provisions of the WTO (see Table 5.4), the Convention on Biological Diversity (CBD), the Global

[104] JI Rules, para 20(b). [105] ibid, para 27.
[106] ibid, paras 34 and 32.
[107] ibid, para 39(c). [108] ibid, para 32. [109] ibid, para 35. [110] ibid, para 39.

Table 5.4: Comparative summary of public participation provisions in select MEAs, the World Bank and the WTO

	Access to Information	Public Participation in Decision-Making	Access to Justice
WTO Negotiations	• Extensive online access to most final WTO documents	• NGOs not permitted to request the floor during negotiations • NGO attendance restricted (in numbers) at Doha Ministerial • NGOs not permitted to attend (as observers) WTO Committee meetings • NGO participation focused on non-official events such as symposium, briefings and workshops	None
WTO Dispute Settlement	• Deliberations of the panel and documents submitted to it shall be kept confidential	• Limited acceptance and consideration of NGO amicus briefs submitted to the Review Panel or the AB • Review Panel proceedings are entirely closed to the public	None—only parties to a dispute may request an appeal
UNFCCC	• Extensive documentation distributed online and at the UNFCCC negotiations • NGOs often gain access to draft negotiating texts as observers to the negotiations	• Unlimited NGOs may attend negotiations, if accredited • NGOs allowed to participate as observers in contact groups • NGOs may address plenary sessions if invited or permitted by the Chair • Sporadic invitations extended to NGOs to make submissions on specific agenda items	None
CDM	• Nineteen types of information required to be made publicly available (see chart above for details)	• Project participants must invite local stakeholders to provide input on the proposed project • Thirty-day public comment on project validation period open to all stakeholders and UNFCCC accredited NGOs	• Stakeholders are not able to trigger a CDM project review • Stakeholders may trigger a 'spot check' of a DOE with a 'written substantiated complaint'

(*Continued*)

Table 5.4: (*Continued*)

	Access to Information	Public Participation in Decision-Making	Access to Justice
		• Verifying DOEs may interview local stakeholders while conducting on-site inspections • Any technical reports commissioned by the EB must be open to public comment for eight weeks	
JI	• Verification of PDD • Reporting of emissions reduced/sinks enhanced • Accreditation of AIEs	• Thirty-day public comment period open to Parties, stakeholders, and UNFCCC accredited observers on PDD and any supporting documentation	None
CBD	• Extensive documentation distributed online and at the UNFCCC negotiations • Clearing House mechanism prioritizes broad participation and easy access • Efforts to translate documents into local languages and disseminate information in culturally appropriate manners	• Unlimited NGOs may attend negotiations, if accredited • Indigenous leaders co-chair certain CBD working groups with government delegates	None
GEF	• GEF policies and GEF Council documents available online • Efforts to translate documents into local languages and disseminate information on GEF projects in culturally appropriate manners	• Unlimited accredited NGOs may attend GEF NGO consultation • Ten NGO representatives may attend each semi-annual GEF Council meeting • Involvement of NGOs and local communities invited throughout GEF project cycle	None
World Bank Inspection Panel	• Extensive documents tracking the progress and details of Bank project investigations available online	• NGOs may be interviewed by the Inspection Panel or International Advisory Panel during a project investigation	• Impacted individuals may bring a complaint directly to the Inspection Panel

Environment Facility (GEF), and the World Bank Inspection Panel (the Panel) each provide useful examples upon which emerging regimes such as that under the CDM and JI could build.

5.1 The World Trade Organization

According to the 2003 Global Accountability Report, the WTO ranked third out of eighteen international organizations and corporations with regard to access to online information.[111] However, while strong access to online information may play an important role in providing civil society information, the WTO experience demonstrates that a large volume of web-accessible WTO documents does not by itself ensure meaningful public participation. The WTO continues to receive harsh criticism for remaining 'opaque in its operations, inappropriately hidden from scrutiny and insulated from external criticism.'[112] NGOs' ability to observe WTO negotiations of particular concern, to submit comments for consideration by WTO Parties, and to provide amicus briefs for WTO disputes remains severely hindered.

Despite these challenges, publicly available WTO documentation and timely access to public information continues to improve. In May 2002, WTO Members revised the document 'derestriction' procedures and reduced the standard release delay from eight to nine months down to six to twelve weeks.[113] However, most negotiating documents and all documents submitted by a Member to a dispute panel remain closed to the public. This limits the ability of NGOs to understand how specific decisions were reached. Most importantly, access to only final decisions provides no opportunity for public participation in the negotiation process of WTO decision-making.

The basis for all NGO engagement in the WTO stems from the general provision in the 1994 WTO Agreement that 'The General Council may make appropriate arrangements for consultation and cooperation with non-governmental organizations concerned with matters related to those of the WTO.'[114] In response to civil society's growing interest in the workings of the WTO, a series of additional provisions has been adopted by the WTO to address the permissible level of NGO engagement in both WTO negotiations as well as dispute settlement proceedings.

Accredited NGOs are permitted to attend the plenary sessions of the WTO Ministerials, but are not offered any opportunity to make statements on the floor

[111] WTO, 'WTO get high marks for accountability, transparency' (11 February 2003) <http://www.wto.org/english/news_e/news03_e/global_account_report_11feb03_e.htm> accessed 28 January 2004.

[112] JW Head, 'Throwing Eggs at Windows: Legal and Institutional Globalization in the 21st Century Economy' (2002) 50 U. Kan. L. Rev. 731, 759.

[113] WTO, 'NGOs: Derestriction process: WTO moves towards a more open organization', <http://www.wto.org/english/forums_e/ngo_e/bernie_derestrictiontext_e.htm> accessed 28 January 2004.

[114] WTO, 'Marrakesh Agreement Establishing the World Trade Organization', Art V:2 (15 April 1994) <http://www.wto.org/english/docs_e/legal_e/final_e.htm> accessed 28 January 2004 (hereinafter 'WTO Agreement').

or co-chair smaller working groups, as they are in the UNFCCC and CBD respectively.[115] NGOs continue to be marginalized by exclusive and closed WTO negotiations. Rather than improve the public participation provisions of the WTO negotiations themselves, the WTO has focused on creating alternative fora for NGOs, mostly outside of the negotiations and apart from the Members themselves.[116] While such measures may meet the WTO's modest 1996 NGO guidelines, which agree to 'to improve transparency and develop communication with NGOs',[117] they fail to provide for meaningful NGO participation in WTO decision-making.

The continued exclusion of NGOs from WTO Committees is a key example of the lack of meaningful participation in WTO substantive arenas. The WTO Committee on Trade and Environment (CTE) is one of the committees to which environmental NGOs have expressed interest in gaining access. However, the CTE refused to expand on the WTO's 1996 NGO Guidelines and stated that 'the primary responsibility for taking into account the different elements of public interest lies with national governments.'[118] While similar reliance on national governments to consider the public's interests and input is not included in any official CDM text, the UNFCCC website points to national governments as the conduit through which NGOs may participate.[119]

The WTO Uruguay Round Dispute Settlement Understanding (DSU) sets the rules for WTO dispute settlements and the mandate for the WTO Dispute Settlement Body (DSB).[120] With regard to public participation in and access to information on WTO disputes, the DSU provides that all 'deliberations of the panel and the documents submitted to it shall be kept confidential'.[121] The lack of breadth of the expertise of the panel members beyond trade policy also raises doubts regarding the panel's ability to appropriately weigh other non-trade issues.[122] Even members of the Appellate Body (AB) have come to openly recognize the non-transparent reputation that the WTO DSU has earned.[123]

115 WTO, 'Registration for non-governmental NGOs', <http://www.wto.org/english/thewto_e/minist_e/min03_e/ngo_acc_e.htm> accessed 28 January 2004.

116 WTO, 'WTO Secretariat activities with NGOs', WT/INF/30 (12 April 2001) <http://www.wto.org/english/thewto_e/minist_e/min01_e/min01_ngo_activ_e.htm>.

117 WTO, 'Guidelines for arrangements on relations with non-governmental organizations' (26 July 1996) <http://www.wto.org/english/forums_e/ngo_e/guide_e.htm> accessed 28 January 2004 ('WTO NGO Guidelines')

118 WTO, 'CTE on: other organizations' <http://www.wto.org/english/tratop_e/envir_e/cte09_e.htm> accessed 28 January 2004.

119 UNFCCC, 'Issues' <www.unfccc.int/issues/mechanisms.html> accessed 15 March 2004.

120 WTO, 'Agreement Establishing the World Trade Organization, Annex 2—Understanding on Rules and Procedures Governing the Settlement of Disputes', Art 14 and Appendix 3, paras 2 and 3. <http://www.wto.org/english/docs_e/legal_e/28-dsu.doc> accessed 28 January 2004.

121 ibid, Appendix 3, para 3.

122 S Shapiro, 'International Trade Agreements, Regulatory Protection, and Public Accountability' (Washington DC 2002) 54 Admin. L. Rev. 435, 451; see also L Wallach and M Sforza, *Whose Trade Organization? Corporate Globalization and the Erosion of Democracy* (1999) 197.

123 J Bacchus, 'Table Talk: Around the Table of the Appellate Body of the World Trade Organization' (2002) 35 Vand. J. Transnatl L. 1021, 1024.

Uncertainty prevails with regards to the status of NGO amicus briefs submitted to the WTO Dispute Panel and AB. Both dispute bodies have held that they possess the authority to accept and consider independent NGO amicus briefs pursuant to DSU Articles 13 and 17, respectively. However, the criteria, which the dispute bodies apply, to determine if they will consider an amicus brief in their deliberations remain vaguely defined and only applied in a cursory manner. The 1998 Dispute Panel, in the *Shrimp–Turtle* case, refused to accept unsolicited information from NGOs, reasoning that it would be 'incompatible with the provisions of the DSU as currently applied.'[124] On appeal, the AB clarified that NGOs could submit information in the form of amicus briefs, but only as part of their own government's submission.[125] Consequently, 'in these cases, NGOs do not participate as independent actors, but rather as "appendages" to the parties' submissions.'[126]

Almost all subsequent NGO requests to submit amicus briefs have been denied by the AB. In *European Communities—Measures Affecting Asbestos and Asbestos-Containing Products (Asbestos)*, the large number of NGOs interested in submitting amicus briefs triggered a 'communication' from the AB on procedures for accepting and rejecting NGO submissions.[127] Of the seventeen NGO requests to submit amicus briefs, the AB denied every one.[128] The more recent AB ruling in *European Communities—Trade Description of Sardines (Sardines)* further perpetuated the ambiguity of the criteria by which the AB accepts or rejects an NGO amicus brief for consideration.[129] The AB reiterated its authority to accept and consider NGO amicus briefs, but rejected the amicus brief in question for failing to 'assist' the AB.[130] The AB's Report offers no guidance on the types of information, which it would find useful and relevant to a dispute.[131]

The opaque nature of the WTO continues to attract extensive scrutiny, especially since the 1999 Seattle WTO Ministerial Meetings. Some progress has been made regarding online access to WTO information and documentation. However, NGOs remain prohibited and severely hindered from observing WTO negotiations of particular concern, submitting comments for consideration by WTO Members, and providing amicus briefs for WTO disputes.

The UNFCCC and CDM websites parallel the extensive online document disclosure of the WTO. However, the transparency and access to the negotiations of the UNFCCC and CDM far exceed that provided to NGOs under the WTO.

[124] WTO, 'United States—Import Prohibition of Certain Shrimp and Shrimp Products (WT/DS58/R): Report of the Panel', para 7.8 (15 May 1998).

[125] WTO, 'United States—Import Prohibition of Certain Shrimp and Shrimp Products (WT/DS58/AB/R): Report of the Appellate Body', para 100 (12 October 1998).

[126] JL Dunhoff, 'Civil Society and the WTO: the Illusion of Inclusion?' (2001) 7 ILSA J. Int'l & Comp. L. 275, 279.

[127] WTO, 'European Communities—Measures Affecting Asbestos and Asbestos-Containing Products (WTO/DS135/AB/R): Report of the Appellate Body', para 52.2 (12 March 2001).

[128] ibid, para 56.

[129] WTO, 'European Communities—Trade Description of Sardines (WTO/DS231/AB/R): Report of the Appellate Body', para 160 (26 September 2002).

[130] ibid.

[131] ibid.

The continued public scrutiny under which the WTO operates serves as an important lesson of the risks and costs of restricting public access to both WTO decision-making and dispute resolution mechanisms.

5.2 NGO Participation and Indigenous Leadership in the Convention on Biological Diversity

The CBD allows indigenous peoples to co-chair certain working groups most relevant to indigenous expertise and interests, an approach considerably stronger than the observer status of NGOs in the UNFCCC negotiations.[132]

The basic participation provisions contained in the CBD text and the Rules of Procedure are nearly identical to those of the UNFCCC.[133] As such, NGOs may participate as observers and without the right to vote in COP meetings unless at least one-third of the Parties present raise objections.[134] However, the CBD text places a greater emphasis on access to information than does the UNFCCC. In addition to the CBD's Article on 'Public Awareness and Education'[135] (similar to UNFCCC Article 6), the CBD also provides an article dedicated to the 'Exchange of Information'.[136] The exchange of information relating to, *inter alia*, research, training, and traditional knowledge is further facilitated by the CBD's 'Clearing House Mechanism' which recognizes that 'broad participation and easy access must be a top priority.'[137] Emphasizing indigenous communities 'unique knowledge and expertise', and the need to ensure the widest indigenous participation possible, the CBD Clearing House Mechanism offers an information database accessible both electronically (by internet and CD-ROM) and in paper-based format.[138]

The most exceptional element of the CBD's public participation procedures is its inclusion of indigenous peoples to co-chair the working group on Article 8(j).[139] Article 8(j) provides that each Party shall:

> Subject to its national legislation, respect, preserve and maintain knowledge, innovations and practices of indigenous and local communities embodying traditional lifestyles relevant for the conservation and sustainable use of biological diversity and promote their wider application with the approval and involvement of the holders of such knowledge, innovations and practices and encourage the equitable sharing of the benefits arising from the utilization of such knowledge, innovations and practices.[140]

[132] The CBD includes 188 countries, and 94 are Party to the CBD's Cartagena Protocol on Biosafety. See CBD, 'Parties to the Convention on Biological Diversity/Cartagena Protocol on Biosafety' (13 December 2002) <http://www.biodiv.org/world/parties.asp> accessed 22 April 2004.

[133] CBD, Convention on Biological Diversity, Art 23 <http://www.biodiv.org/doc/legal/cbd-en.pdf> accessed 15 January 2004 (hereinafter 'CBD text'); CBD, Rules of Procedure for Meetings of the Conference of the Parties to the Convention on Biological Diversity (Annex to Decision I/1, as amended by Decision V/20), Arts 6, 7, 29, and 30 <http://www.biodiv.org/doc/handbook/cbd-hb-03-en.pdf> accessed 15 January 2004.

[134] ibid.

[135] CBD text, n 133 above, Art 13.

[136] ibid, Art 17.

[137] CBD, 'Clearing House Mechanism' <http://www.biodiv.org/chm/default.aspx> accessed 15 January 2004.

[138] ibid.

[139] The International Indigenous Forum on Biodiversity (IIFB) helps facilitate the indigenous position of working group co-chair.

[140] CBD text, n 133 above, Art 8(j).

While such a central role clearly positions indigenous communities to more directly control the process and procedure of the working group, the right to vote remains reserved for governmental Parties only. However, a similar leadership role held by a non-governmental 'observer' to international negotiations is unprecedented.[141]

Not only have the Parties to the CBD supported the inclusion of indigenous peoples at the head of the Article 8(j) working group, but several CBD decisions have also emphasized the importance of extensive indigenous participation at all stages of CBD's negotiation and implementation, both at national and international levels.[142] For example, CBD decisions have repeatedly encouraged Parties to include indigenous representatives on their governmental delegations to the CBD negotiations.[143] In addition to the co-chair role of indigenous people in the Article 8(j) working group, CBD decisions have also supported further indigenous participation in the working group on Access and Benefit Sharing.[144] At the 2002 CBD meeting (COP6 held in The Hague, 7–19 April 2002) the final agreement urged Parties and governments to:

> develop, implement and evaluate, in cooperation with indigenous and local communities, strategies aimed at promoting awareness and enhancing access by indigenous and local communities to information on issues relating to Article 8(j) and related provisions of the Convention.[145]

While the place of indigenous peoples continues to develop in response to the often aspirational provisions laid out in the CBD decisions, the central role that indigenous peoples must play for the successful and sustainable implementation of the CBD has clearly been acknowledged.

The CDM and JI rules have taken limited steps to proactively include indigenous peoples and local communities beyond adopting a broad definition of the term 'stakeholders' to include 'groups or communities affected, or likely to be affected' by the proposed CDM and or JI activity.[146] The collaborative approach to including an indigenous leader as a co-chair of the Working Group on Article 8(j) offers an example of how indigenous expertise and input can play a role in the development of international environmental policies likely to impact indigenous communities directly. Despite calls from the International Forum of Indigenous

141 Perhaps with the exception of the International Labor Organization which allows the right to vote to labour constituencies. See Constitution of the International Labor Organization, 62 Stat 3485, TIAS 1868, 15 UNTS 35, adopted 9 October 1946, Arts 3 and 4.

142 CBD, Decision V/16—Article 8(j) and related provisions, Annex 1, para 1 <http://www.biodiv.org/decisions/default.asp?dec = V/16> accessed 15 January 2004.

143 ibid, para 18; CBD, Decision IV/9—Article 8(j) and related provisions, para 3 <http://www.biodiv.org/decisions/default.asp?lg = 0&dec = IV/9> accessed 15 January 2004.

144 CBD Decision V/16, n 142 above, para 5.

145 CBD, Decision VI/10—Article 8(j) and related provisions, para 27 <http://www.biodiv.org/decisions/default.aspx?m = COP-06&id = 7184&lg = 0> accessed 27 May 2004.

146 CDM Rules, n 36 above, para 1(e); JI Rules, n 102 above, para 1(e).

Peoples and Local Communities on Climate Change for an augmented role for indigenous peoples in the CDM and the creation of a UNFCCC working group on indigenous peoples and climate change, indigenous peoples are not explicitly mentioned in either the CDM or JI rules.

5.3 NGO Consultations and Project Involvement in the Global Environment Facility

Established in 1991, the GEF has contributed $4 billion in grants to over 1,000 environmental development projects in more than 140 countries.[147] The GEF recognizes that '[a]s informed and effective advocates, nongovernmental organizations (NGOs) have had a role in shaping the GEF and its agenda from the very beginning.'[148] Accordingly, the GEF sets a strong example of a model, which allows NGO input from the early policy-setting stages up through the implementation and monitoring of specific projects.[149]

Not surprisingly, the GEF, as the designated financial instrument for both the UNFCCC and CBD, provides for somewhat similar NGO participation rules. The GEF Rules of Procedure allow the GEF Chief Executive Officer (CEO) to, 'in consultation with the Council, invite representatives of other organizations and entities, including NGOs, to attend or observe the Council meetings.'[150] The GEF accreditation process for NGOs is fairly similar to that of the UNFCCC.[151] Once an NGO has completed the form and fulfilled the basic criteria of demonstrating interest and competence in one of the GEF's issue areas, the NGO is entitled to attend the GEF NGO Consultations and potentially the GEF Council Meetings.

A one-day NGO Consultation session precedes each semi-annual GEF Council meeting. The NGO consultation offers NGOs an opportunity to raise concerns on particular projects and to report back on GEF-related initiatives led by their respective organizations.[152] At each GEF Council meeting, ten NGOs are permitted to attend the proceedings.[153] While GEF-accredited NGOs may receive

[147] GEF, 'What is the Global Environment Facility?' <http://gefweb.org/What_is_the_GEF/what_is_the_gef.html> accessed 6 January 2004.

[148] GEF, 'Non-governmental Organizations' <http://gefweb.org/Partners/partners Nongovernmental_Organ/partners-nongovernmental_organ.html> accessed 6 January 2004.

[149] GEF, 'Regional NGO Focal Points' <http://gefweb.org/participants/Focal_Points/NGO/ngo.html> accessed 6 January 2004.

[150] GEF, Rules of Procedure, para 22; see also para 2(n): ' "Attend" means that a representative shall be permitted to be present in the Council meeting room and, at the invitation of the Chair, may address the Council;' and para 2(o):' "Observe" means that a representative shall be permitted to observe the Council proceedings from a viewing room. At the invitation of the Chair, an observer may address the Council.'

[151] GEF, 'Accreditation Application Form' <http://gefweb.org/Partners/partners-Nongovernmental_Organ/Accreditation_Process/GEF_Accreditation_Form.doc> accessed 6 January 2004.

[152] GEF, 'GEF Meetings' <http://gefweb.org/Outreach/Meetings_Events/meetings_events.html> accessed 6 January 2004.

[153] GEF, 'Regional NGO Focal Points' <http://gefweb.org/participants/Focal_Points/NGO/ngo.html> accessed 6 January 2004.

information as members of the GEF NGO network, limited information is available on the web for the interested public to learn about the NGO's priorities or to read reports back on the NGO consultations.[154]

The GEF's *Public Involvement* document reiterates the lessons progressively implemented through the CBD decisions, emphasizing that 'Effective public involvement is critical to the success of GEF-financed projects. When done appropriately, public involvement improves the performance and impact of projects'.[155] The GEF Instrument stipulates that 'projects shall provide for full disclosure of all non-confidential information, and consultation with, and participation as appropriate of, major groups and local communities throughout the project cycle.'[156] Furthermore, the GEF's *Public Involvement* document details a list of project benefits resulting from robust public participation.[157] While the UNFCCC does not provide for a similar NGO consultation prior to decision-making, extensive CDM and JI project documents are made available and open to limited comment periods prior to implementation of the project.

5.4 Access to Justice: Direct Access to Recourse provided by the World Bank Inspection Panel

The Panel is unique in providing direct access for local and indigenous communities to bring a complaint against the World Bank (the Bank) for failing to adhere to its own policies. The following section therefore focuses only on the Panel, without addressing the Bank's more general public participation policies. As one of the few such mechanisms which provides for direct access to redress to local stakeholders, the Panel offers a useful model for consideration in the further elaboration of CDM and JI Rules regarding access to justice. While the Panel often receives international recognition for its level of responsiveness to local concerns, fundamental problems remain which prevent the Panel from fully serving the needs and rights of the communities that Bank projects most directly impact.[158]

The Panel's Operating Procedures define the required form and scope of a request to the Panel for an investigation, who can submit a request, as well as the procedures and time lines, which the Panel, Board of Executive Directors and

154 See <http://www.gefweb.org> accessed 6 January 2004.

155 GEF, 'Public Involvement in GEF-Financed Projects' (June 1996) <http://www.gefweb.org/Operational_Policies/Public_Involvement/public_involvement.html> accessed 6 Januray 2004 (hereinafter 'GEF Public Involvement').

156 GEF, Instrument, para 5 <http://www.gefweb.org/Documents/Instrument/instrument. html> accessed 6 January 2004.

157 GEF Public Involvement. See <http://www.gefweb.org/Operational_Policies/Public_Involvement/public_involvement.html>.

158 Since its creation in 1993, the Panel has received twenty-seven requests for inspections of Bank-funded development projects. Of those twenty-seven requests, the Panel recommended investigation of thirteen, and the Board granted approval for investigation for nine complaints; see The World Bank, *The Inspection Panel Annual Report 2002* (Washington, DC, 2002) 45.

Requester must respect for the request to proceed.[159] The Panel's Operating Procedures highlight that:

> The Panel was established for the purpose of providing people directly and adversely affected by a Bank-financed project with an independent forum through which they can request the Bank to act in accordance with its own policies and procedures.[160]

The Panel is the first body to receive a complaint. Prior to registering the complaint, the Panel first 'decides whether the Request is prima facie not barred from its consideration.'[161] For example, complaints regarding projects which are not funded by the Bank, would present a prima facie bar to Panel investigation. The actual registration of the Request is recognized by the Panel as a 'purely administrative procedure.'[162] The Panel then sends the complaint to Bank Management for their initial review.[163] The Bank Management in turn has twenty-one days to provide the Panel 'with written evidence that it has complied, or intends to comply with the Bank's relevant policies and procedures.'[164] The Panel then has another twenty-one days to both consider the Management's Response and to establish if the Requesters are eligible to bring a Request.[165]

A Request may be filed by:

(i) any group of two or more people in the same country as the project;
(ii) a duly appointed local or foreign representative acting on explicit instructions as the agent of adversely affected people; or
(iii) an Executive Director of the Bank in special cases of serious alleged violations of the Bank's policies and procedures.[166]

At the close of the twenty-one days, the Panel decides whether or not it will recommend an investigation of the project.[167] The final decision of whether or not to investigate ultimately rests with the Board.[168] Following the Board's approval, the Panel undertakes a full investigation for which there are no time limitations.[169] After the Panel conducts an investigation the Report is sent to the Executive

[159] See The World Bank, *The Inspection Panel Operating Procedures*, for the complete Panel provisions relating to procedures; see also The Inspection Panel, Investigation Report: Chad-Cameroon Petroleum Development and Pipeline Project (Loan No 4558-CD); Petroleum Sector Management Capacity-Building Project (Credit No 3373-CD); and Management of the Petroleum Economy Project (Credit No 3316-CD), iii (17 July 2002) (hereinafter 'Panel, Panel Investigation Report—Chad).

[160] The World Bank, *The Inspection Panel Operating Procedures* (Washington, DC, 1993) 2 (hereinafter 'Bank, Operating Procedures').

[161] Panel, Panel Investigation Report—Chad, n 159 above, III.

[162] ibid.

[163] Bank, Operating, Procedures, n 160 above, para 18(d).

[164] ibid, para 27.

[165] ibid, para 30.

[166] ibid, para 4.

[167] ibid, paras 30–2.

[168] ibid, para 39; the 181 Member Governments of the International Bank for Reconstruction and Development (IBRD) are represented by a twenty-four member Board of Executive Directors. The voting is weighted according to the shares the members hold in the IBRD. The IBRD combined with the International Development Association (IDA) make up the Bank's public sector lending. Source: D Clark, *A Citizen's Guide to the World Bank Inspection Panel* (The Center for International Environmental Law, 1999) 2.

[169] Bank, Operating Procedures, n 160 above, para 45.

Directors and President of the Bank as well as to the Bank Management.[170] Upon receiving the Panel's Investigation Report, the Bank Management has six weeks to respond to the Panel's findings.[171] The Board concludes the investigation with a final decision on measures to be taken in light of both the Panel's findings and the Management's Recommendations.[172] Both the Panel's Report and the Management's Recommendations are made publicly available after the Board's decision.[173] Following its Investigation Report, the Panel's mandate excludes it from any further action or consideration of the Management's Recommendations.[174]

The Panel's mandate establishes clear limitations on the allowable scope of the Panel's review of any Bank project.[175] The Panel's role is to check if the Bank has complied with its own policies and rules.[176] The Panel is not in a position to review the adequacy of the Bank's policies themselves. The Panel's findings are as strong as the Bank's rules will permit.

The Panel's experience demonstrated three key strengths, which should be built into any project-based accountability mechanism under consideration by CDM and JI. The Requesters to the Panel gained meaningful consideration of their rights and concerns thanks to (1) the ability for local impacted communities to bring a claim directly to the independent Panel, (2) the extensive on-site fact-finding investigations performed by the Panel itself as well as the International Advisory Group (IAG) in certain claims, and (3) the measures taken to make information available in a timely and culturally appropriate manner. The Panel's integrity and effectiveness have benefited immensely from the increased incorporation of these three core elements into the functioning of the Panel.

Local communities' and civil society's direct access to the independent Panel is imperative in ensuring that people impacted by a project are empowered to raise concerns regarding the protection of their rights in light of the implementation of a development project. Direct access also significantly contributes to the legitimacy of the claims brought to the independent body. A review body lacking such institutional independence would risk findings which reflected conflicts of interest and failed to appropriately weigh the concerns of the injured party.

The ability of the Panel to conduct its own on-site investigations, and also to draw from the findings of the IAG in certain reviews, is fundamental to the Panel's fact-finding mandate. The Bank's Operational Procedures limited the Panel's function to that of considering the facts. The CDM and JI should consider the advantage of both clearly defining the scope of the review body, and also appropriately enabling the body to perform its fact-finding duties.

The Panel continues to lead in the arena of accountability mechanisms. However, it falls short of providing an adequate remedy for local and indigenous

[170] Bank, Operating Procedures, n 160 above, para 53. [171] ibid, para 54.

[172] ibid, para 55. [173] ibid, para 56.

[174] ibid, para 2(d); see also D Clark, 'The World Bank and Human Rights: The Need for Greater Accountability' (2002) 15 Harv. Hum. Rts. J. 205, 218.

[175] Bank, Operating Procedures, n 160 above, para 2. [176] ibid, para 1.

communities adversely impacted by Bank-supported projects. The gap that remains between the Bank's policies relating to local and indigenous peoples and the Panel's ability to ensure remedial actions in response to legitimate complaints stems from the absence of sufficient oversight and enforcement following the conclusion of a Panel investigation.[177] While serving as a learning tool for other lending institutions is not formally included in the Panel's mandate, the instructive potential of the Panel's experience is increasingly informally recognized as crucial for the successful development of new compliance mechanisms.[178] Failure to recognize the Panel's significant handicap due to weak enforcement mechanisms following inspection may expose CDM and JI to international criticism and distrust.

6 Conclusion

The CDM and JI rules continue to evolve. Experience thus far, in particular with regards to the CDM, has demonstrated a strong willingness on the part of the EB and the UNFCCC secretariat to ensure a transparent and participatory process. While the public comment periods remain brief and notification not always consistent, the trend is clearly responsive to the concerns of and contributions from stakeholders and Parties. Overall, the CDM rules lay a promising foundation.

Several initiatives have emerged which highlight key strengths and weaknesses of these two Kyoto Protocol mechanisms. Through their pursuit of strategically using publicly available CDM information and expanding and supplementing the existing rules, CDM Watch, the CDM Gold Standard, and the International Forum of Indigenous Peoples and Local Communities on Climate Change provide important examples of different approaches to strengthening public participation in the CDM and JI.

First, at the time of the agreement on the Marrakesh Accords, a new NGO was established in an effort to facilitate the flow of publicly available CDM information in a timely manner so as to better position interested stakeholders to engage in decision-making on proposed CDM projects, entities, or methodologies. The CDM Watch website serves as an information clearinghouse on CDM projects and policy developments.[179] In addition to the ongoing efforts undertaken by CDM Watch to alert NGOs worldwide of opportunities for public comment periods on CDM-related issues, CDM Watch also recently published a stakeholder toolkit.[180] Similar to the website, the CDM Toolkit seeks to lower the barriers to public engagement in the CDM by explaining CDM terminology and procedures in plain language terms to provide interested stakeholders with the necessary tools to become

[177] Clark, n 174 above, 206. [178] Shihata, n 2 above, 264.

[179] See <http://www.cdmwatch.org> accessed 4 January 2004.

[180] CDM Watch, 'The Clean Development Mechanism (CDM) Toolkit: A Resource for stakeholders, activists and NGOs' <http://www.cdmwatch.org/files/CDM%20Toolkit%20English%20final.pdf> accessed 20 January 2004.

informed about CDM projects and to exercise the available public comment periods as appropriate. CDM Watch also plays an important coordinating role within the international network of environmental NGOs (Climate Action Network). By bringing key advocacy opportunities to the attention of interested NGO stakeholders, groups are better able to prioritize their CDM-related work, and the international Climate Action Network as a whole is better able to coordinate the available capacity of staff to review and comment on technical (and often lengthy) CDM documentation.

Secondly, in October 2002, the World Wildlife Fund (WWF) launched the 'Gold Standard' for CDM and JI projects.[181] The Gold Standard establishes a set of criteria against which project developers can measure their performance in, *inter alia*, creating tangible sustainable development goals and ensuring adequate public participation.[182] Specifically, the Gold Standard places an emphasis on the importance of local stakeholder engagement, pointing to the need for culturally appropriate channels for communicating proposed project information to impacted local communities.[183] As many NGOs have pointed out over the course of the development of the CDM and JI rules, while the internet may serve as an excellent resource for some NGOs and developed country individuals, it fails to provide appropriate information to local communities who may be impacted by the project.[184] Building on the criteria elaborated under the WWF Gold Standard, WWF seeks to increase the environmental and social benefits of CDM and JI projects, which will in turn bolster investor certainty in these projects.[185]

Thirdly, the First International Forum of Indigenous Peoples and Local Communities on Climate Change was held at the UNFCCC negotiations in Lyon, France in September 2000.[186] Since the Lyon negotiations, the International Forum of Indigenous Peoples and Local Communities on Climate Change has repeatedly emphasized that:

> Indigenous peoples demand that the principles of transparency, prior informed consultation and consent, independent third party verification and monitoring, benefit sharing, risk reduction, appeals mechanism and compensation be guaranteed. Furthermore, we emphasize the need for these principles to be applied in culturally and linguistically appropriate manners.[187]

While the positions of the International Forum of Indigenous Peoples and Local Communities on Climate Change have not yet resulted in explicit incorporation of indigenous provisions in the CDM, the official UNFCCC recognition of indigenous peoples as one of the five categories of UNFCCC observers at least is a

[181] World Wildlife Fund, 'The Gold Standard: Quality Standards for CDM and JI Projects' (October 2002) <www.worldwildlife.org/climate/cop8_standards.pdf> accessed 5 January 2004.

[182] ibid. [183] ibid. [184] Baumert and Petkova, n 29 above.

[185] World Wildlife Fund, n 181 above.

[186] International Forum of Indigenous Peoples on Climate Change, 'Declaration of the First International Forum of Indigenous Peoples on Climate Change', Lyon, France (4–6 September 2000) <http://www.treatycouncil.org/new_page_5211.htm>.

[187] ibid.

step in the right direction towards acknowledging the fundamental role indigenous and local communities could play in ensuring that proposed CDM projects meet their ultimate objective of promoting sustainable development.[188]

In conclusion, the CDM and JI public participation rules are basically up to par with international standards as evidenced by the efforts of CDM Watch to coordinate scarce resources and make the best use possible of the CDM rules as they currently stand. At the same time, as illustrated in the previous comparative section, the CDM and JI rules still leave room for improvement. The CDM Gold Standard demonstrates one approach to rewarding enhancement of the rules, including increased local community involvement. As these two mechanisms are in the early stages of operationalization, it remains to be seen the extent to which the public participation rules will be further improved upon.

188 Kyoto Protocol to the UNFCCC, 10 December 1997, 37 ILM 32, Art 12.

PART III

ARTICLE 6 OF THE KYOTO PROTOCOL: JOINT IMPLEMENTATION

6

Joint Implementation: History, Requirements, and Challenges

*Charlotte Streck**

1 Introduction

In the context of the Kyoto Protocol, Parties with defined emissions limits can participate in emissions trading through trading in Assigned Amount Units (AAUs) as defined under Article 17 of the Kyoto Protocol or through the participation in Joint Implementation (JI) project activities. Participation in JI project activities implies the transfer of AAUs which have been converted to Emission Reduction Units (ERUs). The Kyoto Protocol limits the availability of this type of emissions trading to Annex B Parties that are also members of Annex I[1] of the United Nations Framework Convention on Climate Change (UNFCCC).[2] Developing country Parties without defined emission limits can participate in emissions trading in a far more limited way through the Clean Development Mechanism (CDM), under Article 12 of the Kyoto Protocol.

Joint Implementation allows project sponsors to invest in projects that reduce greenhouse gas (GHG) emissions vis-à-vis an established baseline scenario and to sell the difference between the baseline emissions and the project emissions with the consent of the government hosting the project. Through its approval, the government agrees to convert AAUs[3] in an amount equivalent to the verified emission reductions to ERUs and transfer them to the registry accounts of the investing country or private entity.

Although the term 'Joint Implementation' itself is not mentioned anywhere in the Kyoto Protocol, Article 6 of the Protocol establishes the basis for what became known as 'Joint Implementation'. From the definition of the mechanism in the

* Charlotte Streck is a legal counsel at the World Bank. The views expressed in the chapter are the views of the author and do not necessarily represent the views of the World Bank.

[1] Annex I countries are those included in Annex I of the UNFCCC and the Kyoto Protocol. The emission caps for individual countries are listed in Annex B of the Kyoto Protocol, see Appendix 1 below.

[2] Arts 6(3), 3(11), Annex B and UNFCCC, Annex I.

[3] The Kyoto Protocol creates 'assigned amounts', which represent the quantity of greenhouse gas emissions a country is allowed to emit in a defined first compliance period (2008–12).

Kyoto Protocol in 1997, it took another four years before the Conference of the Parties to the UNFCCC (COP) could reach an agreement on the implementation guidelines and on the participation requirements, which define the rules under which JI projects can be designed and implemented. Since 2001, however, dozens of projects, which claim to be eventually eligible under Article 6 of the Kyoto Protocol, have been initiated.[4] A number of countries have entered into bilateral agreements on the implementation of JI projects and a number of multilateral banks, as in the case of the International Bank on Reconstruction and Development (IBRD) and the European Bank on Reconstruction and Development (EBRD),[5] have engaged in the emerging market for JI projects.

This chapter gives an overview of the background to the mechanism referred to as Joint Implementation. It analyses the conceptual differences between CDM and JI, and summarizes the participation requirements as well as the project cycle for JI projects. From there it will describe the contractual arrangements which, so far, have been chosen in order to secure the generation and transfer of ERUs generated by JI projects. In this context, it will also stress the role and the challenges for JI host countries in the emerging carbon market. Of particular interest in this context is the impact of the European Emissions Trading Scheme (EU ETS) for JI in those countries which recently have acceded the European Union (EU) or which will accede before the end of the first commitment period established by the Kyoto Protocol in 2012. This chapter will conclude with a summary of strategic priorities for JI countries.

2 Background

A JI type of mechanism has been discussed in the context of the international climate change negotiations since 1992. Throughout these discussions the benefits of the mechanism have been disputed[6] and an agreement was not reached until the third session of the COP and the final negotiations on the Kyoto Protocol.

The negotiations on a JI mechanism have their roots in the Articles 3(3) and 4(2)(a) and (d) UNFCCC. Article 3(3) UNFCCC recognizes the cost-effectiveness of global reductions in GHG emissions. It thus opened the way for flexibility in the achievement of such emission reductions, which became an important issue in the negotiations, which led to binding emission targets and the Kyoto Protocol.[7] In

[4] The precise number is unknown as JI projects do not require registration until 2008.

[5] The Netherlands EBRD Carbon Fund was established on 27 October 2003, see <http://www.ebrd.org/country/sector/energyef/carbon/main.htm> accessed 29 February 2004.

[6] R Loske and S Oberthür, 'Joint Implementation under the Climate Change Convention' (1994)6(1) International Environmental Affairs 45–58.

[7] A Michaelowa, 'Project-based Instruments: Economic Consequences of the Kyoto and the Buenos Aires Framework and Options for Future Development' in K-L Brockmann and M Stronzik (eds), *Flexible Mechanisms for an Efficient Climate Policy* (Heidelberg, Physica, 2000) 40.

accordance with Article 4(2)(a) UNFCCC Parties to the UNFCCC may implement 'policies and measures jointly with other Parties and may assist other Parties in contributing to the achievement of the objective of the Convention . . . ' and Article 4(2)(d) mandates the COP to take decisions regarding criteria for joint implementation as indicated in Article 4(2)(a).

At the first session of the COP, the Parties to the UNFCCC decided to establish a pilot phase for Activities Implemented Jointly (AIJ) among Annex I Parties and, on a voluntary basis, with non-Annex I Parties that so request.[8] The COP decided to create the AIJ as a pilot through a simple decision under the UNFCCC. Other than deemed necessary by many which argued for adopting a project-based GHG mitigation mechanism involving developing countries in the context of a strong supporting framework, preference was given to the idea of gaining experience through a voluntary pilot programme.[9] Consequently, participation in AIJ projects was voluntary, funds needed to be additional to Official Development Assistance, and no credits were granted to any Party as a result of any reduction in GHG emissions stemming from AIJ projects. The project eligibility requirements for AIJ projects already reflected what later would become the participation requirements for CDM and JI projects. To be eligible under AIJ, projects needed to result in 'real measurable and long-term environmental benefits related to the mitigation of climate change that would not have occurred in the absence of such activities',[10] to secure prior government 'acceptance, approval or endorsement',[11] and be 'compatible with and supportive of national environment and development priorities.'[12] Emissions reductions generated through AIJs could not be used for the fulfilment of commitments of Annex I Parties under Article 4(2)(b) of the Convention, but they could contribute to the achievement of the objective under the Convention and to the fulfilment of commitments under Article 4(5) of the Convention.[13]

Before the final review of AIJ,[14] the issue of JI came to the table in 1997, when the Treaty, which later became the Kyoto Protocol, was negotiated. While the

[8] FCCC/1995/7/Add.1, Decision 5/CP.1.

[9] C Jeepma and W van der Gaast, *On the Compatibility of Flexible Instruments* (Kluwer, Dordrecht, 1999); C Carraro (ed.), *International Environmental Agreements in Climate Change* (Kluwer, Dordrecht, 1999).

[10] Decision 5/CP.1 FCCC/1995/7/Add.1, para 1(d).

[11] ibid, para 1(c).

[12] ibid, para 1(b).

[13] Decision 5/CP.1 in FCCC/1995/7/Add.1, para (b) of the recitals.

[14] Between 1997 and 2001 the Subsidiary Body for the Scientific and Technological Advise (SBSTA) and the Subsidiary Body for Implementation (SBI) synthesis reports evaluating the experience gained with AIJ and submitted it for consideration to the COP. These reports showed that the bulk of AIJ took place between Annex I Parties, with host countries being EITs. Eighty-five out of 152 activities were carried out among Annex I Parties. AIJ: Fifth Synthesis Report, FCCC/SBSTA/2001/7, para 6. In analysing the different type of projects, it should be borne in mind that out of all projects in EITs, fifty-five were small activities of a similar type implemented by Latvia, Estonia, and Sweden. The percentage of non-Annex I Parties which participated in AIJ, significantly increased over time. In 1997, only three non-Annex I Parties had AIJ, of which one country in the Latin American and Caribbean region, Costa Rica, hosted eight of ten activities of non-Annex I Parties. AIJ: Synthesis Report, FCCC/SBSTA/1997/12, para 6(a). Four years later, the majority of all host countries were non-Annex I countries. AIJ, Fifth Synthesis Report, FCCC/SBSTA/2001/7, Annex.

industrialized countries showed significant interest in including the concept in the Treaty, there were major differences as to when and how to include non-Annex I Parties as well as to the rules which would govern the inclusion of a JI mechanism in the context of the Protocol. While the negotiation group around the US Government[15] favoured a less stringent approach without any restrictions on the participation in the mechanism, the EU favoured an approach with clearly defined monitoring and verification requirements and which would exclude sinks.[16] Developing countries generally opposed any participation in JI, their concerns reached from a feared 'neo-colonialism', to concerns that Annex I countries would be let 'off the hook', and the incomplete evaluation and analysis of AIJ.[17] The consolidated negotiations text brought to the table in Kyoto by the Chairman therefore foresaw JI with developing countries only after the final evaluation of AIJ.[18]

Lesson from the AIJ pilot phase were, on one hand, the importance of transparency, capacity building, and reporting,[19] and on the other hand, the benefits of flexibility, which can produce a number of different financing arrangements and involve different project participants.[20] AIJ provided an important source of experience upon which to build the rules and guidelines for JI and CDM, and the decisions establishing AIJ where the precursor for the provisions on JI among Annex I countries as well as for the CDM. Final negotiations in JI were driven by the proposal for a mechanism especially for developing countries and the progress made on what became the CDM. The acceptance of such mechanism which included developing countries in mitigation efforts, made it easy to agree on a similar mechanism for Annex I countries. The agreement was eventually secured by modelling JI closely after AIJ, with little restrictions but also very little guidance on the implementation of the mechanism in the actual text of the Kyoto Protocol.

The adoption of the Kyoto Protocol did not mark the end of the agreements necessary to operationalize the idea of JI. The protocol, adopted under the Framework Convention, is by itself more a framework than an agreement ready for immediate implementation. Parties still needed to agree on details on how to administer, use, and implement the flexible mechanisms, additionally to develop the compliance system outlined in the Protocol, and provisions for the land-use change and forestry sector, methodologies for estimating emissions and removals,

[15] As in many other treaties, States sharing common interests have formed bargaining groups in the context of the UNFCCC in order to organize themselves to up-hold their negotiating positions. In the context of the Kyoto Protocol, the US formed part of the so-called 'Umbrella Group', which consisted of the US, Japan, Canada, Australia, Norway, New Zealand, Russia, and Ukraine.

[16] See for a summary of the negotiations history: S Oberthür and HE Ott, *The Kyoto Protocol* (Heidelberg, Springer Verlag, 1999) 153.

[17] ibid.

[18] FCCC/AGBM/1997/7.

[19] For example, 'Evaluation of (non-sink) AIJ-Projects in Developing Countries (2000)', Wuppertal Papers on behalf of the Deutsche Gesellschaft für Technische Zusammenarbeit (GTZ) GmbH, No 100, January 2000, [33]

[20] K Danish, E Brenes, and J Rotter, 'Legal Dimensions of AIJ Project Development Activities', in R E Dixon, *The U.N. Framework Convention on Climate Change Activities Implemented Jointly (AIJ) Pilot: Experiences and Lessons Learned* (Kluwer, Dordrecht 1999) 281–307.

and reporting obligations. After several years of complicated negotiations and the pulling out of the US of the process, it was the successful outcome of COP7, which was held from 29 October to 9 November, 2001 in Marrakesh, which provided the international community with the rulebook needed to implement JI and CDM projects, the so-called 'Marrakesh Accords'.[21]

The Marrakesh Accords contain a number of guidelines for the implementation of Article 6 and recommend them for adoption by the first COP of the UNFCCC serving as the Meeting of the Parties to the Kyoto Protocol (COP/MOP) after the entry into force of the Kyoto Protocol.[22] Yet, compared to the modalities governing the CDM, the JI text included in the Marrakesh Accords is still vague and incomplete. Since JI activities do not officially start before 2008, a number of specific guidelines still remain to be established. The Article 6 Supervisory Committee to be established to monitor JI activities, will eventually help to interpret the Accords. However, it will not be put into place before the entry into force of the Kyoto Protocol.[23] However, JI implementation has started already and the absence of the Committee is counterbalanced through an active engagement of the host as well as the acquiring country Party in the process of implementing JI projects.

3 Conceptual Differences Between CDM and JI

In the context of the Kyoto Protocol, Parties with 'quantified emission limitation and reduction commitments' (QELRC)[24] can participate in International Emissions Trading (Article 17), in JI (Article 6) or cooperate with developing country Parties in the implementation of CDM projects. The JI mechanism is similar to the CDM as both mechanisms are based on GHG emission reductions or removals through project-based activities. Emission reductions generated by JI or CDM need to be 'additional' to any that would occur in the absence of the project activity.[25] Both mechanisms generate credits than can be used for the purposes of meeting Annex I Parties' emission reduction obligations. In both cases approval by the host country is required[26] and private sector entities may be authorized to participate in project activities.[27]

However, there are significant differences between the two mechanisms. In fact, both mechanisms evolved out of non-overlapping negotiations and the language and the structure of Articles 6 and 12 are different and were not drafted in parallel. Most importantly, Article 6 describes a mechanism to be executed by and between

[21] See FCCC/CP/2001/13/Add.2 <http://www.unfccc.int> accessed 18 April 2004.

[22] FCCC/CP/2001/13/Add.2. The Decision on Implementation Guidelines for JI Projects is contained in the draft Decision 16/CP.7.

[23] ibid, para 3: The COP/MOP '*decides* to establish the Article 6 supervisory committee, at its first session, to supervise, *inter alia*, the verification of ERUs generated by Article 6 projects' (emphasis added).

[24] Art 3(1) Kyoto Protocol.

[25] Arts 6(1)(b) and 12(5)(c) Kyoto Protocol.

[26] Arts 6(1)(a) and 12(5)(a) Kyoto Protocol.

[27] Arts 6(3) and 12(9) Kyoto Protocol.

Annex I Parties to meet their obligations under Article 3 of the Protocol, whereas Article 12 shows an attempt to give developing country Parties the possibility to benefit from the economic instrument laid out by the Protocol.

The fact that JI can only be implemented between Parties with QELRCs leads to important conceptual differences in the treatment of CDM and JI projects. Whereas the integrity of the JI depends on the host countries' ability to accurately establish and account for its AAUs, the integrity of the CDM depends on the rules governing the independent verification and certification of emission reductions. Countries participating in JI are ultimately responsible for achieving their emissions targets, and JI projects do not change the overall amount of GHG emitted under the cap of the established QELRCs. JI transactions do not result in additions to the amount of AAUs allocated to Annex I countries, since, other than in the CDM, the transferring Party has AAUs from which to make a corresponding reduction. Therefore the Protocol's accounting provisions included in Article 3(10) and (11) require that the transfers of ERUs receive identical accounting treatment as emissions trading; indeed from an accounting perspective there is no difference between the project-based ERU transfers and AAU transfers based on Article 17.[28]

In the CDM, the host as well as the acquiring country have an incentive to transfer more, rather than fewer, emission reductions.[29] For the same conflict of interest situation the CDM host country is also not authorized to undertake verification (as in the case of what is described as Track One JI).[30] In JI, where the host country is ultimately responsible for all project activities, the host country has the right to verify emission reductions as being additional as soon as it has the proper monitoring and accounting framework in place.[31] If the Annex I country hosting a JI activity does not meet the full set of accounting criteria, it nevertheless can participate in JI, in which case the rules and procedures are modelled after the CDM. But even then, the country needs to meet the reporting requirement of the Protocol in order to issue and transfer ERUs.[32] Additionally, ERUs cannot be used by a Party to meet its targets under the Kyoto Protocol, in case a participating country's compliance is in doubt.[33] In this case ERUs can be transferred but cannot be used for compliance purposes until the doubts are lifted.

Rules guiding the independent verification are therefore key tools to ensure that the CDM does contribute to the global effort to mitigate climate change.

[28] A Petsonk, DJ Dudek, and J Goldman, *Market Mechanisms & Global Climate Change: An Analysis of Policy Instruments, Environmental Defense Fund in Cooperation with the Pew Center on Global Climate Change* (Washington, DC, 1998) 18.

[29] ibid 20. [30] See ch 8 below.

[31] In the same context, difficulties in measurements of activities involving carbon sinks have restricted the use of forestry projects in the CDM to afforestation and reforestation activities, whereas the government guarantee and AAU backing in JI has given the COP the confidence to allow ERUs to be generated from all forestry activities.

[32] FCCC/CP/2001/13/Add.2, draft Decision -/CMP/1 (Art 6), Annex, para 25.

[33] Art 6(4) Kyoto Protocol, FCCC/CP/2001/13/Add.2, draft Decision 16/CP.7, Annex, para 29.

Consequently, Article 12(7) states that the COP/MOP 'shall' at its first session elaborate modalities and procedures with the objective of ensuring transparency, efficiency, and accountability through independent auditing and verification of project activities.[34] The need for such procedures is lessened in the case of JI and Article 6 gives the COP/MOP a less prominent role, when it authorizes the body to develop measures in the form of guidelines.[35] Although 'guidelines' generally refer to non-binding, non-regulatory rules, it is expected that the JI guidelines will widely be considered as binding and mandatory.

On the other hand, the COP has authorized a 'prompt start' of the CDM, which makes it possible that Certified Emission Reductions (CERs) obtained during the period between 2000 and 2008 could be used to assist in achieving compliance during the first commitment period.[36] ERUs on the other hand can only be generated and transferred for emission reductions generated in the first commitment period, namely between 2008 and 2012.[37]

In summary, the conceptual difference between JI and CDM leads to a more stringent system with respect to the monitoring and verification of emission reductions generated by CDM projects. In the case of the JI the mechanism is generally burdened with less regulation, as the overall cap of emissions of industrialized countries would remain unchanged in case of more lax baselines. However, the responsibility, involvement and the accountability of the country hosting the JI activity is key and shifts the safeguards for the integrity of the mechanisms to the insurance that the country meets a set of reporting and accounting requirements.

4 Participation Requirements and Project Cycle

ERUs generated through JI project activities are nothing but another form of AAUs, which can be traded under Article 17 of the Kyoto Protocol. The differences between the two 'Kyoto currencies' consist in the confirmation that each ERU corresponds to a real and measurable verified emission reduction generated by an approved JI project activity. Before the conversion from an AAU to an ERU confirms the successful completion of the JI Project Cycle, the JI project needs to be designed, developed, approved, and implemented.

4.1 Participation in a JI Project

In order to be eligible for participation in a JI project the host country as well as the acquiring country Party needs to fulfil a number of requirements. These eligibility

[34] The COP has adopted with the Marrakesh Accords the necessary guidelines and modalities and will submit the Accords to the COP/MOP for adoption at its first session.

[35] See Art 6(2) Kyoto Protocol.

[36] Art 12(10) Kyoto Protocol.

[37] Another difference consists in that CDM projects have to pay a certain amounts of CERs to the benefit of adaptation projects and an administration fee.

Table 6.1: Comparison of the Kyoto Protocol Flexible Mechanisms

	International Emissions Trading (Article 17)	**Joint Implementation (Article 6)**	**Clean Development Mechanism (Article 12)**
Implementation of a Project	Not project based	Project based mechanism	Project based mechanism
Achievement of ERs	No additional ERs	Emission reduction activity (additional and measurable)	Emission reduction activity (additional and measurable)
Participants	Only Annex I Parties	Only Annex I Parties	Non-Annex I Parties (+ Annex I Party)
Verification of a Project	n.a.	Independent validation (determination) possible	Independent validation mandatory
Verification of ERs	n.a.	Independent verification of ERs possible	Independent verification and certification mandatory
Host Country Verification	n.a.	Host country can validate Projects (determination) and ERs (Track One JI)	No host country verification possible
Authorized Private Sector Participation	Yes	Yes	Yes
Host Country Approval	n.a.	Required	Required
Integrity Guaranteed	Through accuracy the accounting for AAUs	Through accuracy the accounting for AAUs	Through independent validation and verification
Land Use and Land Use Change Activities	n.a.	Full range of project activities eligible	Limited to Afforestation and Reforestation
Prompt Start (2000)	No	No	Yes
Rules	n.a.	Guidelines	Mandatory modalities and procedures

requirements focus on the need to ensure certainty about a country's ability to account for its emissions. Without information on what a country's emissions are, the risk of overselling credits that do not represent genuine reductions or surplus credits would be too high.[38] In order to establish adequate safeguards, the criteria determining eligibility for participation in AIJ projects have been incorporated in Article 6 of the Kyoto Protocol, supplemented by criteria determining that the use of JI shall be supplemental to domestic policy and measures resulting in emission

[38] J Lefevere, 'Greenhouse Gas Emission Allowance Trading in the EU: a Background', 27 September 2002, Foundation for International Environmental Law and Development (FIELD) 9.

reductions in Annex I countries.[39] Joint Implementation is designed to allow the participation of private entities in the implementation of projects and it is widely expected that the private sector will eventually become the main driver of the market. The international rules allow legal entities to participate in the flexible mechanisms, but this is done under the responsibility of the countries that authorize them to do so.[40] Private sector participation poses a problem where the country is ultimately liable for its compliance with the targets established under the Kyoto Protocol.[41] In order to enable the Parties to control JI activities, all Parties involved in the project have to give their approval. This means that both the host country government as well as the government, which acquires the ERUs or in which the legal entity that acquires the ERUs is headquartered, have to approve the JI project.

The eligibility requirements for Annex I Parties to acquire and transfer ERUs are contained in the Marrakesh Accords, and require the participating countries to:

(a) be a Party to the Kyoto Protocol;
(b) have assigned amount calculated and recorded in accordance with relevant guidelines and decisions;
(c) have in place a national system for the estimation of emissions by sources and removals by sinks of all greenhouse gases;
(d) have in place a national registry;
(e) have submitted annually the most recent required inventory;
(f) have submitted the supplementary information (eg on sinks) on assigned amount and makes any adjustments and recalculations required.[42]

While countries can engage in JI transactions before they meet all these criteria, as long as the Parties involved are Parties to the Kyoto Protocol and have approved the project, they need to develop a registry and the capacity to execute and register transactions involving AAUs and ERUs before the actual transfer of ERUs can take place. Before starting projects using the procedures for Track One JI, Annex I Countries would also need systems in place, which are prerequisites for the transfer of ERUs.

4.2 Project Development

Preparation of a JI Project involves a number of steps additional to the normal due diligence of an investment project. The preparation of the JI project can be undertaken either by the project sponsor before an entity has been identified which

[39] Article 6 (1)(d) Kyoto Protocol.

[40] Neither the Kyoto Protocol nor the Marrakesh Accords give any guidance on what this authorization should entail. It is therefore the prerogative of the country to decide on the criteria and the format of the authorization.

[41] See Art 6(4) Kyoto Protocol.

[42] FCCC/CP/2001/13/Add.2, draft Decision 16/CP.7, Annex, para 21.

would be interested in acquiring the credits or it can be done jointly by the project sponsor and interested investor or carbon purchaser.

Once the potential project has been identified, project participants select an appropriate baseline methodology for the establishment of a credible baseline for the project. The established baseline scenario together with the Monitoring Plan, which summarizes the data requirements and monitoring methodology to facilitate periodic verification of emission reductions, provide the basis for establishing emission reductions to be achieved throughout the project. The guidelines included in the Marrakesh Accords specify that the baseline for JI projects shall be established on a project-specific basis and/or using a multi-project emission factor, in a transparent manner and taking into account relevant national and/or sectoral policies and circumstances.[43] Additionally, many project participants prepare a report including spreadsheets which estimate the flow and volume of emission reductions to be achieved throughout the project and which form the basis for negotiations of the contractual terms of the project.

Rules on methodologies for setting baselines and developing Monitoring Plans are not included in the Kyoto Protocol and can be found only in a rudimentary form in the Marrakesh Accords. For JI, however, the approval of baseline methodologies through the CDM Executive Board can serve as relevant guidance on the evolving standards for baseline setting.[44]

For all projects, documentation must be provided indicating results of any environmental and social reviews at the national and local governmental level following the guidance provided by the UNFCCC. Similarly, stakeholder consultations are to be held, and the appropriate documentation shall be made publicly available, directly or through the UNFCCC Secretariat.[45]

The Monitoring Plan will be included as part of the Project Design Document (PDD). The Plan has to ensure transparent control procedure and quality assurance within the project boundary during the entire crediting period. The PDD is developed on the basis of the baseline assessment, the due diligence of the project, host country requirements and the environmental and social assessment.

4.3 Validation, Determination

The PDD constitutes the basis of the validation of the project. Validation means the determination of whether a project meets the relevant requirements of Article 6 of the Kyoto Protocol. Successful validation of a project is a prerequisite for the project to start generating ERUs. The validation step in JI is referred to as 'determination' in the Marrakesh Accords.

[43] FCCC/CP/2001/13/Add.2, draft Decision 16/CP.7, Annex, Appendix B.

[44] The CDM Executive Board has set up the Methodologies Panel to assess and approve baseline methodologies. The Panel has assessed a number of baseline methodologies, information on which can be found at the Panels website <http://cdm.unfccc.int/EB/Panels/meth>. The site also lists the Methodologies already endorsed and approved by the Executive Board.

[45] FCCC/CP/2001/13/Add.2, draft Decision 16/CP.7, Annex, para 32.

4.4 Negotiation and Execution of the Project Agreements

After successful validation/determination of the project, either the host country or the project sponsor enters into a JI project agreement with the entity that wishes to acquire the credits generated by the project. The most common JI agreement takes the form of a purchase and sale contract, the so-called Emission Reductions Purchase Agreement (ERPA). It governs the transfer of verified emission reductions and ERUs[46] against a purchase price paid on the transfer of the legal title to the emission reductions or on the receipt of the ERUs in the registry accounts of the purchasing party. Although the model of the purchase contract is most widely used, it is not the only form of a JI agreement. Such agreement could also take the form of a loan under which repayments are made in ERUs. The rights to the emission reductions can also accrue to the investors in the projects, which may choose to use the emission reductions for compliance purpose or further trade.

Since the project is implemented in an environment still characterized by significant regulatory risk, a robust contractual framework governing the implementation of the project is key. It needs to give necessary security and detail to ensure a fair allocation of responsibility and risk, while being sufficiently flexible to allow adaptations and concretization of the terms over time.[47]

4.5 Verification

Article 6(1)(b) of the Kyoto Protocol provides that a project must provide a reduction in emissions by sources, or an enhancement of removals by sinks, that is additional to any that would otherwise occur. Whereas Article 12 establishes third party verification, Article 6 provides only that the COP/MOP may further elaborate guidelines and modalities governing JI project activities. This meaningful silence in Article 6 has opened the way to include the possibility of host country verification in JI projects (Track One JI).

Consequently, the Marrakesh Accords provide for two possibilities for verifying emission reductions generated by JI projects. They can either be verified by the host Party if it meets the eligibility requirements (Track One JI) or through a verification procedure similar to the verification of emission reductions under the CDM under Article 6, Supervisory Committee. Once the host Party qualifies for Track One JI, the country may verify reductions from an Article 6 project as being additional.[48]

Some authors have interpreted the fact that a host government can verify emission reductions generated by JI projects as a waiver on the requirement of additionality of the project. They claim that additionality for Article 6 projects is not mandatory because the host country can decide how many AAUs it converts

[46] For details on the definition of the commodity purchase, see ch 3 above.

[47] See chs 16 and 20 below.

[48] See ch 8 below.

into ERUs and transfers for a JI project.[49] However, these authors overlook the fact that Article 6(1)(b) of the Kyoto Protocol clearly states that a JI project is to provide 'a reduction in emissions by sources, or an enhancement of removals by sinks, that is additional to any that would otherwise occur.' This means that JI host countries can verify emission reductions applying national criteria, guidelines and standards, however, this does not exempt such projects from the requirement to confirm that the emission reductions actually have occurred.

4.6 Transfer of Emission Reduction Units

With its approval of the project the host country has agreed to undertake the conversion and transfer of ERUs on the basis of the verification document. It marks and labels AAUs with a project identifier and converts it into ERUs.

Unlike in the case of the CDM, where the host country involvement is limited to the approval of the project, JI host countries need to monitor the project implementation as well as the verification of the emission reductions generated over the time. It is always the final action of the host country of transferring the ERUs from its registry to the acquiring Annex I entity, which successfully concludes the process of generating and acquiring ERUs. In the case of Track One JI, it is the host country which also verifies the emission reductions and therefore establishes the standard of verification.

Necessary conditions for such successful participation in the market is that JI countries maintain their eligibility to participate in the Kyoto mechanisms. This is of utmost importance taking into consideration that a country acquiring ERUs from a country, which is not in compliance with the Protocol, will not be able to use these ERUs for compliance purposes.

5 Contracts Involving JI Host Countries

Although many of the countries in Europe today have climate change programmes which include an engagement in JI, neither the AIJ national programmes nor more modern JI programmes have so far led to the establishment of authorizing legislation under which the projects could be implemented. Solid contractual arrangements are key where a detailed international and domestic legal framework is outstanding. Robust agreements can also make detailed legal provisions unnecessary, provided the interests of the project participants, as well as the integrity of the mechanism, are sufficiently protected. Whereas AIJ contracts typically involved full project financing,[50] the absolute majority of JI carbon contracts are limited to the procurement of emission reductions or ERUs. Today, most contract types are

[49] Danish, n 20 above, 304. [50] ibid.

either negotiated on a contract by contract basis or included in the bidding documents in an emission reduction tender.[51]

These contracts are generally concluded between the acquiring party and the project sponsor. In a great majority of JI contracts concluded over the last four years, the host country is not involved as a party. This experience differs from early expectations, when most observers had expected a more direct involvement of the host country. Such direct involvement was deemed necessary taking into account the commitment of the host country for approved projects: the country has to transfer the ERUs and will ultimately be liable for its compliance under the Kyoto Protocol. However, today the project developer who is in control of the GHG mitigating activity and responsible for securing financing of the project, is considered the most appropriate partner for receiving payments for the achievement of GHG emission reductions. The payments for emission reductions complement the financial package of the project.

The absence of the host country as a party in the project agreements, however, does not limit the importance of the country in the project cycle. The host country needs to allocate AAUs (which will eventually become ERUs) to project participants according to the terms and conditions of the project agreement.[52]

According to the Marrakesh Accord, the host country has to approve the project.[53] The international guidelines do not give more detail on the format of the approval, however, the approval should include the following elements:

(a) The approval of the project for the purposes of JI;
(b) The authorization of the private sector entities to participate in the project;
(c) The establishment of the legal title to the emission reductions sold under the project;
(d) The commitment to the transfer of ERUs in an amount equivalent to the verified Emission Reductions generated by the project.

Approval of JI projects requires clear and standardized procedures. Countries that do not have transparent and predictable approval procedures will discourage investment from potential JI project participants. Generally, a conductive environment in recipient countries determines the attractiveness of JI host countries: JI investments are likely to go to countries with working national institutions and a stable overall macroeconomic environment.[54]

The participation in Track One JI requires the compliance with additional reporting and accounting requirements.[55] In case a country qualifies for Track One JI, it needs to develop its own rules on verification of emission reductions and

[51] See ch 17 below.

[52] E Petkova and KA Baumert, *Making Joint Implementation Work: Lessons from Central and Eastern Europe* (World Resources Institute Climate Notes, Washington DC 2000) 4.

[53] FCCC/CP/2001/13/Add.2, draft Decision 16/CP.7, Annex, para 33.

[54] For an analysis of the institutions required in JI countries: see Petkova and Baumert, n 52 above.

[55] FCCC/CP/2001/13/Add.2, draft Decision 16/CP.7, Annex, para 21.

removals generated by JI projects. On the basis of this decision it establishes the amount of AAUs that it agrees to convert into ERUs and to transfer to the acquiring project participant.

The requirement of host country approval gives host countries bargaining power to establish priorities for specific project classes or to demand a share in the ERUs generated by the projects. Since a Letter of Approval is not an adequate instrument to define the broader context for the cooperation between the Parties, many host countries have decided to enter into framework agreements with sovereign carbon purchasers governing the selection and approval of the project. Many Economies in Transition (EITs) have entered into Memoranda of Understanding (MoU) with countries that have an interest in purchasing ERUs from JI projects in EITs.[56] The World Bank's Carbon Finance Business has developed Host Country Umbrella Agreements (HCAs), which serve a similar purpose.[57] Memoranda of Understanding, as well as HCAs typically outline the basic requirements for projects, the commitment of the host country to promote the development of JI, to approve individual projects and to transfer ERUs to the accounts of the partnering country, and dispute settlement. Unlike the Letter of Approval, which is issued from a designated JI focal point (in most cases the Ministry of Environment), MoUs or HCAs usually go through a national ratification process becoming national law in the country. Many carbon purchasers appreciate the additional security this instrument gives over a Letter of Approval.[58]

Another key element of these agreements is the guarantee of the host government to back the emission reductions generated before 2008 with AAUs (early credits). The Marrakesh Accords have confirmed that ERUs can only be earned for the period between 2008 and 2012, and provide that 'projects starting as of the year 2000 may be eligible as Article 6 projects if they meet the requirements of the guidelines for the implementation of Article 6 of the Kyoto Protocol as set out in the annex below and that ERUs shall only be issued for a crediting period starting after the beginning of the year 2008.'[59] This provision puts JI at a disadvantage against the CDM, which allows project participants to receive credits starting in the year 2000. Participants in JI projects have realized early that the proceeds from the sale of ERUs within the commitment period are usually not sufficient to make JI projects profitable. They therefore explored various approaches to increase the payment flows to these projects. Today the most

[56] Bulgaria has entered into Memoranda with Austria, Denmark, Italy, and the Netherlands; the Czech Republic has entered into Memoranda with Austria, Denmark, France, Germany, and the Netherlands; Romania has entered into Memoranda with Austria, Denmark, Italy, the Netherlands, Sweden, and Switzerland. For more info: Point Carbon, Carbon Market Europe, 5 March 2004 <http://www.pointcarbon.com> accessed 7 March 2004.

[57] At the time of the printing of this book, the World Bank had entered into HCAs with Romania, Bulgaria, the Czech Republic, and Poland.

[58] A Letter of Approval as a unilateral instrument under international law, and is equally binding, however, it hardly receives the attention and the support of a negotiated and ratified agreement.

[59] FCCC/CP/2001/13/Add.2, draft Decision 16/CP.7, para 5.

commonly used approach is what is referred to as 'early crediting', based on the mechanism defined in Article 17 of the Kyoto Protocol. Article 17 defines the rules of International Emissions Trading. The requirements to transfer AAUs under Article 17 are the same as the requirements of requiring or transferring ERUs under Article 6 with the exception that under Article 17 a Party is not allowed to make transactions which would result in its holdings of AAUs dropping below the required levels of its mandatory compliance reserve level of AAUs held (the so-called commitment period reserve).[60] A country can freely decide how to manage and allocate its AAUs. In case it has large amount of surplus AAUs, it may use such AAUs in order to back the verified emission reductions generated before 2008 by a JI project. Thereby the transfer of its AAU corresponds to real and measurable emissions reductions, while at the same time helping to optimize the revenue stream the project sponsor can obtain for their project. In order to back the 'early credits' with AAUs, the country allocates some of its AAUs from the first commitment period to projects started before the beginning of this period. This has the effect of using up a portion of the host country's assigned amount headroom for the commitment period but at the same time reduces actual emissions by facilitating the early start of projects. The same project receives AAUs under an Article 17 transaction before it becomes a JI project for the period between 2008 and 2012. Whether a country wishes to engage in such transaction to underwrite bona fide emissions reductions generated by a project prior to 2008 is a matter within the sovereign discretion of an Annex I Party.

Selling countries can either engage in project activities directly and actively structure the sale of AAUs or it can agree to simply back the activities implemented by private entities. In the latter case, the project entity receives the proceeds from the sale of the emission reductions. In the former, the country can structure the sale of the AAUs in a variety of ways, including direct or forward sales, debt-swaps or options.

Countries can also allocate a certain number of AAUs to promote renewable energy or energy efficiency projects in a pilot programme.[61] It can allocate the AAUs to project proponents, which could sell these AAUs under a forward purchase thereby increasing the profitability of the project.

5.1 Role and Challenges for Host Countries

While JI projects can be implemented among any Annex I countries, transition countries are expected to play a key role in the implementation of JI. The Russian Federation and almost all of the Eastern European EITs have large amounts of surplus AAUs available. This is because the energy consumption in most EITs currently generates emissions that lie well below their 1990 baseline emissions, which establish the reference emissions for the QELRCs set forth in the Kyoto

[60] FCCC/CP/2001/13/Add.2, draft Decision 18/CP.7, para 6.

[61] See also J Palmisano, 'Early Crediting for JI Projects Is a Useful Policy Instrument', Evolution Markets—Executive Brief, 16 November 2001.

Protocol. In Russia alone, a drop in emissions of a total of an estimated 712 million metric tonnes of CO_2 equivalents, or 30 per cent, for the period 1990–94, was not caused by active climate policy but rather stems from the breakdown of the Soviet economic activity.[62] The difference between the 1990 levels and the additional AAUs freed-up through the downturn of the economic activity in Russia and EITs is commonly referred to as 'hot air'.

The existence of hot air places most EITs in a position where they could sell their surplus quotas on the market. However, estimates show rapid economic growth for a number of EITs, which would mean that the countries would need at least parts of their surplus AAUs for their own purposes. Even so, under the most optimistic scenarios most of the EIT countries will have GHG emissions in 2012 which lie well below their 1990 levels. Additionally, given the prevailing inefficiencies in large parts of their energy and industrial sectors, transition countries should also have scope to reduce emissions further at relatively little cost.[63] The comfortable positions of EITs under the Kyoto Protocol and the large amounts of surplus AAUs[64] make them the favourite countries for the implementation of JI projects.[65]

Above and beyond the approval of JI projects, JI host countries will have to develop a strategy vis-à-vis their participation in Article 17 transactions and the sale of AAUs as well as a management system for their AAUs. With regard to the participation in Articles 17 and 6, they will need to establish appropriate institutions, which assume these management responsibilities. Coordination of the activities as well as the clear and transparent allocation and accounting of AAUs allows a country to maximize its benefits from the engagement in the Kyoto Mechanism. The activities that need to be managed and coordinated by the host country include:

(a) The commitment of AAUs through International Emissions Trading under Article 17 of the Kyoto Protocol, which includes:
   a. Forward sales of AAUs;
   b. Early crediting for JI projects;
   c. Transactions implemented under a scheme which 'greens' AAUs;[66]

62 A Moe, 'Russian quota sales: Not just a lot of hot air' available at <http://program.forskningsradet.no/petropol/hotair.php3> accessed 29 February 2004; D Victor, N Nakićenović, and N Victor, 'The Kyoto Protocol carbon bubble: implications for Russia, Ukraine and Emission Trading', IR 98-094, Luxemburg estimates the total size of the 'carbon bubble' of 344 millions of metric tons of carbon (MtC), where Russia and Ukraine account for 225 MtC.

63 D Victor, N Nakićenović, and N Victor, 'The Kyoto Protocol Emission Allocations: Windfall Surpluses for Russia and Ukraine' (2001) 49(3) Climate Change 263–77, W Chandler, *Energy and Environment in Transition Economies* (Boulder, Colo, 2000).

64 Which vary significantly among the countries. Russia, Ukraine, and Poland have the largest amounts of hot air with Hungary or Slovenia not having the allocation that would allow with the current projections of growths to sell surplus AAUs.

65 At the time of writing, the World Bank has entered into HCAs with Bulgaria, Romania, the Czech Republic, and Poland, and MoUs haven been concluded between Switzerland, Denmark, Austria, the Netherlands, Romania, Bulgaria, and the Czech Republic, to quote only a few.

66 See ch 8 below.

(b) Approval of JI Projects to be implemented under Articles 6 of the Kyoto Protocol;
(c) Maintaining the eligibility requirements for the participation in Articles 6 and 17;
(d) Oversight on the compliance with the QELRCs;
(e) Activities of the UNFCCC focal point and key UNFCCC negotiators of the country;
(f) If applicable, the implementation of the Emissions Trading Scheme of the European Union;
(g) If applicable, the issuance of Renewable Energy Certificates.

The country possessing a surplus of AAUs will spread its risk through the involvement in a number of activities. The terms of conditions of these activities will differ. While it may be wise to sell an initial amount of AAUs before the Kyoto Protocol has entered into force, a number of AAUs should be retained allowing the country to buffer economic growth and at the same time taking advantage of future price increases. The activities funded from the proceeds can range from emission-reduction investment to awareness raising, demand side management, or institutional funding.

A successful AAU management system ensures that the country is enabled to explore all investment opportunities; it can promote energy efficiency and an increasing energy mix in a country. It also attracts foreign direct investments while ensuring long-time reductions in GHG emissions of a country. It also supports the Kyoto Protocol and builds infrastructure for international emissions trading.

6 JI and the European Emissions Trading Scheme

In order for the EU Member States to meet their international obligations to reduce GHG emissions by eight per cent of its 1990 level by 2012 under the Kyoto Protocol, the EU is implementing an EU-wide Emissions Trading Scheme (EU ETS). The Emissions Trading Directive was published on 25 October 2003 in the *Official Journal* as 'Directive 2003/87/EC of the European Parliament and of the Council, of 13 October 2003 establishing a scheme for greenhouse gas emission allowance trading within the Community and amending Council Directive 96/61/EC' and establishes a scheme which allows the trade of authorizations to emit a certain amount of greenhouse gases in the European Union (EU Allowances) from 1 January 2005.[67] In its original form,[68] the scope of the Directive was limited to trading within the EU, and did not link the EU ETS to

[67] See ch 23 below. [68] [2003] OJ L 275/32.

emission reduction credits generated under the Kyoto Protocol. To remedy this, the Commission of the European Communities released on 23 July 2003 the first draft of a 'Proposal for a Directive of the European Parliament and of the Council, amending the Directive establishing a scheme for greenhouse gas emission allowance trading within the Community, in respect of the Kyoto Protocol's project mechanisms' (Linking Directive).[69] On 12 September 2004 the Council approved the final text of the Linking Directive.[70]

Establishing the link between CDM and JI on the one hand and EU ETS on the other is the core objective of the Linking Directive. The Proposed Directive foresees the possibility of direct use of CERs and ERUs for compliance purposes under the EU ETS. The scope of regulation however is broader: The Linking Directive does also regulate the approval and accounting of JI projects in the EU. The approval of JI projects in EU Member States raises a few very technical questions concerning the compatibility with the EU ETS. Most of these problems are related to the so-called double counting issue, that is, how to avoid that the same emission reductions result in the issuance of ERUs and at the same time free up EU Emission Allowances. With respect to the double counting problem, three different cases of JI projects in EU Member States must be distinguished:

(i) JI project activities that are undertaken at sources covered by the EU ETS, for example, the refurnishing or fuel switch in a power plant (above 20 MW);

(ii) JI project activities that have no direct link to sources covered by the EU ETS but lead to emission reductions at such sources, for example, the installation of a wind park or the improvement of energy end use efficiency (so called 'Indirect Projects');

(iii) JI project activities at sources that are not covered by the EU ETS.

The agreed version of the Directive leaves it up to the operator whether they wish to engage in JI, selling ERUs or whether they wish to sell EU Allowances for any project activity that reduces CO_2 emissions, provided that the operator cancels an EU allowance from his account for each ERU issued for the project. This proposal also allows the export of ERUs outside of the EU (namely to Japan, which has expressed a strong interest in participating in the JI projects in Eastern Europe). Whether such conversion will be attractive to operators remains to be seen.

With respect to installations that are not covered by the EU ETS, but whose emissions are linked to any emissions of installations covered by the EU ETS (indirect coverage) the agreed text of the Linking Directive allows the implementation of JI projects in the renewable energy and demand side sectors, as long as the governments of Member States cancel an amount of EU Allowances equivalent to the amount of ERUs issued for JI projects. This means that the Member State's government in preparing the National Allocation Plan needs to establish a reserve

[69] COM(2003)403 final. [70] 2003/0173 (COD) Pre-cons 3671/04.

of EU Allowances corresponding to the expected amount of ERUs issued and transferred to JI projects. The Allowance reserve needs to be taken into account when allocating the Allowances to the installation covered by the EU ETS (a larger reserve resulting in the allocation of fewer allowances).

7 Outlook

In the recent years, the engagement in JI has seen several ups and downs. While the possible participation of JI host countries in the trade of AAUs and Track One JIs opens the door to flexible contractual arrangements, not all countries have been welcoming hosts of JI projects. Unclear responsibilities, the lack of a strategy and transparent approval procedures have impeded the early carbon market in many JI countries. Insecurities involving the accession of many potential JI host countries to the EU and the link between the JI and the EU ETS have created further hurdles for the market. However, conceptually JI comes with reduced risks compared to the CDM, since an agreement with the host country can mitigate the risk associated with the baseline, additionality, verification, and approval of the project.

It is therefore key that, in order to maximize the efficient participation and the benefits offered in global Emission Trading and JI, that all Annex I countries develop a strategy on how to manage their AAUs. Such a strategy could look at the following points:

7.1 Assessment of the Kyoto Targets

The first step required of each Annex I country is to determine how to achieve the QELRCs. In order to do this, National Inventories and Registries need to be established[71] to calculate net anthropogenic emissions and the countries, AAUs. Once a National Registry is established and the country has calculated its AAUs, it will know how many AAUs are spare for JI investment or trading (assuming they can be greened), or what projects or legislation need(s) to be initiated to reduce its emissions to achieve its target under the Kyoto Protocol.

7.2 Management of Surplus AAUs

It is expected that EITs will have significant volumes of spare, or additional, AAUs once actual emissions are calculated by the National Registry. How each country manages their additional AAUs is a matter of national policy. It is therefore recommended that each country should consider developing a national policy on how to manage additional AAUs. For a national policy to take best advantage of JI

[71] FCCC/CP/2001/13/Add.2, Decision 19/CP.7 Modalities for the accounting of assigned amounts under Article 7, paragraph 4, of the Kyoto Protocol.

and International Emissions Trading opportunities, it may look at implementing monitoring, compliance, and enforcement systems, analyse what needs to be done to become linked into the EU ETS, and identify what steps need to be taken to foster additional investment.

7.3 Development of a National Strategy

A national policy on how to mitigate climate change and make optimal use of carbon finance should identify which sectors ought to be prioritized, or would benefit most from carbon finance.

To this end, a national policy may consider:

- sectors that contribute most to national emissions, and which therefore offer a wide variety of opportunities for prospective investors who could provide modern, clean technology;
- inefficient sectors, or sectors that produce powerful GHGs such as CH_4, N_2O, HFCs, PFCs, and SF_6;
- national development and sustainable development priorities.

7.4 Monitoring Mechanisms

Because ERUs generated through JI projects can be 'guaranteed' through holding spare AAUs as security, ERUs are likely to receive a premium in the carbon market over CERs. To ensure the integrity of the projects, any JI projects done under the Track One will be dependent on national monitoring, verification, compliance, and enforcement mechanisms. The credibility of these mechanisms will also be significant when attempting to attract funds through sales of greened AAUs.

7.5 Fostering Additional Investment

Additional investment can be attracted through providing security and incentives to prospective investors, and through actively seeking out new markets for Emission Reductions. Security can be provided through a transparent and efficient regulatory framework, which includes the monitoring, compliance, and enforcement mechanisms. The underlying legal system supporting any investment should also be transparent and efficient. National legislation may be reviewed and amended where necessary to ensure new units and other interests of prospective investors will be recognized and sufficiently secure.

7

GHG Emissions Trading Registries

*Anthony Hobley and Peter Hawkes**

1 Introduction

In their simplest form National Greenhouse Gas Emissions Trading Registries (National Registries) are electronic databases for recording and tracking Kyoto Units (that is, Assigned Amount Units (AAUs),[1] Removal Units (RMUs),[2] Certified Emission Reductions (CERs),[3] and Emission Reduction Units (ERUs)[4]) and regional/domestic units (for example, EU Allowances).

Such National Registries are in fact critical to any sophisticated system of emissions trading whether International Emissions Trading (IET) pursuant to Article 17 of the Kyoto Protocol or regional trading such as will take place under the EU-wide greenhouse gas (GHG) Emissions Allowance Trading Scheme (EU ETS).

The Marrakesh Accords in Decision 19/CP.7 set out the key functional requirements necessary for National Registries to satisfy the requirements of the Kyoto Protocol. The functions to be performed by National Registries are to ensure the accurate accounting of the issuance, holding, transfer, acquisition, cancellation, and retirement of all Kyoto Units as well as the carry-over between Kyoto commitment periods of ERUs, CERs, and AAUs (that is, banking). In addition, National Registries may perform the same or similar functions for units issued in any domestic scheme (for example, UK ETS) or regional scheme (for example, EU ETS).

The specifications in the Marrakesh Accords provide for National Registries to have similar functional specifications to financial securities registries. They must be secure, be in the form of standardized electronic databases, contain common

* Anthony Hobley and Peter Hawkes are members of Baker & McKenzie's Global Clean Energy and Climate Change Practice Group.

[1] AAUs are the emission allowances assigned under the Kyoto Protocol and equate to the amount of GHGs an industrialized country can emit over a certain commitment period.

[2] RMUs are credit units issued by parties to the Kyoto Protocol equivalent to the net removals of GHGs from a party's activities under Art 3 Kyoto Protocol.

[3] CERs are units created in accordance with the clean development mechanism (CDM) provisions of Art 12 Kyoto Protocol.

[4] ERUs are the carbon credits arising from joint implementation (JI) projects pursuant to Art 6 Kyoto Protocol.

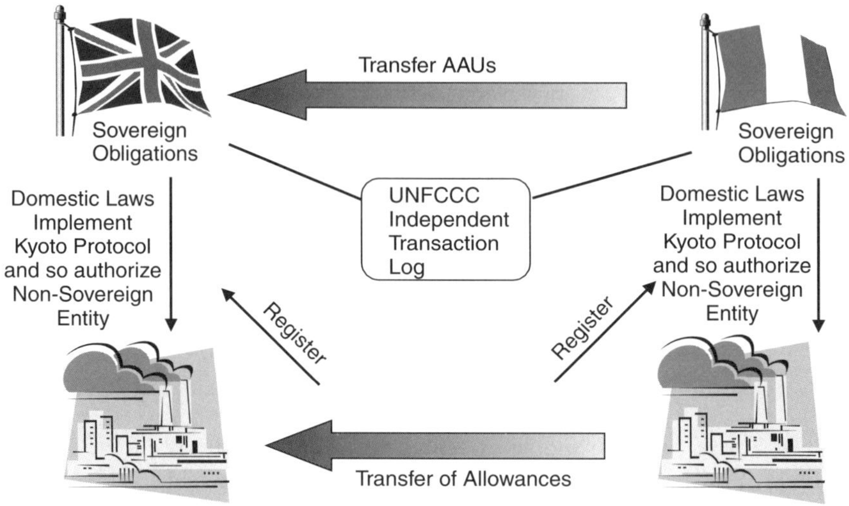

Figure 7.1. International Emissions Trading post-2008

Notes: Article 17 of the Kyoto Protocol envisages non-sovereign entities participating in international emissions trading provided that they are authorized to do so by the sovereign state. The EU ETS will operate in a similar way and is discussed later in this chapter.

data elements, be connected with other National Registries and provide both secured and unsecured access where appropriate. Figure 7.1 illustrates how the Kyoto Protocol envisages that non-sovereign entities (companies or individuals) may participate in IET. The requirement for these non-sovereign entities to be authorized by an eligible Party to the Kyoto Protocol before they can participate in IET is discussed later in this chapter.

2 The Nature of National Registries—Registries not Registers or Trading Platforms

It is important to clearly understand what National Registries are and what they are not.

When reading this chapter it must be remembered that National Registries, and the registries systems they form collectively, are accounting tools. They ensure the accurate recording and tracking of units (Kyoto or otherwise) representing the equivalent of one tonne of carbon dioxide emissions. This is necessary to ensure compliance with emission limitation or reduction obligations either imposed on Parties listed in Annex B to the Kyoto Protocol (Annex B Parties) or entities within those countries. This type of compliance scheme is also a trading scheme in recognition of the belief that emissions trading enables overall compliance to be achieved at lowest cost. That said, National Registries and registries systems do not enforce contracts. Similarly, they do not contain any price or contractual information. As such, this chapter does not address the development of trading platforms and clearing houses that may assist the development of a liquid and transparent market in emissions trading. Instead, it focuses on the importance of National Registries to emissions trading schemes.

Neither are National Registries inventories of GHG emissions. Such inventory information will be gathered elsewhere and input into the National Registries at the time of compliance so that holdings of EU Allowances and Kyoto Units can, at specified times, be automatically matched off against verified emissions. For example, this will allow the National Registries within the EU ETS to alert the relevant regulators when there is a non-compliance so the regulators may know when, and at what level, to impose the penalties provided for under the European Directive establishing the EU ETS (EU ETS Directive).[5]

3 Kyoto Registries

3.1 Annex I Registries: The Marrakesh Accords and the New Delhi Decisions

A country with quantified emission reduction targets included in Annex B of the Kyoto Protocol is only eligible to transfer and/or acquire Kyoto Units if it is in compliance with certain eligibility requirements. The eligibility criteria are established in the Marrakesh Accords. They are broadly similar for the three flexible mechanisms but there are some significant differences. These criteria are most severe in relation to an Annex B Party's eligibility to participate in IET.[6]

If a Party should fail to meet one or more of the relevant eligibility criteria for JI, CDM, or IET this would result in a suspension of the eligibility of that Party to use those mechanisms.[7]

On the assumption that the Kyoto Protocol will enter into force, the draft Decision contained in Decision 19/CP.7 of the Marrakesh Accords will therefore require each Annex B Party to establish and maintain a National Registry to ensure the accurate accounting of the issuance, holding, transfer, acquisition, cancellation, and retirement of all Kyoto Units and the carry-over of ERUs, CERs, and AAUs.[8] A National Registry must be in the form of a standardized electronic database that conforms to the 'technical standards' to be adopted by the Conference of the Parties to the UN Framework Convention on Climate Change acting as the

[5] Dir 2003/87/EC. [6] See ch 22 below.

[7] Para 3 of the Annex to the draft Decision contained in Decision 18/CP.7.

[8] Para 17 of the Annex to the draft Decision contained in Decision 19/CP.7.

Meeting of the Parties to the Kyoto Protocol (COP/MOP) for the purposes of ensuring the accurate, transparent, and efficient exchange of data between:[9]

- National Registries;
- the CDM Registry; and
- the UNFCCC Independent Transaction Log (UNFCCC ITL).[10]

The general design requirements for these 'technical standards' referred to above are set out in Decision 24/CP.8 of the Delhi Decisions and are discussed further below.

Each Annex B Party must designate an organization as its registry administrator to maintain the National Registry. Any two or more Annex B Parties may voluntarily maintain their respective National Registries in a consolidated system, provided that each National Registry remains distinct.

The 'Registry Requirements' contained in the Annex to the draft Decision contained in Decision 19/CP.7 then specify:

- The accounts that each National Registry must have;
- The procedure for the issuance of ERUs, AAUs, and RMUs;
- The procedure for transfer, acquisition, cancellation, retirement and carry-over of Kyoto Units;
- Transaction procedures and the creation of the UNFCCC Independent Transaction Log; and
- Reporting requirements and what is meant by 'publicly accessible information'.

Each National Registry must contain the following accounts:

- At least one holding account for the Annex B Party;
- At least one holding account for each legal entity authorized by the Annex B Party to hold Kyoto Units under its responsibility;
- At least one cancellation account for each commitment period for the purposes of cancelling Kyoto Units on the basis of land use, land-use change, and forestry activities;[11]
- One cancellation account for each commitment period for the purposes of cancelling Kyoto Units following a determination by the compliance committee;[12]
- At least one cancellation account for each commitment period for the purposes of cancelling Kyoto Units for any other reason;[13]
- One retirement account for each commitment period.

[9] Decision 19/CP.7.

[10] Para 19 of the Annex to the draft Decision contained in Decision 19/CP.7.

[11] See ibid para 12(d). [12] See ibid para 12(e). [13] See ibid para 12(f).

The UNFCCC Secretariat will establish and maintain the UNFCCC Registry, which will verify the validity of transactions involving Kyoto Units.

The UNFCCC ITL will act as a central communications hub through which National Registries will exchange data. Automated checks conducted by the UNFCCC ITL will identify any discrepancies in proposed transactions. If a discrepancy is identified then the initiating National Registry must terminate the proposed transaction.

A periodic reconciliation process will also be conducted by the UNFCCC ITL in order to identify (and if possible, resolve) any inconsistency between the information held in a National Registry and the information recorded by the UNFCCC ITL.

As mentioned above, Decision 24/CP.8 contains the general design requirements of the technical standards for data exchange between National Registries under the Kyoto Protocol. These general design requirements are based on principles of accuracy, transparency, efficiency, security, and consistency. They stipulate the required message sequence types (see Table 7.1), serial numbers of Kyoto Units (see Table 7.2), account numbers (see Table 7.3), and transaction numbers (see Table 7.4) as well as the reconciliation and automated checks processes of the UNFCCC ITL.

Table 7.1: Minimum standardized message sequence types for registry systems

Transactions

1. Issuance of units in a National Registry or the CDM Registry
2. Internal transfer of units (a) from the CDM Registry pending account to another account or (b) from an account to a cancellation or retirement account
3. External transfer of units to a National Registry
4. Carry-over of units, as appropriate, to the subsequent commitment period

Other activities

5. Reconciliation of data between National Registries and the UNFCCC ITL
6. Testing of connections between National Registries
7. Notification of change to online status of the UNFCCC ITL
8. Notification of change to offline status of the UNFCCC ITL

Table 7.2: Elements of serial numbers

Element	*AAU*	*RMU*	*CER*	*ERU*
Originating Party identifier	yes	yes	yes	yes
Issuance commitment period	yes	yes	yes	yes
Unit type	yes	yes	yes	yes
LULUCF activity	no	yes	yes	yes
Project identifier	no	no	yes	yes
Unique number	yes	yes	yes	yes

Table 7.3: Elements of account numbers

Element	*Holding account*	*Cancellation account*	*Retirement account*
Party identifier	yes	yes	yes
Commitment period	no	yes	yes
Account type	yes	yes	yes
Unique number	yes	yes	yes

Table 7.4: Elements of transaction numbers

Originating Party identifier
Commitment period
Date
Transaction type
Unique number

Source: Decision 24/CP.8

The Subsidiary Body for Scientific and Technical Assessment is in the process of elaborating the detailed functional and technical specifications required in order to facilitate the implementation of the technical standards in all National Registries in a compatible manner. It is expected that this work together with the implementation and testing of the UNFCCC ITL will be completed before the tenth session of the Conference of the Parties in December 2004.

3.2 The CDM Registry

The CDM Registry is to be established under the authority of the CDM Executive Board (EB) to ensure the accurate accounting of the issuance, holding, transfer, and acquisition of CERs by Parties that are not listed in Annex B to the Kyoto Protocol (non-Annex B Parties).[14]

The structure and format of the electronic database of the CDM Registry must conform to the technical standards (discussed above) adopted by the COP/MOP for the purpose of ensuring accurate exchange of information between the National Registries, the CDM Registry and the UNFCCC ITL.[15]

The CDM Registry will contain at least the following accounts:[16]

(a) a pending account into which the CERs are issued before being distributed to the relevant Party or organization accounts;

(b) at least one account per non-Annex B Party hosting a CDM project or which has requested an account;

[14] See para 1 of Appendix D to the draft Decision contained in Decision 17/CP.7.

[15] See ibid para 2. [16] See ibid para 3.

(c) at least one account for the cancellation of Kyoto Units where accreditation has been withdrawn or suspended; and

(d) at least one account to hold and transfer CERs to cover administrative expenses and assist in adaptation costs but which may not purchase CERs.

Each account in the CDM Registry shall have a unique number, which clearly indicates the Party or organization for which the account is maintained. In addition, each CER shall have its own unique number, which identifies the commitment period for which the CER is used, the non-Annex B Party which hosted the CDM project, and the originating project activity.[17]

Non-confidential information held by the CDM Registry must be made publicly available via the internet.[18]

Following verification and certification of a project, the EB will instruct the registry administrator of the CDM Registry to issue the specified number of CERs (individually numbered) into the pending account. The registry administrator must then promptly forward a proportion of CERs for administration and adaptation into the appropriate management account and forward the remaining CERs to the relevant Parties' and project participants' accounts in accordance with their request.

The EB has not yet established the CDM Registry. During its ninth meeting the EB addressed the possibility of developing an Interim CDM Registry in order to issue CERs on an interim basis as requested and then, once the UNFCCC ITL was operational such issuance would be confirmed. However, this path was not followed. Instead the EB agreed to issue a public call to Parties and organizations for input to the development of the CDM Registry (in particular for the functional or technical specifications, other documentation, and programming codes). In addition, the EB requested the UNFCCC Secretariat to begin development work on the CDM Registry, together with its general work on National Registries. In particular the EB requested the preparation of a draft functional specification for the CDM Registry.

Work has been ongoing on the functional specification of the registry and, at its twelfth meeting, the EB discussed the proposed modalities for the development of the CDM Registry and the responses received from its call for input. The EB anticipates that, depending on the progress of accreditation and registration of projects, the first requests for issuance of CERs are expected in the second or third quarter of 2004.

At the time of writing a number of open issues remained for the EB to resolve, including:

(a) whether holding accounts for Annex B Parties and their entities may be established, in particular so that these Parties and entities may hold CERs in

[17] See para 5 of Appendix D to the draft Decision contained in Decision 17/CP.7.

[18] See ibid para 9.

their own accounts *prior to* the establishment of the UNFCCC ITL and National Registries;

(b) whether CERs held in non-Annex B Party accounts and their entities' accounts may be carried over to a subsequent commitment period;

(c) whether CERs held in non-Annex B Party accounts and their entities' accounts can be transferred to holding accounts in National Registries or other holding accounts in the CDM Registry;

(d) whether non-Annex B Parties or their entities can acquire CERs in their accounts in the CDM Registry other than receiving them from the pending account.

3.3 Authorization of Entity Level Trading

It is important to remind ourselves that the Kyoto Protocol is an international treaty between sovereign States and as such only Parties to the Protocol have any direct rights or obligations under it. These rights and obligations include the right of those Annex B Parties to participate in IET pursuant to Article 17 of the Protocol.

IET allows (but does not require) Annex B Parties to buy and sell Kyoto Units. However, Article 17 does not permit companies and individuals of these countries to participate in IET. It does, however, provide that Annex B Parties may authorize non-sovereign entities to take part in IET. This can be done on a case-by-case basis such as authorizing a company to buy or sell a quantity of Kyoto Units. What is more likely, however, is for parties to authorize all, or at least a certain category of, entities to participate in IET through a general piece of legislation. This broad approach is particularly likely to occur within Annex B Parties that develop an emissions trading scheme as a policy tool for achieving their emission limitation or reduction commitment under the Kyoto Protocol. The EU ETS and the proposed Canadian Federal ETS[19] are good examples of this.

4 Linking Registries—IET Made Real

As described above, each Annex I Party is required to have a National Registry in place before it can take part in IET. Indeed, it is the system of National Registries that will be at the heart of IET by making the holding, accounting, and transfer of Kyoto Units a reality. The UNFCCC ITL will link all the National Registries together and act as communications hub for the registries system under the Kyoto Protocol.

[19] <http://www.nrcan-rncan.gc.ca/lfeg-ggef/English/principles_en.pdf>.

This has been taken a step further in the development of the EU ETS. Each Member State's EU ETS registry must be capable of functioning as that Member State's National Registry pursuant to the Kyoto Protocol.[20] In addition, each Member State's National Registry will also be linked to the Community Independent Transaction Log (CITL), which will act as a communications hub for the EU ETS. The CITL will also be linked to the UNFCCC ITL so that transactions in the EU ETS that are also transactions under the Kyoto Protocol will be checked and processed appropriately.

Once these National Registries, the UNFCCC ITL and CITL and the communication links between them are all in place then, subject to having satisfied the Kyoto Protocol eligibility criteria discussed earlier, the Annex B Parties themselves (that is, their governments) will be able to transfer Kyoto Units to and from their respective government accounts in their respective National Registries.

However, something more will be required for entities (companies and individuals) to take part in IET. Remember for entities to take part in IET they must be authorized by their respective governments. Therefore they will have to be authorized by their governments to hold, receive, and transfer Kyoto Units. In addition, Annex B Parties will have to accommodate this necessity by allowing non-sovereign entities to open accounts within their National Registry for this purpose.

The proposed EU ETS Registries Regulation[21] in effect provides such a blanket authorization to all legal persons within the EU to hold (on a restricted basis) AAUs in the form of EU Allowances. The Regulation will ensure the creation of the EU ETS Registries System (discussed below) and provide the legal authorization to hold and transfer EU Allowances within the twenty-five member countries of the EU (EU25). However, there is no such blanket authorization within the EU25 to hold and transfer Kyoto Units. Even though from 2008 an EU Allowance is also an AAU (more on this later), this only creates a limited legal interest in the AAU in which it is embedded (rather like a licence to occupy a property for specific purpose for a specific period, that is, five years limited interest in that property which must be surrendered at the end of the period). This means the holder of the EU Allowance can hold it and use it in accordance with the EU ETS legal framework but is not totally free to use it as an AAU. It cannot sell it to a government or entity outside of the EU (such as Canada) for use as an AAU. For this to be permitted by the EU, there would probably need to be an agreement in place between Canada and the EU (according to Article 25 of the EU ETS Directive) for the mutual recognition of allowances as well as a physical link between the registries of the two countries as described above.

[20] Art 6(1) of Decision 280/2004/EC.

[21] Draft COMMISSION REGULATION (EC) No . . . /2004 of 21 December 2004 for a standardised and secured system of registries pursuant to Directive 2003/87/EC and Article 6(1) of Decision 280/2004/EC, COM (2004).

But can an EU Member State authorize entities within it to hold, receive, and transfer Kyoto Units? If so, would this mean that an entity in the EU could transfer an EU Allowance to an entity in Canada such that when it arrives in Canada it would be an AAU? At this stage the answer is not clear. Industry has argued that in the absence of such mutual recognition between emissions trading schemes the ability to make such transfers would be unfair as it would decrease the supply of EU Allowances and so put up the cost of compliance. EU Allowances are governed by the rules of the EU ETS, which are not likely to allow this in the absence of a two way mutual recognition agreement. However such an authorization by the UK Government would allow Kyoto Units to be transferred to the National Registry of any other country, which has also authorized its entities to hold, receive and transfer Kyoto Units.

5 The EU ETS Registries System

The proposed EU Registries Regulation imposes obligations on Member States and the European Commission for the establishment of a standardized and secured system of National Registries in the form of standardized electronic databases containing common data elements and the CITL (operating together as the 'EU ETS Registries System') to track the issue, holding, transfer, acquisition, surrender, cancellation, and replacement of EU Allowances and to ensure that there are no transactions incompatible with obligations resulting from the Kyoto Protocol, and to provide for public access and confidentiality as appropriate.

Essentially, the EU ETS Registries System will consist of twenty-six Registries (one National Registry for each Member State and one Community Registry) and the CITL through which all transactions will be communicated and recorded, by which certain transactions will be checked, and by which certain transactions will not be allowed to be completed.

Each Member State and the Community acting through the Commission must establish a Registry by 30 September 2004 in order to allow sufficient time for testing before 1 January 2005 (the first day of the 2005–7 period). The Commission is also required to establish the CITL on behalf of the Community by the same date.

In accordance with Article 19(1) of the Directive, a Member State or the Community may operate and maintain its Registry in a consolidated system with one or more other Member States or the Community provided that each Registry remains distinct.

In accordance with Article 6(1) of the EU Monitoring Decision,[22] a Registry must be capable of functioning as the relevant Member State's National Registry, or the Community's Registry, pursuant to the Kyoto Protocol.

[22] Decision 280/2004/EC.

In essence the proposed Registries Regulation sets out the functional and technical specifications required of each Registry and the CITL, and prescribes how they will operate together as the EU ETS Registries System.

5.1 The Architecture of the EU ETS Registries System

When one looks at each EU Member State's obligations under both the Marrakesh Accords (encapsulated into EU law through Article 6 of the EU Monitoring Decision[23]) and Articles 19 and 20 of the EU ETS Directive one can begin to determine the likely structure of the EU ETS Registries System. Such a system is likely to be made up of individual National Registries in each Member State. Each such National Registry will be capable of handling exchanges according to both the requirements of both the EU ETS and the Kyoto Protocol.

Article 20 of the EU ETS Directive requires that the EU ETS Registries System must contain the CITL administered by the Commission to check all transactions. It makes technical sense that all communications between each such National Registry within the system should be affected by using the CITL as a hub at least for the 2005–7 period. IET will use the UNFCCC ITL as a hub in the same way. This means that all messages and notifications between each National Registry within the EU will pass through the CITL and from 2008, where there is a transfer between National Registries, through both the CITL and the UNFCCC ITL for checking and subsequent onward transmission to the receiving National Registry (see Figure 7.2 below). In essence the CITL will act as the 'gatekeeper' for the

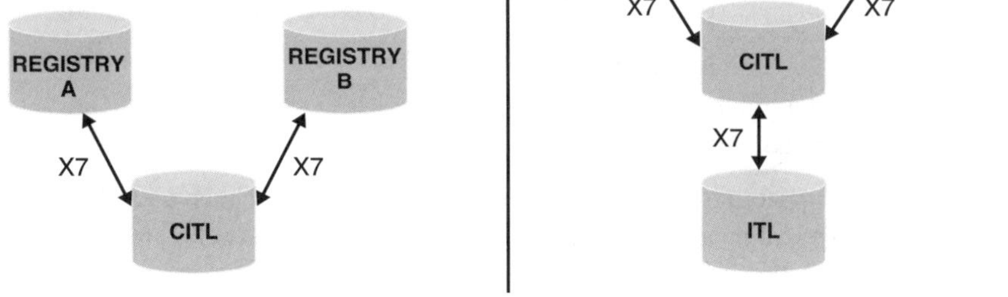

Figure 7.2. Option for Interface CITL/UNFCCC ITL Hubs[24]

[23] Decision 280/2004/EC.

[24] Env Finance Supp, November 2003, XIII.

integrity of the EU ETS and the UNFCCC ITL will perform a similar function for IET under the Kyoto Protocol. There will therefore be no need for individual links between each National Registry. This will have several benefits including lower cost, lower complexity, greater degree of security and ease of management of these links. The CITL will be linked to each National Registry within the EU ETS and to the UNFCCC ITL. A message arriving for a Member State National Registry from a non-Member State National Registry, or the CDM Registry, will be passed on by the UNFCCC ITL to the CITL, which will in turn re-route the message to the receiving Member State National Registry. Naturally messages between non-Member State National Registries will simply pass through the UNFCCC ITL and not transit via the CITL.

5.2 Internal Architecture of National Registries

Each National Registry will be required to contain at all times the following types and number of accounts:

- at least one national holding account;
- one operator holding for each installation;
- at least one person holding account for any legal person within that Member State who wishes to open such an account.

Each National Registry will also be required to contain 'period specific' national retirement accounts and national cancellation accounts (which from 2008 will be the same accounts required pursuant to Decision 19/CP.7 discussed above).

Registry Administrators[25] must assign a unique account identification code to each account in its National Registry in accordance with the requirements of the proposed Registries Regulation. These will largely be consistent with the requirements established under the international rules discussed earlier in this chapter.

The National Registries will also be required to contain the following tables:

- one verified emissions table;
- one surrendered allowances table; and
- one compliance status table.

The CITL shall contain one national allocation plan table for each Member State.

The purposes and functions of these various accounts and tables are elaborated upon throughout this chapter.

[25] Under the Registries Regulation the Registry Administrator means the competent authority or person, designated by the Member State, that operates and maintains the registry in accordance with the requirements of the Regulation.

5.3 Communications and Data Exchange Standards between National Registries

For the EU ETS Registries System to function as a coherent whole each National Registry must be capable of exchanging data via either or both the CITL and the UNFCCC ITL (and in due course both the CDM Registry and the National Registries of non-EU Annex B Parties).

There are of course two key elements to achieve such a capability. The first being physical links between the various National Registries and both the CITL and UNFCCC ITL and the second being standardized communication and data exchange standards which will allow each National Registry to read and understand data sent through to it from any other part of the EU ETS Registries System.

Article 6(1) of the proposed Registries Regulation requires the Central Administrator[26] to establish communication links between each National Registry and the CITL as soon as the relevant National Registry has completed certain testing and initialization procedures. Article 7 also requires the European Commission to put in place a communication link between the CITL and the UNFCCC ITL promptly after the UNFCCC ITL has been established.

Article 7 also requires all communications in the Registries System to be made in accordance with the provisions of the Regulation. Article 6(3) requires that each Registry shall, when carrying out the processes prescribed under the Regulation, follow the format and complete sequence for message exchanges for that type of process as set out in the functional and technical specifications for data exchange standards for registry systems under the Kyoto Protocol, as elaborated by the Secretariat to the UNFCCC.

5.4 Registry Processes

Article 3 requires that each Registry be capable of carrying out the types of processes concerning EU Allowances or Kyoto Units as set out in Annexes VIII, X, and XI. These include the issuance and creation of EU Allowances, surrender and cancellation of EU Allowances, retirement and cancellation of Kyoto Units, and transfer of such units.

Annex VII prescribes the information which will need to be put into the National Registry to initiate any particular process, and any such process will have a unique identification code in accordance with the criteria set out in Annex VI.

Article 3 in effect establishes the requirements to affect each type of process. These requirements include the information required to initiate a particular type of process;[27] the unique process identification number;[28] the automated checks

[26] 'Central Administrator' is defined under the Registries Regulation as meaning 'the person designated by the Commission pursuant to Article 20 of Directive 2003/87/EC to operate and maintain the Community independent transaction log.'

[27] Annex VII.

[28] Annex VI.

performed by a National Registry before continuing a process;[29] and that the message exchange sequence where each type of process must follow the functional and technical specifications for data exchange standards for registry systems provided for under the Kyoto Protocol.

Before the link between the CITL and the UNFCCC ITL is established, all processes except for external transfers will be considered complete and irreversible when:

- the CITL informs the initiating National Registry that it has not found any discrepancies in the proposal sent by the initiating National Registry; and
- the initiating National Registry has sent confirmation to the CITL that it has updated its records in accordance with the proposal.

For the process of external transfers (that is, transfers of units out of one National Registry to an account in another National Registry) of EU Allowances or Kyoto Units the procedure for confirming the finality is the same as described above except that it is the acquiring National Registry which is informed by the CITL that there are no discrepancies, and it is the acquiring National Registry which must inform the CITL that it has updated its records in accordance with the proposal of the initiating National Registry.

After the link between the CITL and UNFCCC ITL is established, all processes shall be considered final when confirmations are sent and received by both the CITL and UNFCCC ITL.

5.5 Opening Accounts

Within the EU ETS Registries System each operator of an obligated installation (Operator) will automatically have an operator holding account opened for it in its National Registry in respect of the greenhouse gas emissions permit ('GHG Permit') issued to it.[30] Its allocation of EU Allowances for that installation will be deposited into this account.

Importantly the EU ETS Directive provides that any legal person within the EU may also hold or transfer EU Allowances.[31] Therefore any person or company may open a Person Holding Account within any National Registry. They may then use such an account to hold, receive or transfer EU Allowances. Whether or not they may also hold, receive or transfer other Kyoto Units in these accounts will be for each Member State to decide. The proposed Linking Directive[32] will de facto

[29] Arts 28 and 29.

[30] Art 15 of the draft Registries Regulation.

[31] Art 12 of the EU Directive provides that any legal person may transfer EU Allowances and Art 19 provides that any legal person may hold allowances.

[32] The full title of the Linking Directive is the 'Directive amending Directive 2003/87/EC establishing a scheme for greenhouse gas emission allowance trading within the Community, in respect of the Kyoto Protocol's project mechanisms'. At the date of writing the Linking Directive has been adopted by the European Parliament and will enter into force when adopted by the European Council and published in the *Official Journal*.

authorize operators to hold CERs and ERUs but no one else. It is to be hoped that for the sake of stimulating the international market in Kyoto Units that EU Member States will be quick to grant such authorization in respect of all Kyoto Units to all persons. Such a step would be a simple, cheap and effective way to give the fledgling Kyoto carbon markets much-needed certainty by making it clear that entities have rights to directly hold and trade Kyoto Units regardless of whether those entities have an emission limitation or reduction obligation.

As discussed above, the EU ETS Directive together with the proposed Registries Regulation (when it enters into force) will together constitute the authorization with respect to EU Allowances and to a limited extent the associated AAUs (see section below on the allocation and issue of EU Allowances) but not other Kyoto Units. However, each Member State in Annex B remains ultimately responsible for both transaction integrity and that the participation by legal entities is consistent with the Kyoto Protocol rules applying to the Annex B Parties themselves. This is made clear in Article 19(3) of the EU ETS Directive, which states that the EU Registries System must ensure that transfers are consistent with the Kyoto Protocol rules. This together with the need to balance transfers of Kyoto Units between Annex B Parties with the corresponding transfers of domestic or regional units suggests that the most logical approach to the design of the EU ETS Registries System is a single National Registry in each Member State that incorporates both the Kyoto Protocol functions and the EU ETS functions.

Account holders in each National Registry will be able to access the accounts within the National Registries through a secured website in much the same way as people routinely access their bank accounts through internet banking. An important question is whether these account holders have the same property rights over their EU Allowances and possibly other Kyoto Units held in their accounts within the National Registries as they do over the money held in their internet-accessible bank accounts.

5.6 The Allocation Process and Creation of EU Allowances

Articles 15 and 16 of the proposed Registries Regulation provide for the allocation and issue of EU Allowances for the 2005–7 and 2008–12 periods respectively. In both cases the proposed Registries Regulation only permits the allocation and issue of EU Allowances in accordance with a National Allocation Plan (NAP), which has not been rejected by the Commission under Article 9(3) of the EU ETS Directive. The procedure to then be followed where no such rejection has taken place is for the Commission to instruct the Central Administrator to enter the NAP table for the relevant Member State into the CITL. Thereafter for the 2005–7 period the Registry Administrator for the relevant National Registry may only then (and by 28 February 2005) issue the total quantity of EU Allowances as set out in the NAP table into the national holding account. The 2005–7 EU Allowances are not backed by AAUs and in effect are created at this point from 'thin air'.

Therefore it is clear from this process that for the 2005–7 period, an EU Allowance will be an electronic unit (numbered similarly to an AAU) clearly identifiable as an EU Allowance by means of an EU Allowance identifier indicating that it may be used by an Operator to satisfy its compliance obligations under the EU ETS.

For the 2008–12 period and all subsequent five-year periods the process is slightly different. By 28 February 2008 (and the same date within the first year of all subsequent periods) the Registry Administrator shall issue the total quantity of EU Allowances as set out in the NAP table into the national holding account by *converting* an equal quantity of AAUs held in the national holding account to EU Allowances.

Therefore for the 2008–12 period and each subsequent five-year period, an EU Allowance will be an AAU that has been converted into an EU Allowance by the addition of the EU Allowance identifier to the unique unit identification number of the AAU (see Figure 7.3 below). Under the Kyoto Protocol and public international law, such Allowances will still be regarded as AAUs. However, according to the European Commission and Member State laws, they will be Allowances and can only be held and transferred by account holders in accordance with the EU ETS Directive and the proposed Registries Regulation.

On 28 February 2005 and each year thereafter the registry administrator shall transfer the number of EU Allowances specified for each installation in the relevant NAP to the relevant operator holding account.

The Registry Administrator may at any time transfer a number of EU Allowances to the operator holding account of a 'new entrant'[33] if instructed to do so by the competent authority.

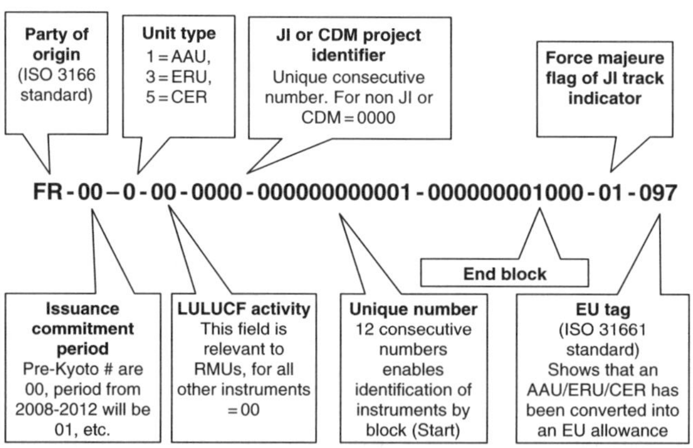

Figure 7.3. Proposed Numbering of EU Allowances[34]

[33] Art 11(3) of the EU ETS Directive states that when deciding upon allocation, Member States shall take into account the need to provide access to allowances for new entrants. For instance, the UK NAP provides for 7.7 per cent of the total Phase 1 allocation to be set aside for distribution for free to installations that commence operations after 31 December 2003 but before the end of 2007.

[34] Env Finance Supp, November 2003, XIII.

In accordance with Article 29 of the EU ETS Directive, *force majeure* EU Allowances will only be issued at the direction of the competent authority of a Member State on the instructions of the Commission during the 2005–7 period and will only be valid for compliance purposes under the EU ETS during that period. They will be entered directly into the section of the surrendered allowance table designated for the relevant installation.

5.7 The Legal Nature and Characteristics of EU Allowances

Embedding the EU Allowances within AAUs in the way described in effect makes it possible for there to be a direct link between each Annex B Party's assigned amount under the Kyoto Protocol and the EU Allowances issued pursuant to the EU ETS. Such a link should ensure a seamless integration between the accounting of units at both international and the domestic/regional level particularly in relation to determining compliance with the Annex B Party's emission limitation or reduction commitment under the Kyoto Protocol.

What this would mean is that such EU Allowances would have different legal characteristics or legal nature under international law than they would under EU law (and hence EU Member States' national laws). Under international law, when an EU Allowance is viewed by an Annex B Party (for example, Spain), EU allowances would be AAUs, which could be used for compliance by that Party towards its international obligations under the Kyoto Protocol. Of course, the rights of such a Party to take EU Allowances for its own purposes would be subject to the property rights of the entity which holds those EU Allowances in its account within that Annex B Party's Registry.

When those same EU Allowances are viewed by a non-sovereign entity, the EU Allowances would be Allowances, which could be used for compliance within the EU ETS or transferred freely to any other account within the EU ETS Registries System.

The interesting point is that an EU Allowance can have these different legal characteristics at the same time at two different legal planes. Of course, in many respects the above is a simplification to illustrate this point. These two sets of legal characteristics are not entirely exclusive of each other. The Annex B Party, which is also a Member State, also recognizes the tagged AAU as an EU Allowance, which it must recognize for satisfying an Operator's compliance obligations under the EU ETS. At the same time, an entity may find an Annex B Party to buy EU Allowances due to the AAU nature of the EU Allowances.

5.8 Project Credits

Once in force, the Linking Directive will amend the EU ETS Directive to allow Operators to use CERs (from 2005) and ERUs (from 2008) to satisfy their compliance obligations. Behind the scenes, a CER/ERU is exchanged for an EU Allowance, which is then automatically surrendered. Accordingly, the proposed Registries Regulation includes provisions in Article 22(1)*bis* for the use of CERs and ERUs by an Operator.

The remaining complicating factor is the issue of supplementarity. The Linking Directive requires each Member State to both set a limit at the Member State level and also at the individual installation level on the number of CERs and/or ERUs which can be used for compliance. The fact that such a limit is to be set will mean that each National Registry and the CITL will need to keep track of the total number of such project credits, which are used for compliance. This may result in a need for a project conversion table within each National Registry or perhaps in the CITL. In any event, there is no such limit for the 2005–7 period.

5.9 Transfers

For those entities participating in the EU ETS and those who wish to trade Kyoto Units, it will be critical to them that they can freely transfer EU Allowances and Kyoto Units both between accounts within the same National Registry and between accounts in different National Registries within the EU ETS with minimum (if any) regulatory oversight or approval required before such transfers can take place. It is also a key issue identified in case law on whether or not such rights constitute a property right or not.[35]

So it is an important question to ask whether or not the proposed Registries Regulation provides a clear legal right to make transfers, which cannot be interfered with by Governments or the Registry Administrators. Article 19(1) of the proposed Registries Regulation provides as follows:

> The registry administrator shall, in accordance with and subject to the provisions of this Regulation, carry out any transfer within its registry requested by an account holder in accordance with the internal transfer process as set out in Annex VIII and carry out any transfer between registries requested by an account holder in accordance with the external transfer process as set out in the functional and technical specifications for data exchange standards for registry systems under the Kyoto Protocol, elaborated by the Secretariat to the UNFCCC.

[35] In the UK, the decision of the Court of Appeal in *Celtic Extraction Ltd and Bluestone Chemicals v Environment Agency* [2001] Ch. 475 is helpful in providing some criteria which can be used to determine whether or not such rights are property interests. In that case the Court was called upon to consider whether or not a waste management licence (in simple terms a right to deposit waste material into a hole in the ground) was property for the purposes of s 436 of the Insolvency Act 1986. The Court developed three tests which need to be satisfied if such a permit can be considered to be property:

1. there must be a statutory framework conferring an entitlement on one who satisfies certain conditions even though there is some element of discretion exercisable within the framework;
2. the permit must be transferable;
3. the exemption or licence will have value.

On its face the use of the word 'shall' imposes a requirement on the Registry Administrator to ensure that the transfer proceeds as requested by an account holder. In practice this should be an automatic process where the account holder accesses its account online and requests a transfer to a specified account either in the same National Registry or another National Registry within the EU ETS Registries system.

Through the careful choice of its words, Article 49 makes it also clear that individual Member States and Registry Administrators should not be able to impose any additional checks or requirements over and above those required by the EU ETS Directive and the proposed Registries Regulation itself, for example, regulatory approval before a transfer can be made or submission of certain specified documents. Such additional requirements would simply be contrary to a Registry Administrator's obligation under Article 49 to carry out the transfer.

5.10 Eligibility and the Commitment Period Reserve

To ensure that transactions within the EU ETS are consistent with the Kyoto Protocol, Article 50 of the proposed Registries Regulation imposes two requirements. Member States must instruct their Registry Administrator *not* to initiate certain transactions if that Member State does not meet the eligibility requirements of the Kyoto Protocol. This may be particularly significant for Cyprus and Malta as they are not (yet) Annex B Parties. However, it is not limited to them and any Member State could, for one reason or another, find that it no longer meets the eligibility criteria. The second requirement is imposed on the Commission, which must notify a Member State when that Member State is approaching a breach of its commitment period reserve.[36] Although not re-stated in the proposed Registries Regulation, Annex B Parties are prohibited from making transfers (or allowing transfers to occur), which would breach their commitment period reserve.

5.11 The Trading and Compliance Cycle

Once an installation's annual emissions report has been verified as satisfactory, the verifier must enter or approve the entry of the annual verified emissions figure into the relevant section of the verified emissions table for that installation for that particular year. According to Article 15 of the EU ETS Directive, this must be done by 31 March each year in order to avoid a block being placed on further transfers of EU Allowances from the relevant operator holding account.

Article 52 of the proposed Registries Regulation provides that an operator shall surrender allowances for an installation by requesting the Registry Administrator

[36] The commitment period reserve is the share of EU Allowances which must be held in a party's registry to reduce the risk of overselling of EU Allowances and consequent non-compliance at the end of the commitment period.

to transfer a specified number of EU Allowances (or CERs or ERUs) from the relevant operator holding account into the national holding account of that National Registry. The Registry Administrator must also enter the number of transferred EU Allowances (or CERs or ERUs) into the section of the surrender allowance table designated for that installation for that year. *Force majeure* allowances will automatically be surrendered when issued in accordance with Article 43 of the proposed Registries Regulation.

Each time an entry is made in the surrendered allowance table or the verified emissions table, the Registry Administrator must calculate the sum of allowances surrendered minus the sum of all verified emissions for that installation. This figure must then be entered into the section of the compliance status table designated for that installation.

From 2008, this calculation shall include the addition of a 'correction factor'. The correction factor shall be zero if the 2007 figure was greater than zero, but shall remain as the 2007 figure if the 2007 is less than or equal to zero. This will ensure that the client's status of each operator is carried through from the 2005–7 period into the 2008–12 period.

On 1 May each year from 2006 each Registry Administrator must notify the compliance status table to the competent authority for that Member State. To ensure that accurate compliance measures are followed in the event that no verified emissions have been entered into the verified emissions table for an installation, the competent authority may estimate a 'substitute emission figure' for the purposes of calculating a figure for the compliance status table. Using the information contained in the compliance status table, the competent authority is able to calculate the relevant excess emissions penalty to be applied to each non-compliant Operator.

Units surrendered during the 2005–7 period are cancelled in the following ways:

- CERs are transferred from the national holding account into the national cancellation account for the 2008–12 period;
- EU allowances and *force majeure* allowances are transferred from the national holding account to the national retirement account for the 2005–7 period.

For the 2008–12 period and each subsequent five-year period, units surrendered are cancelled in the following ways:

- Surrendered EU Allowances are converted back into AAUs by the removal of the EU Allowance identifier from the unique serial identification code of each AAU.
- A number of Kyoto Units equal to the total number of EU Allowances, CERs and ERUs surrendered are transferred from the national holding account to the national retirement account for the relevant period.

In this way it is extremely easy for Member States that are also Annex B Parties to reattribute part compliance with their obligations under the Kyoto Protocol against the verified annual emissions from their installations caught by the EU ETS.

5.12 Banking through the Cancellation and Borrowing and Replacement Procedure

'Banking' is the term used to describe the right to carry over EU Allowances or units over from one compliance year or period to the next for use in that subsequent year or period. Within the context of the EU ETS when one talks about 'banking' one is talking about the carry over of units from one period to the next. This is because EU Allowances are not referenced to a year or vintage but instead are referenced to a period, that is, 2005–7, 2008–12, 2013–17, and so on. EU Allowances issued within a period can be used for compliance against any year within that period, even the year before they were issued. This is possible because EU Allowances will be issued on 28 February each year, which is prior to the requirement to surrender EU Allowances against emissions for the previous year on 30 April.[37] This is often referred to as 'borrowing' forward.

However, whilst such borrowing of EU Allowances from a following year's allocation to the next *within* a period is permitted by the EU ETS Directive, this is not the case from one period to the next. This is because the EU ETS Directive makes it clear that only EU Allowances issued within a particular period can be used for compliance within that period.

How does 'banking' work? Article 13(2) of the EU ETS Directive together with Article 60 and Annex IX of the proposed Registries Regulation essentially creates a cancellation and replacement procedure, which, for all practical purposes, amounts to banking.

Basically, the Registry Administrator will on 1 May in the first year of a period, cancel all EU Allowances held in its National Registry which were issued during the previous period and which have not been surrendered or cancelled. It shall then re-issue an equal number of EU Allowances for the current period. However, there are some differences in how this will work between the 2005–7 period and the 2008–12 period, and thereafter between all subsequent periods. One key issue is that for the 2005–7 period EU Allowances are not based on AAUs (they are merely numbered similarly to AAUs) whereas it is at the current time proposed that for the 2008–12 period and all future periods the EU Allowances will be an AAU with an EU tag. This makes the cancellation process for 2005–7 EU Allowances slightly different to that for EU Allowances from all subsequent periods. This may of course need to be re-visited if the Kyoto Protocol fails to come into force.

[37] See Figure 7.3 above for the proposed numbering of an EU Allowances.

As between the 2005–7 period and the 2008–12 period it is an option for Member States as to whether or not they will allow 'banking'. Therefore the Registry Administrator will only issue the replacement EU Allowances for the 2008–12 period if instructed to do so by the competent authority. At the present time it looks like most, if not all, Member States are not going to permit such banking, so the procedure for banking between the first two periods may well be academic.

For all subsequent periods Member States must allow such banking, that is, they must issue new EU Allowances on a 'one for one' basis to the holders of the EU Allowances from the previous period which have been cancelled under this procedure.

5.13 Availability and Reliability of Registries

Article 68 of the proposed Registries Regulation sets out the requirements of which the Central Administrator and each Registry Administrator has to ensure reliability, and continued operation of the CITL and each National Registry. Article 68(a) provides that the Central Administrator and each Registry Administrator must take all reasonable steps to ensure that the registry is available for access by account holders twenty-four hours a day, seven days a week, and that adequate procedures for safeguarding data are in place.

However the Central Administrator and Registry Administrators are allowed to suspend access to the CITL or National Registry if there is a breach of security of the CITL or National Registry, which threatens the integrity of either the CITL, National Registry or the EU ETS National Registries system; and the back-up facilities are similarly affected. In any event, such prior notice as is reasonably practicable must be given of any such suspension.

5.14 Liability of the CITL and National Registries for Errors and Mistakes

The National Registries and the CITL themselves do not have legal personalities and cannot therefore be sued. Instead liability, to the extent that it is not successfully disclaimed, will lie between the Registry Administrator (or the Central Administrator) and the relevant competent authority (or the Commission).

The proposed Registries Regulation says very little about the liability or otherwise of the Central Administrator or Registry Administrators for errors or mistakes of the CITL or individual National Registries. Indeed, Article 288 of the EC Treaty prohibits the Community and its institutions from excluding liability for non-contractual liability. As a result, the Regulation does not contain a 'no liability' article with respect to the Central Administrator or CITL. The Commission probably also felt that the delegated powers they had under Article 19 of the EU ETS Directive to legislate on emissions trading registries did not extend to interfering with the laws on liability in Member States and so did not seek to cover this directly in the Regulation.

Therefore the proposed Registries Regulation does not either exclude such liability or directly provide for it. Ultimately such questions may have to be resolved based on existing principles of liability under the relevant laws of each Member State. However, Articles 15(4) and 19(4) allow the Registry Administrator to require operators in respect of operator holding accounts and all other persons in respect of Person Holding Accounts to agree to comply with reasonable terms and conditions addressing the issues set out in Annex V to the proposed Registries Regulation. One of the items listed in Annex V is liability. The list of issues which theses terms and conditions may cover also includes: the account holders' obligations; the Registry Administrator's obligations; and warranties and indemnities. It is highly likely that in any dispute between an account holder who has suffered a loss as a result of a mistake or error by a Registry Administrator and that Registry Administrator, the respective duties and obligation set out in these terms and conditions will play an important part in determining questions of ultimate responsibility and liability.

Of course the nature and extent of any liability that may attach to a Registry Administrator or the Central Administrator will depend on the relationship between it and the person bringing a claim. For example, if a Member State appoints a company to be the Registry Administrator of its National Registry and the Commission similarly designates a company to be the Central Administrator of the CITL, liability of those companies may well be addressed in the relevant outsourcing contract.

The types of claims that could be brought against a Registry Administrator, Central Administrator, competent authority, Member State, or the Commission will depend on the law of the relevant Member State. Even if the Registry Administrator is a company, it will essentially be exercising a public function and although this may be regulated by contract, such contracts may in many Member States be subject to public administrative law principles. Such principles are dependent on the law in each Member State and can only be analysed in depth on a case-by-case basis.

In addition, whether an account holder, having accepted the terms and conditions based on the criteria contained in Annex V of the Regulation, can successfully bring a claim against a Registry Administrator will depend on the robustness and legality under the particular local law in each case of any disclaimer contained within these terms and conditions. As an example, under Italian civil law, contractual liability may be addressed in the terms and conditions of a contract with the general limits posed by article 1229 of the Italian Civil Code, which states:

(i) any agreement which in advance excludes or limits liability of the debtor for fraud or gross negligence; and

(ii) any agreement which in advance excludes or limits liability of the debtor in cases where the acts of the debtor or of its auxiliaries constitutes violation of a duty arising from public policy is null and void.

Additional and specific rules are established (Section XIV*bis* of the Italian Civil Code) for contractual liability in consumer contracts.

Another example is shown by a brief overview of what may be the case in the German system. In Germany the legal construction for the operation of the National Registry may in theory be based on public law, on civil law, or on a mix of public law and civil law. The above only applies to contracts governed by civil law. As far as public law applies, the question is: *how far a restriction of liabilities is possible, even though the account holders are somewhat forced to use the Registry*. This question may also be relevant for the determination of what may be meant by the term 'reasonable' with regard to section 307 Civil Code.

In Germany there are restrictions for contract terms and conditions of standardized contracts.[38] These rules are more restricting for consumer contracts than for contracts with companies. For example, the user of a standardized contract may not exclude liability for gross negligence in a contract with a consumer.[39] Standardized contracts are subject to a control of the content even if they are concluded between companies. Any terms and conditions of such contracts are invalid if they are unreasonable.[40]

Whether an account holder can successfully bring a claim directly against the Central Administrator will depend on the circumstances and the relevant Member State law. For example, in the UK tortious and public law claims may arise.

The relationship between a Registry Administrator and the Central Administrator is currently unclear. What if an account holder suffers a loss because of a mistake by the Central Administrator? Who does he have an action against? What if he recovers his losses from the Registry Administrator? Can the Registry Administrator recover its losses from the Central Administrator?

According to Article 8(3) of the proposed Registries Regulation, the Member State and the Community retain (via their respective Competent Authorities) ultimate responsibility and authority for the operation and maintenance of their Registries whether or not they have outsourced their running.

One issue which is of particular interest is the extent to which liability could arise where an account holder suffers a loss if a transfer is terminated because the Member State in whose Registry the EU Allowances are held is out of compliance with the eligibility criteria under the Kyoto Protocol. This potentially raises an issue under international law of 'creeping' or 'indirect' expropriation. Under international law, foreign nationals may claim compensation from a state where the host state has deprived the foreign investor of the value of its investment. This principle is derived from the right of the individual to hold property. In the EU this right has been enshrined in Article 1 of Protocol 1 to the European Convention on Human Rights. However, as the case law in this area demonstrates, the Courts have given a wide latitude to States to interfere with the rights of property holders in certain circumstances particularly where the State is acting in the public interest.

[38] §§. 305ff. Civil Code. [39] §. 309 Civil Code. [40] §. 307 Civil Code.

Whether the actions of a Member State in limiting the movement of EU Allowances due to that Member State's international obligations amount to expropriation of those rights is a complex question. International law will take into account the legitimacy of the aims of the Member State in imposing those limitations and the actual effect that this has on the foreign investors.

It will not be possible for those operating and maintaining the EU ETS Registries System to avoid all liability. However, liability can be minimized, particularly through:

(a) the terms and conditions governing participation in the National Registries;
(b) the legal nature of the relationships between Registry Managers, the Central Administrators, competent authorities, Member States and the Commission; and
(c) the effect of any current or future Member State law.

To fully understand the complexities of the liability regime in each jurisdiction it would be necessary to carry out a review of the relevant laws in each Member State, to see the final implementing regulations in each Member State and how these deal with liability and of course the precise terms and conditions which account holders must agree to.

5.15 Public Access to Information and Reporting

Public access to publicly available information contained in the EU ETS is clearly set out in Article 9 and Annex XVI of the proposed Registries Regulation.

Certain information is required to be updated on a weekly basis by the Central Administrator and Registry Administrators. This information relates to identifying the holder of each account within the EU ETS Registries System. In addition, for operator holding accounts, further information is updated on a weekly basis relating to identifying the installation that each account relates to and the number of EU Allowances allocated to that installation.

The following additional information for each operator holding account must be displayed one year after the relevant year:

- The verified emissions figure for that installation;
- The EU Allowances surrendered by that operator (these are identified by serial identification codes);
- A symbol identifying whether the installation is (or is not) in breach of its obligation under the EU ETS Directive to surrender EU Allowances equal to its annual verified emissions.

Registry Administrators are clearly instructed as to the information that must be made available on a National Registry's public website and how frequently that information must be generated. Following is a list of such publicly available information:

- Certain information for each project identifier against which the Member State has issued ERUs is displayed in the week after the issue has taken place;
- The total quantity of Kyoto Units held in each account is available five years after the relevant year;
- The total quantity of AAUs issued is available one year after the relevant year;
- The total quantity of ERUs issued is available one year after the relevant year;
- The total quantity of Kyoto Units acquired from other registries and the identity of the transferring accounts and registries is available five years after the relevant year;
- The total quantity of RMUs issued is available one year after the relevant year;
- The total quantity of Kyoto Units transferred to other registries and the identity of the acquiring accounts and registries is available five years after the relevant year;
- The total quantity of Kyoto Units cancelled on the basis of activities under Article 3(3) and (4) of the Kyoto Protocol is available one year after the relevant year;
- The total quantity of Kyoto Units cancelled following a determination of the compliance committee under the Kyoto Protocol that a Member State is not in compliance with its commitment under Article 3(1) of the Kyoto Protocol is available one year after the relevant year;
- The total quantity of Kyoto Units or EU Allowances cancelled is available one year after the relevant year;
- The total quantity of Kyoto Units and Allowances retired is available one year after the relevant year;
- The total quantity of ERUs, CERs, and AAUs carried over from the previous commitment period is available one year after the relevant year;
- The total quantity of EU Allowances from the previous commitment period that have been cancelled and replaced in a given year shall be available from 15 May of that year;
- The current holdings of Kyoto Units in each account are available five years after the relevant year;
- A list of persons authorized by the Member State to hold Kyoto Units must be updated weekly;
- The number of *force majeure* allowances and the installation, identification number of the relevant installation to which those allowances have been issued must be displayed in the week after those *force majeure* allowances have been issued;

- The number of EU Allowances allocated and transferred to each operator holding account pursuant to the decision taken under Article 11 of the EU ETS Directive together with the installation number of the installations to which the EU Allowances are transferred must be displayed in the week after such a transfer is made;
- The number of CERs and ERUs used by an operator for compliance purposes, together with the installation identification number *and* the serial identification code of each such CER and ERU must also be displayed. Note however that the proposed Registries Regulation is silent on when this particular type of information must be available.

Similarly, the Central Administrator must make all relevant information available to the public concerning each completed transaction via the CITL public website five years after the relevant year.

Concern has been raised by industry about the level of transparency of transfers of units within the EU ETS Registry System. Some fears were held that if information was made available in real time (or even perhaps at a maximum of one year delay), with some effort it would be possible to determine the trading strategies of various account holders within the EU ETS system. You may now consider with the delay placed on the availability of unit holding and transaction information how a satisfactory balance has been achieved between providing relevant publicly accessible information and maintaining confidentiality as appropriate in a way that is consistent with existing EU and international laws including the Aarhus Convention and the Kyoto Protocol.

Registry Administrators and the Central Administrator are prohibited from releasing additional information held in the National Registry or the CITL to the public.

6 Registries and Market Development Issues

As discussed earlier in this chapter, registry systems are not trading platforms and one of the main purposes of their design is to ensure compliance with emission limitation and reduction commitments. For example, it is recognized that the objective of the implementation of the Directive establishing the EU ETS is to promote reductions of GHG emissions in a cost-effective and economically efficient manner. That said, it is still necessary to consider, particularly given the creation of the EU ETS and the ability for legal entities to participate in emissions trading, a number of market development issues that are important for a liquid, transparent market in emissions trading to evolve.

In late 2003 a number of entities keen to promote the EU ETS trading market sent a collective memorandum to the UK Government's Department of Environment, Food and Rural Affairs (DEFRA) summarizing these issues. These issues are equally applicable to the development of other emissions trading markets under the Kyoto Protocol.

A central theme to these issues is the need for legal certainty in relation to an allowance (the product being traded) and how the trading of allowances fits within the existing legal regime.

In order to promote a trading market for the purposes of reducing GHG emissions, the key requirement is establishing absolute legal certainty in relation to the legal nature of an allowance and its existence over a period of time. What is an allowance as a matter of law? How does a person own it? How long does a person own it for? Can an allowance be revoked?

It naturally follows from the legal nature of an allowance that certainty is also required in relation to the taking of a security interest over an allowance. Further, the order of priority of interests in the enforcement of a security interest over an allowance must be made clear. Governments may also need to establish a system whereby security interests in allowances can be recorded and registered.

The accounting treatment of allowances is yet to be finalized by the International Accounting Standards Board. In addition, the taxation treatment of transactions and allowances at national level, and in particular for cross-border transactions, remains unclear. It is also not clear to what extent the trading of allowances falls within the regulation of a commodity or a security or a financial instrument under existing European or national laws.

Certainty should also be provided as to how allowances will be treated upon insolvency of a company. Will allowances be treated as assets of the company? Will a security interest over an allowance be valid and enforceable against an insolvency official?

At the time of writing, DEFRA was drafting guidance on these market development issues and we understand that other Member States and the Commission were being engaged to consider and discuss these issues and the possibility of harmonization in some areas.

7 Conclusions

As discussed in this chapter, an efficient and reliable registries system is critical to ensure compliance with emission limitation and reduction commitments. Without an appropriate accounting tool it would not be possible to reconcile each Operator's (or Annex B Party's) EU Allowance (or Kyoto Unit) holdings against its verified emissions for the relevant compliance period.

At the very same time, an efficient and reliable registries system is crucial if a liquid and transparent emissions trading market is going to develop and prosper. The integrated nature of the compliance and trading aspects of the Kyoto Protocol and the EU ETS cannot be ignored and the development of National Registries lies at the heart of determining the potential for success of Operators, Annex B Parties and, indeed, market players.

8

Track One JI and 'Greening of AAUs': How Could It Work?

*Jari Väyrynen and Franck Lecocq**

1 Introduction

Articles 6 and 17 of the Kyoto Protocol offer seemingly clear-cut options for countries with economies in transition (EITs). On the one hand, Article 6 Joint Implementation (JI) is a project-based mechanism by which Emission Reduction Units (ERUs) are transferred in exchange for emission reductions achieved through a project. On the other, trading under Article 17 offers countries the possibility to sell Assigned Amount Units (AAUs) on the market without any reference to an underlying project or activity, as long as they meet minimal requirements for trading.[1]

The frontier between the two universes, however, is more porous than it appears at first glance. First, the so-called Track One JI essentially allows the buyer and the seller to agree on the definition and validation of the additionality requirement for a particular project, without the two stage third-party validation procedures of the Clean Development Mechanism (CDM) or of Track Two JI.[2] Pushed to the limit, Track One JI is akin to AAU allowance trading with a project fig leaf.

At the other end of the spectrum, the fact that most EITs are likely to have emissions far below their targets set by the Kyoto Protocol, has prompted some potential buyers to seek 'greening schemes', whereby the proceeds from the sales of AAUs would be earmarked for environmentally friendly activities. The contract could then stipulate that the sale of AAUs would be effective only if these activities were implemented or, by some definition, generated a certain amount of emission reductions, no different than in a project-based transaction.

* Frank Lecocq is an economist with the Development Economics Research Group of the World Bank. Jari Väyrynen is an environmental specialist in the World Bank's Carbon Finance Business. The views expressed in this chapter are the views of the authors and do not necessarily represent the views of the World Bank. The authors would like to thank Robert Marschinski who provided research support for this chapter. The entire section 1 is based on World Bank Prototype Carbon Fund, *Track One Joint Implementation—Current Status of EIT Countries in Meeting the Eligibility Requirements*, 4 June 2003. Significant contributions to the chapter were made by Mr Jan Pretel, Head of Climate Change Department of the Hydrometerological Institute of the Czech Republic who was working with the Prototype Carbon Fund as a short-term consultant at the time.

[1] See ch 22 below.

[2] See chs 9 and 10 below on the CDM and ch 6 above on JI.

The objective of the present chapter is to explore the grey (and rich) area between 'pure' JI and 'pure' AAU trading. Section 2 is dedicated to Track One JI, and section 3 focuses on 'greening schemes'. Both sections are highly speculative since these hybrid instruments are only in their infancy, and since there is still considerable uncertainty over the fate of the Kyoto Protocol. Yet careful reading of the Marrakesh Accords,[3] as well as lessons learned from early project activities, provides some insights on the risks and potential benefits of using Track One JI and 'greening' schemes, and some indications as to how EITs can best make use of the palette of instruments at their disposal.

2 Track One JI

2.1 What is Track One JI and Who is Eligible to Use It?

The Marrakesh Accords define two different tracks for JI, depending on which entity verifies the ERUs:

- In Track One JI, verification is made by the host Party if it meets the eligibility requirements;[4]

OR

- In Track Two JI, verification is made through procedures under the Article 6 Supervisory Committee ('CDM like' procedures).[5]

Under Track One the host Party may verify that emission reductions (ERs) from a JI project are additional, no independent verification is required. Upon such verification, the host Party may issue the appropriate quantity of ERUs. In contrast to Track Two JI, participants in Track One JI project justify their own choice of baseline and monitoring methodology, and do not need the approval of the Article 6 Supervisory Committee for the methodology. Furthermore, under Track One JI, no independent determination (equivalent to validation in the CDM) is required. Instead, project approval, baseline setting and monitoring are governed by the 'national guidelines' of the host Party.[6] The key differences between Track One and Track Two procedures are summarized in Table 8.1.

However, the Annex I host Party can only start using Track One JI procedures if it meets the following eligibility requirements:

(a) It is a Party to the Kyoto Protocol;

(b) Its assigned amount has been calculated and recorded in accordance with relevant guidelines and decisions;

[3] Found in FCCC/CP/2001/13/Add.1-3.

[4] FCCC/CP/2001/13/Add.2, p 13, para 23.

[5] FCCC/CP/2001/13/Add.2, section E, pp 13–16.

[6] See ch 6 above for JI participation and eligibility requirements.

Table 8.1: Comparison between Track One and Track Two JI

	Track One	**Track Two**
Project design (baseline/ monitoring plan requirements)	National guidelines, taking into account Appendix B of the Marrakesh Accords	Article 6 Supervisory Committee (to be established once the Kyoto Protocol enters into force)
Checking project design against requirements and project approval	In accordance to national guidelines	Determination / validation by independent entity Verification by independent entity
Verification and issuance of ERUs	In accordance to national guidelines	Issuance by Article 6 Supervisory Committee

(c) It has in place a national system for the estimation of anthropogenic emissions by sources and anthropogenic removals by sinks of all greenhouse gases;

(d) It has in place a national registry;

(e) It has submitted the most recent required inventory;

(f) It submits the supplementary information (eg on sinks) on assigned amount and makes any adjustments and recalculations required.[7]

How likely is it that EITs will meet these requirements in the short term? At time of writing, the record is mixed. All European Union (EU) accession and candidate countries, plus Ukraine, had ratified the Kyoto Protocol by early 2004. (The Russian Federation has not ratified at time of writing.) It is unlikely that all the requirements for accounting AAUs will be met before 2005, even in the EU accession and candidate countries. National systems for estimation of anthropogenic greenhouse gases (GHGs) emissions by sources and removals by sinks are already in operation in all EU accession and candidate countries. National registry systems are not yet developed in any of the EITs, but the EU accession (and possibly also the candidate) countries should have their registries in operation from the beginning of the European Emissions Trading Scheme (EU ETS)[8] in 2005. It is hard to estimate when national registries might be in place in the rest of the EITs. Finally, most EITs are also slowly but surely meeting the rather complex requirements for full-time series of national GHG inventories and inventory systems for both future years and recalculations for past years. This task has been difficult to complete because of lack of resources.

[7] FCCC/CP/2001/13/Add.2, p 12, para 21.

[8] Directive 2003/87/EC of the European Parliament and of the Council of 13 October 2003 establishing a scheme for greenhouse gas emissions allowance trading within the Community and amending Council Directive 96/61/EC [2003] OJ L275/32.

From this assessment, it is clear that the EU accession process has helped the EU accession and candidate countries to intensify their efforts and to meet the eligibility requirements for Track One JI and Allowance Trading. Preparedness and national circumstances greatly improved in the last one or two years in this group of countries. It appears that the Czech Republic, Latvia, Poland, Slovakia, Hungary, Estonia, and Slovenia are best positioned to meet all the eligibility requirements by or during 2005. Bulgaria, Lithuania, Romania, and Croatia are slightly less advanced in their efforts, but should be able to meet the requirements in the 2006–7 time frame. Finally, it seems unlikely that the Russian Federation and Ukraine would meet all requirements in the very near future.

However, it may be that the EU ETS makes much of this a moot point, as it appears that the inclusion of the energy sector in the EU ETS will make JI less interesting for countries joining the EU. While projects covered by the EU ETS can still be implemented as JI as long as an EU Allowance[9] is cancelled from the operator's account for each transferred ERU,[10] the EU Allowance market may be a more lucrative option for many project sponsors.

For the installations not covered by the EU ETS (the so-called indirect projects) the final rules on how to handle the double counting issues are yet to be agreed.[11] What is clear is that such projects will have to be taken into account in the National Allocation Plans one way or the other.[12] One proposal is to create a reserve pool of Allowances to cover such JI projects. This would imply that an equivalent number of Allowances that would have been otherwise allocated (that is, in the absence of such reserve pool) are not allocated to the installations whose emission are indirectly reduced by such JI projects. However, since the total number of Allowances to be allocated in any given EU country's National Allocation Plan (NAP) is not strictly stipulated (although each NAP does have to be approved by the EU Commission), it is unclear whether this reserve pool system will be implemented.

In any case, the requirement that the *acquis communautaire* be taken into account in the NAPs (and in any JI project baseline) reduces the number of EU Allowances that will be allocated in the EU ETS. These factors mean that there may be more interest in AAU trades at the government level. Indeed, given the complexities associated with dealing with JI projects in the EU ETS context, some potential JI host countries may consider stopping JI project activities and focusing on Article 17 International Emission Trading or some form of 'greening' of AAUs instead.

[9] Allowances under the EU ETS are referred to as EU Allowances throughout the chapter in order to distinguish this form of allowance from AAUs.

[10] Proposal for a Directive of the European Parliament and of the Council amending Directive 2003/87/EC establishing a scheme for greenhouse gas emission allowance trading within the Community, in respect of the Kyoto Protocol's project mechanisms, hereinafter the 'Linking Directive'.

[11] Such a situation might occur, for example, if a significant renewable energy JI project that is not covered by the EU ETS, would displace coal generated electricity and thus reduce the activity level of a coal plant. This scenario would free up some EU Allowances from the coal plant for trading under the EU ETS as well as generate ERUs for the JI Project that could be imported to the EU trading system.

[12] See n 10 above, Linking Directive, para 8, and para 3.2 of the explanatory memorandum attached to the Linking Directive.

2.2 Possible Baseline and Monitoring Methodologies for Track One JI

The Marrakesh Accords state that under Track One JI, verification of emission reductions should be guided by national guidelines.[13] Yet the Accords also define principles for baseline setting and monitoring that even Track One JI must follow (note that the paragraph quoted does not refer to either Track One or Track Two JI):

28. A Party hosting an Article 6 project shall make publicly available, directly or through the secretariat, information on the project in accordance with the reporting guidelines set out in Appendix B below and the requirements contained in decision/CMP.1.

Appendix B referred to in the quote above lays out the basic requirements for baselines and monitoring plans. It is not clear from this paragraph, however, whether 'reporting guidelines' refers to the whole Appendix B or just to parts of it. In any case, Appendix B specifies that the baseline for JI projects shall be established on a project-specific basis and/or using a multi-project emission factor, in a transparent manner and taking into account relevant national and/or sectoral policies and circumstances.

Appendix B further states that project participants shall include a monitoring plan as part of the Project Design Document (PDD). The monitoring plan has to ensure transparent control procedure and quality assurance within the project boundary during the entire crediting period.

What do these requirements seem to imply for baseline setting under Track One JI?

The Marrakesh Accords do not introduce significant methodological differences between CDM projects and Track Two JI projects. Methodologies developed by various JI project developers so far appear to be generally consistent with the respective modalities and methodologies proposed and approved for CDM projects. Indeed, many of the methodological innovations in the CDM may well be applicable in JI and vice versa. On the other hand, it appears clear that the host Party's authority to 'validate' projects and verify emission reductions includes the authority to approve baseline and monitoring methodologies that may be significantly different from those methodologies that are approved for the CDM or may be approved for Track Two JI.

The methodological and regulatory latitude that the Marrakesh Accords delegate to host Parties for Track One JI should obviously take into account the host Party's priorities, legislation, and policies. One plausible assumption is that the host Party may wish to lighten the regulatory burden for JI projects as much as possible in order to maximize the project leveraging potential of carbon

[13] FCCC/CP/2001/13/Add.2, p 13, para 23.

finance and hence emission reductions. This desire may be constrained by two considerations:

- Host Parties with a relatively tight AAU reserve position may want to make sure that their Track One JI projects actually reduce the emissions they claim and thus will not interfere with the plans regarding the use of their (excess) AAUs. Such Parties may want to build an 'insurance' to the methodologies used that would guard against crediting too many ERs to projects;
- Host Parties with plenty of excess AAUs and no obligation or plan to allocate allowances under the EU Emissions Trading Scheme or other similar systems may want to promote Track One JI as a financing mechanism, which channels money directly to projects without passing through the government's budget. In this case, the methodology may also serve as an indirect instrument of allocating these resources.

Depending on such domestic conditions, as well as preferences of the buyer, host Parties may want to explore the following generic approaches to Track One JI methodologies:

- *Standard baselines and multi-project emission factors for technologies or sectors* set at a conservative level to limit as much as possible the risk of selling ERs/ERUs (backed by AAUs) that have not really occurred. Such standards could be set without much research into their accuracy, because the government (as opposed to the project sponsors) would not lose out if standards are too conservative: real ERs that were not credited to the project could still be sold by the government as AAUs. Too lax standard would naturally lead to the reverse situation, and possibly concerns of unfair State Aid under the EU rules.
- *Domestic version of internationally approved Track Two JI methods.* In this approach more attention would be given to project by project analysis and their independent evaluation. However, to reduce transaction costs (both time and money), all project development steps would be performed domestically by the private sector and/or by government agencies. The system would still be Track One JI as the host Party would verify or confirm the verification of ERs (as opposed to the Article 6 Supervisory Committee).
- *Negotiated baselines.* EIT Governments could negotiate baselines and a performance monitoring system with project participants. This method would allow more room to use Track One JI as an incentive mechanism for investments and to meet national priorities in the relevant sectors of the economy.
- *AAUs as subsidies/grants.* The government could set aside AAUs for distribution to projects and use this pool in lieu of (or to replace) other subsidy/grant schemes. The pool could be distributed either by using some criteria or through a tender process in which investment projects would compete with each other for AAUs by bidding for the lowest number needed to implement

the investment project. In this approach in particular, EU regulations on State Aid would have to be considered carefully.

While the cap-and-trade system established for Annex I countries allows some flexibility for baseline setting in JI, establishing counterfactual baselines in JI is complicated by the fact that JI governments have an incentive to move the entire economy towards a lower carbon intensity to free up more AAUs: the baseline tends to gradually disappear or is dynamic as the Government subjects all GHG emitting activities to increasingly stringent regulations in an effort to improve the country's emissions inventory. This is even more the case for EU accession countries, which are required to implement the EU ETS and various other EU Directives, and for countries that are likely to get close to their emission target by 2012. Thus, the baseline for JI projects is likely to get increasingly unclear, making CDM-like baseline methods more difficult to implement.[14]

2.3 Risks and Benefits of Moving to Track One JI

Track Two JI procedures are cumbersome and there is ample reason to try to move to Track One JI procedures as soon as possible. Relying on national guidelines for verification of ERs, issuance of ERUs and for other aspects of project development could greatly simplify the process and lead to significant reduction in transaction costs both in terms of time and money.

Viewed from a macro-level perspective, that is, the overall Kyoto Protocol framework, the risks of moving to Track One JI and host country verification of ERs are minimal. Annex I Parties operate under an emissions cap established by the Kyoto Protocol. It is therefore reasonable that the host country, which is eventually liable for meeting its commitments under the international regime, assumes the responsibility for managing its Assigned Amounts and verifies ERs under the streamlined version of JI.

However, at the micro/project level, there is still a need to maintain sufficient credibility that real and measurable ERs have been created.

Another issue that countries wishing to move to Track One JI need to consider is the need to maintain the Compliance Period Reserve included in the Marrakesh Accords.[15] For many EITs this is not a serious concern as they have significant surplus of AAUs. However, there are a few EITs (most notably Croatia, the Czech Republic, and Slovenia) that are likely to be fairly close to their Kyoto target by the end of 2012. For these countries, it is particularly important to manage their AAU assets carefully.

[14] Please note that countries that recently joined are bound by the EU Directive as amended by the Linking Directive, which specifies that the *acquis communautaire* should be taken into account in the establishment of baselines for project activities. See n 10 above, Linking Directive, para 9, and paras 3.2 and 4.1 of the explanatory memorandum attached to the Linking Directive. See also ch 29 below.

[15] FCCC/CP/2001/13/Add.2, p 54, para 6.

Obviously, these uncertainties and options available need to be reflected in the contractual arrangements. Any JI project agreement should provide for flexibility to move to Track One JI procedures. They can, for example, include a provision which makes it possible to waive the requirement of independent verification in the event that the host country is eligible for Track One JI. It is equally important that project agreements that have been or will be concluded before the EU Linking Directive[16] enters into force deal with the situation created by the new regulatory environment within the EU. Examples of such flexibility could include emission reduction purchase agreements that would allow for the possibility to deliver and accept EU Allowances instead of ERUs or somehow index the price paid for the ERUs to the prices in the EU Allowance market.

2.4 The Disappearing Line between Track One JI and Article 17 International Emissions Trading?

Both Track One JI and Article 17 International Emissions Trading operate in the context of the overall national cap set by the Assigned Amount and the host country needs to fulfil the same set of eligibility requirements in order to participate in either Track One JI or in Article 17 International Emissions Trading. The only additional requirements that a country which wishes to participate in Track One JI needs to fulfil, compared to Article 17 International Emissions Trading, are:

- Emission reductions generated by a project activity which reduces GHG emissions or enhances the removal of GHG by sinks;[17]
- Verification of (actual) emissions reductions by the host Party;[18]
- Issuance and transfer of ERUs instead of transfer of AAUs;[19]
- Reporting requirements for JI.[20]

In other words, under Track One JI a concrete project activity still needs to be implemented. In contrast, International Emissions Trading is linked to achievement of emission reductions for the economy as a whole.

However, the possibility to apply flexible approaches to determining which reductions in GHGs are verified as being additional, as well as the other far-reaching rights of the host country under Track One JI, may make it difficult to draw the line between Track One JI and Emission Trading under Article 17. The border blurs even further where the transfer of AAUs between two Parties under Article 17 is linked to concrete reduction activities on the ground.

[16] See n 10 above.

[17] See document FCCC/CP/2001/13/Add.2, p 13, paras 23 and 24.

[18] ibid.

[19] See Annex to Decision 19/CP.7 Modalities for the accounting of assigned amounts under Article 7, para 4, of the Kyoto Protocol generally, and in particular para 29 of the Annex.

[20] ibid, para 28 and Appendix B.

In these instances the countries need to take a decision on whether to follow the JI approach or whether they would implement certain project-related activities under a 'greening AAU' approach. Given the extra requirements of Track One JI and the limits that the EU ETS may set on for the supply of such credits from EU accession and candidate countries, some form of 'Greening of AAUs' approach under Article 17 seems an increasingly attractive option. This is discussed further in the following section.

There is, however, one important benefit of JI compared to Article 17 International Emissions Trading. The EU Linking Directive is likely to recognize ERUs as units directly convertible to EU Allowances. This means that private sector entities subject to liability in the EU ETS will likely find ERUs much more interesting than AAUs, no matter how much greening they undergo. The demand for ERUs may therefore exceed supply taking into account that EU ETS allocations are likely much stricter in the 2008–12 period—the period for which ERUs can be issued. This may also lead to higher unit prices for ERUs than AAUs.

3 Greening of AAUs

Aggregate GHG emissions from EITs are consistently projected to be below their Kyoto targets during the period 2008–12, although the exact amount of 'excess' AAUs, the so-called 'hot air', varies depending on the estimate (Table 8.2). Country projections, although less abundant, suggest that most EITs should have 'hot air', with the possible exception of Croatia, the Czech Republic, and Slovenia. The Russian Federation represents the bulk of the volume (from 39 per cent to 54 per cent), followed by Ukraine (25 per cent to 31 per cent) and Poland (5 per cent to 10 per cent).

From a legal point of view, all the AAUs are the same, regardless of the performance of the country relative to its target. If the conditions for trading under Article 17 of the Kyoto Protocol are met,[21] 'excess' as well as 'normal' AAUs can be traded. In fact, it is very difficult to distinguish between them *ex ante*, because trading will take place during the first commitment period, that is, before the emissions during that period are measured (emissions inventories are typically available with a one-to-two year delay).

From a political economy point of view, on the other hand, the presence of 'hot air' presents a serious challenge to the Kyoto Protocol.[22] On the one hand, buying what has been dubbed as 'paper tonnes' might not be palatable with public opinion. As a result, major stakeholders, governments and firms are under pressure not to purchase 'hot air'. But too high a price for carbon is equally unacceptable to governments and firms, especially since the US will in all probability remain out of the Kyoto Protocol. In this context, and given the uncertainties which surround the future price of carbon, foreclosing the possibility to tap into the 'hot air' reserve

[21] See section 1 above. See also ch 22 below.

[22] T Koch and A Michaelowa, ' "Hot Air" Reduction through non-Quantifiable Measures and Early JI' 5(2) *Joint Implementation Quarterly* 9–10 (1999).

Table 8.2: Three Emissions Projections for Economies in Transition

Country	Base Year Emissions		Kyoto Target (Relative to Base Year)	2001 Emissions		Projected 2010 AAUs Surplus		
	Year	($MtCO_2e$)		($MtCO_2e$)	(Relative to Base Year)	(MtCO2e), based on National Communications**	DoE*** (2003)	IEA**** (2002)
Bulgaria	1988	144.4	−8%	65.7	−54%	25.3	63.5	54.1
Croatia	1990	31.6	−5%	26.9	−15%	−4.3	1.7	−2.1
Czech Republic	1990	192.0	−8%	148.1	−23%	46.2	20.5	−0.7
Estonia	1990	43.5	−8%	19.4	−55%	21.4	15.5	16.8
Hungary	1985–7	112.5	−6%	79.0	−30%	19.2	22.5	11.2
Latvia	1990	29.1	−8%	11.5	−61%	13.2	12.3	13.0
Lithuania	1990	51.5	−8%	18.8*	−64%	—	23.7	24.9
Poland	1988	564.4	−6%	382.8	−32%	125.5	126.9	72.0
Romania	1989	264.3	−8%	148.2	−44%	—	86.9	65.6
Russian Federation	1990	3,040.1	0%	1,909.0*	−37%	656.0	631.9	753.1
Slovakia	1990	72.2	−8%	50.1	−31%	12.6	13.5	6.4
Slovenia	1986	20.1	−8%	19.3*	−4%	−1.9	−1.8	−4.6
Ukraine	1990	919.2	0%	453.2*	−51%	—	347.5	376.3
TOTAL						**1,337.4****	**1,364.7*****	**1,385.8******

Method: Projected emissions are computed by applying the emissions growth rate from 2001 to 2010 given in the selected projection to the 2001 emissions level (all gases included) as reported to the UNFCCC. **The resulting AAUs surpluses differ from those in the sources mentioned above because all gases are included, and because we use the latest emissions data from the UNFCCC.**

Sources: Base year data, latest emissions inventory and national communications are from UNFCCC website (in Common Reporting Format) as of 20 February 2004. Other projections are from International Energy Agency, 2002. World Energy Outlook 2002 (IEA, Paris and U.S. Department of Energy, 2003); International Energy Outlook 2003 (DoE, Washington, DC.).

* As of 22 February 2004, latest emissions inventory was 1999 for Russian Federation, 1998 for Lithuania and Ukraine, and 1996 for Slovenia. To estimate 2001 emissions, the variation rate of fossil-fuel CO_2 emissions between latest inventory and 2001 (IEA, 2003). CO_2 Emissions from Fossil-Fuel Combustion 1971–2001 (IEA, Paris) was applied to total GHG emissions in latest inventory.

** No projection was available in the National Communications for Lithuania, Romania and Ukraine. Average from International Energy Agency (IEA), 2002. World Energy Outlook 2002 (IEA, Paris and US Department of Energy, 2003). International Energy Outlook 2003 (DoE, Washington, DC.) were used for these countries when computing the total 1,337.4 $MtCO_2e$ figure.

*** We use reference case US Department of Energy, 2003. International Energy Outlook 2003 (DoE, Washington, DC.) projections. DoE provides projections only for energy-related CO_2. We applied the projected growth rates from 2001 to 2010 to the 2001 data for all gases. Also, DoE provides projections only for two regions: Former Soviet Union (Estonia, Latvia, Lithuania, Russian Federation and Ukraine) and Eastern Europe. Country-Level projections are computed by applying the growth rate of the corresponding region to their 2001 emissions.

**** We use reference case International Energy Agency, 2002. World Energy Outlook 2002 (IEA, Paris) projections. IEA provides projections only for fossil-fuel CO_2. We applied the annual projected growth rates from 2000 to 2010 to the 2001 data for all gases. Also, IEA provides projections only for two regions: Russian Federation and other Transition Economies. Country-Level projections are computed by applying the growth rate of the corresponding region to their 2001 emissions.

is also a difficult decision to make. On the seller side, the existence of 'excess' AAUs also creates uncertainty because the demand for it is unclear. This raises questions about the potential gains that could be achieved through the Kyoto Protocol, and thus weakens support for ratification.

The idea of 'greening' AAUs, that is to somehow tie part of the revenues from the sale of 'excess' AAUs to activities beneficial to the environment, may constitute a good way out. By ensuring the environmental integrity of the AAUs, it provides political ground for the buyer, while increasing demand for sellers, and providing additional certainty to the market. In fact, the Government of Canada and some EU Member States have announced that they would not purchase 'hot air' unless it is somehow 'greened'.[23] The issue is of course for buyer and seller to agree for a mutually acceptable definition of 'greening', and to create an implementation scheme, which is both credible and transparent.

3.1 Possible Definitions of 'Greening'

'Greening AAUs' schemes, also known as Green Investment Schemes (GIS), are typically defined as mechanisms through which revenues from the sales of AAUs are earmarked for environmentally related purposes in seller countries.[24] This definition is very broad, and still leaves a wide range of options open depending on the activities that are deemed eligible, and on when exactly an AAU is considered 'greened'. Although there can be a priori as many answers to these questions as there are pairs of interested buyers and sellers, two main forms of 'greening' can be distinguished.

3.1.1 Hard greening

The first—which we call 'hard' greening—considers that an AAU is greened when the activities financed through the proceeds of the sale have generated one unit of emission reductions compared with what would have happened otherwise. There can be many variations around this theme (one might require more or less emission reductions per AAU sold), but the central idea remains that the yardstick for measuring greening is the amount of emission reductions generated.

Under this definition, 'greening AAUs' is akin to Track One JI.[25] But unlike traditional JI, it allows for the crediting of emission reductions generated out of

[23] Canada, *Climate Change Plan* (2002) states that 'One issue that has received considerable attention is the possibility of buying surplus permits from Russia and eastern European countries, where economic decline over the past decade has lowered their emissions below their Kyoto target. The Government of Canada has already indicated that it would buy such permits only if the selling country agrees to invest the proceeds from the sale in projects and activities that contribute to emissions reductions. This is called "greening" surplus permits and ensures that these purchases result in real environmental benefits that contribute to addressing climate change' (p 43).

[24] See W Blyth, *Green Investment Schemes: Options and Issues* (IEA, Paris, 20 November 2003), and K Tangen, et al., *A Russian Green Investment Scheme: Securing Environmental Benefits from International Emissions Trading* (Climate Strategies, 2002) <http://www.climate-strategies.org/gisfinalreport.pdf>.

[25] If a country meets the requirements for trading AAUs under Art 17, it also meets the requirements for Track One JI.

the first commitment period, either before 2008 ('early crediting'), or after 2012 ('late crediting').

- 'Early crediting' is already being used by the World Bank Prototype Carbon Fund and the Dutch ERUPT programme along JI projects: it consists of rewarding emission reductions realized before 2008 with AAUs. The amount of excess AAUs decreases (Figures 8.1 and 8.2), but in exchange for the early financing of a project. Monitoring and verification techniques can be applied to the project so that the AAUs are transferred only after the emission reductions are verified.
- With 'late crediting', AAUs would be delivered for emission reductions that the project is expected to generate after 2012. In this case, obviously, the emission reductions cannot be verified before the AAUs are transferred. However, some form of mixed monitoring system can be imagined. For example, effective emission reductions could be measured up to 2012, and the fact that the project is operational at that date, and deemed likely to keep generating emission reductions in the future, could be considered sufficient to trigger payment for the AAUs corresponding to the emission reductions expected beyond 2012. In this case again, the supply of 'excess' allowances for the first commitment period diminishes. However, the baseline for the second commitment period also diminishes (see section 3.2.5).

As Figures 8.1–8.3 illustrate, 'hard' greening can both be a complement to a JI project (that is, JI for 2008–12 plus early and late crediting in the form of AAUs), or a stand-alone activity (in which case the emission reductions realized between 2008 and 2012 are credited with AAUs under the greening scheme and not with ERUs).

Impact of Early/Late Crediting on National Emissions Balance

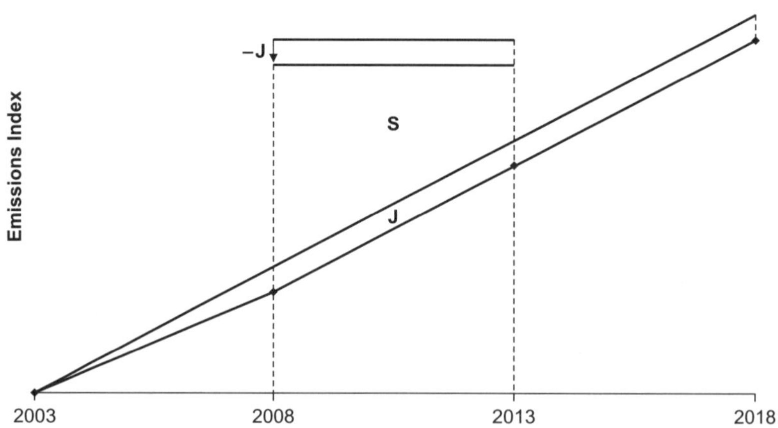

Figure 8.1. Standard JI Project

The buyer pays for the emission reductions generated by the project during the first commitment period (J) in exchange for an equivalent amount of AAUs turned into ERUs. The amount of 'excess' AAUs (S − J + J) does not change.

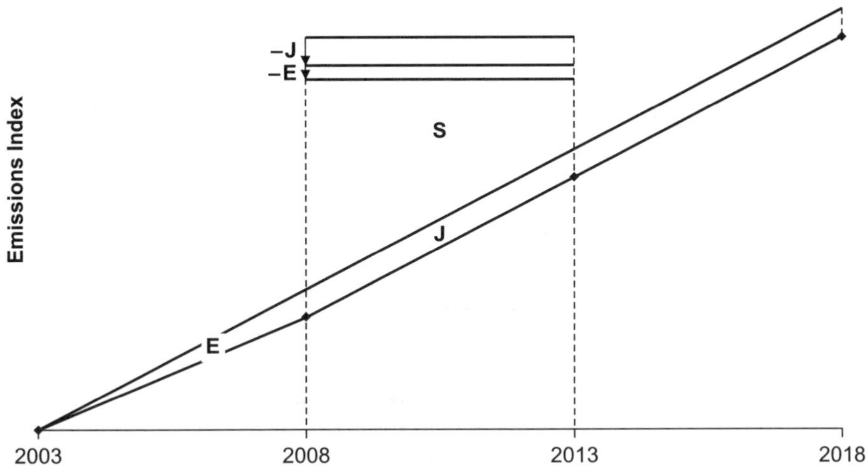

Figure 8.2. Early crediting

The buyer pays for the emission reductions generated by the project before 2008 (E) in exchange for an equivalent amount of AAUs. The amount of 'excess' AAUs is diminished by E (S − J − E + J).

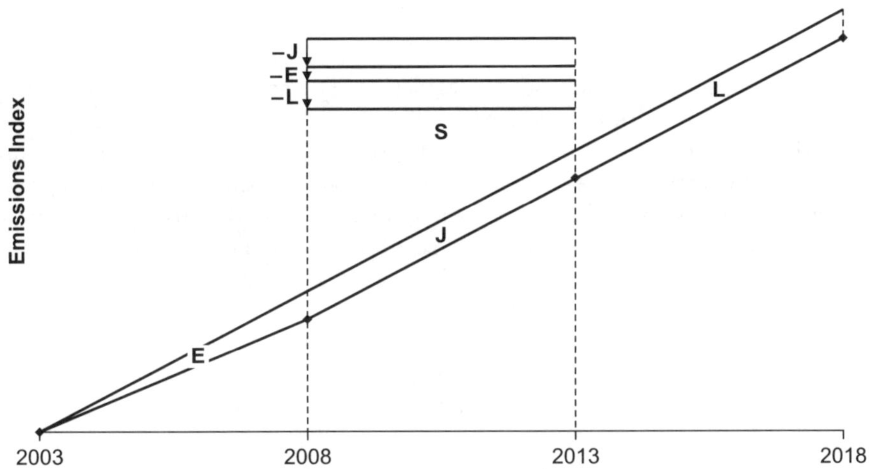

Figure 8.3. Late and early crediting

The buyer pays for some of the emission reductions generated by the project after 2012 (L), in exchange for an equivalent amount of AAUs: the surplus AAUs is diminished by E + L (S − J − E − L + J). On the other hand, an amount of L has been reduced from the baseline during the second commitment period.

3.1.2 Soft greening

A second possibility is to define 'greening' as the effective implementation of certain pre-defined activities, such as, for example, implementation of a demand side management programme, dismantling of energy subsidies, or capacity building activities related to climate change.

This form of greening does not overlap with any of the existing instruments, because it is based on the implementation of pre-defined policies, and not on the effective measure of emission reductions. This allows for supporting a much wider range of policies and measures, which might not be easily packaged into JI projects, but could still significantly affect emissions in the medium and long run.

3.1.3 On the interpretation of 'earmarking'

Central to the definition of 'greening' is the notion of earmarking the proceeds from the sale of AAUs for 'environmentally friendly' activities. One possible interpretation is that *all* the proceeds from the sales of AAUs should be earmarked this way. Such a position is consistent with the overall objective of establishing a 'greening' scheme, which is to ensure the environmental integrity of the sales of AAUs.

Central to the definition of 'greening' is the notion of earmarking the proceeds from the sale of AAUs for 'environmentally friendly' activities. One possible interpretation of that provision is that *all* the proceeds from the sales of AAUs should be earmarked. This would be consistent with the overall objective of establishing a 'greening' scheme, which is to ensure the environmental integrity of the sales of AAUs. However, one must bear in mind that in JI or CDM transactions, only *part* of the proceeds from the sales of ERUs or CERs support mitigation. If average abatement costs are lower than the price of carbon, the seller makes a profit out of the sales, which can be used without restriction. This supplemental income from carbon finance is a critical incentive for sellers.[26]

In this context, if all the proceeds from the sale of AAUs had to be earmarked for environmentally friendly activities in 'greening' schemes, there would be no profit associated, and this would constitute a major disincentive. If the buyer still wants to make sure that no environmentally harmful activities are funded through the scheme, one solution might be to negotiate two lists of activities to be funded with the proceeds of the sale of AAUs. The first would include the activities necessary for greening (either hard or soft). The second—which would be larger—would set criteria on how to spend the surplus (that is, the proceeds from the sales minus the costs of the activities necessary to consider that the AAUs are greened).

3.2 Risks Associated with 'Greening AAUs'

3.2.1 Delivery risk

As in any project-based mechanism, the contract would have to allocate the risk of non-performance between the buyer and the seller. In the case of 'hard' greening, this is the risk that the project is not being implemented, or that it does not generate as many emission reductions as expected. In the case of 'soft' greening, it

[26] By contrast to, for example, the incremental cost approach followed by the Global Environmental Facility, where only the mitigation costs are covered.

is the risk that the activities are not being implemented, or clearly failed before the end of the commitment period (for instance, a Demand Side Management programme is abandoned after a few years). How these risks are shared will determine, inter alia, the balance of upfront versus 'on delivery' payment, the nature of the liabilities incurred by buyers and sellers, and the penalties attached, and the price at which the 'greened' AAUs are paid.

3.2.2 *Non-transferability of AAUs*

Greening AAUs entails a transfer of AAUs from one country to another. A prerequisite is thus that both countries meet the requirements for trading under Article 17.[27] This introduces significant risk, since in case the seller does not meet these requirements at the end of the commitment period (because it never met them, or because it failed to maintain the proper standards), the transfer of AAUs cannot occur at all; presumably at a time where fallback supply options are either non-existent—because it is too late—or very expensive.

In addition, this risk is relative to the whole country. Unlike project risk, it thus cannot be mitigated by developing a portfolio of activities, unless this portfolio is across several countries. To hedge against it, the seller could assume the risk and liability unilaterally, but the question would remain whether the buyer would find this credible. Alternatively, the buyer must either place call options somewhere else, or rely on specific cross-country insurance schemes (for example, two economies in transition backing each other up for AAUs in case of default), which do not exist yet.

3.2.3 *Challenge to the environmental credibility of the 'greening'*

Buyers and sellers incur the risk of having the environmental credibility of the greening process challenged at any point during the process. This risk is not specific to 'greening', but it might be more acute in this case because no third-party institution is called to validate the environmental integrity of the process. This risk can be mitigated in two ways.

First, both for hard and soft greening (or some combination thereof), credible argument must be made that the activities financed under the project would not be financed otherwise. Establishing the baseline might be tricky, especially in the case of policy measures, but some relatively low cost forms of assessment can be imagined (for example, control group with countries at same stage of economic development, short-list of eligible sectors where the rest of the funds can be spent, and so on).

Secondly, the monitoring of the earmarked funds must be transparent and credible, which has both contractual and institutional implications on the way the proceeds of the sales are collected, stored, disbursed, and accounted for.

[27] See ch 22 below.

3.2.4 Overselling AAUs and falling short in 2012

Given the unpredictability of emissions even in a short time period, there is a risk that the 'excess' AAU budget might not be as large as expected. Therefore, a risk for sellers is to commit too much of their AAU budget in advance and fall short in 2012, when no credits are left on the market, or only very high prices.

This risk is not specific to 'greening'. It also exists for AAU trading, and for JI projects. It is also not homogenous for all forms of 'greening'. In the case of 'hard' greening during the commitment period, emission reductions presumably make for the transferred AAUs (see Figure 8.1 above). But in the case of soft greening, or of early or late crediting, the amount of 'excess' AAUs is reduced (Figures 8.2 and 8.3 above). Lastly, it diminishes with time, as emission levels during the commitment period are projected with greater and greater accuracy.

One way to mitigate this risk is to sell only a limited fraction of the expected AAU budget through 'greening schemes'. But it is very often possible to come up with a scenario, however implausible, in which the amount of 'excess AAUs' is very low or even negative. Politically, these scenarios are all the more difficult to refute as they usually derive from very high economic growth assumptions.

Ultimately, the seller must confront the trade-off between:

- Selling early and getting the resources, but risking paying some money back in case emissions are higher than expected; and
- Not selling, but possibly not being able to sell at all (or at very low price), in the future (in case the uncertainties around current and future international regimes for climate mitigation continue).

This requires strong planning, taking into account all instruments (and not only greening) that the country has at its disposal. Institutionally, this calls for housing all the entities dealing with the various Kyoto instruments (and where relevant the EU ETS) under the same roof, so that decisions may be taken in a coordinated way.

3.2.5 Weakening the country's negotiation position for post-Kyoto

Sellers engaging in 'greening AAUs' scheme might also feel that they are weakening their negotiating position for negotiations on second commitment period targets. This is a 'race to the bottom' argument where all countries limit their mitigation efforts in the hope to gain as much as possible in the next round of quota allocation. This risk is especially important if grandfathering is expected to be the underlying principle for allocating emissions quota in the future.

This risk is not specific to 'greening schemes', but it might be particularly important in the case of 'late crediting', because the country states that it is also reducing emissions beyond 2012.

This risk can be mitigated in several ways. First, buyer countries can pledge not to use the 'good work' by EITs against them. One prerequisite is that the specific policies and measures that were adopted under the 'greening scheme' are recorded in a very transparent way. It must also be recalled that the emission reductions generated (directly or indirectly) by the activities supported under the 'greening scheme' will lower the country's emission trajectory for the second commitment period (Figure 8.3 above).

However, as in the previous case, the bottom line is that the seller will have to weigh the potential gains from the greening schemes against the potential risk of a weaker negotiating position. If the activities financed under the scheme also have large economic and environmental benefits irrespective of the amount of emission reductions they generate, the choice becomes easier.

4 Conclusions: Risks and Benefits of Using Track One JI and Greening of AAUs Approaches

Track One JI and 'greening AAUs' present economies in transition with a wide range of tools to maximize the development benefits of participation in the Kyoto Protocol. Track One JI allows to sidestep the procedural constraints associated with Track Two JI. 'Greening' AAUs provides a way to finance beneficial policies and measures which cannot easily be framed as a project, and to sustain the flow of investment in projects generating emission reductions beyond 2006–7—at which time JI project flows are likely to decrease substantially because there will not be enough time to produce a sufficient amount of ERs before 2012—by resorting to 'late crediting'.

Compared to Article 17 International Emissions Trading and greening of AAUs, the attractiveness of moving to Track One JI is limited for EU accession countries. The EU Emissions Trading System is likely to limit the applicability of JI in order to avoid double counting. Such limits, while beneficial for the efficient implementation of the EU Scheme, may exclude small renewable projects with the EU from the benefits of carbon finance available to JI projects in other EITs.[28] EU accession countries therefore seem to have an interest in focusing on the requirements and institutions needed for EU ETS and Article 17 International Emissions Trading.

For other EITs (Bulgaria, Romania, Russia, Ukraine), Track One JI provides an interesting option for ensuring maximum benefits for both the global climate and the national economy by reducing transaction costs significantly compared to Track Two JI. Furthermore, demonstrating Track One JI in these countries will provide a transparent and cost-effective process for assuring buyers of AAUs that seek some form of 'greening' that the host country has the capacity to review and account for actual emissions reductions achieved. While the modality and

[28] This conclusion assumes limited price feed-through of carbon allowances onto electricity prices in Europe. If fossil-fuel electricity producers face steeper production costs because of the EU ETS, then small-scale renewables, even out of the system, would benefit.

definition of 'greening' of AAU sales will vary from buyer to buyer and be different in each transaction, host countries that have demonstrated capacity for Track One JI will be at a considerable advantage in being able to administer such contracts and assure the buyer that ERs resulting from the investment of proceeds of AAU sales will be monitored, verified and properly accounted for.

From all this, only one thing is clear and applicable to all potential JI host Parties: they need to have a comprehensive strategy on how to manage their AAU asset and which mechanisms they want to use for the maximum benefit of the global climate and their domestic economies. Countries will vary in their strategies with regards to the respective roles that Track Two and Track One JI, Article 17 International Emissions Trading and Greening of AAUs under it, as well as the EU Trading Scheme play. Formulating such a strategy may not be easy, however, especially given the blurred borderlines between the three mechanisms.

PART IV

ARTICLE 12 OF THE KYOTO PROTOCOL: CLEAN DEVELOPMENT MECHANISM

9

CDM Project Cycle and the Role of the UNFCCC Secretariat

*Maria Netto and Kai-Uwe Barani Schmidt**

1 Introduction

The Kyoto Protocol to the United Nations Framework Convention on Climate Change (UNFCCC) provides for three 'market-oriented mechanisms': Joint Implementation (JI), Emissions Trading (ET), and the Clean Development Mechanism (CDM). One purpose of these three mechanisms is to maximize the cost-effectiveness of climate change mitigation by providing an opportunity to reduce greenhouse gas (GHG) emissions abroad at lower cost.

In particular, with the CDM, Parties to the Convention aimed at breaking new ground in promoting international investment in climate friendly technologies for the benefit of developing countries.

The CDM aims to help developing countries (non-Annex I Parties) achieve sustainable development and the ultimate objective of the Convention, while helping industrialized countries (so-called Annex I Parties) meet their emissions targets under the Kyoto Protocol.

Under the CDM, industrialized countries may implement projects in developing countries that aim at reducing GHG emissions. As governments establish emissions targets for their private sector, companies will have an incentive to be engaged in the CDM.

The incentive to invest in CDM project activities, comes from the opportunity of crediting emission reductions from projects in developing countries, in the form of certified emission reductions (CERs) which may be used by industrialized countries towards meeting their emissions targets under the Kyoto Protocol. CERs may be traded on carbon markets. In other words, this means that part of a return of an investment into a project in a developing country, which reduces emissions beyond what would have been the case without the CDM, can be generated by the CERs.

* Maria Netto and Kai-Uwe Barani Schmidt are program officers with the UNFCCC Secretariat in Bonn, Germany. The views expressed in this chapter are the views of the authors and do not necessarily represent the views of the United Nations and the UNFCCC Secretariat.

The Kyoto Protocol has not yet entered into force. The rules for entry into force of the Kyoto Protocol require 55 Parties to the Convention to ratify (or approve, accept, or accede to) the Protocol, including Annex I Parties accounting for 55 per cent of that group's carbon dioxide emissions in 1990.[1] As at 15 April 2004, 122 countries had ratified the Kyoto Protocol[2] accounting for 44.2 per cent[3] of emissions from Annex I Parties.

Although the Kyoto Protocol has not entered into force, Parties to the Convention, through the 'Marrakesh Accords',[4] agreed to facilitate *the prompt start of the CDM* by adopting its 'Modalities and Procedures'[5] and by establishing the CDM Executive Board.

The CDM has thus been operational since November 2001.

For the UNFCCC Secretariat the operationalization of the CDM meant an important change from its role of supporting Parties' negotiations of legal text for modalities, to a more interactive role in supporting the implementation of CDM. This challenge includes the need to service different actors, from the private and public sectors, interacting under the CDM.

This new role implied, *inter alia*, the need to:

- Maintain a neutral position while supporting different interactions amongst diverse actors;
- Implement an electronic workflow management system involving actors scattered/spread around the world;
- Maintain a reliable and accurate information system;
- Respond to outside demands in an even more expedient manner;
- Implement procedural steps with a short turnover time; and
- Create a number of operational procedures.

This chapter will provide an overview of CDM actors and steps of the CDM project cycle and the role of the Secretariat in its support thereof.

2 CDM Actors

The CDM provides for a unique partnership where both intergovernmental process and private sector play an important role. While the intergovernmental

[1] Art 25 Kyoto Protocol. [2] See <http://unfccc.int/resource/kpstats.pdf>.

[3] See UNFCCC website: <http://unfccc.int/resource/kpthermo_if.html>.

[4] See Report of the Conference of the Parties at its seventh session, Decision 17/CP.7, FCCC/CP/2001/13/Add.2.

[5] See Decision 17/CP.7 and its Annex 'Modalities and procedures for a clean development mechanism', FCCC/CP/2001/13/Add.2.

process and the Executive Board of the CDM are given a regulatory and supervisory role, the operational tasks are the responsibility of project participants and designated operational entities (DOEs)—mainly from the private sector. As will be seen from the description below, the structure of the CDM is decentralized and private sector oriented.

2.1 COP/MOP (COP)

The Kyoto Protocol stipulates that the CDM shall be subject to the authority and guidance of the Conference of the Parties serving as the meeting of the Parties to the Protocol (COP/MOP) and be supervised by an Executive Board of the CDM.[6]

The COP/MOP is thus the overall governing body of the CDM. Under the prompt start phase of the CDM (before the Kyoto Protocol enters into force) the Conference of the Parties of the Climate Change Convention (COP) shall however assume the responsibilities of the COP/MOP.[7]

The COP/MOP (COP)[8] reviews annual reports of the Executive Board and provides guidance to the Board on its operations.[9] In addition, the COP/MOP (COP) shall assist in arranging funding of CDM project activities, as necessary, and review the regional and sub-regional distribution of DOEs and distribution of CDM project activities.

2.2 Executive Board

In the Marrakesh Accords, the COP agreed that an Executive Board should be established and that until the entry into force of the Kyoto Protocol this Board should act as the Executive Board of the CDM.[10]

The Executive Board is comprised of ten members and ten alternate members. The Board shall be composed by one member and one alternate member from each of the regional groups of the United Nations, two members and two alternate members from Annex I Parties,[11] two members and two alternate members from non-Annex I Parties and one member and one alternate member from small island developing States.

6 See Art 12(4) Kyoto Protocol.

7 FCCC/CP/2001/13/Add.2, Decision 17/CP.7, para 2.

8 In the following the bracketed (COP) shall indicated that the referred issue is under the responsibility of the COP until the COP/MOP convenes for its first session.

9 Please refer to the COP/MOP section on the UNFCCC CDM website for the list of decisions on guidance by the COP to the Executive Board, <http://cdm.unfccc.int>.

10 FCCC/CP/2001/13/Add.2, Decision 17/CP.7, para 4.

11 The annex to Decision 17/CP.7 on modalities and procedures of a CDM stipulates that when referring to 'Parties' it refers to Parties that have ratified the Kyoto Protocol. Decision 17/CP.7 stipulates however that during the period before the entry into force of the Kyoto Protocol Parties to the Convention may nominate members and alternates for the Board. Once the Kyoto Protocol has entered into force, members and alternates from Parties that did not ratify would have to be replaced.

Although members are to be nominated by constituencies and elected by the COP/MOP (COP) they shall act in their personal capacity.[12] This in fact constitutes one of the innovations of the CDM; the members of the intergovernmental body (the Board) are not to represent individual interests of governments, or have any pecuniary or financial interests on projects or companies involved in the implementation of the CDM.

The Board operates by following rules of procedure adopted by the COP.[13] These rules of procedure cover a number of issues including the Board's voting rules and quorum, terms of members and alternates and their re-election, cause for termination of membership, chairmanship, avoidance of conflict of interests by Executive Board members, number and form Board meetings, confidentiality provisions, attendance by observers to Board meetings, transparency provisions and public availability of information, and use of outside expertise.

As mentioned above, the Executive Board was established to supervise the CDM and undertake mainly a 'regulatory' function under the CDM. As we will see in detail in the description of the project cycle below, the Board is responsible for:

- Making recommendations to the COP/MOP (COP) with regards to modalities and procedures for a CDM;
- Developing guidance to project participants and DOEs on the implementation of the CDM;
- Implementing the accreditation process of DOEs, including overseeing and developing procedures for accreditation, reviewing accreditation standards and making recommendations to the COP/MOP (COP), accrediting companies as DOEs and making recommendation on suspension/withdrawal of DOEs;
- Approving baseline and monitoring methodologies necessary for assessing and quantifying emission reductions by proposed project activities;
- Registering CDM project activities;
- Issuing CERs and maintaining a CDM registry for tracking CERs;
- Maintaining and making publicly available a repository of approved rules, procedures, methodologies and standards relating to the CDM.

2.3 Panels

The modalities and procedures for a CDM stipulate that the Board may 'establish committees, panels or working groups to assist it in the performance of its functions. The Executive Board shall draw on the expertise necessary to perform its

[12] FCCC/CP/2001/13/Add.2, Decision 17/CP.7 on modalities and procedures for a CDM, para 8(c) of the Annex.

[13] FCCC/CP/2002/7/Add.3, Decision 21/CP.8, Annex I.

functions, including from the UNFCCC roster of experts. In this context, it shall take fully into account the consideration of regional balance.'[14]

During its first two years of operations, the Board has drawn, on many occasions, on the work by expert panels.

Panels operate following general guidelines and specific terms of reference and guidance developed by the Board.[15] Panels may be created for temporary and specific tasks, as was the case for the panel to develop recommendations on simplified modalities and procedures for small-scale CDM project activities, or for filling a standing function, such as the panels on accreditation and methodologies.

Like Board members and alternates, panel members are to act in their individual capacity. The panels' recommendations to the Board are publicly available, unless they address confidential or proprietary information. The terms of reference of the panels specify their size, terms, mandates, competence criteria for their experts and whether experts are entitled to receive fees for specific work. Experts from panels are selected from public calls for candidates in accordance with the competence criteria set out in the terms of reference. Panels may, in addition, work with the assistance of outside experts that undertake specific tasks such as desk reviews of proposed methodologies or assessments of companies applying for accreditation.

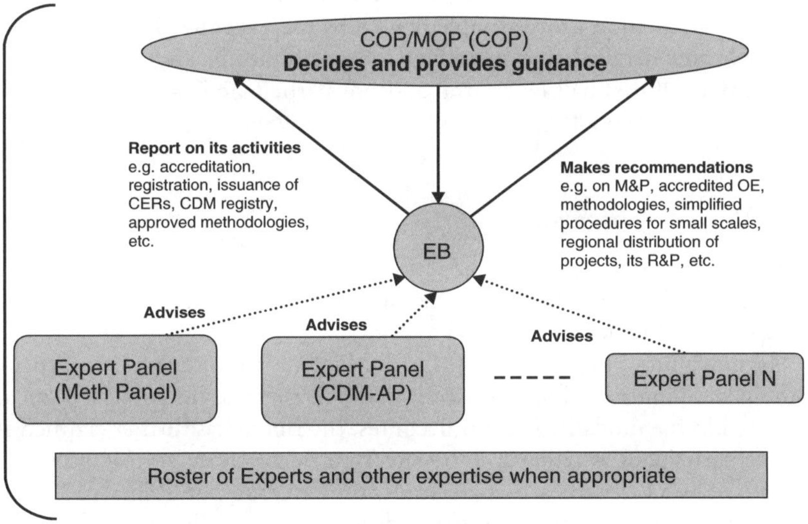

Figure 9.1: EB—Institutional set-up/Decision-making

Source: UNFCCC Secretariat.

14 FCCC/CP/2001/13/Add.2, Decision 17/CP.7 on modalities and procedures for a CDM, para 18 of the Annex.

15 General guidelines for panels as well as the specific terms of references of different panels and guidance by the Board are available under the section on Panels on the UNFCC CDM website: <http://cdm.unfccc.int/>.

2.4 Designated Operational Entities

A DOE, under the CDM, is either a domestic legal entity or an international organization, which is accredited and designated, on a provisional basis until confirmed by the COP/MOP (COP), by the Executive Board to perform three key functions:[16]

- Check for completeness and forward to the Board proposed new methodologies;
- Validate and subsequently request registration of a proposed CDM project activity using an approved methodology;
- Verify emission reductions of a registered CDM project activity, certify as appropriate and request the Board to issue CERs accordingly.

The role of the DOEs is another innovative tool of the CDM. The companies accredited as DOEs are those expected to check the adequacy and adherence of project activities to the CDM modalities of the DOEs. A DOE request to register a project activity or to issue CERs will be implemented unless a request for review is triggered within a set time frame. These companies are part of the institutional infrastructure of the CDM and play a key role in the credibility of system.

The delegation of responsibility by the COP/MOP (COP) to DOEs implies that the accreditation process implemented by the Executive Board needs to be a thorough and detailed one including the possibility of spot checking. It also implies that DOEs shall be neutral and not participate in the design of project activities and methodologies.

2.5 Project Participants

The Kyoto Protocol stipulates that participation under the CDM may involve private and/or public entities, and is subject to whatever guidance may be provided by the Executive Board.[17]

The modalities and procedures for a CDM refer to the term 'project participants' without defining it. Taking into consideration the roles of project participants described in the modalities and procedures, the Board has further clarified that a project participant is either a Party involved or a private and/or public entity authorized by a Party to participate, under the Party's responsibility, in CDM project activities. Project participants take decisions on the allocation of CERs from the CDM project activity under consideration.[18] Project participants propose and implement a given CDM project activity.

[16] Entities applying for accreditation (applicant entities) may carry out some activities as DOEs, including submitting proposed new methodologies, for the purposes of witnessing whether they are qualified to undertake functions of DOEs.

[17] See Art 12(9) Kyoto Protocol.

[18] Please refer to the CDM–PDD Glossary of Terms available under the documents section on the UNFCCC CDM website: <http://cdm.unfccc.int/>.

They may also propose to the Board new baseline and monitoring methodologies to account for emission reductions. This is one of the important innovations of the CDM—methodologies for accounting for reductions are developed from a bottom-up approach.

2.6 Designated National Authorities

The CDM modalities and procedures state the need for a country to establish a Designated National Authority (DNA) in order to participate in CDM project activities.[19] The main tasks to be carried out by the DNA for the CDM are defined in the CDM modalities and procedures. The DNA shall,[20] among other things:[21]

- Give a written approval to the project participants of voluntary participation of the Party to the CDM project activity;
- In the case of a host Party, to include in the written approval a confirmation that the CDM project activity assists it in achieving sustainable development.

2.7 Public

Public transparency and scrutiny play an important role in ensuring the CDM credibility. All information, which is the basis for decision-making, is publicly available. In addition, in two stages of the CDM project cycle, public comments are sought:

- When designing a project activity, project participants are to call for comments by local stakeholders;[22]
- Before validating a project activity, a DOE shall make the project design document of a proposed project activity publicly available and receive public comments on validation requirements from Parties, stakeholders, and UNFCCC accredited non-governmental organizations.[23]

[19] Please refer to paras 29 and 37(a) of the Annex to Decision 17/CP.7 on modalities and procedures for a CDM, FCCC/CP/2001/13/Add.2. [20] Please refer to para 40(a), ibid.

[21] Updated information on role/functions of DNAs is made available in DNA section: <http://cdm.unfccc.int/>.

[22] In that regard, the Executive Board has further clarified that 'an invitation for comments by local stakeholders shall be made in an open and transparent manner, in a way that facilities comments to be received from local stakeholders and allows for a reasonable time for comments to be submitted. In this regard, project participants shall describe a project activity in a manner which allows the local stakeholders to understand the project activity, taking into account confidentiality provisions of the CDM modalities and procedures.' Please refer to the section on clarifications/guidance: <http://cdm.unfccc.int/>.

[23] Detailed 'procedures on public availability of the CDM project design document and for receiving comments as referred to in paragraphs 40 (b) and (c) of the CDM modalities and procedures' have been developed by the Executive Board. In accordance with these procedures DOEs shall make PDDs available through a link to the UNFCCC CDM website. Please refer to the section on procedures: <http://cdm.unfccc.int/>.

In addition to these opportunities, the public, including Parties, NGOs, stakeholders, and institutions are able to make comments on the documents on which the Executive Board agreed to seek public comments. Current calls for input and comment can be found through the main page of the UNFCCC CDM website. Currently public input is sought in different areas of work of the Board such as:

- Work plans for specific tasks and/or development specific technical recommendations/guidance;
- Proposed new methodologies;
- Applicant entities and accreditation procedures.

2.8 Secretariat

The UNFCCC Secretariat services the COP/MOP (COP) and the Executive Board, performing functions assigned to it accordingly. In this role the Secretariat is to directly or indirectly support and link all actors described above (Figure 9.2). It does so by providing means for actors to communicate and interact in a cost effective manner, for rules to be known as adopted by the COP/MOP and the Executive Board and for the Board, panels, experts, and entities to be able to work professionally.

The Secretariat maintains a web-based information system providing up-to-date information on the implementation/operation of the CDM (such as, Board/panel meeting reports, relevant decision/guidance of COP/MOP (COP), guidance of the Board, procedures, calls for public input, repository of approved methodologies, modalities for different types of project activities (for example, small scale, afforestation and reforestation), registered project activities).

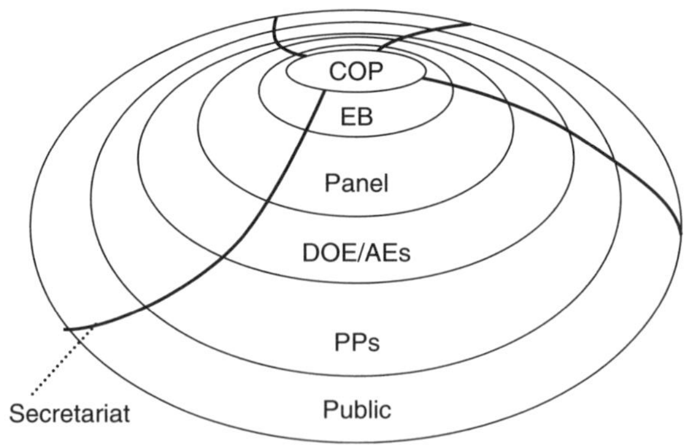

Figure 9.2: CDM actors and the role of the Secretariat
Source: UNFCCC Secretariat.

The Secretariat is expected to play a neutral role in servicing the CDM actors and not make recommendations or judgements on specific CDM project activities.

3 CDM Project Cycle

3.1 Introduction to the Cycle

The CDM project cycle is, in essence, very similar to the project cycle of any other project. There is a design phase, a phase where the project design is confirmed to meet preset design features, a phase of implementation when monitoring is performed, a phase when performance/output of the project is verified and certified to meet requirements and finally a phase when a document is issued to confirm existence of the output.

The negotiations over the design of the CDM, where initial positions ranged from the view that most of the decisions should be taken by the supreme body of the intergovernmental process (COP/MOP (COP)) to a completely private sector system, led to the novel design of an intergovernmental mechanism, where the infrastructure includes private/public sector companies (designated operational entities (DOEs)) which take decisions unless objected to, within a short period of time, by a ten-member Board. Figure 9.3 below relates steps to actors of the project cycle. It illustrates that the majority of steps are within the control of private sector actors and that steps where the Executive Board gets involved are 'automatic'

		CDM 'infrastructure'	
	Project Participant (PP)	**Operational Entity (DOE)**	**Executive Board (EB)**
Project Design	Provide info		
Validation/Registration		Confirm that requirements are met	* (*within* 8 *weeks*)
Monitoring	Implement		
Verification & Certification		Certify ERs	
Issuance of CERs			* (*within* 15 *days*)

***An automatic step unless a review is requested.**

Companies (private/public)

Written letter of approval from DNAs involved are required prior to request for registration.

Figure 9.3: CDM project Cycle Roles and responsibilities

Source: UNFCCC Secretariat.

unless a review is triggered by three of the ten Board members or by the governments which are project participants.

The project cycle is, in its major steps, the same for all types of CDM project activities. Differences exist in modalities for small-scale and afforestation and reforestation CDM project activities such as different requirements for validation and monitoring plan and deadlines. This chapter describes the project cycle as described in the CDM modalities and procedures without highlighting these differences for small scale and afforestation and reforestation CDM project activities. These differences are described in the modalities for these types of CDM project activities.[24]

There are two inputs to the project cycle required for it to work as sketched out above and explained in more detail below:

- The DOEs are required to be accredited by the Executive Board and designated by the supreme body;
- Methodologies need to be approved by the Executive Board for determining the reductions in emissions that are additional to any that would occur in the absence of the certified project activity and monitoring of such reductions.

3.2 Accreditation of Operational Entities

The Executive Board has the mandate to accredit entities and recommend to the COP/MOP (COP) their designation as operational entities. The Executive Board has established an accreditation procedure[25] which describes the process of accreditation covering application, assessment of whether applicant entities meet the accreditation standards and requirements, and recommendation of a decision to the Executive Board by a panel of accreditation experts.

The accreditation is valid for three years with no scheduled assessment of the entity during this period. The DOE can seek re-accreditation, during or after the expiration of a three-year term. The DOE may be assessed on the basis of 'spot checks', which can be triggered in accordance with the accreditation procedure. The accreditation, provided by a single body—the Executive Board—is sector specific[26] without any geographical limitation. In accordance with a phased accreditation approach, a DOE may, for a period of time, be limited to perform validation/registration functions before it will be also accredited to perform the verification function.

The role of the Secretariat in this process is to provide logistical, managerial, and substantive support to the work of the Board, the panel, assessment teams and

[24] The 'Simplified modalities and procedures for small-scale CDM project activities' and 'Modalities and procedures afforestation and reforestation project activities under the CDM' are available in the documents section: <http://cdm.unfccc.int/>.

[25] Please refer to the section on procedures: <http://cdm.unfccc.int/>.

[26] See list of sectoral scope(s) for accreditation: <http://cdm.unfccc.int/DOE/scopes.html>.

applicant entities, as applicable. For example the Secretariat:

- Maintains a web-based database and communication system permitting the Board, applicant entities, panel members and assessment teams, distributed globally, to communicate and manage their work flows;
- Maintains a roster of accreditation assessment team experts;
- Prepares background documentation to assist decision making by the Board and the panel related to the accreditation process;
- Recommends options for and implements procedures related to the accreditation process;
- Organizes physical meetings of the Panel;
- Coordinates and logistically supports the work of assessment teams on-site;
- Assumes observer functions at assessments;
- Administers contracts and fee payments.

3.3 Approval of New Methodologies

As referred above, one innovative feature of the CDM is that the modalities provide for a bottom up approach using the large potential of private sector knowledge and ideas with regard to the development of methodologies for determining the reductions in emissions that are additional to any that would occur in the absence of the certified project activity (baseline methodologies) and monitoring of such reductions (monitoring methodologies). The Executive Board approves such methodologies based on a recommendation by a panel of methodology experts (also known as the Meth Panel).[27]

Project participants may put forward methodologies by submitting to a DOE a proposal supported by information on the planned project which intends to make use of the new methodology. After an assessment of the proposal, involving two external experts hired by the Meth Panel, the public in general providing comments and the possibility for feedback by project participants on a draft recommendation of the Meth Panel, the Panel makes a final recommendation to the Executive Board for its consideration. If the Board approves the methodology it is included in the repository of approved methodologies against which DOEs can validate requests for registration by project participants of projects as CDM project activities.[28]

In addition to this bottom up approach, the Board may undertake efforts to provide consolidations of existing approved methodologies to streamline and ease their use or to develop methodologies.

[27] The Board may establish additional panels for treating methodological issues for specific project types such as small-scale or afforestation and reforestation.

[28] Please refer to the section on procedures (<http://cdm.unfccc.int/>) for detail on 'Procedures for submission and consideration of a proposed new methodology'.

The role of the Secretariat in this process is to provide logistical, managerial, and substantive support to the work of the Board and the Meth Panel. For example the Secretariat:

- Maintains a web-based database and communication system permitting the Board, DOEs, Panel members and experts, distributed globally, to communicate and manage their work flows;
- Maintains a roster of experts to undertake desk reviews;
- Prepares background documentation to assist decision making by the Board and the Panel;
- Recommends options for and implements procedures related to the process of approving new methodologies;
- Organizes physical meetings of the Panel;
- Administers contracts and fee payments.

4 The Project Cycle Steps

This section describes the five steps of the CDM project cycle, and how the UNFCCC Secretariat is involved in each of these steps.

4.1 Design of a CDM Project Activity

The actors in the project design phase are the project participants. Based on the modalities and procedures for a CDM and guidance by the Executive Board the project participants have to design the project activity such that it could be validated and registered as a CDM project activity. One of the most difficult design tasks for a project participant is to ensure that the proposed project activity uses approved methodologies for baselines and monitoring, and applies it correctly to the proposed activity. Project participants shall submit information on their proposed CDM project activity using the CDM–PDD. Project participants are guided by the modalities and procedures for a CDM[29] as well as by the Executive Board.[30]

If in the design phase project participants conclude that a new methodology is required they may propose it as described above.

The role of the UNFCCC Secretariat is to ensure that information on approved methodologies and guidance for design of project activities is easily available in a cost effective manner. Although the Secretariat is to provide information on the

[29] In particular paras 43 to 52 (regarding baselines), 53 to 60 (regarding monitoring) and Appendix B of the Annex to Decision 17/CP.7 on modalities and procedures for a CDM, document FCCC/CP/2001/13/Add.2.

[30] For more detail please refer to the section on clarifications/guidance: <http://cdm.unfccc.int/>.

generic level of design it shall not provide input on particular design or eligibility questions.

4.2 Validation/Registration

The DOE is the key actor in this phase. Validation is the process of independent evaluation of a project activity by a DOE against the requirements of the CDM on the basis of a CDM–Project Design Document (CDM–PDD).[31] The sections 'Validation and Registration' as well as 'Verification' of the modalities and procedures for a CDM describe the key functions of DOEs.

After being contracted by project participants, a DOE is to check that validation requirements[32] have been met after reviewing the CDM–PDD and any supporting documentation prepared and submitted by the project participants (including ensuring that the baseline and monitoring methodologies used are approved by the Board).

The DOE is to make the CDM–PDD publicly available,[33] receive public comments by Parties, stakeholders, and UNFCCC accredited non-governmental organizations and make them publicly available.

After the deadline for receipt of comments, the DOE is to make a determination as to whether the project activity should be validated. The DOE is to inform project participants of its determination on the validation of the project activity. If the DOE determines the proposed project activity to be valid, it is to submit to the Board a request for registration[34] in the form of a validation report. This shall include the CDM–PDD, the written approval of the host Party, and an explanation of how it has taken due account of comments received. Before submitting the request for registration, the DOE has to have received the written approval(s) from DNAs of Parties involved.[35] The validation report is to be made publicly available.

Registration is the formal acceptance by the Executive Board of a validated project as a CDM project activity. 'The registration by the Executive Board shall be deemed final eight weeks after the date of receipt by the Executive Board of the request for registration, unless a Party involved in the project activity or at least three members of the Executive Board request a review of the proposed CDM project

[31] See latest version of the CDM–PDD in the documents section: <http://cdm.unfccc.int/>.

[32] Validation requirements are stipulated in para 37 of the Annex to Decision 17/CP.7, FCCC/CP/2001/13/Add.2. See also guidance provided by the EB at its eighth meeting available in the reference section: <http://cdm.unfccc.int/>.

[33] The Board has elaborated 'Procedures on public availability of the CDM-PDD and for receiving comments as referred to in paras 40 (b) and (c) of the CDM modalities and procedures (version 03)'. The current version is available in the reference section: <http://cdm.unfccc.int/>.

[34] Detailed 'procedures for registration of a proposed CDM project activity' and level of fees for project registration have been elaborated by the Executive Board. Please refer to section 4.1 above.

[35] The modalities and procedures for a CDM do not specify whether the written approval by DNAs shall be obtained before or after making a PDD publicly available (paras 40(a) and (f), 37(a) and 28 to 30 of the Annex to Decision 17/CP.7).

activity'.[36] The request for review 'shall be related to issues associated with the validation requirements'. The registration as a CDM project activity is the prerequisite for the verification, certification and issuance of CERs related to that project activity.

The role of the UNFCCC Secretariat is to ensure that information on, *inter alia*, public availability of CDM–PDDs, validation reports, requests for review and registrations is provided through the UNFCCC CDM website in an accurate and cost effective manner. The Secretariat serves as a communication channel between DOEs and the Executive Board in case guidance is required by DOEs in order to perform their functions. For example, the Secretariat:

- Maintains a web-based database and communication system permitting the Board, DNAs, DOEs and project participants, distributed globally, to communicate and manage their work flows;
- Prepares background documentation to assist decision making by the Board;
- Recommends options for and implements procedures related to the process of validation and registration;
- Administers registration fee payment.

4.3 Monitoring

Monitoring is implemented by the project participants in accordance with the monitoring plan contained in the registered CDM–PDD. The modalities and procedures for a CDM provide guidance for the establishment of a monitoring plan which is to be based on an approved methodology.[37] The preparation by project participants of a monitoring report in accordance with the monitoring plan (contained in the CDM–PDD) is a pre-condition for verification and certification by a DOE and the issuance of CERs by the Executive Board.

The UNFCCC Secretariat has no particular role in this step.

4.4 Verification/Certification of the CDM Project Activity

Verification is the periodic independent review and ex-post determination by the DOE of the monitored reductions in anthropogenic emissions by sources of GHGs that have occurred as a result of a registered CDM project activity during the verification period. Certification is the written assurance by the designated operational entity that, during a specified time period, a project activity achieved the reductions in anthropogenic emissions by sources of GHGs as verified.

Like in the validation/registration phase, the DOE is the key actor in this phase. As a first step, the DOE is to make the monitoring report received from the project

[36] Detailed 'Procedures for review as referred in paragraph 41 of the CDM modalities and procedures' have been elaborated by the Executive Board. Please refer to the section on procedures: <http://cdm.unfccc.int/>.

[37] Please refer to paras 53 and 54 of the Annex to Decision 17/CP.7.

participants publicly available and proceed with verification steps. The DOEs have to establish a verification report which shall be made publicly available[38] and based on this report, the DOE shall:

> certify in writing that, during the specified time period, the project activity achieved the verified amount of reductions in anthropogenic emissions by sources of greenhouse gases that would not have occurred in the absence of the CDM project activity. It [The DOE] shall inform the project participants, Parties involved and the Executive Board of its certification decision in writing immediately upon completion of the certification process and make the certification report publicly available.[39]

The role of the UNFCCC Secretariat is to ensure that information on, *inter alia*, public availability of monitoring plans, validation reports, and certification reports is provided through the UNFCCC CDM website in an accurate and cost effective manner. The Secretariat serves as a communication channel between DOEs and the Executive Board in case guidance is required by DOEs in order to perform functions. For example, the Secretariat:

- Maintains a web-based database and communication system permitting the DOEs and project participants, distributed globally, to communicate and manage their work flows;
- Recommends options for and implements procedures related to the process of verification and certification.

4.5 Issuance

The certification report submitted to the Board by a DOE constitutes the request for issuance of CERs which is considered final after fifteen days, unless a Party involved in the project activity or at least three members of the Board request a review of the proposed issuance of CERs. Like in the 'automatic' step at registration, the scope of such a request has been limited. The review is limited to issues of fraud, malfeasance or incompetence of the DOE.

Once the issuance is considered final and upon instruction by the Executive Board, the CDM registry administrator issues the CERs within the CDM registry established and maintained to ensure the accurate accounting of the issuance, holding, transfer and acquisition of CERs by non-Annex I Parties. At this stage a share of the CERs issued is withheld to cover administrative costs and to assist developing country Parties that are particularly vulnerable to the adverse effects of climate change to meet the costs of adaptation.[40]

The role of the UNFCCC Secretariat is to ensure that information on, *inter alia*, requests for issuance, issuance and related project information is provided through

[38] ibid, para 62. [39] ibid, para 63.

[40] For more detail on share of proceeds and registration fee refer to section 4.1 above, and see: <http://cdm.unfccc.int/>.

the UNFCCC CDM web-based information system in an accurate and cost effective manner. The Secretariat provides operational support with regard to the implementation in a timely manner of the review procedure. The Secretariat presently assists the Board in designing and developing the CDM registry. The Board has not yet taken a decision on how the CDM registry is operated and what exact role the Secretariat is to play in this regard.

5 Conclusion

In summary, the CDM project activity cycle provides for a number of novelties for a multilateral mechanism. It is implemented to assist countries meet legally binding commitments by linking their interest to the interest of other countries to achieve sustainable development.

In designing the mechanism, issues important for private sector involvement are addressed, such as the:

- 'Bottom up' approach in defining criteria and rules (project participants can make proposals for methodologies);
- Involvement of private sector (project participants, designated operational entities);
- Decentralized decision-making (for example, unless a request for review is triggered, the decision to request registration of a project activity by one of the many DOEs is implemented by the Secretariat);
- Decision-making is based on expert panels established through a public call for experts and bearing in mind the need for incorporating knowledge from all regions of the world.

By being assigned the task to act as secretariat to the CDM Executive Board, the UNFCCC Secretariat was requested to service a day to day process of implementation and decision making in addition to its role to service a negotiation process. The Secretariat is also assuming a new UN role in supporting a globally operating accreditation body. The new roles provided new challenges with regard to the:

- Treatment of public as well as confidential information of a competitive nature (for example, the accreditation process);
- Need to avoid conflict of interest situations;
- Need to ensure fair and equitable treatment of all actors and not to create competitive advantages by, for example, providing information/advice to one private sector actor and not to another;
- Need to respond to shorter deadlines/turnover times.

The CDM is a challenging experience for all actors involved and may possibly provide a useful experience for other multilateral agreements.

10

Keeping it Clean—Safeguarding the Environmental Integrity of the Clean Development Mechanism

*Ernestine Meijer and Jacob Werksman**

1 Introduction

The Kyoto Protocol[1] to the UN Framework Convention on Climate Change (UNFCCC),[2] when in force, will create legally binding caps on the greenhouse gas (GHG) emissions of developed (Annex I) Parties.[3] The Clean Development Mechanism (CDM) of the Kyoto Protocol was included in the Protocol to accommodate Annex I governments' demands for flexibility in the implementation of these commitments. The CDM is one of three 'flexibility mechanisms' built into the Kyoto Protocol, in addition to international emissions trading among Annex I

* Ernestine Meijer is a senior research fellow, Center on Environmental and Land Use Law, New York University School of Law, and Jacob Werksman is Adjunct Professor of Law, New York University School of Law. The views expressed in this chapter are the views of the authors and do not necessarily represent the views of New York University School of Law.

[1] Kyoto Protocol to the United Nations Framework Convention on Climate Change 1997, UN Doc FCCC/CP/1997/L.7/Add.1 reprinted in (1998) 37 ILM 22 (not yet in force). For background on the negotiation of the Protocol and the design of the CDM, see S Oberthur and H Ott, *The Kyoto Protocol: International Climate Policy for the 21st Century* (Springer Verlag, 1999). 'The Clean Development Mechanism: Unwrapping the Kyoto Surprise' in *Review of European Community and International Environmental Law* 7:2 (Autumn 1998) 147–58; R Estrada-Oyuela, 'A Commentary on the Kyoto Protocol' in G Chichilnisky and G Heal (eds), *Environmental Markets, Equity and Efficiency* (Columbia University Press, 2000).

[2] United Nations Framework Convention on Climate Change, 1992, reprinted in (1992) 31 I.L.M. 849.

[3] The Kyoto Protocol specifies legally binding commitments by most industrialized countries (Annex I countries) to reduce the level of their GHG emissions compared to the level of 1990. For the first commitment period (2008–12) an average reduction of at least 5 per cent of the overall emissions of such gases below 1990 levels is aimed for. The reductions of GHGs agreed to are different for each country. The countries of the EU, for example, agreed to a reduction of 8 per cent and the US to 7 per cent, while the Russian Federation agreed to stabilize at the 1990 level and Iceland was allowed to have its emissions grow by 10 per cent. The amount of GHGs these countries are allowed to emit is called their 'assigned amount'. The maximum of this assigned amount is called the 'emissions cap'. Those countries that are not Annex I countries (non-Annex I countries) have not agreed to a GHG stabilization or reduction.

Parties, and a project-based offset trading regime, also among Annex I Parties, known as Joint Implementation.[4] The CDM is unique among these mechanisms, in that it allows an Annex I Party to increase its cap by obtaining emissions credits generated by investments in a CDM project based in an uncapped, developing (non-Annex I) Party. These emissions credits are called Certified Emissions Reductions (CERs) and are denominated in tonnes of carbon equivalent. The CDM thus provides capped Parties with geographical flexibility, by allowing them to meet their commitments through projects in developing countries where opportunities for emissions reductions are likely to be cheaper. Proponents of the CDM have noted that in addition to making compliance with the Protocol more cost effective, CDM projects can lead to improvements in local air quality and energy efficiency, and to increased investment in, and technology transfer to, developing countries. By doing so, the CDM could provide a modest response to the more mainstream trend in foreign direct investment that has been criticized for leading more energy intensive and polluting manufacturing sectors to move 'south' where labour, environmental standards, and other production costs are often lower.

However, the CDM, by allowing capped Parties to receive credit for emission reductions taking place in uncapped countries, raises concerns about the overall environmental performance of the Kyoto Protocol, as measured in terms of trends in global emissions of GHGs. If all countries included in Annex I of the Protocol were to ratify the Protocol and remain within their caps, without resorting to CERs, the Protocol would lead to an approximate 5 per cent reduction in their 1990 contributions to GHG emissions by the end of the Protocol's first commitment period (2012).[5] Each additional 'carbon-tonne' of CER transferred under the CDM will entitle an Annex I Party to an equivalent increase in emissions within its territory, while remaining in compliance. The more successful the CDM is at generating CERs, the more an Annex I Party can use those CERs to increase its emissions above its cap, and the more the overall environmental performance of the Kyoto Protocol depends upon the environmental integrity of the CDM. Environmental integrity is the demonstrated ability of the CDM to approve projects and to certify emissions reductions that are widely perceived as additional to what would otherwise have occurred, and support projects that contribute to long-term reductions in concentrations of GHGs in the atmosphere.[6]

Before moving to an assessment of the potential environmental integrity of the CDM, it is important to not belabour the contrast between 'capped' and 'uncapped' Parties in the context of an international emissions trading market. Without the CDM, compliance with the Protocol's caps would depend solely

[4] See, for a general overview of project based regimes, *Greenhouse Gas Emissions Trading and Project-based Mechanisms* (OECD, 2004).

[5] Art 3(1) Kyoto Protocol. This rough estimate was included in the Protocol to provide the reader with a benchmark of the Protocol's overall contribution to combating global warming, which might otherwise have been lost in the details. It does not take into account the significant potential emissions reductions lost to the system caused by the US not ratifying or executing the Protocol.

[6] See Art 12(2) Kyoto Protocol.

upon the performance and good will of its Annex I Parties, backed by the Protocol's Compliance System.[7] Although that system is stronger than those attached to many other international treaties, it suffers from many of the inherent weaknesses of international compliance systems and cannot guarantee the performance of capped Parties.[8] It should, however, be noted, that there may be a negative feedback loop between the Protocol's Compliance System and the CDM's integrity. The more successful the Compliance System is in policing an Annex I Party's domestic compliance, the greater the pressure on the CDM to lower the cost of domestic compliance by generating higher volumes of CERs. On the other hand, some might argue that the CDM, by decreasing compliance costs, and by promoting investment in and technology transfer to developing countries will have a long-term positive impact on the emissions trends, despite the gaps it may open up in the Protocol's formal, legal structure by shifting focus from capped to uncapped Parties.

At the heart of concerns surrounding the CDM's integrity is that, in order for a CDM project to generate CERs, the project's proponents must demonstrate a 'counterfactual', that is, they must show that the emissions reductions resulting from a particular investment are 'additional' to reductions that would have occurred in the absence of that investment. This 'additionality' is measured by comparing the lower emissions that result from the investment against a scenario of the emissions that would likely have occurred had the investment not taken place (known as the baseline scenario). If additionality is not demonstrated to a high degree of certainty, the integrity of the CDM as a source of emissions credits will be undermined. If, for example, the CDM certifies emissions reductions generated by the introduction of improved energy efficiency technologies that would have become widely used in the host country within the time frame of the crediting period, these reductions are not likely, 'in fact', to have been additional. Such predictive errors, if they are unchecked and widespread, could translate, through CERs, into increased emissions in Annex I Parties and net global increases in emissions. While we can never be fully certain that a CER was 'in fact' additional (because the additionality calculation is based on a counterfactual), we can be very confident that a CER, once it enters the Kyoto system, will be used by an Annex I Party to increase its emissions beyond its cap.

The role of the CDM is further complicated by the creation of a 'market' in which the participants on both sides of the transaction, rather than competing, have short-term interests in common. The non-Annex I Party hosting the CDM project and the Annex I Party receiving the CERs share a short-term interest in

[7] The text of the Kyoto Compliance System is contained in the Annex to Decision 24/CP.7, FCCC/CP/2001/13/Add.3, taken by COP7. COP7 recommends that the first session of the Conference of the Parties serving as the Meeting of the Parties to the Protocol (COP/MOP) adopt the Annex.

[8] X Wang and G Wiser, 'The Implementation and Compliance Regimes under the Climate Change Convention and its Kyoto Protocol' (2002) 11(2) RECIEL 181–98.

applying higher emissions baseline scenarios to any project, in order to demonstrate a greater level of additionality in emissions reductions, leading to the certification of a larger quantity of CERs. If, in the longer term, the market is flooded with cheap CERs, returns on investment could drop, and host countries may develop an interest in producing fewer, higher quality CERs. Concerns about inflated baselines and an overproduction of CERs are not necessarily based on a presumption of bad faith. The counterfactual baseline scenarios at the heart of each project will depend upon highly subjective judgments. If those exercising these subjective judgments are consistently rewarded when projects generate higher levels of CERs, a biased trend could result. Bad faith collusion between investors and hosts, in manipulating the CDM rules does, nonetheless, remain a possibility. Although the Protocol's Compliance System will promote compliance with the CDM rules this system too is overseen by the Parties themselves.

The design of the CDM and the Protocol's other flexibility mechanisms is based, in part, on existing domestic regulatory schemes, but none of these provided precedents that could answer all of the CDM's integrity-related concerns.[9] At the time the Protocol was negotiated existing models were all 'cap-and-trade' systems; there were no precedents for an emissions trading scheme based on exchanges between 'capped' and 'uncapped' sources. There was no international or domestic precedent for a market based regulatory mechanism in which governments play the roles of ultimate buyer, seller, and regulator. Finally, the CDM is unique in that it seeks to integrate into the state-based compact of the Kyoto Protocol, a role for the private sector, in order to tap its natural role in the market, and its expertise and relative independence from governments in providing project assessment and monitoring. The engagement of the private sector necessarily triggers the application of domestic regulatory frameworks, as well as private international and domestic laws generated by commercial contracts. While the legal relationships that engage private entities can 'thicken' the amount of law applicable to a CDM transaction, these relationships can also cloud the traditional state-to-state lines of accountability under public international law.

The designers of the Kyoto Protocol were well aware of these concerns and challenges. In response, they have built into the CDM's Project Cycle an impressive set of safeguards designed to ensure clear and consistent standards for calculating the environmental additionality of CDM projects, processes for encouraging technical assessments, transparency and public scrutiny of CDM operations, and mechanisms for rejecting or discounting CERs when problems emerge. A unique set of institutions has been created to oversee these procedures. These institutions include as their members government representatives, individual experts, and independent auditors. The CDM's designers have generated a creative but untested mix of relationships between states, private

[9] J Werksman, 'Compliance and the Kyoto Protocol: Building a Backbone into Flexible Regime' in J Brunnee and E Hay (eds), *Yearbook of International Environmental Law* 9 (Oxford University Press, 1999) 48–101.

Examples of Existing and Proposed Domestic Cap-and-Trade Systems

Ontario SO_x NO_x Trading Scheme[10]

Ontario's emissions trading system is of the 'cap, credit and trade' variety, a hybrid that incorporates features of pure 'cap-and-trade' with those of a 'baseline-and-credit' system.

A certain number of entities is subject to an emissions cap for SO_x and NO_x. These caps are lowered gradually to achieve provincial commitments to reduce emissions. Other uncapped emitters can achieve emissions reductions. These are calculated through a baseline (that is, the emission rates in the year prior to when the credit creating action was taken) and credits system. Credits will result from actions taken at the project level. Credits cannot be created by a company that reduces its emissions by going out of business or by cutting back on production. The Ministry of Environment needs to approve the credits before they can be registered. The Emission Reduction Credits (ERCs) only exist when they are created and registered.

The capped entities can use the ERCs from the uncapped ones only to fulfil a part of their allowances. The maximum allowable use of ERCs to an emitter will be limited to 33 per cent of allowances used for NO_x and 10 per cent of allowances used for SO_2.

UK Emissions Trading Scheme[11]

The UK Greenhouse Gas Emissions Trading Scheme 2002 equally is a combination of 'cap-and-trade' and 'baseline-and-credit'. The scheme allows for different kinds of participants:

1. Entities participating voluntarily. A baseline and a target of absolute emissions reductions is set for them for each year. When the target is reached in a year, an incentive payment is paid.
2. Companies, which already had emissions targets under Climate Change Agreements (CCAs). As the targets under these CCAs are often relative (efficiency targets), these entities can participate in an emissions trading scheme to fulfil their commitments, but the scheme is controlled by a 'gateway'.
3. Non-capped entities. These can undertake emissions reduction projects and sell the reduction credits. A project must be additional (baseline + credit). The reduction credits will be received after verification of the actual reduction.

Capped entities can use the reduction credits generated by the non-capped ones to reach their target.

Proposed US Climate Stewardship Act of 2003[12]

The Climate Stewardship Act of 2003 (S. 139) is a plan proposed by Senators McCain (Rep., Arizona) and Lieberman (Dem., Connecticut) to cap US GHG emissions.

[10] Ontario Regulation on Emissions Trading, made under the Environmental Protection Act, entered into force 31 December 2001. See also: <http://www.ene.gov.on.ca/envision/env_reg/er/documents/2001/RA01E0020-C.pdf> accessed 31 January 2004.

[11] UK Greenhouse Gas Emissions Trading Scheme 2002. See also: <http://www.defra.gov.uk/environment/climatechange/trading/ukets.htm#Rules> accessed 31 January 2004.

[12] Proposal for Climate Stewardship Act of 2003 (S.139). US Senate site: <http://thomas.loc.gov/cgi-bin/query/D?c108:1:./temp/~c1080hNGaA> accessed 31 January 2004. See also <http://web.mit.edu/globalchange/www/MITJPSPGC_Rpt97.pdf> accessed 31 January 2004.

The Act caps sectors at their 2000 emissions levels in Phase I of the program, running from 2010 to 2015, and to their 1990 emissions levels in Phase II starting 2016. Credits generated outside the capped sectors may be used by capped entities for staying within their cap. These 'outside credits' may come from CO_2 sequestration in soils and forests, reductions in emissions from uncapped sectors, allowances acquired from foreign emissions trading systems, and from a special incentive program for automobile manufacturers.

actors, and international institutions designed to promote environmental integrity.

This chapter will briefly introduce the actors in the CDM Project Cycle and then review the nuts and bolts of the Project Cycle, flagging the actors' roles and relationships. It then turns to a more detailed look at the framework of rules built into the CDM designed to promote the CDM's environmental integrity, and raises questions that the CDM's rules and practices have yet to answer. It will conclude with remarks on whether a mixed market mechanism like the CDM, and the direct engagement of the private sector in its operations are likely to strengthen or to confuse the legal basis for the implementation of the Kyoto Protocol.

2 Overview of the CDM

Article 12 of the Kyoto Protocol sets out the basic provisions of the CDM, but leaves out many details of the CDM's operation. This is in part because negotiators ran out of time in trying to complete a mechanism that came as a last minute 'surprise' to many observers, and in part to accommodate delegations' diverging perspectives on how the CDM would function. Article 12 anticipates that the Parties to the Protocol, once it has entered into force, will, through the Conference of the Parties serving as the Meeting of the Parties (COP/MOP), take additional decisions to bring the CDM into operation. Even as the Protocol was adopted, negotiators anticipated that the COP/MOP would have to elaborate, amongst other things, 'modalities and procedures with the objective of ensuring transparency, efficiency and accountability' of the CDM.[13]

At the seventh session of the Conference of the Parties to the UNFCCC (COP7) in Marrakesh, nearly four years after the Protocol was adopted, the first set of decisions on the institutions and procedures for the functioning of the CDM was taken. These decisions form part of the Marrakesh Accords, which include modalities for the 'prompt start' of the CDM that will allow it to begin operations prior to the entry into force of the Protocol. The Marrakesh Accords include draft decisions that will be forwarded to the first meeting of the COP/MOP for formalization.[14]

[13] Art 12(7) Kyoto Protocol.

[14] Decision 17/CP.7 on 'Modalities and procedures for a clean development mechanism as defined in Article 12 of the Kyoto Protocol' and its annex on 'Modalities and procedures for a clean development mechanism as defined in Article 12 of the Kyoto Protocol', hereinafter the 'CDM Modalities and Procedures'.

As a first basic step towards ensuring the environmental integrity of the CDM, Article 12 provides that Annex I countries may only use the CERs of CDM projects to contribute to compliance with *part of* their emissions caps.[15] This provision was included at the insistence of the European Union and many developing country delegations, including small island States, that were sceptical about trading between capped and uncapped countries. These delegations shared a concern that the CDM would take the pressure off Annex I countries to cut domestic emissions, and could restrict the development choices of developing countries by pressuring them to limit their emissions. Since then, delegations have been unable to agree how a limitation on the use of the CDM could be quantified, agreed, and enforced. Instead, Annex I Parties have agreed to include, as part of their regular reports to the COP/MOP information that demonstrates (presumably in qualitative terms) how their use of the CDM and the Protocol's other flexibility mechanisms is 'supplemental to domestic action.'[16] This effort to cap the quantity of CERs available for use in Annex I countries does, however, reveal a continuing level of concern over whether the CDM will deliver a net gain or a net loss to efforts to combat climate change.

3 CDM Project Cycle

3.1 Actors in the CDM Project Cycle

As mentioned before, an untested mix of private and public, existing and newly created entities play a role in the CDM Project Cycle. The main institutions involved are the Conference of the Parties serving as the Meeting of the Parties (COP/MOP) and the Executive Board (EB), both of which will be formally established when the Kyoto Protocol enters into force. Until then the Conference of the Parties to the Convention (COP), and the interim Executive Board established by the COP, are operating the CDM on a pilot basis.[17] The Protocol's COP/MOP, which will consist of all the Parties to the Kyoto Protocol, will provide overall authority and guidance to the CDM. The EB, composed of ten members (four from Annex I and six from non-Annex I Parties) is in charge of the regular supervision of the CDM. The EB will be assisted in some of its activities by at least three panels of experts. Since the Kyoto Protocol has not entered into force, these entities still lack a solid legal basis, but a temporary solution has been found to nevertheless enable the prompt start of the CDM: the COP of the UNFCCC acts as the

[15] Decision 15/CP.7, seventh preamble: '[t]he use of the [flexibility] mechanisms shall be supplemental to domestic action and that *domestic action shall thus constitute a significant element of the effort made by each Party included in Annex I* to meet its quantified emission limitation and reduction commitments under Article 3, paragraph 1.' (Emphasis added.)

[16] Annex to Decision 15/CP.7, paras 1 and 2. Some have argued that the main meaning of the term 'supplemental' strongly suggests that the percentage of the cap met by the flexibility mechanisms could not exceed 49 per cent.

[17] Art 12 Kyoto Protocol.

COP/MOP and in this capacity it has, on an interim basis, adopted the CDM Modalities and Procedures and elected the members of the EB.[18]

Other entities involved in the Project Cycle are project developers, host country governments, Designated Operational Entities, and the purchasers of CERs. A CDM project is initiated by a project developer. Either public or private entities can develop CDM projects. The project is hosted by the non-Annex I Party in whose territory it will be based. So-called Designated Operational Entities (DOEs), which can be domestic legal entities and international organizations, carry out monitoring functions at several stages of the Project Cycle. They validate a proposed CDM project and verify and certify during the lifetime of the project, that emissions reductions have occurred and can thus generate CERs. The EB accredits DOEs on the basis of a set of requirements agreed by the COP/MOP,[19] while the COP/MOP is responsible for their final designation as DOEs.[20] The CERs resulting from a CDM project can be purchased by private and public

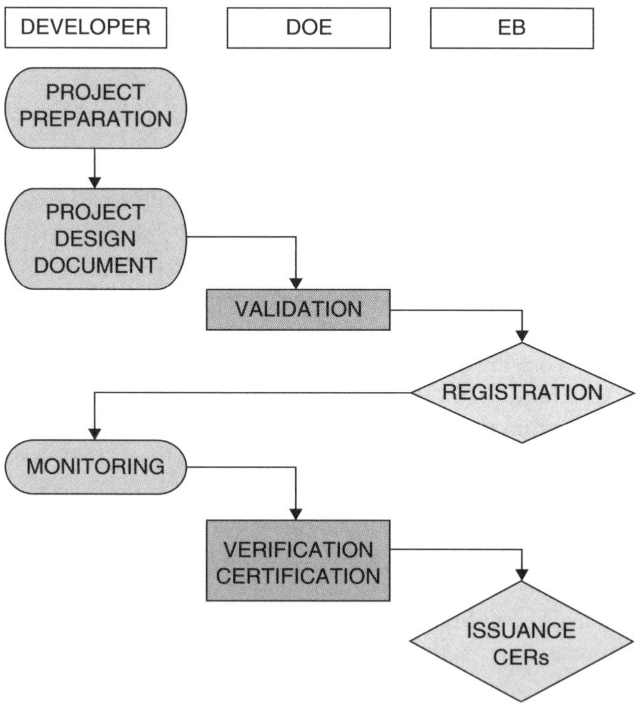

Figure 10.1: CDM Project Cycle

[18] Decision 17/CP.7, Arts 1–2.
[19] ibid, Art 6(b).
[20] Annex to Decision 17/CP.7, para 3(c).

entities, but are used by an Annex I Party, at the end of the Protocol's commitment period (2012) to demonstrate its compliance with its commitment.

The CDM Modalities and Procedures[21] spell out the criteria and the procedures by which projects can generate CERs, and are referred to as the CDM Project Cycle.[22] The steps of the Project Cycle are described here in more detail, focusing on aspects linked to the CDM's environmental integrity.

3.2 Project Eligibility

The CDM rules exclude some projects from eligibility. Out of long-term concern of some environmentalists that the efforts to combat global warming not lead to proliferation of nuclear energy, Annex I countries are to refrain from using emissions reductions generated from nuclear facilities to meet their reduction commitments.[23] Remaining concerns about the methodological challenges of calculating the long-term contributions of projects based on the uptake of carbon dioxide by plants, led Parties to agree that land use and land-use change projects are only eligible as CDM projects in as far as they are limited to afforestation and reforestation. Thus the more typical CDM projects are likely to be projects involving energy efficiency technologies, renewable energy sources (wind, solar, biomass, and so on), or projects switching from high carbon intensive fuel to lower carbon intensive fuel.[24]

3.3 Project Additionality

The project proponents must assess the project's additionality, and set this out in the Project Design Document (PDD).[25] This assessment must demonstrate that

[21] Decision 17/CP.7 of 10 November 2001 and Annex to Decision 17/CP.7 (including appendices).

[22] For small-scale projects, a more simplified Project Cycle has been developed in order to limit transaction costs and to help make these projects compete financially with larger scale investments. The small-scale projects eligible for a simplified regime are:

(i) Renewable energy project activities with a maximum output capacity equivalent of up to 15 MW (or an appropriate equivalent);
(ii) Energy efficiency improvement project activities which reduce energy consumption, on the supply and/or demand side, by up to the equivalent of 15 MW/hours per year;
(iii) Other project activities that both reduce anthropogenic emissions by sources and directly emit less than 15 kilotonnes of carbon dioxide equivalent annually.

(Decision 17/CP.7, Art 6(c) and Annex to Decision 17/CP.7, paras 5(e), 45(d), and 55).

While this streamlined Project Cycle for small-scale projects is not the focus of this chapter, it reflects an interesting trade-off between environmental integrity and other development objectives. Because smaller projects generate lower volumes of CERs they pose lower risk in terms of increasing global emissions. The benefits of encouraging investments in smaller communities and economies, and of engaging these countries in the climate change process for smaller projects, were considered worth any additional environmental risks.

[23] Decision 17/CP.7, fifth preamble.

[24] Examples from the UNDP, *Clean Development Mechanism Manual*, ch 2. See <http://www.undp.org/seed/eap/html/docs/cdmchapter2.pdf> accessed 20 February 2004.

[25] On the website of the Convention a model PDD is provided: <http://cdm.unfccc.int/Reference/Documents/cdmpdd/English/cdmpdd.pdf> accessed 16 December 2003.

the project will in fact achieve GHG emission reduction, when compared to the situation in the absence of the project.[26] Each PDD must describe:

- the *baseline scenario* from which this additionality is measured (the scenario that reasonably represents the GHG emissions that would occur in the absence of the proposed project activity),[27]
- the *project boundary and any adjustments for leakage*. Each project must calculate all 'greenhouse gas emissions under the control of the project participants that are significant and reasonably attributable to the CDM project activity'. These must then be adjusted for the 'net change of . . . greenhouse gas emissions which occurs outside the project boundary, and which is measurable and attributable to the CDM project activity'.[28]

Requiring project proponents to estimate and discount for potential 'leakage' helps to address concerns raised by environmentalists that CDM investments could displace, rather than replace GHG-intensive activities. For example, leakage can result if fossil fuel saved through a CDM project results in increased use of that fossil fuel elsewhere.[29] All methodologies used to calculate additionality have to be approved by the EB,[30] which is assisted in this function by the 'Panel on guidelines for methodologies for baselines and monitoring plans' (the 'Meth Panel').[31]

3.4 Monitoring Plan

The PDD also has to include a detailed monitoring plan, designed to transparently and verifiably collect all relevant data for the establishment of the baseline, of the project's emissions and of possible leakage.[32] Before the drawing up of the PDD, a new methodology for monitoring is submitted to the EB for its approval. The procedure for approval is similar to the one for the establishment of a baseline methodology and the EB is assisted, again, by the Meth Panel.[33] If an already existing and approved monitoring methodology is used, it does not need the (renewed) approval of the EB.

3.5 Environmental Impact Analysis/Assessment

In addition to calculating the project's impact on GHG emissions, the project developer or the host country government must also carry out an 'analysis of the

[26] See Annex to Decision 17/CP.7, paras 37(d), 43, 44.

[27] Annex to Decision 17/CP.7, para 44. See on baselines: *Emission baselines, estimating the unknown*, OECD and IAEA (OECD, 2000).

[28] Annex to Decision 17/CP.7, paras 51–2.

[29] See B Schlamadinger and G Marland, *Land Use and Global Climate Change: Forests, Land Management and the Kyoto Protocol*, prepared for the Pew Center on Global Climate Change (June, 2000) 7 and 32.

[30] Annex to Decision 17/CP.7, paras 5(d), 37(e) and 38.

[31] Under the CDM Modalities and Procedures, the EB has been given the right to establish committees, panels or working groups for assistance. See ibid, para 18. The Meth Panel was established during the fourth meeting of the EB. See report CDM-EB-04 of 10 June 2002.

[32] Annex to Decision 17/CP.7, para 53.

[33] ibid, paras 5(d), 37, 38, and 54.

environmental impacts of the project activity, including transboundary impacts, in accordance with procedures as determined by the host Party.' If those impacts are considered significant by the project participants or the host country, an environmental impact assessment must be undertaken 'in accordance with procedures as required by the host Party.'[34] The results of the environmental impact assessment must be included in the PDD.[35] The CDM Modalities and Procedures do not provide for the situation that the host country does not have any laws on environmental impact assessment.

3.6 Public Participation of Local Stakeholders in Project Design

Under the CDM Modalities and Procedures, the project developer must make sure that local stakeholders ('the public, including individuals, groups or communities affected, or likely to be affected, by the proposed clean development mechanism project activity') participate in the process of designing the project.[36] The proposed project is to be made available to the public for comments and a summary of the comments, as well as a report on how due account was taken of these comments is to be part of the PDD.[37] It was strongly argued during the negotiation of the Protocol, that public scrutiny and participation would improve the environmental integrity of CDM project design and decision-making.[38]

3.7 Validation

Once all the conditions described above are fulfilled, the PDD, together with the written approval of the host country (including a confirmation that its project participation is voluntary and that the project activity assists it in achieving sustainable development)[39] and an affirmation that the money used by an Annex I country for the creation and implementation of the project is not the result of a diversion of that country's Official Development Assistance (ODA),[40] is submitted by the project developer to a DOE for validation.[41] The affirmation regarding diversion of ODA is meant to address two related concerns. The first concern, raised by many NGO observers, relates to the CDM's environmental integrity. Financial assistance for development projects that lower host country GHG emissions predates the Kyoto Protocol and therefore, should form part of the baseline of emissions reduction efforts of the host country, and should not be considered additional. The second

[34] ibid, para 37(c) and Appendix B, Art 2(e). [35] ibid.

[36] ibid, para 1(e). The decision only mentions 'local stakeholders' here (para 37(b)), whereas it mentions 'stakeholders and accredited NGOs' in the international consultation procedure (para 40(c)). From that it can be inferred that local stakeholders do not need to be NGOs or accredited to the UNFCCC in order to participate in these consultations. [37] ibid, para 37(b) and Appendix B, Art 2(g).

[38] See, Rio Declaration on Environment and Development, Principle 10.

[39] Decision 17/CP.7, fourth preamble, Annex to Decision 17/CP.7, para 40(a).

[40] ibid, seventh preamble paragraph and Appendix B, Art 2(f).

[41] ibid, Annex, paras 27(a) and 35.

concern, raised by developing country governments, was that donor interest in promoting the CDM as a source of emissions credits for their own use, should not lead to a diversion of ODA from existing priorities, such as public health or education.

3.8 International Public Comment Period

Once the PDD is complete, the DOE assigned to validate the project must make it available to the public in an international consultation procedure.[42] Within thirty days Parties, stakeholders, and non-governmental organizations (NGOs) accredited by the UNFCCC can provide their comments to the PDD. The DOE makes the comments publicly available, and, when submitting the project for registration, explains how it has taken due account of these comments.[43] Again, public scrutiny is assumed to lead to improved project design.

3.9 Registration

Registration is the formal acceptance by the EB of a validated project as a CDM project activity. A project must be registered before verification, certification, and the issuance of CERs related to that project activity, can take place.[44] A request for registration is considered granted and the registration final, unless three or more members of the EB or a Party request review of the proposed CDM activity. The review can only deal with 'issues associated with the validation requirements', including the additionality assessments and other requirements associated with the project's environmental integrity. The outcome of the review is communicated to the project participants and the public. If the EB rejects the project, it is possible for the project developer to revise the proposed activity in order to have it revalidated and re-registered, taking into account the remarks made in the review report. Pending a review a project cannot be registered.

3.10 Implementation and monitoring

All of the phases described above precede the actual implementation of the CDM project.

In the implementation phase, the project is carried out. The monitoring plan that was part of the PDD (and that was accepted by the EB) takes effect in this phase. As mentioned before, the monitoring plan establishes a set of methods to measure the GHG emissions of the project activity in relation to the baseline, within the project boundary and taking into account possible leakage and environmental impacts.[45] Based on the monitoring, GHG reductions can be calculated and submitted for verification as CERs.[46]

[42] Decision 17/CP.7, fourth preamble, Annex to Decision 17/CP.7, para 40(b).
[43] ibid, para 40(f). [44] ibid, para 36. [45] ibid, para 59.
[46] ibid, paras 53–8.

3.11 Verification and Certification

Verification and certification of emissions reductions achieved by the CDM project is carried out by a DOE. In order to avoid conflicts of interests, except in exceptional circumstances, the DOE that validated the project cannot also be responsible for verifying and certifying emissions reductions.[47] The DOE verifies the information given by the project developer (if necessary through on-site visits or data from other sources), and generates a verification report that certifies in writing the amount of emissions reductions that would not have occurred without the project.[48]

3.12 Issuance of CERs

The EB ultimately issues the CERs. This issuance is automatic unless at least three members of the EB or a Party involved in the project activity requests a review.[49] At this stage in the CDM Project Cycle, the EB's discretion is limited. The scope of the review is limited to 'issues of fraud, malfeasance or incompetence of the designated operational entities . . . '[50] Combined, these rules place a considerable procedural momentum behind the approval of the CERs by the EB, and leave considerable discretion and authority with the DOE. This momentum will, presumably, only increase with the volume of projects seeking CER issuance, and the technical and political resources necessary to challenge any particular issuance are strained.

Once this process is complete, the EB instructs the CDM registry administrator to issue the required quantity of CERs to the EB's account. A part of the CERs transferred to the EB's account are then transferred to a special account used to cover administrative expenses and to meet costs of adaptation projects in those states that are particularly vulnerable to the adverse effects of climate change.[51] The remaining CERs are forwarded to the accounts of Parties and project participants as requested by them.[52]

4 Institutions and Procedures Safeguarding the CDM's Environmental Integrity

4.1 The Parties to the Protocol

At the core of the CDM Project Cycle are the Parties to the Protocol: developing country Parties who will benefit from CDM project activities, and the Annex I

[47] ibid, para 37(e) [48] ibid, paras 61–3. [49] ibid, para 65. [50] ibid, para 65.

[51] ibid, para 66(a). The share reserved for adaptation projects is set at 2 per cent of the CERs issued for a CDM project activity (Decision 17/CP.7, Art 15(a)). CDM projects in the least developed country Parties are exempt from sharing proceeds to assist with the cost of adaptation projects. This makes them more attractive for investors. [52] ibid, para 66(b).

Parties that can use CERs generated by those projects to contribute to compliance with their emissions caps. Participation by Parties in the CDM is voluntary[53] and limited to Parties to the Kyoto Protocol.[54] Furthermore, each Annex I Party is required to meet basic eligibility requirements before it can use CERs generated by the CDM.[55] These eligibility requirements are primarily related to the calculation of the Party's initial cap, the submission of emissions inventories, and the national registries that will be used to track the trade taking place through the flexibility mechanisms.[56] This information is essential for the smooth functioning of the Protocol, and for ensuring, for example, that CERs are not double counted as they move from Party to Party. Compliance with these requirements will be reviewed by the Enforcement Branch of the Protocol's Compliance System, which will have the power to suspend an Annex I Party's eligibility to use the CDM if that Party fails to meet these requirements.[57] The suspension can be reviewed and eligibility reinstated through an expedited procedure, and need not interfere with CDM project-level activities. But if a suspended Party is a major, upfront investor in a project, its suspension could put the financial situation of the project at risk.

Non-Annex I host countries are not, however, under the jurisdiction of the Protocol's Enforcement Branch and cannot be suspended from CDM eligibility. The monitoring of CDM project performance will thus be left to the DOEs, and to the performance requirements set out in the CDM Modalities and Procedures, including the public participation mechanisms and the environmental impact assessment. Compliance with these requirements will depend primarily on host country legal and regulatory systems.

4.2 The Executive Board

As has been described, the EB plays a central role in supervising the CDM Project Cycle, and much of the CDM's integrity will depend upon the EB's performance. The EB, for example, will establish the methodologies for calculating environmental additionality, it is responsible for the accreditation of the DOEs, for ensuring access to information and the transparency of the CDM's operations, and for the final issuance of CERs. For these reasons, the integrity and legitimacy of the EB's decision-making will be essential to the integrity, legitimacy, and environmental additionality of CERs.[58] General, political oversight of the EB will be provided by the COP/MOP. The Compliance Committee has no jurisdiction to review decisions made by the EB.

[53] Art 12(5)(a) Kyoto Protocol, Annex to Decision 17/CP.7, para 28. [54] ibid, para 30.

[55] ibid, para 31. [56] ibid.

[57] Decision 24/CP.7 (Procedures and Mechanisms relating to compliance under the Kyoto Protocol).

[58] See on the importance of impartiality and independence as a basis for legitimacy D Bodansky, 'The Legitimacy of International Governance: A Coming Challenge for International Environmental Law?' 93 *American Journal of International Law* 596 (1999), and L Gormley and J De Haan, 'The Democratic Deficit of the European Central Bank' 21 E.L. Rev. 95 (1996).

The CDM's designers have made significant efforts to ensure that the EB's decisions are made in an apolitical and professional manner. Although nominated by Parties and elected by the COP/MOP, EB members are not intended to serve as political representatives, but are instead required to '[p]ossess appropriate technical and/or policy expertise and shall act in their personal capacity.'[59] In a step unusual for a treaty-based body of this nature, the EB members are required to take a written oath of service declaring they have no pecuniary or financial interest in any aspect of a CDM project activity or any DOE.[60] Members may be suspended by the EB 'for cause including, *inter alia*, breach of the conflict of interest provisions, breach of the confidentiality provisions, or failure to attend two consecutive meetings of the executive board without proper justification.'[61]

These safeguards should help address concerns that the CDM could develop a culture of approval, similar to that observed in development banks, where members tend not to challenge projects based in other countries, for fear that projects in which their own countries have an interest will be challenged in retaliation. A seminal review of the World Bank published in 1992 that led to major reforms showed that complex safeguards designed to ensure the social and environment integrity of World Bank projects, were routinely ignored under the pressure to approve loans.[62] While the EB in the CDM's Project Cycle will not be under pressure to 'move money' in the same way as the Bank, it will be under pressure to approve projects and to issue CERs.

However, some of the CDM's decision-making safeguards may be difficult to apply in practice. During the pilot phase of the EB, many of the members of the EB elected to serve in the interim EB also participated in its design, and continue to represent their governments in the ongoing negotiations on the design of the CDM, and other aspects of the Protocol. While these individuals are undoubtedly experts in the treaty texts they themselves designed, it is open to question whether it is reasonable to expect them to wear both hats, of independent decision-maker in the CDM, and diplomatic representative in the COP, comfortably. It might be unlikely for a diplomatic representative sitting on the EB to have a direct financial or pecuniary interest in a project of the kind that the CDM rules prohibit, but there can be little doubt that these individuals have a professional interest in seeing their interpretation of the Kyoto Protocol prevail, and that they have an interest in having projects that are advantageous to their own countries, approved. The risk remains, therefore, that the CDM's project and CER approval process will be vulnerable to 'political horse-trading.'

Nevertheless, individual steps of the CDM Project Cycle, described above, will be subject to a high degree of transparency and scrutiny, in particular by professionals

[59] Annex to Decision 17/CP.7, para 8(c).

[60] ibid, para 8(e) and (f) and Rules of procedure of the Executive Board of the Clean Development Mechanism, Rule 10. No similar oath is required, for example, of Council Members of the Kyoto Protocol's Compliance Committee that will assess and enforce Parties' compliance with the Protocol or the Global Environment Facility which will operate as the Protocol financial mechanism.

[61] Annex to Decision 17/CP.7, para 10.

[62] World Bank, *Effective Implementation: Key to Development Impact* (Wapenhans Report, 1992) 14.

from the private sector and by the NGO community, which will make political interference more difficult. EB meetings are open to accredited NGO observers, and proceedings are webcast live and archived on the internet.

4.3 Expert Panels

The independence of the EB's decision-making is enhanced by the guidance of external technical experts. The EB has created Panels on methodologies, on accreditation, and on small-scale projects, to provide this guidance. While these panels tend to be chaired by members of the EB that also serve as country representatives, they also include in their membership international civil servants and representatives from the private, NGO, and academic sectors.[63]

The recommendations of the panels will play a very important role in the EB decisions. Already, the panels have developed and recommended at least thirty-six new methodologies to the EB. Eight of these methodologies have had the final approval of the EB, six of which were accepted after earlier (partial) refusals. When reviewing the different recommendations of the Meth Panel and the subsequent decisions by the EB, it becomes clear that the EB closely follows the Meth Panel's recommendations.[64]

4.4 Private Entities

Article 12 of the Kyoto Protocol provides that CDM projects may involve private entities. As the CDM rules and practice have evolved, private entities have begun to play two main roles: as project developers, and as DOEs responsible for validating projects, and verifying and certifying emissions reductions. A private entity involved as project developer, can only participate in the project with the authorization of a Party. In addition, a private entity participating in a CDM activity may only acquire or transfer CERs if the Party, which had authorized the participation, is eligible to do so. The participation of an authorized private entity does not add to or remove from the Party's responsibility for the fulfilment of its Kyoto obligations.[65]

This set of relationships reflects a Party's ambivalence towards the role of private entities in implementing sovereign obligations. The private entity is empowered to participate, but can only do so with the authorization of a Party. There is no requirement that the entity have any pre-existing link to its sponsor government

[63] Under the CDM Modalities and Procedures, the EB has been given the right to establish committees, panels or working groups for assistance. See Annex to Decision 17/CP.7, para 18. The Meth Panel was established during the fourth meeting of the EB. See report CDM-EB-04 of 10 June 2002.

[64] The Meth Panel recommends that a proposed methodology is either accepted (A), kept under consideration (B) or is not approved (C). In almost all cases documented on the UNFCC website, the EB follows the Meth Panel. See <http://cdm.unfccc.int/methodologies/process?cases = A>, for the A-cases, <http://cdm.unfccc.int/methodologies/process?cases = B> for the B-cases and <http://cdm.unfccc.int/methodologies/process?cases = C> for the C-cases, accessed 6 May 2004.

[65] Annex to Decision 17/CP.7, paras 33 and 36.

and, presumably a Party could for example authorize the participation of a company registered in another Party, or even a company registered in a non-Party. The entity's interests are linked to the performance of its sponsor government—if a Party becomes ineligible to acquire and transfer CERs, so do the entities it has sponsored. On the other hand, the rules clarify that the authorization by a Party of an entity does not result in devolution of that Party's sovereign commitment under the Protocol. It remains unclear whether, under private international law or under domestic law, a government could be held accountable for the actions of a private entity that it has agreed to sponsor. In other words, does the act of authorization result in a relationship of principal and agent between the state and the entity, such that the sponsoring government could be held liable in case of a breach of contract or other law?[66]

Private entities acting as DOEs play a crucial role in promoting the CDM's integrity by validating CDM projects, preparing them for registration, and verifying emissions reductions for issuance by the EB as CERs.[67] To be eligible as a DOE, an entity must demonstrate that it meets a long list of requirements compiled by the secretariat from the Marrakesh Accords. A DOE must be either a domestic legal entity or an international organization and employ a sufficient number of persons competent in the areas of work in which it operates.[68] It must demonstrate financial stability, insurance coverage and resources and sufficient arrangements to cover possible legal and financial liabilities.[69] A DOE also has to document its systems of legal accountability and liability, described in rather opaque terms as 'procedures for handling complaints, appeals and disputes'.[70] In addition, a DOE has to show for each project that it has no conflict of interest with the project participants.[71] As of the date of writing this chapter, twenty-three entities, all of them private entities, twenty-one from developed countries, have entered the accreditation process.[72]

The DOE must establish its bona fides on the basis of recommendation by the Accreditation Panel (AP).[73] This Panel is advised by Accreditation Teams (AT), which carry out the more detailed assessment work, including desk reviews, on-site assessments, and spot checks. The name and sectoral interests of each applicant DOE are posted on the Secretariat's website and Parties, stakeholders, and accredited NGO observers are invited to comment on the merits of the application. As of the date of this chapter only six brief comments were observable with regard to the

[66] If the relationship between the private entity and the sponsoring State would be one of principal and agent, a situation could arise where a private DOE, acting as the agent of country A, would have to validate a project initiated by a private project developer, acting as the agent of country B. In essence, this could be seen as country A validating the project of country B. That would be quite an untenable situation and relatively unique in international law.

[67] Annex to Decision 17/CP.7, para 27.

[68] Not every DOE is designated to work in *every* project area. Some will be designated only for certain specific areas.

[69] Annex to Decision 17/CP.7, Appendix A, para 1.

[70] ibid, Art 1(a).

[71] ibid, para 27(d).

[72] See <http://cdm.unfccc.int/DOE/CallForInputs> accessed 6 May 2004.

[73] Establishment of this panel is based on Annex to Decision 17/CP.7, para 18 *juncto* 25.

twenty-three DOE applications.[74] Two of the applications had been approved.[75] When the Protocol enters into force the COP/MOP will formally designate the DOEs. The COP/MOP has the additional task of ensuring a sufficient regional and subregional distribution of DOEs by promoting accreditation by the EB of DOEs from developing country Parties.[76]

As has been mentioned, the credibility of the DOE function is further safeguarded by rules that, for example, prohibit (unless the EB decides otherwise) the DOE that was in charge of the validation of a project in its design phase, from being involved in the later verification and certification of emissions reductions from that project.[77] DOEs are held accountable by the COP/MOP for complying with applicable CDM Modalities and Procedures and relevant decisions of the COP/MOP and the EB.[78] Accountability is partly achieved through the inclusion of provisions on suspension or withdrawal of designation in the Modalities and Procedures. If the EB finds, after review of a CDM project activity, that a DOE no longer meets accreditation standards or other provisions of COP/MOP decisions, it may recommend to the COP/MOP that the designation of an Operational Entity is withdrawn or suspended. Withdrawal or suspension directly takes effect after the recommendation.[79] At the moment there are no procedures that would allow the DOE to formally challenge and reverse these decisions.

DOE accountability is also achieved through financial penalties. If significant deficiencies are identified in a validation, verification or certification report for which a DOE was responsible, the EB can issue a review by another DOE to find out the magnitude of the deficiencies. If this review shows that excess CERs have wrongly been issued, the responsible DOE has to pay for the cost of the review and it has to buy or transfer as many CERs as the excess amount to the EB's cancellation account.[80] In addition, a DOE has to comply with the applicable laws of the Party hosting the CDM project activity.[81] These may include contract, customary, and statutory law.

In cases involving private investors and DOEs foreign to the host country, regional or bilateral investment treaties between the host country and the entity's home country may apply. This raises a series of interesting questions about the interaction between CDM instruments that promote objectives of economic liberalization, prohibit non-discrimination between foreign and domestic investors and among foreign investors, and regulate expropriation of investments.[82] It is unclear, for example, whether CDM rules that favour investors that have been

[74] See <http://cdm.unfccc.int/DOE/CallForInputs> accessed 6 May 2004. This may be due, in part, to the lack of information posted on each applicant. The website only indicates the name of the entity and the sectors in which it hopes to operate.

[75] DNVCert and JQA. See <http://cdm.unfccc.int/DOE/list> accessed 6 May 2004.

[76] Annex to Decision 17/CP.7, para 3(c), Appendix A.

[77] ibid, para 37(e).

[78] ibid, para 26.

[79] ibid, para 21.

[80] ibid, para 22.

[81] ibid, para 27(c).

[82] J Werksman, K Baumert, and N Dubash, 'Will International Investment Rules Obstruct Climate Protection Policies?' 3 *International Environmental Agreements: Politics, Law and Economics* **3**(1)59–86 (Kluwer, 2003).

authorized by Protocol Parties could be viewed as discriminatory. The legal situation is further complicated by the direct involvement of the EB in the regulation of private entities. By what procedural means or legal authority would penalties or costs imposed by the EB be enforced against entities? Does the home or host country step in to enforce these decisions? What role would international law play in mediating any disputes that might arise between the enforcing state and the regulated entity, or between the EB and the regulated entity?

Although almost all CDM projects will have one DOE involved in the validation of baselines, and another in the certification of emissions reductions, the CDM Modalities and Procedures do not pre-determine who will be contracting and paying for these services. Generally, it can be assumed that the DOE contract will be financed by one or more of the project proponents, as negotiated between them. There are, as yet, no standard terms of reference to govern these contracts. The CDM Modalities and Procedures do, however, require a DOE to demonstrate that it has 'no real or potential conflict of interest with participants in the CDM project . . . '[83] and that staff and senior management are not involved in any commercial, financial or other processes, which might influence their judgment.[84] This would seem to suggest that a DOE cannot itself be an investor in, or purchaser of, the project, and that the fees it receives under the contract cannot be linked to the project's performance. If DOE service contracts are developed so as to protect against such conflicts of interest, and are well-monitored and enforced, this should help to promote the CDM's integrity.

4.5 Public Participation

Public scrutiny of CDM decision-making will play a central role monitoring the environmental integrity of the CDM and its projects. As has been described, the CDM at several stages has possibilities for public participation. The project developer must allow local stakeholders to participate in the design of the project, and indicate how he has taken due account of these comments in the PDD. A DOE must undertake a similar international consultation procedure prior to validation of the PDD. The DOE accreditation process also allows the possibility for public comments. Although each instance of public participation undoubtedly helps attain a higher level of integrity of the CDM, the requirements are somewhat skeletal compared to many domestic 'notice and comment' rules. For example, although local stakeholders can participate in the design of the project, the CDM rules do not indicate how stakeholders are to find out about a plan to develop a CDM project and about possible implications of the project.

In order to provide for more standardized rules in respect of public participation, the CDM could refer to emerging regional and international standards in the area of environmental decision-making. Instruments such as the UN/ECE Aarhus

[83] Annex to Decision 17/CP.7, para 27(d). [84] ibid, Appendix A, para 2(a)(ii).

Convention on Access to Information, Public Participation in Decision-making and Access to Justice in Environmental Matters[85] and the EC Directive on Public Access to Environmental Information,[86] which contain more-or-less generally accepted requirements for access to information, public participation, and access to justice, could be taken as a starting point for the public participation requirements of the CDM. Besides enhancing environmental integrity, there also is a more market-oriented argument for introducing such public participation provisions in the CDM: they are more uniform, and more detailed, standardized rules on public participation could help prevent a 'race to the bottom' towards (low cost) projects in countries with no or sub-standard public participation legislation.

4.6 Environmental Impact Assessment

The same arguments in favour of harmonized rules apply to CDM requirements on environmental impact analysis/assessment (EIA). As was noted in the description of the CDM Project Cycle, when a project participant or the host country finds that a project may have a significant environmental impact, an EIA must be undertaken by the project developer, 'in accordance with procedures as required by the host Party.'[87] But the more extensive the rules for the EIA, the more significant the chances are that the CDM project will not even materialize and/or the more costly the CDM project may become because of environment-driven adjustments. Since a host country generally benefits from a CDM project, the absence of uniform EIA rules for the CDM may create an incentive for the host country to not insist on a thorough EIA, in order to make its country attractive for CDM projects. The CDM's geographical flexibility should not become a means of channelling projects to host countries with the lowest environmental standards. Again, existing regional and international standards, such as the Espoo Convention on Environmental Impact Assessment in a Transboundary Context,[88] or the World Bank's operational procedures on environmental assessment,[89] could prove useful in this regard by establishing uniform principles for EIAs.

5 Concluding Observations

The CDM's revolutionary experiment in international environmental policy-making has already generated more detailed rules, institutions, and procedures

[85] UN/ECE Convention on Access to Information, Public Participation in Decision-making and Access to Justice in Environment Matters, 38 ILM 517 (Aarhus, 25 June 1998).

[86] Directive 2003/4/EC of the European Parliament and of the Council of 28 January 2003 on public access to environmental information and repealing Council Directive 90/313/EEC [2003] OJ L41/26.

[87] Annex to Decision 17/CP.7, paragraph 37(c).

[88] 30 ILM 802 (Espoo, 28 February 1991).

[89] World Bank Operational Policy/Bank Procedures 4.01, available at <http://wbln0018.worldbank.org/Institutional/Manuals/OpManual.nsf/944eea1d5fb31d95852564a30060b223/9367a2a9d9daeed38525672c007d0972?OpenDocument> accessed 20 July 2004.

than is typical of an international environmental agreement. And yet its innovations raise more questions, particularly for lawyers, than its designers have been able, thus far, to answer through practice. Given the importance of certainty and predictability to the attraction of investment capital and the creation of new markets, these questions may plague the CDM for some time.

Clearly, the most important step towards the CDM's long-term future will be the entry into force of the Protocol. At least until the Kyoto Protocol enters into force, the pilot phase EB, where so much of the CDM's credibility and legitimacy rests, is operating under the somewhat attenuated authority of the UNFCCC COP. Even after the Protocol enters into force, and the present arrangements are blessed by the Protocol's COP/MOP, the legal basis for EB decisions will remain unclear. The authority of treaty bodies such as the COP/MOP and the EB to make decisions with binding legal effect has been frequently questioned.[90] The legal route to challenging or confirming the EB's authority is similarly unclear.

When the CDM was first designed, governments and observers seemed to have placed a great deal of trust in the private sector as a source of finance, innovation, and, crucially, independent monitoring and evaluation of CDM activities. While the private sector remains the main source of foreign direct investment in climate related activities in the developing world, public perceptions of corporations' capacities to independently audit this industry were severely shaken by scandals such as the one involving ENRON and its accountants. The fate of the CDM and the environmental integrity of the Kyoto Protocol may thus depend upon an arm's length partnership between States and private entities that combines the strengths of an innovative market, and of a more traditional regulator.

[90] R R Churchill and G Ulfstein, 'Autonomous Institutional Arrangements in Multilateral Environmental Agreements: A Little-Noticed Phenomenon in International Law' (2000) 94(4) *American Journal of International Law* 623–60.

11

Meeting Participating Country Responsibilities under the CDM: Designating a National Authority

*Robert O'Sullivan and Charles Cormier**

1 Introduction

The Kyoto Protocol's[1] Clean Development Mechanism (CDM) assists developed countries reach their quantified emissions limitation and reduction commitments in a cost effective manner, and at the same time assists developing countries by promoting and enabling sustainable development through the transfer of clean technology and responsible private investment. As such, the CDM provides an opportunity for developing countries to participate in the global effort to reduce global greenhouse gas (GHG) emissions and help mitigate climate change.

Participation in the CDM is voluntary,[2] but in order for a developing country to host a CDM project, the host country is required to:

(i) Ratify the Kyoto Protocol; and
(ii) Designate a National Authority for the CDM.[3]

If the host country decides to allow private or public entities to participate in the CDM within its jurisdiction, the host country is also required to authorize such participation.[4]

* Robert O'Sullivan is a junior professional associate at the World Bank's Legal Department. Charles Cormier is a senior environmental specialist at the World Bank's Environmental Department. The views expressed in this chapter are the views of the authors and do not necessarily represent the views of any organization they may be associated with. Many of the example DNAs referred to in this chapter are still under development and subject to change. The authors would also like to thank Charlotte Streck and Eric Haythorne for reviewing and commenting on this chapter.

[1] FCCC/CP/1997/L.7/Add.1 Decision 1/CP.3, 'Adoption of the Kyoto Protocol to the United Nations Framework Convention on Climate Change,' Annex, reprinted in 37 ILM 22 (1998), *not yet in force* (hereinafter the Kyoto Protocol, or the Protocol).

[2] FCCC/CP/2001/13/Add.2, Decision 17/CP.7/CMP.1 'Modalities and procedures for a clean development mechanism', Annex, para 28.

[3] ibid, paras 29 and 30.

[4] ibid, para 33.

This chapter analyses what is required to create an efficient and transparent Designated National Authority (DNA) within host countries under the Kyoto Protocol. It focuses on what a host country DNA is required to approve, how this is done, and the outcome of the approval process. Finally administrative structures and funding a DNA are discussed.

2 Participating Country Responsibilities: Designating a National Authority

The Marrakesh Accords state that Parties participating in the CDM shall designate a national authority for the CDM. The purpose of the DNA is to provide:

> written approval of [1] voluntary participation from the designated national authority of each Party involved, including [2] confirmation by the host Party that the project activity assists it in achieving sustainable development.[5]

The requirement to provide written approval that a Party voluntarily participates in the CDM was included in the text to emphasize that the CDM does not create binding obligations on developing countries who participate in it.[6] The second requirement is more substantive as it asks the host country to determine whether or not a project will further its national sustainable development objectives. To do so most host countries intend to identify project specific sustainable development criteria and apply these criteria in an approval procedure that is used to assess potential CDM projects.

2.1 Project Approval

2.1.1 Project approval: scope of review

The scope of review in the approval of CDM projects required by the DNA is best understood in the context in which the CDM was created. The CDM was created to engage developing countries in the global effort to stabilize anthropogenic emissions of GHGs through participating in Joint Implementation (JI)-type projects.[7] As JI operates within the context of industrialized (Annex I) countries, it was created in a regulatory context where the Party achieving emission reductions has to account for its overall emissions and is therefore liable for any Emission Reduction

[5] ibid, para 40(a).

[6] Juan Pedro Searle, member of the delegation from the Republic of Chile to COP7.

[7] The CDM evolved from a Brazilian proposal for a 'Clean Development Fund' whereby Annex I Parties would pay a penalty for non-compliance with their emission reduction commitments into the Fund, which in turn would fund mitigation projects in non-Annex I countries (S Thorne and E Rovere, *Criteria and Indicators for Appraising Clean Development Mechanism (CDM) Projects* (Helio International, October 1999). During the rounds of climate change negotiations this proposal was developed into the Clean Development Mechanism.

Units (ERUs) it transfer to another Annex I country.[8] Developing countries have not assumed a legally binding commitment to reduce their GHG emissions and forgo the possibility to follow a 'first track CDM' procedure under which the host country would gain eligibility to certify emission reductions generated in their territory. Since CDM host countries are not liable for any over-verified emission reductions generated by projects implemented in their territory, without independent verification there would be an incentive by Project Participants to boost emission reductions generated by those projects.[9]

To protect the credibility of the CDM and avoid this conflict of interest, the responsibilities of assessing baseline and monitoring plans against approved methodologies, leakage, project validation, additionality requirements, and verification and certification of emission reductions was delegated to an 'independent auditor'—referred to as a Designated Operational Entity (DOE).[10] Final approval of new baselines and monitoring methodologies was delegated to the Executive Board (EB).[11] These rules governing the CDM have been referred to as the 'international criteria' of the CDM.

The responsibility of ensuring the CDM furthered the core objective of promoting sustainable development was assigned to the country hosting the CDM project. Accordingly the Marrakesh Accords state that DNA is required to confirm that the project 'assists [the host country] in achieving sustainable development'—it does not require the DNA to review the other key aspects of the CDM that are delegated to other entities. Similarly it is not mandated that the DNA would assess the financial viability of projects. The project participant, not the DNA, is responsible for ensuring that the project is financially viable and that other aspects of the project—such as checking technological specifications and necessary licences—comply with the legal requirements of the host country.

This of course does not mean that a DNA can never peruse aspects of the project outside the requirements of Marrakesh Accords, but rather that when approving a project, the DNA is not required to confirm all aspects of the project. The DNA is not liable if it approves a project that fails to be registered by the CDM EB, or encounters financial difficulties. Approval can be limited to approving voluntary participation and sustainability, and does not amount to a general approval of the project, or to a representation or guarantee that the project will be approved by the validating DOE and registered by the EB.

[8] Each time a JI host country issues an Emission Reduction Unit (ERU) for a JI project it hosts, it must deduct an equivalent amount of Assigned Amount Units (AAUs) or Removal Units (RMUs) from their national registry. See generally, FCCC/CP/2001/13/Add.2, Decision 19/CP.7 on the 'Modalities for the accounting of assigned amounts under Article 7, paragraph 4, of the Kyoto Protocol'. For a discussion within this text, see chs 6 and 7 above on JI.

[9] See chs 6 and 7 above for a more detailed discussion of the JI mechanism.

[10] See note 2, paras 27 and 40. See also Art 12(5) Kyoto Protocol. The concepts of leakage and additionality are discussed in ch 10 above on CDM concepts, requirements, and the project cycle.

[11] See note 2, para 5(b) of the Annex to Decision 17/CP.7.

Duplicating approval processes for identical issues covered elsewhere may lead to inconsistencies between the project as approved by the DNA, and the project as validated by the DOE and as authorized by the EB. Arguments used to support an expansion of the DNA's role include arguments that the DNA's preliminary analysis of financial viability, or of the international criteria will increase the probability of final approval[12] and ensure only high quality projects go before the EB.[13] While a preliminary review of international criteria may help a project solidify its Project Design Document (PDD),[14] requiring the DNA to review these criteria will cause a significant increase in costs to the DNA and the project participant that are likely to outweigh the benefit gained. As transaction costs in the CDM are already high, requiring the project to undergo duplicate approval of international criteria may act as a significant deterrent to the development of CDM projects, and the ability of the CDM to act as a catalyst to finance sustainable development. If a host country wants to become involved in developing baselines or monitoring plans, or assisting project participants develop these, a more effective method of doing this may be via a broader CDM capacity-building programme rather than as part of the DNA's approval procedures.

2.1.2 Project approval: sustainable development objectives

Article 12 of the Kyoto Protocol states that one of the objectives of the CDM is to 'assist parties not included in Annex 1 to achieve sustainable development'. During the negotiation of the Kyoto Protocol several attempts were made to agree upon standard indicators for sustainable development. Consensus was not reached, and it was agreed that sustainable development criteria and objectives should be determined on the national level.[15]

The term 'sustainable development' is open to many different interpretations, but most, if not all interpretations agree on a few defining principles that make up the 'magic triangle' of economic, social, and ecological sustainability.[16] Procedural components of sustainable development such as transparency, community participation and access to justice are also often recognized. Many countries already include principles of sustainable development in national laws or policy. For example,

[12] P Manso, 'Establishing a National Authority for the Clean Development Mechanism: the Costa Rican Experience', in *Greenhouse Gas Emission Trading & Project-based Mechanisms* (OECD, 2004) 43.

[13] See UNDP, UNEP, and Kingdom of Morocco, Ministry of Land-Use Management, Water and the Environment, Secretary of State for the Environment paper, *Clean Development Mechanism CDM-Morocco, Strategy-Organization-Procedures* (published as part of the CDM Capacity Development Project CD/CDM Morocco) 16.

[14] The PDD is a standard document used to submit a project for validation by a DOE and registration by the EB.

[15] S Thorne and S Raubenheimer, *Sustainable Development (SD) Appraisal of Clean Development Mechanism (CDM) Projects—experiences from the South*SouthNorth *(SSN) project* (South*South*North) 12, available at <http://www.southsouthnorth.org>.

[16] C Sutter, *Sustainability Check-Up for CDM Projects: How to Assess the Sustainability of International Projects under the Kyoto Protocol* (Swiss Agency for Development and Cooperation, 2003) 26.

sustainable development criteria is often incorporated into environmental laws, as part of environmental impact assessment processes, or as discrete national policies. In the case of South Africa, it is incorporated into the Constitution in the Bill of Rights.[17] To date, most, if not all of these laws or policies focus on sustainable development at the macro level and do not contain project specific criteria. To gauge effectively whether or not individual CDM projects contribute to sustainable development, criteria focusing on project level assessment may need to be developed.

To date, host countries have followed similar approaches to assess whether or not a CDM project fulfils national sustainable development criteria. Chile adopts the simplest approach when it states that a project meets national sustainable development objectives if it complies with national laws, drawing attention to national environmental impact assessment laws.[18] The approach adopted by Morocco, Argentina, and Peru builds on this by combining reference to national legislation with compliance to general sustainable development indicators or objectives.[19] Other countries complement compliance with national laws with specific, project level sustainable development criteria. Colombia, India, South Africa, and Uruguay[20] and a number of other countries[21] have, or are planning on adopting this approach. Countries that do not have an existing set of sustainable development priorities or strategies may find the sustainable development frameworks provided by international agencies, such as the United Nations Commission on Sustainable Development, a useful starting point.[22]

[17] Art 24(c)(iii) of the Constitution of the Republic of South Africa 1996 (Act No. 108 of 1996).

[18] Personal communication with the DNA of Chile by the authors. DNA contact information is listed on the website of the UNFCCC Secretariat at <http://unfccc.int>.

[19] Information on the DNA for Peru and Argentina collected from communication with each of these DNAs. See n 13 above for information on the Moroccan DNA. DNA contact information is listed at <http://unfccc.int>.

[20] The criteria in India and South Africa are still under development. Uruguay has published its assessment criteria (C Sutter and L Santos, *Host Country Approval for CDM Projects in Uruguay: Application of a Sustainability Assessment Tool* (Swiss Federal Institute of Technology and Ministry of Housing, Territorial Regulation and Environment, Uruguay, 2003), as has Colombia (Republic of Colombia, Ministry of Environment, Colombian Office for the Mitigation of Climate Change, *Criterios y Procedimientos Para La Aprobación Nacional de Proyectos de Captura de Gases de Effecto Invernadero Eligibles al Mecanismo de Desarrollo Limpio, Propuesta en Discusion*, (Criteria and Procedures for the National Approval of Projects for the Reduction of Greenhouse Gas Emissions eligible for the Clean Development Mechanism, Proposal under discussion, 12 February 2003).

[21] It is also likely that many of the twelve Countries participating in the UNEP CD4CDM project will adopt this approach. These countries are Bolivia, Cambodia, Côte d'Ivoire, Ecuador, Egypt, Guatemala, Jordan, Morocco (which, as noted above, already applies general criteria), Mozambique, Philippines, Uganda, and Viet Nam. For more information on this project see <http://www.cd4cdm.org/>.

[22] The UNCSD indicator set is a useful framework, which has the advantage of international support and field tests through the cooperation of national governments. The UNCSD has developed a comprehensive set of indicators (totalling more than 100), which are listed on the website <http://www.un.org/esa/sustdev>. Other frameworks, particularly those of the US Interagency Working Group for Sustainable Development Indicators ('IWG SDI'), offer a practical, decision-making oriented approach, which can incorporate the most detailed breakdown of aggregate indices. Further useful information on sustainable development assessment frameworks can be found at The UN System-Wide Website on Sustainable Development <http://www.unsystem.org/agenda21>.

Countries that have adopted specific sustainable development criteria have followed two different models. Colombia has taken a hierarchical, criteria-based approach to project approval and has developed a set of sustainable development principles, requirements, criteria, and indicators to be used in CDM projects in the power (supply and demand), transportation, and residue[23] sectors.[24] 'Principles' occupy the highest level, followed by 'Requirements' and 'Criteria' at the next level. Requirements are necessary conditions for observing the principles, and Criteria correspond to specific objectives, which lead to furthering a principle without being mandatory measures of sustainable development on their own. 'Indicators' occupy the third level, and are qualitative and quantitative parameters assessed in relation to the requirements and criteria.[25] Other countries have been even more

> **Box 1. Colombia's Proposed Approval Procedures**
>
> Colombia also has an optional preliminary assessment followed by a two-step approval procedure. During the project preparation stage, the Colombian Office for the Mitigation of Climate Change (COMCC), located within the Ministry of Environment, may conduct a preliminary assessment and offer technical recommendations if requested to do so by the project participant. The COMCC may also issue a preliminary letter of no objection stating that the project does not demonstrate characteristics that conflict with any CDM requirements. The preliminary comments, recommendations, and letter of no objection are designed to be independent of the formal approval process. Their aim is to help improve the quality of CDM projects and reduce the perception of regulatory risk when parties are seeking financing for a project.
>
> The first step in formal project approval involves an initial technical opinion, which is given by the COMCC. The opinion includes a recommendation whether the project should be approved or rejected. This opinion is then sent to the National Inter Sector Technical Committee for Climate Change, which has twenty days to object to the recommendation. After twenty days, if the project is accepted by the Ministry of Environment, the Minister of Environment sends the project participant a letter of approval. The COMCC reserves the right to send comments to the designated Operational Entity responsible for validating the project. The comments may include comment on project issues not related to the sustainable development criteria assessed by the COMCC, such as methodological or legal issues. [see note 20 above]

In addition, the International Institute for Sustainable Development (IISD) provides selected examples of Indicators <http://www.iisd.org>. For CDM-specific assessment of sustainable development criteria, see Sutter, n 16 above.

23 Residues include recovery activities, such as recovery and use of sanitary filling gases, and recovery and usage of gases from sewage water treatment plants.

24 See Republic of Colombia, Ministry of Environment, Colombian Office for the Mitigation of Climate Change, n 20 above.

25 As a example, one of the General Principles states: 'The project should contribute to the state policy and maintain coherence with the national, regional, local and/or sector plans, programs and projects.'

prescriptive in their application of sustainable development criteria, requiring numerous detailed, specific criteria to be met.

Another project based approach developed by Dr Christoph Sutter,[26] and applied in Uruguay,[27] involves the development of a 'Multi-Attributive Assessment of CDM Projects' methodology. This methodology aims to assess the potential effects a prospective CDM project will have on sustainable development. The analysis consists of the following five steps: identify sustainability criteria; define indicators; weigh criteria; assess the project against these criteria; and finally aggregate and interpret the results.[28]

2.1.3 Project approval: procedure

General principles of procedural or administrative law call for transparent approval procedures that produce consistent, timely, and unbiased decisions capable of being reviewed. Applied to the CDM, these principles suggest the DNA should: avoid possible conflicts of interest that may create bias; operate transparently; and state clearly who participates in the decision making; how long the decision making takes; who is responsible for the final decision and under what authority; and how final decisions may be appealed.

A generic approval procedure, or approval procedure template that takes these considerations into account is described below and outlined in Figure 11.1.[29] The generic procedure contains a preliminary, optional assessment of the DNA followed by a complete assessment that draws on input from a wide range of stakeholders. Conflicts of interest and bias are minimized by limiting the roles the DNA assumes, and by the DNA operating transparently through publishing its organizational structure, rules, procedures, submissions received, and reasons for decisions.[30]

Contact with the DNA during the early stages of project planning can be useful. A preliminary assessment, such as a non-binding letter of no objection, will

There are a number of criteria under this principle, one of which states: 'The project has mechanisms addressed to improve the coverage or the quality of home utility services in a low-income local community or communities.'

The indicators under criteria are: 'The project presents technical-economic alternatives and offers compensations in order to improve the coverage or the quality of sustainable sanitation and drinking water home utility services in low-income local community or in communities influenced by the project' and 'The project presents technical-economic alternatives and offers compensations in order to improve the coverage or the quality of sustainable electric power home utility services in low-income local community or in communities influenced by the project.'

[26] See n 16 above. [27] Sutter and Santos, n 20 above. [28] ibid, 31.

[29] It must be emphasized that many different approval procedure options are available. The aim of the generic approval procedure is to reflect one option for applying generally accepted principles of administrative law to the DNA's approval procedure.

[30] See UNDP, UNEP, The World Bank, and World Resources Institute, *World Resources 2002–2004: Decisions for the Earth: Balance, Voice and Power* (World Resources Institute, 2003) 108, where information disclosure is identified as a tool for strengthening corporate integrity and accountability.

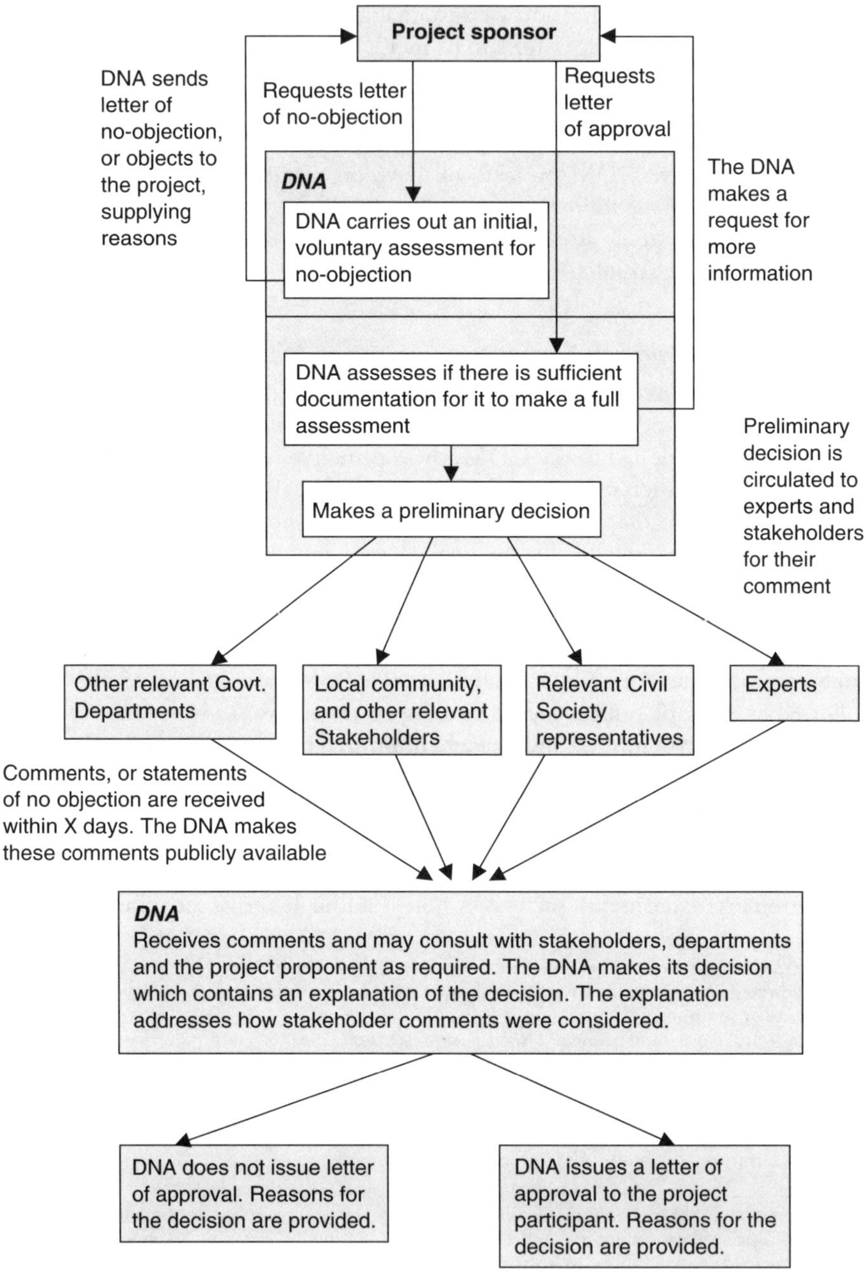

Figure 11.1: Project approval procedure template

provide comfort to investors and project participants and further enhance the ability to obtain project finance, including finance associated with the sale of Certified Emission Reductions (CERs).

A preliminary assessment will not be relevant if a DNA intends to approve a project during the early stages of project planning. Such approval could be based either on preliminary project documentation or specific sustainable development documentation prepared prior to preparation of the PDD, as the PDD is not available at the early development stage of a project. The preliminary no-objection is included in the template as it is likely that many DNAs will require an abridged or complete PDD to approve a project.

In order to perform a complete assessment of any proposed project, the DNA will need information to judge whether the project conforms with the host country's sustainable development objectives. This may include specific documentation on how sustainable development criteria are met, information already included in the host country's environmental impact assessment, or other measures that are adopted. Alternatively, to bring the DNA's documentation into parity with documentation required under international rules, the DNA may consider using the PDD, or an abridged version of the PDD to assess sustainable development requirements.

Once the request for approval is received, the DNA will need to check the documentation to ensure it is complete, and notify the project participant whether it will be able to make a full assessment based on the documents supplied. The DNA can then assess the information and make a preliminary decision either to approve the project, reject the project, or defer making a preliminary decision and request more information or clarification from the project participant. Once a preliminary decision is made, the decision, and reasons supporting the decision can be shared with experts and stakeholder representatives,[31] with a request for assessment and either positive approval, or approval by not objecting.[32] A time limit for receiving a reply is set, after which the stakeholder would be deemed to accept the preliminary decision. After receiving feedback from stakeholders, the DNA would then

[31] In the spirit of Principle 10 of the Rio Declaration on Environment and Development (31 ILM 874 (1992)), incorporating relevant stakeholder consultation in sustainable development assessment decision making will increase the DNA's integrity and provide additional security to project developers. Support for including public participation can also be drawn from research into effective environmental impact assessments. See M A Bekhechi, 'Legal and regulatory framework for Environmental Impact assessment in African countries' in B Chaytor and K Gray (eds.), *International Environmental Law and Policy in Africa* (Kluwer Academic Publishers, 2003) 276. For a more detailed discussion on public participation in the CDM and JI, see ch 5 above.

[32] How the full assessment is carried out will depend in part on the structure of the DNA. If the DNA is established as a self-contained office already incorporating a wide range of stakeholders, a decision may be reached with only a modest amount of external consultation. If the DNA office is smaller, and contains a largely external decision making process, the approval procedure will necessarily reflect this. As most DNAs will initially be small offices, the template proposed assumes the latter structure. Even so, a self-contained office may adopt a similar approach, with the proposed process contained within a single office rather than spread externally.

Box 2. Morocco's Proposed Approval Procedures

Morocco developed its approval procedure in conjunction with the UNDP and UNEP.[33] The DNA for Morocco will approve projects in a two-step procedure. The DNA first performs a preliminary evaluation based on a 'project idea note', followed by an in-depth evaluation based on the PDD. The aim of the preliminary evaluation will be to verify documentation, and harmonize information so that the DNA can easily check whether the project will lead to measurable GHG reductions. The preliminary evaluation will also verify whether the project will conform to national sustainable development criteria, and will not have any external negative effect.

The second step based on the PDD will review whether the project meets international criteria and Moroccan sustainable development criteria. It appears the DNA will assess most issues covered in the PDD, including the project's financial viability, baseline, and environmental and socio-economic impacts. The DNA will also assess whether there is voluntary participation, there are real, long-term benefits related to the mitigation of climate change, and whether the emission reductions are additional to any that would occur in the absence of the project. The initial assessment is expected to take no more than two weeks, and the in-depth assessment a maximum of four weeks.

Project review in Morocco is undertaken by the 'CDM National Council', which is comprised of representatives from a number of different ministries, departments, members of civil society and other experts.[34] Decisions taken by the Council are adopted by the majority of members present at the relevant meeting.

make its final decision.[35] Feedback received, and reasons for making the final decision can be made publicly available.

Finally, procedures for reviewing or appealing a DNA's approval may need to be determined, as will the term of such approval, and what grounds, if any, exist for revoking the approval.

If project approval procedures become too complex or onerous, there is a risk that the costs of meeting approval requirements will outweigh the benefits project developers expect to receive from the CDM. This is confirmed by market analysis that has found efficacious project approval by the DNA has a significant influence

[33] See n 13 above.

[34] Ministerial representatives include representatives from: Ministry of Foreign Affairs and Cooperation; Ministry of the Interior; Ministry of Finance and Privatization; Ministry of Commerce, Industry and Telecommunications; The Prime Minister's Delegated Ministry of Economic and General Affairs and Economical Improvement; Ministry of Agriculture and Rural Development; Ministry of Equipment and Transport; Ministry of Energy and Mines; and The High Commissariat of Water, Forestry and Combating Desertification. There is also a representative from each of the General Confederation of Moroccan Enterprises; the Moroccan Center of Export Promotion; the Renewable Energy Development Center; and the Moroccan Center for Cleaner Production. Civil society is represented through non-govermental organizations involved in combating climate change, or initiating sustainable development programmes.

[35] There are a number of different options available here. If a simpler model is preferred, once a request for approval is received, a DNA could send out the request and information to a committee of relevant stakeholders and convene a meeting of this committee to assess approval in a single step.

on the appeal of CDM opportunities in a host country.[36] However, operating efficient regulatory authorities 'is difficult in countries where state authority is weak, budgets are constrained, and technical capacity is low'.[37] Countries with weak state authority, constrained budgets and low technical capacity tend to be countries where CDM investment would be particularly beneficial. For these countries to maximize opportunities under the CDM, it is particularly important that project approval processes are clear, efficient and transparent. Each host country will therefore need to address any legal, political, and administrative issues surrounding how best to establish project approval regulations in national law.[38]

2.2 Outcome of Project Approval: Letter of Approval

The outcome of the DNA's assessment is written approval stating the host country is voluntarily participating in the CDM, and that the Project will assist the host country achieve sustainable development.[39] As discussed below, the host country may also delegate authorization responsibilities to the DNA. If this is done, the letter may also state the DNA authorizes the participation of private entities participating in the project for the purposes of the Marrakesh Accords.[40]

2.2.1 Additional statements

Project sponsors have indicated the letter of approval can increase the chances of attracting project finance if the letter provides additional comforts to sponsors and helps protect the interests of project developers. Some CDM investors therefore want to see a range of additional confirmations included in a Letter of Approval. Additional statements requested (but not always included) include statements that:

(i) The establishment of one of the project participants as the focal point for the project with respect to all communications for the purpose of the United Nations Framework Convention on Climate Change (UNFCCC), in particular with regard to instructions regarding allocations of CERs at issuance;

[36] Carbon Market Analyst, 22 December 2002, Point Carbon. See also, UNDP, *The Clean Development Mechanism: A Users Guide* (UNDP, 2003) 14. This is consistent with analysis of regulation generally; see also World Bank, *Doing Business in 2004: Understanding Regulation* (World Bank, IFC, OUP, 2003), ch 7, and Committee for Economic Development, *Modernizing Government Regulation: The Need for Action*, Policy Statement by the Research and Policy Committee of the Committee for Economic Development, (Committee for Economic Development, 1998).

[37] See n 30 above, 108.

[38] Generally, there are four types of instruments that give effect to regulation—specific legislation; presidential or cabinet decrees; contracts (including regulatory licences); and administrative procedures. See LJ Guasch and P Spiller *Managing the Regulatory Process: Design, Concepts, Issues, and the Latin America and Caribbean Story*, World Bank Latin American and Caribbean Studies, Viewpoints, (World Bank, 1999) 42–48.

[39] See n 5 above.

[40] See the definition of 'Authorization of a private and/or public entity to participate in a CDM project activity' in the Clean Development Mechanism Guidelines for Completing the Project Design Document (CDM-PDD), the Proposed New Methodology: Baseline (CDM-NMB) and the Proposed New Methodology: Monitoring (CDM-NMM), Version 01, adopted by the EB on 1 July 2004.

(ii) The DNA acknowledges the project participants' right, title and interest in and to all of the GHG emission reductions generated by the Project (and any CERs which are created out of the Project);

(iii) The DNA will cooperate to facilitate the CDM process and give assistance, where necessary, for the issuance and transfer of CERs to the Project Participants; and

(iv) Any public funding of the Project does not result in a diversion of official development assistance (if applicable).

The DNA may also wish to safeguard its own interests by minimizing or removing the potential for assuming liabilities others may wish to impose on the DNA. This could be done by including in the letter of approval statements that:

(i) The approval addresses only those issues it is required to address under the Kyoto Protocol, and does not represent the DNA's approval of the validity of the baseline calculations or the conformity of the project to any other CDM validation requirements;

(ii) The letter does not constitute a guarantee that the project developer will be issued CERs by the Executive Board;

(iii) The letter does not guarantee the project has, or will receive the necessary licences and approvals under national law; and

(iv) The developer is responsible for compliance with CDM requirements.

Even though liability is not expected to arise in the absence of such statements, these statements may be useful if a project participant requests the DNA to make guarantees or representations relating to any of the above points, or if the host country or DNA wants to make it explicit that it is not making these representations.

3 Ancillary Roles of the DNA

If the DNA assumes the role of CDM advocate, or a co-developer and promoter of national projects, it will be difficult for the DNA to provide unbiased appraisals of the same projects for submission to the international community. Host countries must be aware of this and ensure that no conflicts of interest arise or are perceived as arising. Promotion of the CDM could, for instance, be delegated to the department that is normally responsible for promoting investment opportunities.[41] A conflict of interest would be avoided if the promoting department was not

[41] This department should be aware that the EB is also responsible for providing information, including providing information on investors seeking projects, and projects available for funding; note 2, para 5(h)(i).

involved in establishing or operating the institutional structure, administration, project approval procedures, or decision-making processes of the DNA, or be responsible for hiring or promoting of any DNA employees. This department would also be able to market and promote these opportunities much more effectively than the DNA. Peru has taken this approach, delegating promotion of the CDM to the government office that promotes foreign direct investment in Peru.[42] This same office may also provide technical advice on baseline and monitoring plan methodologies if the host country sees this as a valuable service.

4 Creating a Host Country DNA: Organizational Structure

Since the DNA acts as a domestic regulatory agency, the Marrakesh Accords do not provide any guidance on *how* to establish a DNA. Generally, since the CDM is a mechanism designed to achieve GHG reductions and promote sustainable development using foreign investment, ministries responsible for environment, forestry, and energy may want to play a role. The ministry responsible for promoting development and investment opportunities is also important to the host country and the CDM. Deciding in which agency to establish a DNA is likely to be influenced by the allocation of existing responsibilities for UNFCCC matters.

Most of the DNAs established in host countries to date integrate these ministries in the DNA, following one of two basic approaches. Some host countries have centralized administration and decision-making authority of the DNA in a single office that draws on relevant ministries, stakeholders, and experts in the decision-making procedure.[43] Alternatively, a contact point for the DNA has been located in a particular department, but relevant ministries, (and perhaps stakeholders and experts) are incorporated into the decision-making structure which becomes the DNA.[44] In either event, the DNA is also often located in the same site as the existing UNFCCC focal point.[45]

4.1 Centralized DNA

In a centralized model, a specific minister, ministry, or department can be designated as the DNA. In perhaps the simplest version of this model, the UNFCCC focal point may be designated the National Authority, with a single individual within that focal point named as responsible for administering the DNA.

[42] Authors' communications with the DNA for Peru.

[43] For example, Peru has adopted this approach.

[44] For example, Argentina has adopted this approach.

[45] Zambia has created its DNA within the Ministry of Environment and Natural Resources, which also acts as the UNFCCC focal point. On an interim basis Cambodia has designated the Cambodia Climate Change Office within the Ministry of Environment as their DNA. The Environment Minister also acts as the UNFCCC focal point. Peru and Morocco have also established their DNAs within their respective UNFCCC focal points.

Incorporating other interested ministries and relevant stakeholders is possible in the process of project approval. When deciding which ministry or department is most appropriate, the host country may consider which sectors are likely to attract CDM projects. For example, the Ministry of Energy may not be the most suitable ministry to house the DNA if the CDM is likely to attract primarily afforestation or reforestation projects. Independence, transparency, and integrity can be readily demonstrated and realized if the DNA operates under an independent governance structure within the ministry housing it.

This model requires minimal cost to establish, allows resources to be focused within a single ministry, draws on pre-existing administrative structures, resources, and capacity, and acts as a clear, well-defined focal point within the host country. The disadvantages are that it may: (a) create the perception of bias in favour of the situs ministry's agenda; (b) create inter-ministerial rivalry over which ministry is the most appropriate location; and (c) if the approval procedure is poorly designed, exclude relevant entities from the decision-making process. This model can be adapted to both large and small volumes of projects. It is a good approach when the host country has limited capacity or resources set aside for the DNA and the host country does not expect a diverse portfolio of CDM projects. Care must be taken though to promote independent, autonomous governance of the DNA within the ministry where it is housed in order to ensure integrity and protect the DNA from bias or improper influence.

4.1.1 Centralized DNA as an independent office

As a variation to this model, a DNA may be created as a discrete, independent government office not located within a particular ministry or department. Such a DNA is autonomously managed and governed, and only accountable to, rather than administered by, a particular minister or ministers.

Requiring ministers to only make decisions regarding the DNA, or give directions to the DNA through parliament can ensure transparent and independent governance. Founding legislation would need to state that the DNA possesses the appropriate authority to issue letters of approval, as well as delegate any authorization responsibilities the DNA may adopt.

The advantages of this approach include: greater DNA autonomy and independence; increased prospects for including private sector and NGO participation in DNA management; the avoidance of potential inter-ministerial rivalry; and the creation of clearer accountability. The disadvantages are that such a DNA will: (a) be more costly, complex, and time consuming to establish; (b) require a high level of host country capacity to provide sufficient skilled personnel without draining one or more government ministries; and (c) produce a powerless DNA unless it is given sufficient decision-making authority. This model is only appropriate if the host country expects a steady flow of CDM projects, there is sufficient capacity within the host country to support an independent DNA, and the host country expects to attract a diverse portfolio of CDM projects.

4.2 DNA as an Inter-ministerial Committee

In this approach, the committee is made up of representatives from any ministry with an interest in CDM projects. A single ministry may contain the secretariat responsible for coordinating other ministries and for acting as the DNA's focal point. In this model, either the principle ministry or the committee itself may act as the DNA.

This approach is adaptable to either a small or large volume of projects. The advantages are that: the host country will be able to draw upon existing experts without relocating them; the approach requires minimal costs to establish and maintain; all relevant ministries are ensured of being involved in decision making; and assessments of projects from different sectors can be readily made without bias or the appearance of bias. Disadvantages, however, are that: (a) delays are more likely as responses will be required from multiple decision makers[46] and cooperation and coordination will be required among multiple ministries, all of which may not be amenable to this; and (b) where and how to establish a focal point will not be readily apparent. This approach can work well if the host country has limited capacity or resources available to allocate to the DNA and expects to receive CDM projects sporadically, across a number of sectors. It is also appropriate if the host country will have problems establishing independent governance within a single ministry, or where inter-ministerial rivalry prevents an effective DNA from being amicably established within a particular ministry.

When deciding how to establish a DNA, each host country will need to estimate the volume and type of CDM projects expected, and the expenditure warranted to establish the DNA. Each host country will need to tailor their DNA to promote transparency, independence, and integrity within its individual political, administrative, legal, and administrative environment. Finally, a degree of flexibility in the design of the DNA may be beneficial, as predicting the evolution of the CDM market is difficult. A design that makes sense to address a small volume of projects may well become unworkable if the volume of projects increases substantially.

Whichever approach is adopted, a clear contact point, transparent, effective, and efficient CDM project approval procedures and the ability to coordinate with relevant governmental ministries and institutions so that Letters of Approval can be issued without delay will all benefit a DNA.

5 Creating the DNA: Funding

A lack of sufficient and sustainable funding of regulatory agencies invariably leads to regulatory inefficiency, and sometimes to the improper use of regulatory power.[47] Secure funding not tied to project approval and able to support staff and housing requirements over the long term will promote efficient operation and help safeguard against corruption.

46 This may offset 'no objection' time limits discussed above.

47 See n 41 above.

6 Conclusions

A global effort reflecting the principle of common but differentiated responsibility is required to address the global issue of climate change. The CDM is the chief method of encouraging developing country participation under the Kyoto Protocol. The purpose of this chapter has been to identify responsibilities host countries assume when designating a National Authority to participate in the CDM. The DNA is responsible for approving CDM projects if the host country is participating in the CDM voluntarily and if the project contributes to the host country's sustainable development objectives. The assessment criteria a DNA applies to determine if a project contributes to the national sustainable development objectives must be accurately identified and efficiently applied. First identifying the potential for CDM opportunities in a host country will keep the size of the DNA in perspective. While promoting opportunities offered by the CDM is key to developing the CDM market, host countries must prioritize separating this task from the regulatory responsibilities of the DNA.

The CDM creates new opportunities for developing countries to gain access to clean and efficient technology, and attract environmentally and socially responsible foreign investment. Realizing these opportunities will be challenging. A number of international difficulties for the CDM, such as the availability of approved baseline and monitoring methodologies, are beyond the control of individual host countries. Other challenges, such as ensuring host countries accurately identify and efficiently carry their responsibilities while minimizing conflicts of interest are within each developing country's control.

12

Benefit Sharing under the Clean Development Mechanism

*Saleemul Huq and Hannah Reid**

1 Introduction

The Clean Development Mechanism (CDM) was set up as an instrument under the Kyoto Protocol of the United Nations Framework Convention on Climate Change (UNFCCC) to provide developed countries that had accepted targets for reducing greenhouse gas (GHG) emissions with flexibility for achieving these targets, by allowing them to take credits from emissions reduction projects undertaken in developing nations. The first objective of the CDM is to reduce GHG emissions and the second objective is to assist developing countries which host CDM projects to achieve sustainable development.[1]

The GHG benefits of each CDM project will be measured according to internationally agreed methods and will be quantified in standard units known as Certified Emission Reductions (CERs). However, the government that is hosting

> **Box 1. The Clean Development Mechanism**
>
> The purpose of the clean development mechanism shall be to assist Parties not included in Annex I in achieving sustainable development and in contributing to the ultimate objective of the Convention, and to assist Parties included in Annex I in achieving compliance with their quantified emission limitation and reduction commitments under Article 3.[2]

* Saleemul Huq is the Director of the Climate Change Programme at the International Institute for Environment and Development (IIED) in London. Hannah Reid is a Research Associate at IIED with the Climate Change Programme. The views expressed in this chapter are the views of the authors and do not necessarily represent the views of IIED.

[1] See L Aukland, PM Costa, S Bass, S Huq, N Landell-Mills, R Tipper, and R Carr, *Laying the Foundations: Preparing the Land-Use Sector—A Quick Guide to the Clean Development Mechanism.* (IIED, London, 2002); and UNFCCC, Kyoto Protocol to the United Nations Framework Convention on Climate Change (1997), <http://unfccc.int/resource/docs/convkp/kpeng.pdf>.

[2] Article 12(2) of the Kyoto Protocol to the United Nations Framework Convention on Climate Change, available at <http://unfccc.int/resource/docs/convkp/kpeng.pdf>.

the CDM project must define its own sustainable development objectives. Designated National Authorities (DNAs) established by each government to examine and approve CDM projects need some way of assessing whether proposed projects and methodologies meet sustainable development objectives.[3]

Over the last few years a number of CDM projects have been established in which the benefits from sustainable development (also called co-benefits or benefit sharing) have been explored. This chapter reviews a number of such efforts in different parts of the developing world and reviews the extent to which benefit sharing has been defined and practised. It will also provide some suggestions for systematizing such co-benefits from CDM projects in future.

The term 'benefit sharing' usually refers to benefits or outcomes from the CDM project that accrues to the host country government or to the local community in the host country. However, sometimes the term is also used to mean sharing of the CER credits generated by the CDM project. This raises equity-related questions such as what share of CERs host countries should receive as well as whether CERs are the property of the host country or not. In this chapter the term 'benefit sharing' is used in the former context only (that is, in the sense of benefits for the host country government as well as local communities).

2 Benefit Sharing in the CDM—A Brief History

The Kyoto Protocol provides legally binding commitments for thirty-nine developed countries to reduce their GHG emissions, thus recognizing the differentiated responsibilities (based on historic accumulated emissions) of countries to take action to combat climate change. The Protocol provides 'flexibility mechanisms' (so named as they allow for flexibility in achieving GHG emission reductions) to help these industrialized countries reach their targets. Such mechanisms include emissions trading (trading of emission allowances between developed nations), joint implementation (transferring emission allowances between developed nations, linked to specific emission-reduction projects), and the CDM (which allows developed nations to achieve part of their reduction obligations through projects in developing countries which sequester CO_2 from the atmosphere). As the only flexibility mechanism to involve developing countries, the CDM provides opportunities to share benefits from projects designed to combat climate change with poorer nations. The idea was that flows of finance and green technologies would enable poorer countries to leapfrog to cleaner and more sustainable development.

[3] A Michaelowa, *Host Country Requirements to Make CDM Process Nationally Efficient*, Paper prepared for UNDP for the Asia Meeting on CDM Operation, Kuala Lumpur, 1–3 April 2003; UNDP and Ecosecurities, *Facilitating CDM Design and Implementation: A Manual for UNDP Country Officers* (Draft, February 2003); and H Reid and S Huq, *Available Approaches and Tools to Assess Sustainable Development for Efficient National CDM Operation*, Paper prepared for UNDP for the Asia Meeting on CDM Operation, April 2003.

The inclusion of the CDM as one of the flexibility mechanisms came as result of political negotiations under the UNFCCC leading up to the Kyoto Protocol to enable the developing countries to participate in the activities under the Protocol (as they did not have to take on any emission reduction targets under the Protocol). The inclusion of the second objective (that is, that CDM projects should also meet sustainable development goals) was due to the desire of the developing countries to ensure that CDM projects were in harmony with their own development priorities. However, during the negotiations no single criterion for what constitutes 'sustainable development' was agreed, so it was left to each host (that is, developing) country to develop its own criteria. This chapter explores some of the experience gained so far in trying to develop and apply such sustainable development criteria to CDM projects. It is not an exhaustive assessment of all (or even a large number of) existing CDM projects, but rather a selection of a few interesting ones which have made an effort to go beyond a very simple approach to meeting sustainable development objectives of the CDM.

3 What is Benefit Sharing in the Context of the CDM?

Sustainable development is a complex concept. It is important for host countries to create a national definition that is compatible with internationally accepted principles governing sustainable development, such as that used in 1987 by the World Commission on Environment and Development, otherwise known as the Brundtland Commission (Box 2), and also one that is nationally appropriate in the context of CDM. Costa Rica has made progress in this regard (Box 3). Internationally accepted principles usually incorporate environmental, economic

Box 2. What is Sustainable Development?

Sustainable Development is development that meets the needs of the present without compromising the ability of future generations to meet their own needs.[4]

Box 3. Sustainable Development Criteria in the context of CDM in Costa Rica

Projects should be compatible with and supportive of Costa Rica's national environment and development priorities and strategies, including biodiversity conservation, reforestation and forest preservation, sustainable land use, watershed protection, air and water pollution reduction, reduction of fossil fuel consumption, increased utilization of renewable resources and enhanced energy efficiency. Projects should enhance the income opportunities and quality of life for rural people, transfer technological know-how and capacity building, and minimize adverse consequences.[5]

[4] World Commission on Environment and Development, *Our Common Future* (Oxford, OUP, 1997).

[5] C Kelly and N Helme, *Ensuring CDM Project compatibility with Sustainable Development Goals*. Center for Clean Air Policy, (Washington DC 2000).

and social dimensions, and some also incorporate institutional (or governance) and technological components.

Host country governments must also determine ways to assess whether sustainable development objectives are met in the context of CDM, usually through the identification and measurement of indicators. The relative importance of such indicators needs to be assessed in the context of CDM, ensuring that they account for temporal variations and cover sustainable development issues at the local, sectoral, national, and global levels. These may not always be in line with each other. For example, while an afforestation project may appear to meet local sustainable development criteria, consultation with the national forestry department may reveal that the area may only have enough water resources for 2,000 hectares more of forest, and land requiring this water has instead been earmarked for small-scale forestry growers. At the same time, certain types of projects could be excluded due to the prevailing lack of international consensus regarding their desirability (for instance, nuclear power projects). If the criteria are too stringent, projects that are intuitively known to meet sustainable development objectives (renewable energy projects, waste to energy projects, energy efficiency projects, sustainable forestry projects) may not be approved. Likewise, single projects may only be able to address a handful of indicators. The real challenge therefore lies in ensuring that the choice and measurement of indicators provide a meaningful assessment of whether or not CDM projects meet sustainable development objectives.

Many groups have compiled lists of sustainable development indicators,[6] and WWF has developed the Gold Standard for this purpose in the context of the CDM (Box 4). Each indicator has a weighting to ascertain the extent to which each sustainable development indicator is met. Such weighting allows for the inclusion of negative impacts on sustainable development. Scores for each indicator vary between +3 and −3. In order to meet the standard, WWF proposes that the total score should be positive and that none of the three sustainable development objectives should have a negative score.

4 Benefit Sharing in Practice

Few projects meet all social, environmental, and economic aspects of sustainable development. However, each project makes a significant contribution to at least

[6] For example see the Commission on Sustainable Development (CSD) indicators for sustainable development <http://www.planbleu.org/indexa.htm>; International Finance Corporation, *Measuring Sustainability. A Framework for Private Sector Investments* (IFC, Washington, DC, 2003); the International Sustainability Indicators Network <http://www.sustainabilityindicators.org/>; the International Institute for Sustainable Development (IISD) and UNEP-Earthwatch Dashboard of Sustainability <http://esl.jrc.it/envind/dashbrds.htm>; the IISD Compendium of Sustainable Development Indicator Initiatives <http://www.iisd.org/measure/compendium/>; S Huq, *Applying Sustainable Development Criteria to CDM Projects: the PCF Experience*, PCFplus Report (Washington, DC, April 2002); and MCSD, *Indicators for Sustainable Development in the Mediterranean Region. Blue Plan*, Mediterranean Commission on Sustainable Development (MCSD, June 2000).

Box 4. WWF Proposed Gold Standard Sustainable Development Matrix[7]

Local/global environmental sustainability

- Contribution to water availability and access locally and regionally (+3 is up to and more than double the number of people with access to water, −3 is no access to water at all);
- Contribution to water quality locally and regionally in the project's area. Water quality will be measured using concentration of main pollutants, including Biological Oxygen Demand and others (+3 is no polluted waters at all, −3 is up to double or more pollutants' concentration in all waters);
- Contribution to local air quality. Air quality will be measured by comparing the concentration of most relevant air pollutants, such as SO_x, NO_x, and particulate matters (+3 is complete elimination of air pollution, −3 is up to double or more air pollution compared with baseline);
- Contribution to local soil condition. Soil condition will be measured by comparing the concentration of most relevant soil pollutants, erosion and the extent of land use changes due to the project (+3 is total elimination of soil pollution/ erosion, −3 is double or more soil pollution/erosion compared with baseline);
- Contribution to local biodiversity. The change in biodiversity is estimated on a qualitative basis considering any destruction or alteration of natural habitat compared to the scenario without the project. A positive change will be given by previously disappeared species re-colonizing the area, a negative change will be given by species disappearing or by introduction of foreign species (+3 is significant restoration and protection of natural habitat within the project area, −3 is destruction or irreversible modification of over 10 per cent of the natural habitat within the project area).

Social sustainability and development

- Employment quality, such as whether the jobs resulting from the project activity are highly or poorly qualified, temporary or permanent in comparison with business as usual. Take temporary and permanent as qualifications for job quality (+3 when all new jobs created are long-term and of better quality, −3 when all new jobs created are long-term and of poorer quality than baseline);
- Livelihoods of the poor: this indicator comprises a number of sub-indicators. Where a sub-indicator is not relevant to the project, it should be ignored. After all the relevant variables have been considered, the total score should be non-negative.

[7] WWF, *The Gold Standard: Quality Standards for CDM and JI Projects*, Final Draft for Consultation and Draft Technical Appendices (December 2002).

- Poverty alleviation will be evaluated by calculating the change in number of people living above income poverty line (+3 is up to double or more people living above poverty line, −3 is up to double or more people living below poverty line);
- Contribution to equal distribution of wealth and additional opportunity for disadvantaged sectors, in particular gender and excluded social groups. The indicator combines quantitative—changes in estimated earned income (normalized to the project's starting year)—and qualitative assessment—improved opportunities (+3 is substantial improvement in both female earned income (over 100 per cent increase) and opportunities, −3 is up to and more than halved female earned income);
- Access to essential services (water, health, education, access to facilities, and so on). Measured by the number of additional people gaining access, which must be directly related to the service and not a spin-off (+3 is increased access to four or more aspects, −3 is decreased access to three or more aspects);
- Access to affordable energy services: the CDM and Joint Implementation provide an important opportunity to improve the coverage of reliable and affordable clean energy services, especially to the poor and in rural areas. Where of a relevant scale, security of energy supply (an indicator of a country's ability to generate the power that is needed for services and the economy in comparison with the baseline), should be taken into account (+3 is 100 per cent or more improvement in coverage and quality of service, −3 is 100 per cent or more worsening in coverage and quality of service).

Economic and technological development

- Employment (numbers), measured by the number of additional jobs directly created by the CDM project (+3 is up to and more than doubled number of jobs, −3 is elimination of all jobs predicted);
- Sustainability of the balance of payments: net foreign currency savings may result through a reduction of, for example, fossil fuel imports as a result of CDM projects. Any impact this has on the balance of payments will be measured (+3 is total avoidance of foreign currency expenditures, −3 is doubled or more net foreign currency expenditures);
- Hard currency expenditures on technology, replicability, and contribution to technological self-reliance: as the amount of expenditure on technology changes between the host and foreign investors, a decrease of foreign currency investment may indicate an increase of technological sustainability. When CDM projects lead to a reduction of foreign expenditure via a greater contribution of domestically produced equipment, royalty payments and licence fees, imported technical assistance should decrease. Similarly a reduced need for subsidies and external technical support indicates increased self-reliance and technology transfer (+3 is a total avoidance of foreign currency expenditures/subsidies with technology, −3 is doubled or more net foreign currency expenditures/subsidies with technology).

one of these arenas, and projects described below place particular focus on the social aspects of sustainable development. Even this small sample of projects demonstrates the wide range of sectors in which CDM can operate. These include waste management, transport, energy, forestry, and industry.

4.1 The Jepirachi Wind Power Project, Colombia

The Jepirachi Wind Power Project, located in Wayúu Indigenous Territory of the north-east Atlantic coast of Colombia, in the Department of Guajira, is one of several projects conducted under the World Bank's Prototype Carbon Fund (Box 5). Once construction is completed the wind generators will deliver around 68.3 gigawatt hours per year to the Colombian National Interconnected System (SIN). The total estimated emission reductions to be achieved by the fifteen windmills are 1.168 million tonnes of CO_2 over twenty-one years. This will be achieved by displacing power that would otherwise have been produced and dispatched by other plants in the Colombian National Interconnected System. The project has preferential access the electricity market (as does energy from thermal and hydropower sources) such that when wind is available the need for energy from less carbon friendly sources will be eliminated.[8]

The Project will contribute to the sustainable development of Colombia in various ways. First, it will demonstrate the potential for wind-based power generation, thereby facilitating investment. Secondly, it will increase the proportion of energy provided to the national grid from a renewable source, thereby increasing national capacity regarding the installation and operation of such technology. These indirect benefits may stimulate the development of more renewable energy sources in Colombia. Thirdly, the project will contribute to an increase in economic

Box 5. The World Bank's Prototype Carbon Fund[9]

The Prototype Carbon Fund (PCF) was established for the purposes of (i) demonstrating how project-based transactions in GHG emission reductions can contribute to the sustainable development of developing countries and countries with economies in transition; (ii) sharing the knowledge gained in the course of the PCF's operations with all interested parties; and (iii) demonstrating how the International Bank for Reconstruction and Development can work in partnership with the public and private sectors to mobilize new resources for its borrowing member countries while addressing global environmental concerns.

[8] Prototype Carbon Fund, *Clean Development Mechanism Project Design Document. Colombia: Jepirachi Wind Power Project*, 8 September 2003, <http://carbonfinance.org/pcf/router.cfm?Page=Projects&ProjectID=3119>.

[9] <http://prototypecarbonfund.org/splash.html>.

activity during the construction period, injecting $21 million into the Colombian economy. Finally, the project is located on land belonging to the indigenous Wayúu community, which is one of the poorest groups in Colombia. According to the census of Wayúu peoples carried out in 1992, there are 128,727 in Colombia. Half of these people live in the Uribia Municipality, which is where the project will be located. Traditional local economic activities include shepherding, fishing, and horticulture. Local people also receive royalties from regional coal mining. The Wayúu will benefit from numerous community-driven activities financed by the project sponsor.

The project developers led an extensive consultation process during project development. This consultation process included national, regional, and local governmental institutions concerned with indigenous peoples, and traditional authorities and communities. The first consultation focused on the installation of wind monitoring devices in 1999, and the consultation process continued throughout the Environmental Impact Assessment process. This involved twenty formal consultation meetings with communities (using translators) and other meetings with government institutions. The consultation process was finalized in June 2002 with an agreement on the Environmental Management Plan (which includes the socio-economic and cultural programmes described below). The Ministry of Interior, Department of Indigenous Communities Affairs, supervised the consultation process.

The Social Management Programme resulting from this consultation process is summarized in Table 12.1. Particular benefits include training to facilitate direct and indirect job creation, refurbishing a local school, cemetery, and health centre (including installation of solar powered refrigeration facilities for vaccines and perishable medicines at the latter), provision of small water storage pits, and establishing a desalination plant (fed by wind power) capable of desalinating two to four cubic metres of water per hour. By targeting water supply, education, and health services, the project addresses priorities identified by the community. There is also an agreement between the project sponsor and the host indigenous community to review the programme two years after its implementation.

In displacing fossil fuel-based power generation, the project will also reduce the levels of pollutants such as NO_x and SO_x, and also ground level ozone, which is thought to have negative health impacts.

These community-driven projects will be monitored in accordance with a set of indicators outlined in the Project Monitoring Plan. The Colombian Government has not set any sustainable development targets for the project so the project developers in collaboration with the community and the PCF have determined appropriate indicators. Independent verifiers will periodically audit the project developer to ensure performance indicators are met. Additional indicators may be provided as the programme is implemented. These will be agreed upon within the third year of operation to reflect the scope and nature of the additional programmes.

Table 12.1: The Jepirachi Wind Power Project Social Management Programme

Programme	Objectives	Project
Information and communication programme	1. To inform communities about the project 2. To establish good relations between communities and the project sponsor 3. To encourage community participation	(a) Dissemination of information on the project (b) Reception and resolution of claims (c) Field visits to follow up the construction process
Employment opportunities programme	Improve community income	(a) Direct employment (recruitment and hiring) (b) Indirect employment (acquisition of raw materials, goods, and services)
Environmental education programme	To promote sustainable development	(a) Dissemination of the Environmental Management Plan to employees and communities (b) Training on environmental project design (c) Ethno-education projects (d) Training on reservoir water management (e) Training on solid waste disposal (f) Training on adequate use of natural resources
Participation and community strengthening programme	1. To strengthen communities 2. To facilitate community access to financial resources	(a) Training on indigenous rights according to Colombian law (b) Training on formulation, implementation and assessment of self management projects to access legal transfers and additional PCF benefits
Information and training for employees programme	To respect community cultural characteristics	(a) Training employees and contractors on the cultural life of Wayúu people
Compensation programme	1. To compensate for the use of land and resources 2. To improve local living standards	(a) Water desalination plant (b) Water storage (c) School rehabilitation (d) Health centre rehabilitation (e) Graveyard rehabilitation
Technology dissemination programme	To disseminate information on new technological advances	(a) Field visits to the wind power plant (b) Dissemination of material on new technologies

4.2 Electricity Generation from Biogas and Bio-diesel in Brazil

Refuse is a huge problem in Rio de Janeiro, and a recent emissions inventory shows that the Jardim Gramacho Landfill at Duque de Caxias is the main source of GHGs (in the form of methane) emitted by the city.[10] A new project at the Jardim Gramacho Landfill is converting this polluting open dumpsite into a sanitary landfill, and illustrating the potential for generating electricity from renewable energy sources. This project is one of several under the South*South*North Project (Box 6). Biogas, produced from the decomposition of organic solid wastes in landfills, and bio-diesel, produced from used vegetable (cooking) oils drives a power generator providing at least partial energy self-sufficiency for operating the landfill site. Remaining fuel needs will come from renewable sources. Technological refinements are still required, for example to ensure the right proportion of bio-diesel and biogas in the fuel mix for the generator, but through the power generation process, methane will be converted to CO_2, thus reducing the GHG effect by a factor of twenty-one. It is projected that the project will reduce emissions by 35,000 tonnes of CO_2 over a ten-year period. Income from the sale of carbon will support the operation of the project, and this process is underway, with the CDM Executive Board (EB) currently undertaking a final review of the project. The electricity generated by the plant is used to clean the water produced at the landfill site, such that only clean water is returned to Guanabara Bay nearby.[11] This has led to the rehabilitation of local mangrove swamps and improvements in associated livelihood opportunities. Employment generation has also occurred through the selective collection of used vegetable oils.

Whilst the project demonstrates the potential that technological solutions can play in improving the local and global environment, those who depend on picking

Box 6. The South*South*North Project[12]

The South*South*North Project is a network of organizations, research institutions, and consultants grouped into one developmental organization with considerable expertise to help public and private stakeholders develop the necessary confidence for dealing effectively with the CDM. South*South*North operates in Brazil, South Africa, Bangladesh, and Indonesia. It seeks to add to the international body of knowledge on the CDM by developing capacity through the actual experience of developing a number of CDM projects. The CDM offers an opportunity to promote sustainable development and to direct the flow of capital, expertise and technology into developing economies through the climate window.

[10] E La Rovère, CBS Dubeux, LBR de Mattos, L da Silva Ribeiro, LB Oliveira, LCS Aquino, MG Buzzatti, and SK Ribeiro, *Inventário de Emissões de Gases do Efeito Estufa do Município do Rio de Janeiro* (SMAC project No 63/99, Rio de Janeiro, 2000).

[11] M Orford, *Climate Change and the Kyoto Protocol's Clean Development Mechanism* (ITDG Publishing, London, 2004).

[12] <http://www.southsouthnorth.org/>.

through refuse to earn their living have expressed concerns that their livelihood will be threatened. Efforts to provide these people with alternative employment opportunities at recycling plants elsewhere have met with limited success, as they feel they would earn less (they can earn as much as twice the minimum wage picking through waste) and their quality of work would deteriorate (they currently see themselves as freelancers or small entrepreneurs, which would change with employment at a recycling plant). The provision of local social benefits for some low-income groups therefore remains a challenge.[13]

4.3 Solar Home Systems in Bangladesh

Bangladesh is one of the poorest countries in the world, with little industry and hence very low emissions. Leveraging investment through the CDM to reduce emissions is therefore challenging.

Approximately 30 per cent of households in Bangladesh are connected to the national grid, and those that are not connected stand little chance of being connected in the near future due to remoteness and prohibitively high costs. Installation of Solar Home Systems can provide electricity in isolated areas where grid electricity is not available. One project aiming to introduce Solar Home Systems to remote areas is owned by Grameen Shakti, which provides micro-credit for the installation of these systems. South*South*North (Box 6) also provides support. Over the next five years the project aims to install Solar Home Systems in 30,000 homes in isolated areas such as coastal districts and flood plain areas. The project therefore targets some of the poorest households in one of the poorest countries in the world. Installation of Solar Home Systems will provide each household with two lights and a plug, and will also save 8,600 tonnes CO_2 per year from substitution of existing energy sources.[14]

Access to light and power will improve rural economies and enable households to increase their income. Light after sunset has allowed carpenters and other skilled people to establish and run small businesses, thus providing employment. Others keep their shops open later due to newly installed lights, and some sell power to neighbouring shop owners. Some rely on the electricity to charge mobile phones, which they can both use to communicate with friends and relatives and rent out for small profits. Women also benefit, for example, in a cluster of 750 Solar Home Systems implemented by the Rural Electrification Board in Narsingdi in Bangladesh, the earnings of women have increased considerably by undertaking sewing contracts.[15] Increased employment opportunities will also arise from servicing and maintaining the Solar Home Systems, and local manufacturers will earn more from supplying system accessories. Grameen Shakti provides several financing

[13] <http://www.southsouthnorth.org/>. [14] ibid.

[15] J Troni, S Agbey, PM Costa, N Haque, M Hession, L Gunaratne, H Rodriguez, and A Sharma, *Moving towards Emissions Neutral Development Project (MEND)*, Final Report (Department for International Development, UK, 2002).

options to support installation and post-sales maintenance, for example through training local youth. People can enjoy entertainment and information from the radio and television, they can use fans, and children can study after dark thus improving their education. Reduced dependence on fuels such as kerosene and diesel reduces indoor air pollution, and frees up precious time previously spent fetching such fuels. Improved security will result from the provision of light during the night in and around homes and shops.[16]

Some poor households struggle to repay the loans, and revenue from the CDM, however small, could provide additional funds to help with this. However, mobilizing this funding is proving difficult. Troni *et al.* note that 'there is no awareness among the financial community of how carbon could help to improve, for example, the repayment terms for solar home system programmes.'[17] In addition, they add that government institutions responsible for implementing CDM policy and processes suffer from inefficiency and inactivity.[18] Technical know-how on CDM project development criteria, focusing on identification, scoping, approval, and so on, needs developing. Mechanisms to appraise CDM projects are absent and independent project monitoring is also problematic (the Ministry of Science and Technology, which is responsible for renewable energy projects such as this, lacks sufficient capacity for monitoring). The level of awareness of opportunities offered by the CDM is low and limited to the Ministry of Environment and Forests, and institutional structures require strengthening to promote inter-ministry dialogue. Knowledge of sustainable development issues is also insufficient. The approval process for CDM projects in Bangladesh is therefore long (12–24 months), lacking in transparency, and there is no fast-track procedure for poverty-focused projects.[19]

4.4 Energy Efficient Housing in South Africa

The introduction of energy efficient housing in ten pilot houses in a low-income housing settlement called Kuyasa, located in the township of Khayelitsha, will save a projected 5,500 tonnes of CO_2 per year. This project is owned by the City of Cape Town, and South*South*North (Box 6) provides support. Whilst small in itself, replication potential is high, and the idea is to expand the project to 2,309 houses, as construction of low-income houses proceeds. The project is therefore playing an important role in laying the foundations for sustainable housing developments in the whole City of Cape Town.[20] The initial ten houses were retrofitted with energy efficient technology, but the expanded project will install energy efficient technologies during construction, thus reducing the installation costs. The project increases the thermal performance of houses by introducing ceilings, energy efficient lighting, and solar water heaters, thus reducing the consumption

[16] See Orford, n 11 above. [17] See Troni *et al.*, n 15 above, 107. [18] ibid.

[19] ibid.

[20] G Morris, S Rosenberg, and M Booi, *Sustainable Energy Service Interventions for Lowincome Housing in Kuyasa, Khayelitsha* (SBE 2003, Technology and Management for Sustainable Building. CSIR, 26–30 May 2003).

of electricity and kerosene and alleviating 'energy poverty'. Energy costs constitute as much as 30 per cent of the running costs of low-income households. The savings in the cost of energy, estimated at R625.83 per household per annum, will provide further benefits to the beneficiaries. The community can thus 'leapfrog' from suppressed energy consumption, into clean technology.

South Africa moved to a system of majority rule when Nelson Mandela came to power in 1994. As part of its commitment to social justice, the new Government established the Reconstruction and Development Programme (RDP), through which it aims to provide low cost housing for its previously disadvantaged citizens. Kuyasa is part of one of these housing schemes.

This project has a particularly high rating in terms of social sustainability and local development. It will create jobs, reduce health risks, and save money. Benefits include a reduction in household dust due to the insulated ceiling, and improvement in the ambient temperature of the house, thereby reducing the need for paraffin stoves and other heat sources which pose fire-related dangers (such as burns and damage to property). Elsie Tabalaza is a 70-year-old pensioner residing in Kuyasa. She is one of the ten beneficiaries of the project pilot phase, and her house has therefore been fitted with a solar water heater, an insulated ceiling, and energy efficient lighting. It is five degrees warmer than her neighbour's house and according to technical research by Agama Green Energy, electricity savings of up to 40 per cent have been experienced. As Elsie explains, 'Things are much better. Since the ceilings have been installed, it is much warmer, and more beautiful than it was before'.[21] The solar powered water heater also reduces the need for heating water on a flame or using an electric kettle, thus further reducing fire risks. A reduction in household smoke and improved local air quality will cause fewer respiratory health problems, such as carbon monoxide poisoning, amongst those living in modified homes and the township. Improved respiratory health profiles will offset the need for local, national government, and state enterprise spending on such diseases.[22] The installation and maintenance of these technologies will also provide employment to and increase capacity and skills amongst local artisans (such as plumbers, electricians, and carpenters). Local and regional job creation will increase further when solar water heaters are manufactured locally. Local capacity regarding the CDM has grown, in part because of training provided by the South*South*North project. Council officials from different line departments who have gained knowledge about CDM can utilize the opportunities that this presents elsewhere.[23]

Installing the new technologies costs R10,998 in each house, and project expansion will involve finding a carbon buyer, and identifying possible sources of bridging finance, such as contributions from the community.[24] One of the remaining challenges is to secure acceptance for the argument that 'suppressed demand' for energy merits status as a CDM project.[25]

[21] See Orford, n 11 above. [22] ibid. [23] See Morris *et al.*, n 20 above. [24] ibid.
[25] See Orford, n 11 above.

4.5 Transport in Yogyakarta City, Indonesia

The Yogyakarta Urban Public Transport Alliance, which includes the Yogyakarta Urban Bus Cooperative, is involved in a project designed to improve the public transport system in Yogyakarta City by using cleaner engines in 200 out of the 500 buses in the area, restructuring urban public transport management, improving routing and schedules, and introducing alternative fuels.[26] Funding under the CDM will help replace diesel engines with liquid petroleum gas engines, and total emissions reductions for the seven-year crediting period would be 9,600 tonnes CO_2. Currently, public transport in Yogyakarta is unsafe, and the bus engines cause significant pollution. Smoke makes city transport unpleasant for residents, and many people use masks, especially those on motorbikes. Improved buses, and better management of transport routes and schedules will reduce local air pollution, and provide a safer cleaner environment in which commuters and others could travel.

4.6 The Plantar Project, Brazil

The Plantar Project, near the city of Curvelo in the State of Minas Gerais, Brazil, has received funding from the World Bank's Prototype Carbon Fund. The project aims to use 23,100 hectares of old eucalyptus plantations, which over the next seven years will produce 9,885,260 tonnes of charcoal, to substitute for the use of coal/coke in the production of 3,802,023 tonnes of pig iron over the next 21 years. Replacing fossil fuels with biomass in this way will reduce CO_2 emissions by 3,070,481 tonnes of carbon during the project lifetime. The charcoal will be produced in a sustainable way, according to the principles and criteria of the Forest Stewardship Council (FSC), which certified the forests in 1998, ensuring the initiative is environmentally, socially and economically sustainable (Box 7).

Box 7. Forest Stewardship Council set of ten principles and criteria for good forest management[27]

1. *Compliance with Laws and FSC Principles*: Forest management shall respect all applicable laws of the country in which they occur, and international treaties and agreements to which the country is a signatory, and comply with all FSC Principles and Criteria.
2. *Tenure and Use Rights and Responsibilities*: Long-term tenure and use rights to the land and forest resources shall be clearly defined, documented and legally established.
3. *Indigenous Peoples' Rights*: The legal and customary rights of indigenous peoples to own, use and manage their lands, territories, and resources shall be recognized and respected.

[26] ibid. [27] <http://www.web.net/fscca/fsc.htm#princi>.

4. *Community Relations and Workers' Rights*: Forest management operations shall maintain or enhance the long-term social and economic well-being of forest workers and local communities.
5. *Benefits from the Forest*: Forest management operations shall encourage the efficient use of the forest's multiple products and services to ensure economic viability and a wide range of environmental and social benefits.
6. *Environmental Impact*: Forest Management shall conserve biological diversity and its associated values, water resources, soils, and unique and fragile ecosystems and landscapes, and, by so doing, maintain the ecological functions and the integrity of the forest.
7. *Management Plan*: A management plan—appropriate to the scale and intensity of the operations—shall be written, implemented, and kept up to date. The long-term objectives of management, and the means of achieving them, shall be clearly stated.
8. *Monitoring and Assessment*: Monitoring shall be conducted—appropriate to the scale and intensity of forest management—to assess the condition of the forest, yields of forest products, chain of custody, management activities and their social and environmental impacts.
9. *Maintenance of Natural Forests*: Primary forests, well-developed secondary forests and sites of major environmental, social or cultural significance shall be conserved. Such areas shall not be replaced by tree plantations or other land uses.
10. *Plantations*: Plantations should be designed and managed consistent with Principles 1–8 and the following criteria. Such plantations can and should complement overall ecosystem health, provide community benefits, and provide a valuable contribution to the world's demands for forest products.

The FSC is one of the largest and most credible groups involved with forest certification. However, certification has been criticized because it is being used for a purpose for which it was not originally designed. The initial evaluation costs are often too expensive for smaller forestry operations, principally in low-income countries, and certification thus favours large producers who spread the costs of certification audits against large production volumes (Box 8). Certification also gives market advantage to already sustainably managed forests rather than encouraging a move towards sustainability particularly in tropical areas.[28] Indeed, more than 90 per cent of certified forests occur in temperate and boreal high-income countries,[29] and only 8 per cent of certified forests are in the tropical low-income countries.[30]

[28] S Bass, K Thornber, M Markopoulos, S Roberts, and M Grieg-Gran, *Certification's Impacts on Forests, Stakeholders and Supply Chains* (International Institute for Environment and Development (IIED), London, UK, 2001).

[29] R Eba'Atyi and M Simula, *Forest Certification: Pending Challenges for Tropical Timber* (International Tropical Timber Organization (ITTO) Technical Series No 19, Yokohama, Japan, 2002).

[30] HT Thang, *Current Perspectives of Sustainable Forest Management and Timber Certification*, Special Paper in congress proceedings, XII World Forestry Congress, 21–28 September 2003, Quebec City, Canada.

Box 8. Challenges faced by community forest enterprises regarding certification[31]

- The high costs of certification for community groups;
- The inaccessibility of both market information and certified forest product markets;
- The inability of forest standards to recognize many (complex) local land use systems, and locally-relevant social issues;
- The lack of links between certification and the (development of) policies to promote community forestry;
- The social and cultural burdens, and the technical challenges, entailed when undertaking the necessary business improvements to support certified forest operations.

The project also provides local environmental benefits including the avoidance of acid rain by the cessation of SO_2 release. A reduction in local air pollution may also lead to improvements in worker health. The project contributes to the recovery of degraded areas, and forests protect the soil, offer shelter to wildlife, and prevent erosion. Whilst the eucalyptus trees are all clones, and thus provide limited gains in terms of biodiversity conservation and associated livelihood benefits, for each 100 hectares of planted forests, at least 20 hectares of indigenous cerrado forests in and around the plantation areas are maintained to preserve the native flora and fauna and to protect local water sources. Biological corridors have been constructed, and effective fire control mechanisms introduced. Plantar will increase investment in biodiversity conservation on its landholdings, and pressure on indigenous forests elsewhere will decrease by the provision of cultivated biomass from forest Plantar plantations.

Forestry also creates many jobs, particularly in rural areas where jobs are scarce. Replacing charcoal blast furnaces with those based on coal reduces employment due to the lack of forestry activity, and low labour needs resulting from greater technological advancement. Job creation helps combat rural exodus and draws additional services such as schools, providing education and development to rural communities. It is estimated that this project will create and maintain around 3,000 jobs during its 21-year lifetime. None of these will involve child labour, and Plantar has certification granted to companies that promote socially sound working conditions (kindergarten, transport facilities, recreation activities, and so on). The quality of work in relation to other employees in the same sector will be better, and particular focus will be placed on the incidence of respiratory diseases among project staff working on carbonization activities. Plantar claims it has provoked no involuntary resettlements and affected no indigenous peoples. Plantar's employees get training in sustainable forest management techniques, appropriate

[31] See n 28 above.

pest control, and general environmental education. Plantar is also sponsoring local agriculture schools and training students in sustainable forestry and agriculture.[32]

Despite these apparent sustainable development benefits, criticism of the Plantar project has been strong.[33] One letter (signed by over fifty Brazilian NGOs, civil society movements, churches, and unions) urging PCF investors not to buy credits from the Plantar project expresses concern over 'the continuing expansion of large-scale eucalyptus monocultures in our states, which result in negative environmental, social, economic and cultural impacts on local communities and stimulate the profit of a few at the expense of the majority of the people.'[34] Greenpeace has questioned the permanence of carbon sequestration and thus the real long-term climate benefits,[35] and others have commented on the disastrous local environmental impacts of eucalyptus plantations, which have caused rivers to dry up, and have minimal biodiversity value.[36] Application of FSC certification has also been challenged by organizations such as SinksWatch[37] and the World Rainforest Movement.[38]

Particular concern has been expressed over the integrity of the project developers who are accused of expelling local people from their land by taking advantage of attractive tax incentives available under the military dictatorship in the past, burning large amounts of native cerrado forests (historically and currently), fraudulently acquiring community land for plantations, poor community consultation, circumvention of Brazilian legislation regarding labour (for example, limiting subcontracting to avoid resultant dangerous working conditions and child labour) and environmental impact evaluation, poor care of past employees, contributed to the local economic crisis (by providing fewer job opportunities than agriculture, and destroying alternative industries reliant on products from the native vegetation),[39] and providing incorrect baseline situations (scenarios that would result if the project does not go ahead).[40]

5 Future Benefit Sharing—Some Suggestions for Systematization

The structure and functions of DNAs will be a key determinant of whether sustainable development objectives accrue in each country. Experience from the most successful host countries to date suggests that it is imperative to have a single unit responsible for the solicitation and approval of projects, and that this unit should

[32] See <http://carbonfinance.org/pcf/router.cfm?Page=Projects&ProjectID=3109>.

[33] See <http://www.sinkswatch.org/projects/plantarmore.html>.

[34] See <http://www.fern.org/pubs/ngostats/Planteng.htm>.

[35] Greenpeace submission on the proposed Plantar CDM project <http://www.prototypecarbonfund.org>.

[36] See <http://www.fern.org/pubs/ngostats/Planteng.htm>.

[37] See <http://www.sinkswatch.org/pubs/SinksWatch%20submission%20regarding%20FSc%20accreditation%20audit%20SCS,%20Plantar%20Minas%20Gerais%20February%202004%20final.doc>.

[38] See <http://www.wrm.org.uy/countries/Brazil/fsc.doc>.

[39] See n 34 above.

[40] See <http://www.cdmwatch.org/plantar.html>.

have full decision autonomy and professional, permanent staff. Because DNAs must examine and approve CDM projects, it is important that they have the capacity to understand and effectively operationalize the accrual of sustainable development objectives.

The lack of guidelines on what constitutes sustainable development and how it should be measured means that CDM projects with sustainable development 'frills' could be priced out of the market. Meeting sustainable development objectives provides additional complications, which may chase away potential investors in CDM projects and lead to a 'race-to-the-bottom' in terms of reducing barriers to investment.[41] This could mean that projects such as the Solar Home Systems project in Bangladesh will not receive the support they deserve. However, experiences in Singapore, where there are stringent hurdles for investors to cross and yet no shortage of investors, compared to Uganda, where there are few obstacles and almost no international investment, suggest that investors are not necessarily discouraged by regulation. This is likely to be because regulation and transparency lowers the risk to investors, and increases the likelihood that investments will have benefits that will last longer than the current policy-maker's term of office. This suggests there could be benefits from systematizing the process of ensuring CDM projects provide sustainable development benefits.

However, other factors, such as investment climate, related investments, and risk also influence where the private and public sectors choose to invest. Many of the poorer regions of the world have to compete with regions that traditionally receive more direct foreign investment (like Singapore) or that have large volumes of CERs to sell (like China, India, Mexico, Brazil, and South Africa). Investors will be less inclined to buy from smaller countries, especially if their approval, and monitoring and evaluation procedures are lengthy, cumbersome, and difficult to implement. Opportunities for creating niche markets for ethically motivated CDM investments exist, and countries with no protocols for sustainable development run the risk of having these issues marginalized.[42] This is the thinking behind the Community Development Carbon Fund (CDCF) launched by the World Bank at the World Summit on Sustainable Development (WSSD) in Johannesburg in 2002. The challenge is to ensure that benefits from the CDM reach the 'poorest of the poor' as opposed to being channelled into less vulnerable communities and countries. Some degree of regulation may help with this, as it is currently a buyer's market for those investing in CDM, which means that sustainable development objectives run the risk of being marginalized unless they are firmly entrenched in national policy and legislation.

Given that the CDM is a subset of foreign direct investments (FDI) for developing countries, albeit unusually one which has an additional sustainable development benefit objective, it is possible that, with special pro-active promotion, some

[41] See for example JA Kim, *Sustainable Development and the CDM: A South African Case Study* (Tyndall Centre for Climate Change Research, Norwich, 2003).

[42] See Huq, n 6 above.

CDM projects could become examples of how to promote FDI in the developing countries (especially the poorest ones, such as the least developed countries) in a manner that will promote sustainable development as well as generate profits for the investors. Thus the experience with CDM can potentially also speak to the larger role of FDI and particularly to the corporate social responsibility (CSR) agenda in FDI (which is a much larger pool of investments than the CDM alone).

Many have commented on the need for considerable institutional capacity strengthening and training of the main participants in the CDM process to ensure potential benefits materialize.[43] This is particularly important among DNA staff, but is also apparent amongst government decision-makers across ministries, potential CDM project owners, project developers and local/regional institutions, both in financial services, and potential suppliers of capacity building services. Most CDM project owners and many key institutions do not yet have sufficient understanding of how the CDM and the Kyoto markets will function and how CERs will be priced, and are unable to realistically estimate the benefits and costs of participating in projects. Without this understanding, the flow of CDM projects may be small and the compliance costs to Annex B countries higher than necessary. Without urgent, well-planned attention to capacity building in CDM host countries, the Kyoto mechanisms cannot make a significant contribution to cost-effective climate change mitigation in the first commitment period and will also fail to unfold their full potential to contribute to sustainable development in host countries. Learning-by-doing is one approach that can help share practical experiences from pilot CDM activities as part of a broader monitoring and evaluation process.

The role of developing simple, but robust, criteria for assessing the sustainable development benefits from potential CDM projects is best taken up by host country governments and their respective DNAs. This should be seen as an opportunity to ensure that proposed CDM projects adhere to some (country driven) criteria that contribute to national sustainable development objectives. It is important to ensure that these criteria are not viewed as a resource-intensive requirement for project development that serves only to increase the burden and costs accruing to potential CDM project developers. It is also important that developing countries share and learn from each other's experiences as they implement CDM projects. It is hoped that this chapter contributes to this growing knowledge base by sharing lessons from some of the more successful existing projects.

[43] See Huq, n 6 above. See also UNDP and Ecosecurities 2003, n 3 above; K Brown and E Corbera, *A Multi-Criteria Assessment Framework for Carbon-Mitigation Projects: Putting 'Development' in the Centre of Decision-making* (Tyndall Centre for Climate Change Research Working Paper, 2003) 29; Y Jin and X Liu, *Prospects of CDM for Promoting Sustainable Development in China—Accelerating Foreign Technology Investment and Technology Transfer*, prepared for the Working Group on Trade and Environment, China Council for International Cooperation on Environment and Development (2000); and Michaelowa, n 3 above.

13

Can Companies or Entities from a Non-Party to the Kyoto Protocol Participate in the Flexible Mechanisms?

*Martijn Wilder**

1 Introduction

The flexible mechanisms of the Kyoto Protocol (the Clean Development Mechanism (CDM), Joint Implementation (JI), and International Emissions Trading(IET)) are intended to provide a cost-effective means to reduce greenhouse gas (GHG) emissions on a global level, by enabling Annex I Parties to meet their Protocol targets by investing in emission reduction projects in other countries or by purchasing Kyoto Protocol credits on an international trading market.

It is clear from the terms of the Kyoto Protocol and the Marrakesh Accords (which flesh out the practical rules of participation in emission reduction projects and emissions trading) that it was generally envisioned that only Kyoto Protocol Parties and the entities they have authorized to participate in projects and emissions trading would be able to participate in the flexible mechanisms. However, as the carbon market gains liquidity, companies from Annex I countries which have not ratified the Protocol (such as Australia and the US) have expressed interest in becoming involved in CDM and JI projects and International Emissions Trading, despite their Governments' prevailing refusal to ratify the Protocol.[1]

The ability of non-Party private entities to participate in the Kyoto Protocol is a politically contentious one. There are three major issues which will be relevant to determine whether a non-Party company can participate in CDM or JI projects and trade Kyoto Protocol credits (such as Certified Emission Reductions (CERs) or Emission Reduction Units (ERUs)):

(i) Do the Kyoto Protocol rules enable non-Party companies to be 'participants' in CDM or JI projects and to trade emission reductions?

* Martijn Wilder is the head of Baker & McKenzie's Global Clean Energy and Climate Change Practice Group. The views expressed in this chapter are the views of the author and do not necessarily represent the views of Baker & McKenzie.

[1] For instance see Environment Business Australia, 'The Business Case for Ratification', accessible at <http://www.environmentbusiness.com.au> accessed 3 May 2004.

(ii) If not, is there any way in which non-Party companies may be able to circumvent such a prohibition?

(iii) Will the domestic legislation and policies of Kyoto Protocol Parties effectively prevent non-Party companies from holding registry accounts to trade emission reductions?

Ultimately, the Kyoto Protocol rules are ambiguous with regards to the ability of non-Party companies to participate in projects and emissions trading. Although the Protocol prevents non-Party governments from directly participating in the flexible mechanisms,[2] it does not address the issue of the ability of non-Party companies to participate with the assistance of a Party to the Protocol. The reality of multinational companies and the ability for non-Parties to establish 'shelf' companies in Party jurisdictions to hold and transfer Kyoto Protocol credits is likely to circumvent any prohibition on non-Party participation in the flexible mechanisms. Due to the difficulty of negotiating amendments to the Protocol and the practice of passing resolutions by consensus of the Conference of Parties to the UNFCCC, it is unlikely that modification of the Kyoto Protocol rules to prevent such circumvention will achieve sufficient political consensus. Therefore, the ability of non-Party companies to participate will depend to a large extent on national policies of Parties to the Kyoto Protocol and the policies of the CDM Executive Board and JI Supervisory Committee with regards to registering projects and developing rules of procedure.

This chapter provides an analysis of the Kyoto Protocol rules as they apply to non-Party companies and other private entities. Future domestic policies and the actions of the Conference of Parties/Meeting of Parties to the Kyoto Protocol (COP/MOP), the JI Supervisory Committee, and the CDM Executive Board are very difficult to predict. Therefore, without concluding whether non-Party companies' participation in the Kyoto Mechanisms will ultimately be possible, this chapter provides comment on the types of potential national policies that may occur to address the participation of non-Party companies and the perspectives of some of the Kyoto Protocol Parties such as Canada and the EU Member States on this issue.

2 Non-Party Participation in the Kyoto Protocol

As a fundamental principle of public international law, an international treaty or agreement cannot impose obligations or confer rights upon a third party to that treaty or agreement unless:

(i) the third party gives its consent (for example, through ratification of the Protocol); and

[2] Arts 6, 12, and 17 of the Kyoto Protocol to the United Nations Framework Convention on Climate Change 1997, FCCC/CP/1997/L.7/Add.1 reprinted in (1998) 37 ILM 22 (*not yet in force*).

(ii) the parties to the treaty or agreement (in this case, Parties to the Protocol) agree that third party states may have rights or obligations under its provisions.[3]

The exceptions to this general principle are where the rights or obligations have become customary international law[4] or where some other agreed arrangement is specifically provided for in the terms of a treaty.[5] Under general international law principles, countries which are signatories to the UNFCCC or the Kyoto Protocol but which have not ratified one or both of them nonetheless have an obligation to refrain from acting in a way which would defeat their object and purpose.[6]

2.1 Participation in the CDM

The Protocol contains no provisions explicitly allowing or disallowing non-Party private entity participation in the CDM although, as discussed below, countries that have not ratified or acceded to the Kyoto Protocol are excluded from directly participating in transferring CERs, ERUs, or Assigned Amount Units (AAUs). Significantly, a number of Parties to the Protocol have articulated a clear intention to exclude companies from non-Parties from participation in the flexible mechanisms of the Protocol.[7]

The specific provisions of the Protocol and the Marrakesh Accords relating to participation in the CDM and the trading and transfer of CERs are as follows:

(i) Article 12 of the Protocol provides the basic outline of the CDM and envisions the participation of private and/or public entities;

(ii) Article 12 also recognizes that CERs accruing from CDM projects shall be used to contribute to compliance with part of an Annex I Party's quantified emission reduction or limitation commitments (QERLC) under the Protocol;

(iii) The Marrakesh Accords provide that a Kyoto Protocol Party may authorize private entities to participate in CDM projects, although it must still remain responsible for the fulfilment of its obligations under the Protocol;

[3] Vienna Convention on the Law of Treaties, Arts 34, 35, and 36.

[4] Rights and/or obligations under an international agreement may over time become accepted as principles of customary international law, through state practice of compliance. This was the case with the Exclusive Economic Zone under the United Nations Convention on the Law of the Sea.

[5] Certain agreements may explicitly provide for non-party participation. For example, under the Basel Convention, which deals with the movement of hazardous wastes across national boundaries and the requirements for disposal of such wastes. This Convention specifically imposes obligations on non-Party States with regards to reporting the migration of hazardous wastes. It also allows Party States to enter into bilateral agreements with non-Party States in regards to waste disposal issues, without the consent of the other Parties to the Convention.

[6] Art 18 of Vienna Convention on the Law of the Treaties 1969, 1155 UNTS 331.

[7] See D Bodansky, 'Implications for U.S. Companies of Kyoto's Entry into Force without the United States', Pew Centre Publication accessible at <http://www.pewclimate.org/docUploads/Kyoto%2DUSBusiness%2Epdf> accessed 3 May 2004.

(iv) The Marrakesh Accords provide that a Party may authorize private entities to participate in emissions trading, and must maintain a public list of such entities;

(v) The participation requirements of the CDM under the Marrakesh Accords provide that for a Party to be eligible to *use* CERs it must be a Party to the Protocol;[8] and

(vi) The participation requirements for emissions trading under the Marrakesh Accords provide that for a Party to be eligible to *acquire and transfer* CERs issued in accordance with the relevant provisions of the Protocol, it must be a Party to the Protocol.[9]

In effect, the Kyoto Protocol and the provisions of the Marrakesh Accords create a new legal right—a CER—which is created for use within the public international law framework of the Protocol by Parties to that framework. Without the Protocol, and in the absence of national legislation recognizing Kyoto Protocol Rights (CERs, ERUs, AAUs, as well as Removal Units, tCERs and lCERs), CERs have no legal existence.

The provisions which exclude Parties from participating apply to sovereign States rather than individual companies. As discussed, there is no specific provision in the Kyoto Protocol or the Marrakesh Accords, which specifically excludes participation by non-Party companies in the CDM. Some commentators have argued that this 'silence' implicitly allows for participation by companies from non-Parties such as Australia or the US. For example, a paper published by the Australian Greenhouse Office (AGO) in 2003 states that '[a]s a non-Party to the Kyoto Protocol, the Australian Government cannot participate directly in the Kyoto mechanisms. However, the rules agreed at Marrakesh in November 2001 do not discriminate between firms from countries that have and have not ratified the Protocol.'[10] The AGO paper argues that there is nothing in the Marrakesh Accords to prevent Parties who have ratified the Protocol from authorizing companies from non-ratifying jurisdictions to participate, and implies that Australian firms will therefore be able to participate. However, a heavy reliance on what the Kyoto Protocol 'does not state' fails to consider broader public international law principles regarding third party rights and obligations under a treaty and may result in an incomplete analysis.

Although it expressly envisions participation of private and public entities as well as national governments, the Kyoto Protocol is essentially an international treaty between sovereign States and the rights and obligations therein apply only to

[8] The Marrakesh Accords, FCCC/CP/2001/13/Add.2, draft Decision-/CMP.1, Modalities and procedures for a Clean Development Mechanism, Art 12, Annex, para 31.

[9] The Marrakesh Accords, ibid (Article 17), Modalities, rules and guidelines for emissions trading under Article 17, Annex, para 2.

[10] See Australian Greenhouse Office paper 'Australian Firms and the Market-based Mechanisms of the Kyoto Protocol' (2003), accessible at <http://www.greenhouse.gov.au/international/kyoto/pubs/mechanisms.pdf> accessed 12 May 2004.

national governments, which have become a Party to the Protocol through an act of ratification or accession. These rights and obligations (including the ability to participate in CDM or JI projects and trade CERs or ERUs) are likely to be passed on to private and public entities through domestic legislation.[11]

Any Kyoto Protocol Party that authorizes private or public entities to participate in CDM activities will remain responsible for fulfilling its overall emission reduction obligations under the Treaty.[12] Entities that have been authorized by a Party can only transfer and acquire CERs if the Party, which has provided authorization for their participation, is in compliance with its obligations under the Protocol (that is, it has an operating registry system in place). This means that companies participating in the CDM will be relying on the continued compliance of the Party, which has authorized them to be able to undertake emissions trading within the Kyoto Protocol registry system.

2.1.1 Non-Party 'participants' in the CDM

The participation requirements of the CDM under the Marrakesh Accords provide that for a country to be eligible to acquire, use, or transfer CERs it must be a Party to the Protocol.[13] The Marrakesh Accords also require every participant in a CDM project to be authorized by a Party to the Protocol,[14] although it is unclear from the text whether this Party must be the national government of the participant.

The Project Design Document Glossary published by the CDM Executive Board[15] defines a CDM project participant as 'either a Party involved or . . . a private and/or public entity authorized by a Party to participate, under the Party's responsibility, in CDM project activities.' The Glossary further states that project participants take decisions on the allocation of CERs from the proposed CDM project. CDM project participants could include the project developer, the purchaser of CERs, and the financier for the project. As part of the project registration process, all project participants must sign a statement to the Executive Board clarifying the way in which they will communicate with the Executive Board and the Secretariat in relation to the project and, in particular, with regard to instructions regarding CER allocation.

If a non-Party company wished to build a CDM project and ultimately receive CERs for that project to sell on the market, it may wish to become a 'project participant'. The details of this participation would become public through the CDM website in the Project Design Document (PDD), which is provided on the CDM

[11] For instance, the New Zealand Parliament has passed the Climate Change Response Act 2002, which establishes the legal and administrative framework for Government to grant AAUs to private and public entities. A summary of the Climate Change Response Act can be found at <http://www.mfe.govt.nz/laws/> accessed 3 May 2004.

[12] Marrakesh Accords, draft Decision -/CMP.1, above n 8, Art 12, Annex, para 33.

[13] See n 9 above. [14] ibid.

[15] Accessible on the CDM Executive Board website <http://cdm.unfccc.int/Reference/Documents/PddGlossary/English/pddgloss> accessed 3 May 2004.

Executive Board website for public comment as part of the validation and registration process.[16] Potentially the public disclosure of non-Party participation in a CDM project could attract criticism from NGOs and certain Kyoto Protocol Parties.

At this stage it is unclear whether the Executive Board will attempt to prevent non-Party participation through the project registration process. The Board is composed of one member from each United Nations regional group, two additional members from Annex I developed countries, two additional members from developing countries and one member from the small island developing States (which are geographically particularly vulnerable to climate change). This composition will ordinarily mean that the Executive Board will contain members primarily from developing countries. It is possible therefore that the Executive Board may be more amenable to participation by non-Party companies, as developing countries do not face the same concern about trade exposure to non-Party competitors due to the costs of compliance that Annex I Parties to the Kyoto Protocol have expressed.[17] Additionally, as a UN body, it is uncertain whether the Executive Board would directly interfere with the political decision of a Kyoto Protocol Party to authorize non-Party companies to participate (for example, by allowing them to hold registry accounts within its national registry) if this was not expressly prohibited by the international rules agreed upon by the Parties.

The ability of non-Party companies to participate in the Kyoto Protocol flexible mechanisms will therefore depend to a large extent on the domestic political policies of Kyoto Protocol Parties rather than on the existing rules of the Kyoto Protocol itself. At this early stage, any analysis about the types of domestic policies that may be implemented to address the issue of non-Party participation in the Kyoto Protocol mechanisms must be merely speculative. These policies may depend to a large extent on the actual costs incurred by domestic industry in Annex I Parties from the implementation of national policies (for example, emissions trading) to meet Kyoto Protocol targets. What is certain, though, is that non-Party participation in the Kyoto Protocol mechanisms will be an issue that the Kyoto Protocol Party Governments will need to consider as the Protocol mechanisms continue to develop and particularly before the CDM Executive Board registry and counterpart national registries are established.

Obviously, if the Kyoto Protocol rules implicitly allowed Parties to authorize non-Party company participation, the decision whether to authorize non-Party

[16] See 'Procedures on Public Availability of the CDM Project Design Document and for Receiving Comments as Referred to in Paragraphs 40(B) and (C) of the CDM Modalities and Procedures (Version 03)' <http://cdm.unfccc.int/Reference/Procedures/public%20availPDD_ver03.pdf> accessed 3 May 2004; PDDs available for public comment are accessible at <http://cdm.unfccc.int/Validation/publicPDD>.

[17] For instance, the New Zealand Government has expressed concerns about potential adverse impacts on it exports in its 'National Interest Analysis: Kyoto Protocol to the UN Framework Convention on Climate Change', <http://www.climatechange.govt.nz/about/kyoto.html>, accessed 3 May 2004.

participants in the CDM would be a largely political one. Under the text of the Marrakesh Accords, it is contemplated that only Annex I Parties will create national emissions trading registries with the ability for companies to hold their own sub-accounts to buy and sell CERs.[18] Annex I Parties may be reluctant to allow non-Party companies a 'free ride' in comparison to their own counterpart companies, which would likely be facing the compliance costs of reducing GHG emissions on a domestic level. In fact, there is a possibility that businesses in Kyoto Protocol Parties (such as those in the European Union) may feel at a competitive disadvantage in comparison with business based in non-Party countries, which are not internalizing the costs of meeting Protocol targets. Such businesses may pressure their governments, not only to prohibit the participation of non-Parties in the Kyoto Protocol flexible mechanisms, but also to take affirmative action to pressure those non-Parties to ratify such as through trade measures. For example, a higher rate of tax could be levied against imports of carbon-intensive products from non-Parties. Such a measure would invoke the interplay between international environmental rules and international trade law, and may well be challenged at the WTO.[19]

For example, Canada has stated that it will allow subsidiary companies from non-Parties to the Protocol (such as American subsidiaries located in Canada) to hold sub-accounts within the Canadian national registry on the same basis as Canadian companies.[20] Canada has argued that a failure to do so may breach World Trade Organization obligations of non-discrimination on the basis of the ownership or location of companies. However, other Parties, such as the European Union, have voiced objection to the prospect of any non-Party participation in the Protocol, citing the possibility that non-Party companies will get a 'free ride', as they will be able to profit from the flexible mechanisms without incurring the costs of binding obligations to reduce their emissions.[21] A consortium of environmental interest groups, including Greenpeace, WWF, Climate Action Network, and CDM Watch, is also pressing Parties to the Protocol to refrain from investing in projects in which the private developer is based in a non-Party country.[22]

2.1.2 Non-party participation in unilateral CDM projects

Prior to the seventh session of the Conference of the Parties to the UNFCCC (COP7) and based on the text of Article 12 of the Kyoto Protocol, it was believed

[18] See n 9 above.

[19] For example, see 'Free Riding on the Climate' by the New Economics Foundation 2003, <http://www.nef.org> accessed 3 May 2004.

[20] See Paul Fauteux, Director General, Climate Change Bureau Environment Canada, 'Party/Non-Party Emissions Trading: Compatibility Now, Convergence Later? (Notes for a presentation)', <http://www.ieta.org/About_IETA/IETA_Activities/SBSTA16/Fauteux.doc> accessed 3 May 2004.

[21] See n 7 above.

[22] See letter written by consortium to Annex I Countries, <http://www.cdmwatch.org/quality-restrictions.php> accessed 3 May 2004.

that a CDM project activity must be undertaken by, or directly involve, an Annex I Party. The draft COP6 negotiating text originally provided that projects undertaken by a non-Annex I Party without any direct Annex I Party involvement ('Unilateral CDM Projects') would not be permitted.[23] However, this bracketed text was later removed and it was recognized by some parties at COP7 that Unilateral CDM Projects would be allowed within the Protocol regime. However, this concept is not specifically referred to anywhere in the Marrakesh Accords, although the text stating that a Party not included in Annex I may participate in a CDM Project Activity if it is a Party to the Protocol has been used to support the concept.

It is interesting to note that the draft 'Linking Directive' for the European Union Emissions Trading Scheme (EU ETS) defines an eligible CDM or JI project activity as a project activity approved by one or more Annex I Parties in accordance with Article 6 or 12 of the Kyoto Protocol and the decisions adopted pursuant to the UNFCCC or the Kyoto Protocol.[24] The EU thereby excludes Unilateral CDM Projects for the purpose of the EU ETS.

Unilateral CDM Projects may enable non-Party companies to acquire CERs without the need to be involved in the project as a participant. For example, a non-Party company could work with a Host non-Annex I country to have the project approved as a Unilateral CDM Project and limit its involvement to purchasing the resulting CERs. In effect, this would require the developing Party company to develop and implement the project, with host Government approval, and then sell the CERs subsequently to a non-Party company through a separate contract. However, as discussed, in order for the transfer of CERs to occur effectively, this would require:

(i) the developing country company to have access to a national registry account, or an account within the CDM Executive Board registry, into which the Executive Board can issue the CERs for the project; and
(ii) the non-Party company to have access to an account within a national registry account of a Party to the Protocol into which the CERs can be transferred.

Some members of the CDM Executive Board have argued that the emissions trading provisions of the Kyoto Protocol only allow for Annex I Parties to trade rights between their national accounts and that therefore developing country Parties are implicitly excluded from emissions trading.[25] However, on the other

[23] See 'Mechanisms pursuant to Article 6, 12 and 17 of the Kyoto Protocol. Modalities, rules, procedures and guidelines under the mechanisms pursuant to Article 6, 12 and 17 of the Kyoto Protocol', FCCC/SB/2000/MISC.4, Add.1, Add.2/Corr.1 and Add.3, <http://maindb.unfccc.int/library/> accessed 3 May 2004.

[24] Proposal for a Directive of the European Parliament and of the Council amending Directive 2003/87/EC establishing a scheme for greenhouse gas emission allowance trading within the Community, in respect of the Kyoto Protocol's project mechanisms.

[25] See Climate Change Program 'Unilateral CDM—Prospect and Pitfalls', <http://regserver.unfccc.int/SEORS/cop9events/FileStorage/FS_442592919> accessed 3 May 2004.

hand, the Marrakesh Accords specifically provide that within the CDM registry, an account will be opened for each CDM host country or developing Party which requests one.[26] It could seem anomalous if the Kyoto Protocol enabled a developing country to hold an account but prohibited it from trading CERs (as in the first commitment period the host country will not have any binding compliance commitments and will therefore not require CERs for compliance purposes). However, the practical administrative limitations of the CDM Executive Board registry account may well mean that it will not be equipped to administer emissions trading on behalf of host countries. This is an issue which the Executive Board is likely to provide future guidance on and will directly impact on the feasibility of Unilateral CDM Projects.

2.1.3 Non-party access to the Kyoto Protocol Trading Registries

CERs are rights created and recognized under international law which only come into existence through the submission of a Certification Report to the Executive Board and electronic issuance through the CDM registry system. Once created, Parties to the Protocol can only use, transfer, and acquire CERs in accordance with the rules of the Protocol and for the purpose of meeting their QELRC. If an Annex I country (such as Australia or the US) has refused to ratify the Kyoto Protocol, there would be little basis upon which it could argue that its companies were entitled to access CERs for international emissions trading purposes, although there would be nothing to stop non-Parties from recognizing Kyoto Protocol credits in a national emissions trading regime.

Figure 13.1 represents the manner in which the registry system will operate.

2.2 Participation in Emissions Trading

To issue or transfer Kyoto Protocol Rights, a company must have access to an electronic account within a national registry established under the Marrakesh Accords. Currently such accounts can only be created by Annex I Parties to the Protocol.[27] As governments will invariably pass down GHG reduction obligations to individual industries and corporations, Kyoto Protocol credits are likely to have value for private entities in Protocol Parties. In most Annex I countries, it is contemplated that Kyoto Protocol credits will be owned by and traded between private and/or public entities.

As registry accounts can only be established by Parties to the Protocol, non-Parties (and businesses incorporated within those non-Parties) will be unable to directly participate in the Protocol framework and thereby acquire, or have transferred to them, Kyoto Protocol Rights.

[26] Marrakesh Accords, draft Decision -/CMP.1, Modalities and procedures for a Clean Development Mechanism, Art 12, Annex, para 5.

[27] See n 9 above.

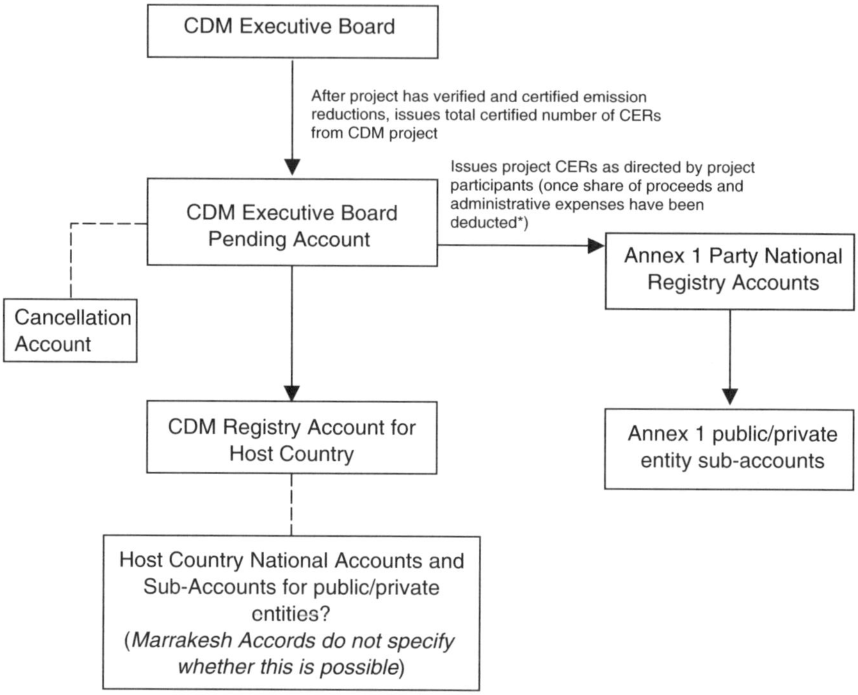

Figure 13.1. Progress of CERs in a CDM Project

* As part of the application to the Executive Board to register a CDM project, project participants are required to provide a signed statement indicating the method by which the Executive Board should communicate with them to determine how CERs are to be allocated after they have been certified and issued.

However, it may be possible for non-Parties, or companies from non-Parties, to enlist the assistance of a Party to the Protocol, or a company holding an account within a Party's national registry, to hold or transfer rights on behalf of the non-Party. As discussed further below, the ability to enlist such assistance will depend largely on the Party's domestic regulation and policies regarding non-Party participation. If the Kyoto Protocol Party is willing to assist, some fee or commission may be imposed by the cooperating Party or public/private entity for this service.

In addition, a subsidiary registered in a Protocol Party jurisdiction could effectively buy, hold and sell Kyoto Protocol credits on behalf of the non-Party parent company. This is not prohibited within the Kyoto Protocol rules. With the realities of multinational companies and globalization, it would in fact be very difficult to prevent Kyoto Protocol non-Parties from indirectly participating in the Protocol flexible mechanisms and speculating on the international emissions trading market. Where those subsidiaries actually have emission reduction obligations under the jurisdiction in which they are based, this would not be considered 'free riding' as they have both the same rights and obligations as their counterpart companies. However, if 'shelf companies' or agent companies are established within Kyoto Protocol Parties simply for the purpose of holding Kyoto Protocol Rights on behalf of a company in a non-Party, this would effectively bypass the provisions in the Kyoto Protocol and Marrakesh Accords prohibiting direct non-Party participation. It could be argued that such measures would not be in furtherance of the objectives of the Kyoto Protocol as they would allow non-Party companies to profit from the CDM and emissions trading without incurring any of the costs of Kyoto Protocol compliance. Whether such circumvention is possible will depend largely on national policies regarding registry participation.

The creation of the registry system at a national level may limit the ability of certain types of companies to participate. For example, it is possible that some national registries will only create accounts for companies with a substantial business in the domestic jurisdiction, to avoid 'shelf' companies being established as subsidiaries of non-Parties to enable non-Party companies to indirectly participate in emissions trading. Alternatively, the creation of registry accounts may require political approval, which may be declined for applications from companies from non-Parties, depending on the form of the regulations and the domestic policy regarding non-Party private entity participation.

3 Structuring Contractual Arrangements to Allow Non-Party Participation

Another means by which a non-Party company could legitimately buy or sell Kyoto Protocol credits would be by entering into forward-looking purchase contracts with a condition precedent that the contract will not enter into force until the country ratifies the Kyoto Protocol. However, such contracts are only likely to be acceptable to a seller of CERs, ERUs, or AAUs if there is a high likelihood that such ratification will occur in the near future. The value of Kyoto Protocol Rights is likely to increase as the first Commitment Period approaches and Parties begin to accumulate rights to meet their obligations. A CDM project proponent or a trader of Kyoto Protocol Rights is therefore unlikely to 'lock' themselves into a contract at a fixed price, which is dependent on a future change of government policy, and thereby be prohibited from selling those rights to other buyers who can use them immediately.

It is important to understand that it is always possible for any country or company to simply acquire contractual 'emission reductions'[28] rather than Kyoto Protocol rights by way of their role in an emission reduction project in a developing country. There is no limitation of any kind stemming from the Protocol that

[28] For example, in its standard CDM ERPA, the World Bank has defined 'Emission Reductions' to mean any right, interest, credit, entitlement, benefit, or allowance to emit (present or future) arising from or in connection with any GHG reduction by the Project and includes any right that may be created under any regulatory or legal regime as a result of the GHG reductions whatsoever.

would prevent non-Party companies from purchasing generic emission reductions, regardless of whether or not those emission reductions have been verified as compliant with the Protocol. In this case, it would only be necessary to ensure that any contracts to purchase such reductions complied with the appropriate governing law relating to the contract and the transaction.

Early trades in carbon rights have already occurred in this manner over the last four years, even by non-Parties to the Protocol. For example, a number of Australian and US companies have entered into transactions to buy and sell emission reductions with total (public) transaction values exceeding US$200 million.[29] In most cases, the emission reductions are transacted on the basis that they are consistent with the rules of the Protocol, but even then, the contracting parties should be well aware that the emission reductions may not ultimately give rise to Protocol compliant rights such as CERs. Whether contractual rights will have value in the future is dependent on government policy changes so that trading is generally of a speculative nature.

There are several means by which a non-Party company could gain value from CERs without directly participating in Kyoto Protocol emissions trading. For example, it could:

(i) use the CERs under domestic legislation or for some other purpose outside the Kyoto Protocol trading regime and request the cooperating Party (or private entity) to cancel them in its account to ensure environmental integrity;
(ii) retain the rights to CERs and require the cooperating Party (or private entity) to hold them on trust in the expectation that the non-Party Government would ratify the Protocol in the future; or
(iii) exercise control over the CERs through cooperation with the Party (or private entity) to cause them to be onsold to another Party with compliance obligations under the Protocol regime.

Any such cooperation would require contractual arrangements to be entered into between the non-Party company and the Party entity in whose national account the CERs are to be held, to ensure that the CERs are controlled, used or transferred as directed and are not otherwise used by that entity.

Such arrangements could include:

(i) An agreement between the non-Party company and a non-Annex I Party (who, for example, has obtained CERs through a Unilateral CDM Project) to hold CERs in its holding account within the CDM Registry;
(ii) A bilateral treaty between a non-Party and an Annex I Party whereby that Party agrees to act as an 'agent' for the non-Party in holding and dealing with CERs in its national registry account; or

[29] Speech by New South Wales Premier Bob Carr at Total Environment Centre, 'In Control Of Carbon' Breakfast Thursday, 20 November 2003, Sydney, Australia.

(iii) Agreements between the non-Party company and a Party country or authorized entities in a Party country with a CDM holding account, granting the non-Party company the right to hold and control CERs held in that account.

In all such arrangements it is likely that some fee or commission would be imposed by the cooperating Party or public/private entity.

4 Potential Future Developments

Once the Kyoto Protocol has entered into force, Kyoto Protocol credits such as CERs or ERUs are likely to be worth substantially more than they are being traded for in the current market, and there may be more incentive for non-Party companies to wish to become involved in CDM or JI Projects and Emissions Trading.

Currently, in the period while the Kyoto Protocol has not entered into force but the more detailed rules on the flexible mechanisms are still developing, the decisions on developing detailed rules and procedures (such as those necessary to decide the issue of the extent to which non-Party private entities are able to participate) are being made by the Conference of Parties to the UNFCCC acting as the Meeting of Parties to the Kyoto Protocol (COP/MOP). If and when the Kyoto Protocol does enter into force (expected to occur upon the ratification of Russia), future decisions on the international rules on participation in the flexible mechanisms will be taken by the COP/MOP, which will exclude countries such as Australia and the US.[30] There is a possibility that COP/MOP (the major Annex I countries being the EU Member States, Japan, New Zealand, and Canada) may take a stronger stance against non-Party private entity participation in the Kyoto Protocol than the current Conference of Parties to the UNFCCC.

5 Conclusion

Non-Party governments will not be able to directly participate in the flexible mechanisms. However, the rules do not exclude the possibility that Parties to the Protocol may authorize project participants from non-Party jurisdictions, or may enable companies from such jurisdictions to hold registry accounts within their national registries. If non-Party companies wished to become involved in a CDM or JI project or trade Kyoto Protocol credits (CERs, ERUs, tCERs, lCERs, AAUs, or RMUs), their ability to do so would depend to a large extent on the policies of the COP/MOP, the CDM Executive Board and of individual Parties to the Protocol with regards to non-Party participation. At present, there is little guidance

[30] These countries are non-Parties to the Kyoto Protocol but Parties to the UNFCCC, so are currently entitled to contribute towards the formation of the Kyoto Protocol rules.

as to how such policies may develop in the future, although, with the realities of multinational companies which can establish a presence in a variety of Party and non-Party jurisdictions, it is likely that indirect participation of non-Party companies in the flexible mechanisms will be possible under the current rules. However, such participation could be seen as contradicting the spirit of the provisions of the Kyoto Protocol establishing the flexible mechanisms, which was to alleviate the cost of meeting Annex I Party obligations through projects that also achieved sustainable development and technology transfer for developing Parties. As the CDM progresses and projects begin operation, it is likely that the CDM Executive Board and the COP/MOP will be called upon to provide guidance on the issues raised in this chapter and the ability of non-Parties to participate in the flexible mechanisms.

PART V

CARBON SEQUESTRATION

14

Generation of Sequestration Credits under the CDM

*Sebastian Scholz and Ian Noble**

Almost six years of assessment and negotiation drew to a close at the ninth session of the Conference of the Parties (COP) held in Milan, Italy, in December 2003 (COP9) when agreement on the definitions and modalities for the use of afforestation and reforestation (A&R) projects in the Clean Development Mechanism (CDM) was reached. The decisions almost complete the Buenos Aires Plan of Action to implement the Kyoto Protocol adopted by COP4 in 1998.[1] This chapter first outlines a brief history of the negotiations to better understand the most recent outcomes of COP9 that are discussed in more detail in the second part of the chapter.

1 Sinks in the CDM—An Eventful Negotiating History

The inclusion of 'sinks' in the Kyoto Protocol was controversial in the lead-up to the drafting of the Protocol at COP3 in 1997 and has remained so. 'Sinks' was a shorthand term adopted to refer to the uptake and loss of carbon from terrestrial vegetation and soils. Some Parties and non-governmental organizations (NGOs) wished to have sinks excluded from the Protocol as a whole, while others saw them as a significant opportunity, or even a necessity, for achieving Kyoto targets. Those opposed to sinks emphasized problems with the measurement of carbon pools in vegetation and soils and problems in ensuring that any losses of sequestered carbon, for which credit had been given, were accounted for and made up in calculating progress towards targets. Issues relating to additionality and leakage in CDM projects were also raised but these did not differ greatly from similar considerations for non-sink activities in the CDM. Those supporting sinks pointed to the massive

* Sebastian Scholz is a PhD fellow at the Center for Development Research, Bonn. Ian Noble is an adviser to the World Bank's Environmental Department. The views expressed in this chapter are the views of the authors and do not necessarily represent the views of their employers.

[1] In addition, COP9 introduced a class of 'small-scale' afforestation and reforestation projects. These projects generate annual Certified Emission Reductions (CERs) of less than 8 kt CO_2. Simplified rules for such projects will be determined before COP10 (see further below).

quantities of carbon exchanged between the atmosphere, vegetation, and soils every year; an amount that exceeds emissions from fossil fuels by thirty fold.[2] They were concerned that by omitting sinks a major exchange of carbon, which could swamp any gains made through actions under the Kyoto Protocol, would be ignored. Eventually, a compromise was reached at COP3. Certain sink activities, afforestation, reforestation, and deforestation, would be included in accounting of Annex I Parties[3] and a wider range of activities would be reassessed after further technical advice had been received from the Subsidiary Body for Scientific and Technological Advice (SBSTA) and the Intergovernmental Panel on Climate Change (IPCC).

In relation to the flexibility mechanisms, sinks projects were allowed[4] in exchanges between Annex I Parties through Joint Implementation (JI) under Article 6 of the Kyoto Protocol. However, Article 12 of the Kyoto Protocol, describing the rules for the CDM, omitted the reference to sinks—a difference in wording that was interpreted by some as excluding sinks projects from the CDM.[5] The negotiating history supports this view in a way: as part of the deletion of all footnotes in a draft text on the CDM in the Kyoto Protocol during COP3, a placeholder for possible further discussion on the inclusion of sinks in the CDM was removed.[6] After this deletion the issue of sinks as a project category in the CDM was not raised again and the SBSTA in June 1998 clarified that 'Article 12 of the Protocol (. . .) mentions certified emission reductions accruing from projects, but it does not refer to sequestration by sinks, as is the case in Article 6'.[7]

However, the issue of sinks in the CDM was brought back on the agenda of the international climate negotiations one year after Kyoto. An initiative headed by developing countries re-introduced a forestry-oriented catalogue of CDM project categories to the negotiating table.[8] Latin-American countries, in particular, had gained beneficial experiences through investments in forestry projects during the pilot phase of 'Activities Implemented Jointly'. A number of industrialized Annex I Parties joined the effort and a vigorous debate ensued. Meanwhile the IPCC, at the request of the Parties, completed a Special Report on Land Use, Land-Use Change

[2] RT Watson (ed.), *Climate Change 2001: Synthesis Report* (IPCC, Cambridge University Press, 2001) and RT Watson and IR Noble, *The Global Imperative and Policy for Carbon Sequestration: The Carbon Balance of Forest Biomass* (Garland Science/BIOS Scientific Publishers, New York, 2004).

[3] Art 3(3) Kyoto Protocol.

[4] The Kyoto Protocol text includes projects 'aimed at reducing anthropogenic emissions by sources *or enhancing anthropogenic removals by sinks* of greenhouse gases in any sector of the economy' (emphasis added).

[5] S Oberthür and HE Ott, *The Kyoto Protocol, International Climate Change Policy for the 21st Century* (Springer, Berlin, 1999).

[6] J Depledge, *Tracing the Origins of the Kyoto Protocol: An Article-by-Article Textual History*, technical paper, prepared under contract to UNFCCC, UN Doc FCCC/TP/2000/2.

[7] SBSTA 1998, *Methodological Issues. Issues related to Land-use Change and Forestry. Note by the Secretariat*, UN Doc FCCC/SBSTA/1998/INF.1.

[8] SBSTA 1999, *Mechanisms pursuant to Articles 6, 12 and 17 of the Kyoto Protocol. Further Proposals from Parties on Issues raised in Decision 7/CP.4, Paragraph 1(a), (b) and (c)*, UN Doc FCCC/SB/1999/MISC.10/Add.3.

and Forestry.[9] This report assessed the feasibility of measuring and monitoring sinks and the contributions that sinks could make to reducing greenhouse gas (GHG) emissions and to wider development goals. Prior to and during COP6 in The Hague in 2000, sinks had become the most contentious issue and were the ultimate cause of the breakdown of this particular meeting. To resolve this critical stalemate of the international climate regime delegates convened to a resumed session of COP6 in Bonn a few months later. Here the deadlock was broken: the negotiations came to a conclusion and agreed on a package of decisions that eventually formed the basis of the 'Marrakesh Accords' at COP7 in 2001, a political milestone of the present climate negotiations. The Marrakesh Accords established the rules for sinks projects for Annex I Parties. Sinks in the CDM were to be limited to A&R projects and the use of sinks projects in developing countries was restricted to meet only 1 per cent of Annex I Parties' Kyoto obligations.[10] This amounts to a maximum of about 30 million tonnes of carbon per year during the first commitment period although the actual limit will be significantly smaller than this as some countries will not use their quota. In addition, the rules governing sinks apply to only the first commitment period and will be revisited before the second commitment period.

COP7 also mandated the SBSTA 'to develop definitions and modalities for including afforestation and reforestation project activities under the clean development mechanism in the first commitment period, taking into account the issues of non-permanence, additionality, leakage, uncertainties and socio-economic and environmental impacts, including impacts on biodiversity and natural ecosystems . . .'[11] by COP9.

COP9 approved additional rules for the inclusion of sinks in the CDM some of which are tailor-made for CDM forestry projects while others have been 'mutatis mutandis' adopted from decisions for other project categories already made in the Marrakesh Accords. Interpretation of the COP9 text is still underway, but it is clear that, similar to other CDM project categories, there are some remaining questions with regard to implementation that will have to be answered by the CDM Executive Board and its Methodology Panel.

Overall, the decisions pave the way for a CDM project category that—designed accordingly—is expected to show substantial developmental benefits to rural communities[12] while providing an attractive supplementary mitigation option for Annex I countries.

[9] RT Watson, IR Noble, B Bolin, NH Ravindranath, DJ Verardo, and DJ Dokken (eds.), *Land Use, Land-Use Change and Forestry. A Special Report of the IPCC* (Cambridge University Press, 2000).

[10] See Decision 17/CP.7, Modalities and procedures for a clean development mechanism, UN Doc FCCC/CP/2001/13/Add.2 (hereinafter Decision 17/CP.7), paras 7(a) and (b).

[11] ibid, para 10(b).

[12] JO Niles, S Brown, JN Pretty, AS Ball, and JP Fay, 'Potential Carbon Mitigation and Income in Developing Countries from Changes in Use and Management of Agricultural and Forest Lands', in IR Swingland, EC Bettelheim, J Grace, GT Prance, and LS Saunders (eds.), *Carbon, Biodiversity, Conservation and Income: An Analysis of a Free-market Approach to Land-use Change and Forestry in Developing and Developed Countries* (London, The Royal Society, 2002).

2 COP9 Decisions: A First Analysis and Interpretation

2.1 Definitions

As the title of the decision on 'Modalities and procedures for afforestation and reforestation project activities under the clean development mechanism in the first commitment period of the Kyoto Protocol'[13] unmistakably clarifies, the provisions made are only valid until the end of the year 2012. However, it provides legal security for projects that are registered before the end of the first commitment period. This implies investment security for project participants even if the decisions concerned are subject to revision for further commitment periods.[14] Crediting beyond the first commitment period is guaranteed depending on the crediting scheme chosen by project participants.

The legally binding definitions of 'afforestation' and 'reforestation' for Land Use, Land-Use Change and Forestry (LULUCF) projects under the CDM are the same as those previously agreed for reporting under the other forestry-related Article 3(3) and (4) of the Kyoto Protocol.[15] The definition of 'reforestation' includes only lands that did not contain forest before 1990. Some parties tried to shift this base year for CDM activities to 2000 on the grounds that data was difficult to obtain for the earlier date and that there was no incentive to reforest areas that were cleared after 1 January 1990 but well before the rules of the Kyoto Protocol could have been anticipated.[16] They did not succeed as there was concern that moving the threshold date from 1990 to 2000 would make lands that were deforested or degraded later than 1990 eligible to receive carbon credits without accounting for the relevant carbon loss. Furthermore, some argued that a shift might be an incentive to clear natural forests between now and the first commitment period in the hope that the

[13] Draft Decision -/CMP.1 2003, Modalities and procedures for afforestation and reforestation project activities under the clean development mechanism in the first commitment period of the Kyoto Protocol, UN Doc FCCC/SBSTA/2003/L.27 (hereinafterDraft Decision -/CMP.1 2003). All references made to specific paragraphs in this chapter refer to this document, its Annex, and Appendices, if not otherwise stated.

[14] ibid, para 3 '. . . the treatment of land use, land-use change and forestry project activities under the clean development mechanism in future commitment periods shall be decided as part of the negotiations on the second commitment period . . .'

[15] Draft Decision -/CMP.1 2001, Land use, land-use change and forestry, UN Doc FCCC/CP/2001/13/Add.1, Annex, Section A. Definitions para 1(b) ' "Afforestation" is the direct human-induced conversion of land that has not been forested for a period of at least 50 years to forested land through planting, seeding and/or the human-induced promotion of natural seed sources; (c) "Reforestation" is the direct human-induced conversion of non-forested land to forested land through planting, seeding and/or the human-induced promotion of natural seed sources, on land that was forested but that has been converted to non-forested land. For the first commitment period, reforestation activities will be limited to reforestation occurring on those lands that did not contain forest on 31 December 1989'.

[16] These lands are often the lands that could most contribute to biodiversity conservation and the rehabilitation of land degradation as they tend to be close to remaining forest patches or in the early, and still reversible, stages of degradation.

date might be shifted again in the future and the land thereby become eligible for reforestation activities.

Although there were some reasons for adjusted definitions and a reforestation threshold tailored to the specific circumstances of CDM host countries most delegates for consistency reasons however decided not to differentiate forest-related definitions for different articles of the Kyoto Protocol.

2.2 (Non)-Permanence and Accounting

Sinks projects differ in one very significant way from projects based on other activities. Whereas a tonne of emissions reductions, once achieved, remains a benefit to the atmosphere,[17] a tonne of sequestered carbon is of benefit to the atmosphere only as long as it remains sequestered. This issue of 'permanence' has been the basis of concern about the feasibility of sink projects, both within Annex I Parties (Article 3(3) and (4)) and in the CDM (Article 12). Annex I Parties measure progress towards meeting their targets by preparing a national inventory, which is required to account appropriately for all afforestation, reforestation and deforestation, and any other land use categories the Party chooses to make use of under Article 3(4). If sequestration is included in one commitment period and that carbon is released in a later commitment period, this should be reflected in the national inventory and reported progress towards a target will be consistent with the changes in atmospheric concentrations. In CDM projects, the question arises as to whether losses of carbon will be adequately reported,[18] and if they are, how they will be attributed within the Annex I compliance system. A number of schemes based on creating temporary credits or around insurance were suggested. Eventually COP9 agreed to two forms of temporary crediting: temporary CERs (tCERs) and long-term CERs (lCERs).[19]

In both cases, a verifier must confirm every five years that any sequestered carbon that has been used to create credits (either tCERs or lCERs) remains sequestered. More precisely, the verifier confirms that the project retains enough sequestered carbon to cover the credits issued. This encompasses projects that incorporate a planned harvesting regime, or which retain a buffer of sequestered carbon to cover unplanned carbon losses. The accounting system monitors sequestration within the project boundary, not the particular trees or particular stands of trees and their soil. If there has been a net loss of sequestered carbon there are a

[17] Unless it leads to emissions elsewhere or in subsequent years — that is, leakage.

[18] This is due to the fact that developing countries are not subject to quantified targets under the Kyoto Protocol.

[19] Draft Decision -/CMP.1 2003, Annex, Section A. Definitions para 1(g) ' "Temporary CER" or "tCER" is a CER issued for an afforestation or reforestation project activity under the CDM which (. . .) expires at the end of the commitment period following the one during which it was issued; (h) "Long-term CER" or "lCER" is a CER issued for an afforestation or reforestation project activity under the CDM which (. . .) expires at the end of the crediting period of the afforestation or reforestation project activity under the CDM for which it was issued'.

series of rules governing the replacement of the issued credits with other credits.[20] For tCERs the expiry, verification and re-issuance of tCER certificates automatically corrects for any losses of carbon from a project. With lCERs, if net carbon stocks decrease between two certification reports, the difference in lCERs needs to be replaced by AAUs, CERs, ERUs, RMUs or lCERs from the same project activity.[21] If carbon stocks increased since the last verification, new lCERs may be issued by the Executive Board.

Sinks projects in the CDM are also subject to a crediting period over which the project can generate new carbon credits, as in other CDM activities. For sinks projects, the agreement between the buyer and the seller can be extended for a crediting period of either non-renewable thirty years, or twenty years, twice renewable to a maximum of sixty years. At the end of each crediting period, the baseline must be re-assessed.

COP9 also introduced a rule that states that at the end of the agreed crediting period all tCERs and lCERs must be replaced by credits from other projects. This means that a liability to replace temporary credits at some date twenty to sixty years hence is created. Technically, this restriction is not necessary from the point of view of maintaining a benefit to the atmosphere, as there is no reason not to extend the five-yearly verifications indefinitely.[22] Some parties felt that such a liability would deter the widespread use of sinks in the CDM, others felt that liabilities at least twenty, and as much as sixty years in the future, would have little practical impact on the use of sinks in the CDM. A number of schemes to deal with liability are under development.

The technical differences between tCERs and lCERs are minor and mainly related to the types of credits that can be used to replace credits at the end of the crediting period and the requirements in reporting to the registry. They will require different contracts between buyers and sellers (five-yearly for tCERs and an initial twenty- or thirty-year crediting period for lCERs) and this may create different perceptions of the contract and commercial arrangements. The choice between tCERs or lCERs has to be made upfront and remains fixed once the project starts.[23]

It appears likely that tCERs will incur greater transaction costs than lCERs mainly because new certificates have to be issued every five years rather than every twenty or thirty years. Even though this is expected to be a simple electronic transaction, the current expectation is that each transaction will attract a considerable administrative fee and possibly additional fees for special funds to assist developing

[20] Draft Decision -/CMP.1 2003, Section K, para 49. [21] ibid, paras 49 and 50.

[22] There appears to be a false analogy here with the crediting period for non-sink projects where the crediting period sets a limit to the period for which the baseline is assumed relevant and the project can be assumed to be delivering new benefits to the atmosphere. The analogous case for sinks would be to stop the issuance of additional credits from a project at the end of the crediting period, while allowing the continuing verification of credits already issued.

[23] Draft Decision -/CMP.1 2003, Section K paras 38 and 39.

nations. Thus, users will have to balance the greater flexibility offered by tCERs in that buyers or sellers can select to disengage every five years, against the increased transaction costs.

The two crediting period options, a single one and another one that can be renewed twice, conceptually repeat the Marrakesh Accords for energy projects. However, the increased length acknowledges the long-term character of forestry projects. Here again legal security for projects that are registered prior to the end of the first commitment period is guaranteed even though there is a possibility for rules and modalities to change for future commitment periods. One of the remaining open questions is the point in time at which the crediting period is supposed to start. The draft decision states it 'shall begin at the start of the afforestation or reforestation project activity.'[24] Dutschke points out that this could mean that emissions from initial land preparation are not adequately taken into account.[25] He sees this doubt reinforced as the initial verification 'may be undertaken at a time selected by the project participants.'[26] On the other hand, chapter four of the IPCC Good Practice Guidance, whose use is 'encouraged', promotes the inclusion of this kind of ancillary emissions in monitoring practice.[27] The question will need to be addressed by the Executive Board in approving methodologies for A&R projects. Credits generated by sinks either in Annex I countries or through the CDM can be used only in the commitment period in which they are certified and cannot be carried over to a subsequent commitment period (that is, no 'banking').[28]

Another important issue is the likely price of temporary credits versus CERs. Chomitz and Lecocq dealt with the basic issue arguing that no one would pay more for a temporary credit today than the difference between the current price for a permanent credit CER and the discounted value of permanent credit price at the end of the contract.[29] If the price of permanent CERs rises faster than the discount rate, then the price of temporary credits should fall to zero. However, supply and demand may affect this conclusion. If many parties are unable to meet their commitment near the end of a commitment period and CERs are in short supply, temporary credits may be a more attractive option. Simple analyses based on a discount rate of 7 per cent suggest that tCERs should be priced at less than 30 per cent of a CER, but the price for an lCER contracted for sixty years could approach that of a CER depending on the buyer's view of the likelihood that the credit would have to be replaced after sixty years and at what price.

[24] ibid, Section G, para 23

[25] M Dutschke, *Growing Trees for Climate — The Milan Climate Conference Paves the Way for Forestry Projects in Developing Countries* (2004), <http://www.biocarbon.net/> accessed 25 January 2004.

[26] Draft Decision -/CMP.1 2003, Section I, para 32.

[27] IPCC Good Practice Guidance for LULUCF, IPCC National Greenhouse Gas Inventories Programme (2004).

[28] Draft Decision -/CMP.1 2003, Section K, para 41.

[29] They assumed this would be five years only. See, KM Chomitz and F Lecocq, *Temporary Sequestration Credits: An Instrument for Carbon Bears* (World Bank Policy Research Working Paper 3181, December 2003).

The value of sinks credits will also spread according to the buyer's assessment of project quality and the associated risks. Project failure (that is, the loss of carbon through deliberate action or accident) will result in an earlier than planned need for replacement with CERs. Liability for tCERs and lCERs is on the side of the buyer (Annex I Party), in whose registries the unit is kept. If after the due date for credit replacement the Annex I Party is out of compliance and therefore not allowed to buy or sell any emission permits, its allowance deficit will increase. How Annex I Parties will pass this liability along to separate entities within their country (for example, private companies with national targets) is yet to be seen.

It is likely that many CDM sink deals will be handled through a third party pooling entity, such as the BioCarbon Fund of the World Bank.[30] Such arrangements allow buyers to spread their risk over a wide range of projects while the sellers get the benefits of the experience accumulated by the pooling entity. Such arrangements also lend themselves to more elaborate financial arrangements for dealing with temporary credits and their replacement.

2.3 Additionality and Baseline Establishment

In the initial phases of planning and technical preparation and the subsequent elaboration of the project specific baseline study, proponents have to address the issue of additionality.

The requirements, as they were adopted by COP9, are very similar to the ones established in the Marrakesh Accords for energy projects.[31] The relevant paragraphs read as follows: An A&R project is additional 'if the actual net greenhouse gas removals by sinks are increased above the sum of the changes in carbon stocks in the carbon pools within the project boundary that would have occurred in the absence of the (. . .) project activity.'[32]

Choosing an adequate baseline methodology and establishing a plausible baseline scenario, especially for A&R projects, is a challenging effort. In principle the same counterfactual reasoning has to be undertaken as in the case of energy projects. However, when assessing different future land-use patterns there is no such incontestable tool as an expansion plan like in the field of energy projects that would represent a reasonable baseline scenario. Baselines for A&R projects have to analyse and project past trends and current situations of prevailing land-use patterns. Changes in land use mostly do not exhibit random phenomena but are rather predictable based on factors that can be observed: historically prevailing patterns of land use, presence of transportation networks, access to markets and so on. Sometimes even land-use plans might exist.[33] The decisions refer to this fact by

[30] See the homepage of the BioCarbon Fund, <http://carbonfinance.org/biocarbon/home.cfm> accessed 4 April 2004.

[31] See Decision 17/CP.7, n 10 above.

[32] Draft Decision -/CMP.1 2003, Section G, paras 12(d) and 18.

[33] D Kaimowitz and A Angelsen, *Economic Models of Tropical Deforestation: A Review* (Center for International Forestry Research, Bogor, Indonesia, 1998) and S Brown, IR Swingland,

pointing out that 'relevant national and/or sectoral policies and circumstances, such as historical land uses, practices and economic trends' have to be taken into account when establishing an A&R baseline.[34]

The current decisions establish three different baseline methodology approaches in paragraph 22 from amongst which project proponents have to choose the most adequate one:

(a) Existing or historical, as applicable, changes in carbon stocks in the carbon pools within the project boundary;
(b) Changes in carbon stocks in the carbon pools within the project boundary from a land use that represents an economically attractive course of action, taking into account barriers to investment;
(c) Changes in carbon stocks in the pools within the project boundary from the most likely land use at the time the project starts.[35]

Option (a) establishes the same 'business-as-usual' baseline methodology approach for A&R projects as paragraph 48(a) of the Marrakesh Accords does for energy projects.[36] Depending on the forest definition as pointed out earlier, in a normal case, some form of grassland or grazing land might be the prevailing form of business-as-usual baseline scenario of projects choosing approach (a). This implies that a kind of grass- or grazing land modelling would be most suitable to assess the development of carbon pools in the area concerned for the baseline scenario.

Option (b), as well, is a direct adaptation from the equivalent decision in the Marrakesh Accords for CDM projects from the energy sector. The difference between options (b) and (c) is not self-evident at first view. Under economic considerations 'the most likely land use at the time the project starts' (option (c)) is actually the one 'that represents an economically attractive course of action' (option (b)). By presenting these two alternatives, the regulations deliberately distinguish between the economically attractive course of action and the most likely land use, given a prevailing scheme of land-use activities in the area concerned. Thus, option (c) can be seen as a more 'pragmatic' version of option (b).

In calculating the baseline of net GHG removals by sinks and the with-project GHG removals by sinks, proponents can deliberately exclude carbon pools in their accounting. In addition, emission sources can be excluded. Both options are subject to the provision that as a result of these exclusions, the overall credits the project will claim (net anthropogenic GHG removals by sinks) are not increased.[37] The logic

R Hanbury-Tenison, G T Prance, and N Myers, 'Changes in the Use and Management of Forests for Abating Carbon Emissions: Issues and Challenges under the Kyoto Protocol', in IR Swingland, EC Bettelheim, J Grace, GT Prance, and LS Saunders (eds.), *Carbon, Biodiversity, Conservation and Income: An Analysis of a Free-market Approach to Land-use Change and Forestry in Developing and Developed Countries* (The Royal Society, London, 2002).

[34] Draft Decision -/CMP.1 2003, Section G, para 20(e).
[35] ibid, para 22.
[36] Decision 17/CP.7, Annex, para 48.
[37] Draft Decision -/CMP.1 2003, Section G, para 21.

behind this paragraph is as follows: the accounting of any sinks project is conservative as long as all (at least as much as technically feasible) sinks are included in the baseline on the one hand and all emissions are included in the with-project scenario on the other, that is, when calculating net GHG removals by sinks. Or stated otherwise, project proponents in the proposed accounting have to maximize sinks in the baseline while they have to maximize emissions in the with-project case and can deliberately exclude (minimize) sinks. This paragraph has some implications for the workload of proponents: for example, where soil carbon is a small, but costly to measure sink, then it will have to be estimated in the baseline, but not measured in the project.

To finally come to the amount of carbon sequestered that a certain project can claim (either tCERs or lCERs), the net GHG removals by sinks have to be calculated. This is done by taking the actual sinks removals of the with-project case and subtracting the sinks removals calculated in the baseline. From this figure in one further step any occurring leakage has to be subtracted to arrive at the 'net anthropogenic GHG removals by sinks.'[38] This amount equals the amount of tCERs or lCERs the project is seeking to get certified.

2.4 Leakage

As mentioned above leakage plays an important role in assessing the amount of credits to claim for certain A&R project.[39] Since very early in the negotiation the topic was one of the major concerns surrounding the issue of sinks as many assumed that the CDM would result in widespread establishment of plantations that would drive people off their land only to deforest areas elsewhere. Another scenario was that avoided deforestation would be accepted as a CDM category and that, having been denied access to one area of forest, local people would be forced to clear forests elsewhere.

COP9 adopted wording for additionality that is consistent with that used for other CDM projects, in that leakage has to be taken into account if it 'is measurable and attributable' to the A&R project activity.[40] Also, only leakage that increases

[38] Draft Decision -/CMP.1 2003, Section A. Definitions para 1(c) ' "Baseline net greenhouse gas removals by sinks" is the sum of the changes in carbon stocks in the carbon pools within the project boundary that would have occurred in the absence of the afforestation or reforestation project activity under the clean development mechanism (CDM); (d) "Actual net greenhouse gas removals by sinks" is the sum of the verifiable changes in carbon stocks in the carbon pools within the project boundary, minus the increase in emissions of the greenhouse gases measured in CO_2 equivalents by the sources that are increased as a result of the implementation of the afforestation or reforestation project activity, while avoiding double counting, within the project boundary, attributable to the afforestation or reforestation project activity under the CDM; . . . (f) "Net anthropogenic greenhouse gas removals by sinks" is the actual net greenhouse gas removals by sinks minus the baseline net greenhouse gas removals by sinks minus leakage'.

[39] ibid. Definitions para 1(e) ' "Leakage" is the increase in greenhouse gas emissions by sources which occurs outside the boundary of an afforestation or reforestation project activity under the CDM which is measurable and attributable to the afforestation or reforestation project activity'.

[40] ibid, Section G, para 12(d).

emissions is to be taken into account, thereby precluding any claims for 'positive leakage' such as where a project may lead to additional forest planting or growth outside the project boundaries. If the CDM does lead to large-scale reforestation projects then estimating leakage will require a new set of techniques and data to estimate the flow-on effects on land-use, including reforestation projects that do not proceed because of the CDM project and additional clearing by displaced landowners. This, along with additionality issues, is likely to be one of the main blocks to such large-scale projects. In small-scale community forest projects any leakage is likely to be local and can be monitored as part of project management. The COP9 text requires that project proponents have such monitoring plans.[41] As Schwarze *et al.* point out, sound leakage management begins during the design of a project with the very thoughtful selection of the project site.[42]

2.5 Social and Environmental Safeguards

According to the Marrakesh Accords the design of A&R projects has to take potential '. . . socio-economic and environmental impacts, including impacts on biodiversity and natural ecosystems . . .' into account.[43] Prior to COP9 it was already clear that this regulation especially would deviate from the strict procedure of interpreting and 'translating' the Marrakesh Accords decisions for energy projects mutatis mutandis for LULUCF projects in the CDM. The provision was introduced to respond to the concerns of certain parties as well as particular NGOs, who expected large-scale plantations with disadvantageous side-effects to become eligible under the CDM. Clearly, ignoring these concerns would have eroded public support and undermined the CDM framework. Thus, parties at COP9 had to deliberate how to flesh-out the basic principles set down in the Marrakesh Accords. There was basic consensus amongst parties that negative social and ecological effects were to be avoided, however, there was disagreement over how to achieve this goal. Tabled were two different options on how to assess potential impacts of A&R projects: in the first case, socio-economic and environmental impacts had to be analysed as part of the project design while in the second case, a full environmental and social impact assessment was required. Host countries were afraid of the potential loss of sovereignty by very narrow impact assessment regulations. They argued that most of them had a much stricter, or at least more specific legislation in place. Therefore the decisions specify that any assessment (socio-economic or environmental) has to be made '. . . in accordance with the procedures required by the host Party'.[44]

[41] ibid, Section H, para 25.
[42] R Schwarze, JO Niles, and J Olander, 'Understanding and Managing Leakage in Forest-based Greenhouse-Gas-Mitigation Projects', in Swingland *et al.*, n 33 above.
[43] Decision 17/CP.7, Annex, para 10(b).
[44] Draft Decision -/CMP.1 2003, Section G, para 12(c).

Both proposals made the attempt to highlight overlaps to other UN conventions like the Convention on Biological Diversity (CBD)[45] or the Convention to Combat Desertification[46] and other multilateral environmental agreements (MEA). This reasonable attempt of cross-referencing to other MEAs was finally overruled as not all Parties to the UN Framework Convention on Climate Change[47] have signed the other agreements concerned and implicitly might become obliged to recognize them.[48]

The COP9 modalities and procedures elaborate on provisions with regard to social and environmental impacts in Section G, paragraph 12(c). According to this paragraph, project participants have to submit documentation on the analysis of socio-economic and environmental impacts to the designated operational entity as a pre-requisite for project validation. This documentation has to include an assessment of '. . . impacts on biodiversity and natural ecosystems, and impacts outside the project boundary . . .' (leakage) of the proposed A&R project. How exactly these regulations can be taken care of is specified in the complementary Appendix B which describes the essential components of the project design document (PDD) of an A&R project. Here, paragraphs 2(a), (b) and (c) are of particular relevance.[49] A detailed description of the planned A&R project activity has to be presented in the PDD. It has to consist of a description of the present environmental conditions of the area and information about socio-economic indicators like '. . . legal title to the land, rights of access to the sequestered carbon, current land tenure and land use.'[50] This description at the time the project starts has to be compared and carefully weighted against potential environmental and socio-economic impacts of the envisaged project activity.[51] If project participants or the host country government consider potential impacts to be 'significant', an impact assessment (environmental and/or socio-economic) has to be carried out in accordance with the host country's procedures. Especially this regulation of not establishing an

45 Convention on Biological Diversity, (1992) 31 ILM 818, 5 June 1992.

46 International Convention to Combat Desertification in those Countries Experiencing Serious Drought and/or Desertification, Particularly in Africa, (1994) 33 ILM 1332.

47 United Nations Framework Convention on Climate Change, (1992) 31 ILM 849, 29 May 1992.

48 However, Parties under the current rules have to take into account other '. . . relevant provisions of international agreements that may apply to afforestation and reforestation project activities under the clean development mechanism' (Draft Decision -/CP.9) which could be seen as an implicit reference to other multilateral agreements.

49 Draft Decision -/CMP.1 2003, Appendix B, para 2(a), 'A description of the afforestation or reforestation project activity comprising the project purpose; a technical description of the project activity, including species and varieties selected and how technology and know-how will be transferred, if appropriate; a description of the physical location and boundaries of the project activity; specification of the gases whose emissions will be part of the project activity'.

ibid, para 2(b), 'A description of the present environmental conditions of the area including a description of climate, hydrology, soils, ecosystems, and the possible presence of rare or endangered species and their habitats'.

ibid, para 2(c), 'A description of legal title to the land, rights of access to the sequestered carbon, current land tenure and land use'.

50 ibid, para 2(c).

51 As pointed out ibid, para 2(j) and (i).

international minimum standard to judge whether certain impacts are acceptable or not has been heavily criticized by environmental NGOs. The economic reasoning behind this criticism is that neither participants nor the host country may have any incentive to carry out such assessment because of increasing project development and transaction costs. Furthermore, the fact that the impact assessment is not subject to specific guidelines was of major concern for NGOs.[52] As mentioned above, this decision was taken because of the concerns voiced by many of the potential host countries.

Some European countries made particular efforts to exclude sinks projects involving genetically modified organisms (GMOs). After a vigorous debate the use of GMOs and potentially invasive species was made subject to the mutual agreement of the host country and the investor (Annex I country). The PDD has to report on the selected 'species and varieties'[53] in order to ascertain that the CER-buyer has sufficient background for the purchase decision.[54]

The time frame for the reception of stakeholder comments after validation was extended from thirty days for non-sinks projects to forty-five days for A&R projects.[55] According to the Marrakesh Accords ' "stakeholders" means the public, including individuals, groups, or communities affected, or likely to be affected, by the proposed clean development mechanism project activity'.[56] In addition, the PDD requires information on the process of stakeholder involvement, '. . . a summary of the comments received, and a report on how due account was taken of any comments received'.[57] This indicates that stakeholders should be involved in a very early stage of project development.[58] Their involvement also facilitates the assessment of leakage as potentially competing land-use forms can be detected and land-use shifts can be anticipated.

[52] One way to address social, environmental, and biodiversity aspects of A&R projects is forest certification (See S Subak, 'Forest Certification Eligibility as a Screen for CDM Sinks Projects' (2002) 4 Climate Policy 335–51). Standards existing today do not explicitly certify for carbon sequestration. However, the different assessments show some overlaps with the aspects to deal with in the PDD. This may not be the case for 'small-scale' A&R projects where the certification process is too costly and sustainability issues are not as much of a concern.

[53] Draft Decision -/CMP.1 2003, Appendix B, para 2(a).

[54] For project developers, both the use of potentially invasive alien species as well as GMOs should be subject to a close examination. If species are considered to be invasive there is a considerable threat to other ecosystems, habitats, and other species by their introduction. For the use of GMOs there have been no long-term studies on their ecological impacts and risks (cross-pollination, transmission of altered genetic material and so on). However, both cases would certainly be considered to have significant impacts and therefore require an environmental impact assessment. Due to these added risks to pass validation the BioCarbon Fund of the World Bank, for example, has no intention to seek projects based on GMOs.

[55] Draft Decision -/CMP.1 2003, Section G, para 15(c).

[56] Draft Decision -/CMP.1, 2002, Art 12, Modalities and procedures for a clean development mechanism as defined in Article 12 of the Kyoto Protocol, Annex: Modalities and procedures for a clean development mechanism, Section A, para 1(e).

[57] Draft Decision -/CMP.1 2003, Appendix B, para 2(n).

[58] In the case of forestry projects in rural and remote areas, stakeholders might also include dwellers that do not necessarily have certified rights on the area. However, project success and credibility will depend on their involvement as well.

2.6 'Small-scale' Afforestation and Reforestation Projects in the CDM

COP 9 introduced a class of 'small-scale' afforestation and reforestation projects in the CDM. The adopted text allows for simplified modalities to be developed for small-scale sink projects. These are defined as projects that result in a removal of less than 8 kilotonnes CO_2 per year. Furthermore, the text points out that small-scale projects have to be developed and implemented by low-income communities or individuals which directly links these projects to an improvement of rural livelihoods. As a consequence of limiting the carbon sequestration rate of small-scale projects, the actual size of eligible projects will vary according to the project type, the species planted, and the natural conditions of the area concerned. For fast growing species, the actual project size will be rather limited. For species or conditions with lower uptake rates the respective size will be relatively larger. Depending on these parameters the maximum project area will vary between approximately 200 and 2,000 hectares, although a project may be made up of separated land units and thus affect a much larger area.[59]

The way the modalities and procedures for small-scale projects will be simplified remains to be determined by the Parties which were invited to submit their views to the UNFCCC secretariat by the end of February 2004. Based on these submissions the secretariat will produce a technical paper to be forwarded to SBSTA 20 in Bonn in June 2004. The SBSTA will then propose simplified modalities for adoption at COP10 at the end of 2004. Nothing concrete can be said on the simplifications COP10 will assign to these projects and therefore final costs compared to 'full' A&R projects remain unclear. However, if these projects are to bear full preparation and implementation costs, their feasibility is doubtful.

3 Conclusions and Outlook

The emerging carbon market has more than doubled since 2002, with more than 70 million tCO_2e traded as of November 2003. Most of the volume was exchanged through project-based transactions and nine out of ten tonnes of emission reductions originate from projects located in transition economies or developing countries. However, a number of developing countries in Africa, as well as the poorest countries in Asia are essentially bypassed.[60] Such distributional bias of the CDM benefits challenges the core idea of the mechanism. In this situation sinks projects can play a crucial role: particularly, developing countries can obtain sustained

[59] See Draft Decision -/CMP.1 2003, Annex, Section A. Definitions para 1(b), 'The "project boundary" geographically delineates the afforestation or reforestation project activity under the control of the project participants. *The project activity may contain more than one discrete area of land*' (emphasis added).

[60] F Lecocq and K Capoor, *State and Trends of the Carbon Market 2003*, PCFplus Research Paper (Washington, DC, December, 2003).

benefit from the emerging market through LULUCF projects as they mainly rely on revenues from the agricultural sector. The World Development Indicators reveal that the first sector contributes substantially to the structure of output in developing countries. Especially in low-income countries agriculture contributes up to 50 per cent of the GDP. Of the world's 1.2 billion extremely poor people, 90 per cent depend on forests and their products for their livelihoods. In Central Africa, for example, 65 million persons live inside or near forests and depend on them for food, energy and medicines, and so on.[61] This explains the fact that even prior to a formalized international agreement on a GHG market, farmers in developing countries were interested in the carbon market which in turn led to a constantly increasing number of carbon contracts. A major feature of the A&R modalities as they emerge after COP9 is their openness for the market actors to develop their own specific contractual agreements. In this regard the established tCERs and lCERs open for a broad scope of contractual solutions between developing countries and Annex I Parties.

An outstanding issue to look at in 2006 again is the revision of the linking of the flexible (Kyoto) mechanisms to the European Emissions Trading System[62] (EU ETS). This will concern A&R credits originating from CDM projects as well. The basic question was how credits from the flexibility mechanisms of the Kyoto regime could be converted into European 'allowances'. The EU therefore adopted a directive to amend the EU ETS to link the scheme to emission reduction units that comply with the Kyoto Protocol (Linking Directive).[63] At present, the 'Linking Directive' excludes LULUCF credits (JI and CDM) from recognition under the Community scheme. But there are some suggestions to allow for a conversion. Schlamadinger and Dutschke, for example, propose converting lCERs into two parallel products: an allowance and a bond to cover liability issues. They point out that the allowance could be traded in the EU ETS as long as the bond is 'covered'. Thereby costs of potential non-permanence and restitution could be internalized.[64] The challenge here is to proactively develop a sustainable design of the emerging market—a first step is to ensure market openness for developing countries.

A more comprehensive interpretation of the COP9 decisions is still underway and there are some remaining issues that have to be answered by the CDM

[61] Food and Agriculture Organization of the United Nations, *Global Forest Resources Assessment 2000—Main Report* (Rome, 2001); The World Bank, *World Development Indicators* (Washington, DC, 2003).

[62] Directive 2003/87/EC of the European Parliament and of the Council of 13 October 2003 establishing a scheme for greenhouse gas emission allowance trading within the Community and amending Council Directive 96/61/EC [2003] OJ L275/32.

[63] On 23 July 2003, the EC Commission released the first draft of a 'Proposal for a Directive of the European Parliament and of the Council, amending the Directive establishing a scheme for greenhouse gas emission allowance trading within the Community, in respect of the Kyoto Protocol's project mechanisms', COM(2003) 403 final.

[64] B Schlamadinger and M Dutschke, *Linking Terrestrial Carbon Sequestration to the EU ETS*, Guest commentary, Carbon Market Europe, 27 February (Point Carbon 2004).

Executive Board. Overall, the decisions offer an encouraging starting point for project development as they provide legal security for project participants. The concrete guidelines that are backed by the international community will give sinks projects a remarkable boost. The decisions pave the way for a CDM project category that, accordingly designed on the one hand, can show substantial developmental benefits to rural communities, while on the other hand providing an attractive supplementary mitigation solution for Annex I Parties.

15

Specific Features of Land Use, Land-Use Change, and Forestry Transactions

*Benoît Bosquet**

1 Introduction

This chapter reviews transactions related to climate change mitigation projects conducted in the sphere of Land Use, Land-Use Change, and Forestry (LULUCF) activities, which are also referred to as 'carbon sinks' as they capture and store carbon out of the atmosphere through the process of photosynthesis. The years 1996–9 saw a flurry of LULUCF transactions.[1] In fact, such transactions, mostly between the US and Canada, dominated the carbon market during that period. However, this flow has significantly declined in the past four years. The US has now largely withdrawn from the international climate change negotiations and the focus of climate change mitigation in that country, if any, is on domestic action. The rest of the market has been characterized by a growing number of smaller transactions, with a few exceptions. Moreover, doubts have arisen about the measurability, additionality, permanence, ancillary social, and environmental benefits of LULUCF projects. In sum, significant constraints exist to the development of LULUCF, in spite of the multiple benefits that LULUCF projects help achieve, including local environmental improvements (soil rehabilitation, biodiversity protection), social benefits for local communities (creation of revenues, transfer of knowledge, and improved capacity to adapt to climate change), and strategic aspects (engagement of the developing world in the fight against climate change). This chapter reviews the market for credits generated by LULUCF project activities, that is, credits of emission removals from LULUCF projects, explains the specific features of LULUCF transactions relative to greenhouse gas (GHG) emission reduction purchases from energy projects, and briefly proposes ways to help lift the constraints on the development of LULUCF transactions.

* Benoît Bosquet is a senior natural resource management specialist at the World Bank's Environmental Department. The views expressed in this chapter are the views of the author and do not necessarily represent the views of the World Bank.

[1] World Bank, *State of the Carbon Market 2004* Report (Washington, DC), www.carbonfinance.org.

2 Market Survey

2.1 The Years from 1995 to 2003

A number of LULUCF projects were conducted as part of the Activities Implemented Jointly Pilot Phase programme under the United Nations Framework Convention on Climate Change (UNFCCC), which started in 1995, though on a voluntary basis and not for transfer of credits among Parties.[2]

According to the World Bank's 2003 carbon market intelligence,[3] seventy-six LULUCF transactions were signed between the beginning of 1996 and the end of 2003 representing a volume of 40 million tonnes of CO_2 equivalents (tCO_2e). Relative to the universe of carbon transactions that took place over the same period, LULUCF made up 21 per cent of the total number of transactions and 23 per cent of the total volume. In fact, in volume terms, LULUCF was the single largest project category, before landfill gas management, cow feed improvement, and hydropower, among others. However, more than three-quarters of this volume were transacted between 1996 and 1999. Over the single period 1996–9, LULUCF accounted for 41 per cent of all the volume sold. This share sharply declined in subsequent years: the share of LULUCF in total volume transacted between 2000 and 2003 was merely 9 per cent, and only 7 per cent when the last two years 2002 and 2003 are considered. Nineteen LULUCF purchases were concluded in 2002 and another twenty in 2003.

These numbers reflect a growing interest for LULUCF in the retail market. One firm, Future Forests, accounted for twelve of the nineteen purchases recorded in 2002 and fourteen of the twenty purchases recorded in 2003.[4] Future Forests' purchases were all for less than 10,000 tCO_2e and geared towards small individual and corporate clients. On the other hand, large-scale transactions have become considerably less common. The large transactions involved sellers predominantly located in the US and Australia. Also, most transactions did not comply with the Kyoto Protocol's eligibility criteria with respect to additionality over a baseline scenario, the date for project start, and the host country having ratified the Kyoto Protocol. The notable exception was the World Bank's Prototype Carbon Fund, which signed purchase agreements for 1 million tCO_2e or more from two projects designed to become eligible to Joint Implementation (JI) and the Clean Development Mechanism (CDM), respectively.[5]

Transaction prices have varied from less than US$0.50 to more than US$17 per tCO_2e, but comparison is difficult as some prices are net and other prices are gross of other revenues, notably timber sales.[6]

[2] Intergovernmental Panel on Climate Change, *Land Use, Land-Use Change, and Forestry* (Cambridge University Press, Cambridge, 2000), <http://www.grida.no/climate/ipcc/land_use/>.

[3] World Bank, *Carbon Market Intelligence Survey* (Washington, DC, 2003), <http://www.prototypecarbonfund.org>.

[4] See <http://www.futureforests.com>.

[5] See <http://www.prototypecarbonfund.org>.

[6] See n 3 above; World Bank, *Carbon Transaction Database* (Washington, DC, 2003) (confidential).

2.2 Outlook

Looking forward from 2004, the LULUCF segment of the carbon market can be characterized as a buyers' market given the clear excess of supply over demand. Judging from the experience of the BioCarbon Fund[7] alone, the number of project proposals received by the World Bank's Carbon Finance Business from across the world is ten times higher than the fund will be able to support, and the discrepancy in volume terms, that is, in number of tonnes of GHG removals, is even greater. The World Bank, through the Prototype Carbon Fund and the BioCarbon Fund, is one of the few buyers of LULUCF credits. So far the major public and private buyers have shied away from acquiring these carbon rights associated with sinks. Notable exceptions to this trend include the few transactions from projects carried out in Australia by Japanese power utility companies such as Tohoku Electric and the European semi-conductor manufacturer STMicroelectronics, which has pledged to become carbon-neutral by 2010. STMicroelectronics is also undertaking a reforestation project in Morocco.[8] Another remarkable exception is the announcement by the Brussels Region Government of Belgium in early 2004 that it had entered into agreements with authorities in the Democratic Republic of Congo and Côte d'Ivoire for two reforestation projects.[9] The Brussels Government's investments would sequester in excess of 1 million tCO_2e purchased at prices between US$10 and US$20 per tonne. These investments are especially noteworthy in Europe, which is generally averse to the use of sinks projects to meet climate change mitigation obligations. The likely implication of the excess supply over demand for sinks credits is that prices for LULUCF assets will be lower than for other carbon rights, except when the buyer of the carbon pays in advance and is the sole or main financier of the project, in which case the price will tend to reflect the cost of the project, as in the STMicroelectronics and Brussels Region Government purchases.

2.3 Reasons for Low Prices

LULUCF transactions are typically more complex than energy transactions. If the rules of the CDM can prove daunting for energy project developers, they are even

[7] The BioCarbon Fund provides carbon finance to demonstrate projects that sequester or remove greenhouse gases in forests, agriculture, and other ecosystems. The BioCarbon Fund aims to deliver cost-effective emission reductions while promoting environmental benefits, the reduction of poverty and opportunities for adaptation to climate change, see <http://www.biocarbonfund.org>.

[8] See <http://www.st.com/stonline/company/environm/report02/socenv03.pdf>.

[9] A Gérard, 'Projets de reforestation au Congo et en Côte d'Ivoire', *Le Soir* (Brussels, 11 February 2004); A Gérard, 'Reboiser pour lutter contre le CO_2', *Le Soir* (Brussels, 8 March 2004); Government of New South Wales, *ST Microelectronics* (2003), <http://www.forest.nsw.gov.au/env_services/carbon/investments/stmicro/default.asp>; Government of New South Wales, *ST Microelectronics Investment in NSW Planted Forests* (2003), <http://www.forest.nsw.gov.au/env_services/carbon/media/releases/STMicroElectronics.pdf>.

more complex for LULUCF project developers. For a long time, the framework for these rules remained uncertain, as only the general types of activities eligible to the CDM were known, namely afforestation and reforestation. The ninth session of the Conference of the Parties to the UNFCCC (COP9) finally provided the details.[10] However, the rules on temporary carbon crediting are judged 'overly bureaucratic' and fraught with 'too much uncertainty' by some practitioners, while the obligation to replace all temporary credits with permanent credits or Assigned Amount Units (AAUs) after a maximum of sixty years has been regarded as an open liability by some potential purchasers of sinks credits.[11]

The market for LULUCF credits, be they Emission Reduction Units (ERUs) from JI projects, temporary and long-term Certified Emission Reductions (tCERs/lCERs) from CDM projects, or simply verified removals from voluntary activities, is characterized by severe restrictions. First, Annex I countries are allowed to fulfil their obligations under the Kyoto Protocol using CDM sinks credits (that is tCERs/lCERs as was established by COP9) up to no more than 1 per cent of their 1990 emissions. This is equivalent to an estimated 5–6 per cent of the level of effort necessary for Annex I countries to meet their targets.[12] Any removal of GHGs achieved through CDM LULUCF projects beyond that level cannot be counted as credits by Annex I countries seeking to meet their obligations, regardless of their climate impact. Secondly, the forthcoming European Union's Emissions Trading Scheme will exclude sinks credits up to 2008 at least, given that the system is intended to help European countries meet their Kyoto targets primarily through gains in energy efficiency within Europe.

The European policy-makers' disinclination towards LULUCF projects as a climate change compliance tool translates the views held by a number of non-governmental organizations (NGOs), including some branches of Friends of the Earth, Oxfam, and WWF. According to these views, sinks offer OECD[13] countries too cheap a 'way out' of their climate obligations by allowing industrialized countries to continue emitting GHGs, they are not permanent, cannot be verified, present serious environmental threats where they are implemented, take land away from local people, and so on.[14] Leading figures in the European Parliament echo those views:

> Sinks are difficult to measure . . . Sinks are not a permanent solution. Also what happens if a forest plantation burns down, are the carbon credits then given back? Verification will be very difficult . . . . Apart from . . . the ecological integrity of the Kyoto Protocol . . . there

[10] See ch 14 above.

[11] M Nicholls, 'Forestry Experts Welcome Sinks Deal', *Carbon Finance* (January 2004).

[12] K Newcombe, *Extending the Carbon Market to the World's Poor*, Paper presented at the ABCDE Conference, Paris, 15 May 2003. See <http://wbln0018.worldbank.org/eurvp/web.nsf/Pages/Paper + by + Newcombe/$File/ABCDE + KEN + NEWCOMBE + REVISED.PDF>.

[13] Organization for Economic Cooperation and Development.

[14] Climate Action Network (CAN), *Two Steps Forward, One Step Back: Kyoto Mechanisms Finalised*, Press Release, CAN (Brussels, 9 December 2003); CDM Watch, *Lettre ouverte aux Ministres de l'Union européenne* (CDM Watch, Seminyak, Indonesia, 20 February 2003); Greenpeace, *Sinks in the CDM: After the Climate, Biodiversity Goes Down the Drain* (Greenpeace International, Amsterdam 2003); WWF,

are also economic arguments. Sinks could become so abundant (after 2012) that the price for carbon would not be around 10 € per ton, but only 2 or 3 €. That would mean the financial incentive to modernise our industry ecologically would be gone. We have to realise that the minus 8 target of Kyoto is only the beginning and much bigger cuts are necessary.[15]

Numerous other NGOs and scientists have refuted these arguments. On the basic issue of permanence, there is no question that the carbon sequestered in vegetation will eventually be released back into the atmosphere. But the process may take years, decades, or centuries. If a tonne of carbon dioxide remains sequestered for a period of time equivalent to the residence time of carbon dioxide in the atmosphere, biological sequestration of that tonne is equivalent to a pure reduction of emissions by one tonne, even though scientists disagree over that 'equivalence time'. Furthermore, trees die but forests can be managed to survive for very long times.

Some of the more specific fears about the dangers of LULUCF have been addressed as well. Measuring and monitoring methods are increasingly well established and tested, each additional project contributing to the general knowledge base.[16] Comparative reviews of energy and LULUCF projects have found that these projects face comparable issues regarding measurement and social and environmental benefits, and that potential problems can be dealt with accordingly by project entities.[17]

Remaining uncertainties can be factored into the carbon contracts governing the relationship between the project participants. For instance, when the precise amount sequestered cannot be predicted but a range of values can, the purchase agreement can simply specify that only the quantity of carbon at the lower bound of the confidence interval will be sold.

Critics are correct to point out that LULUCF projects will generate cheaper credits than energy projects. However, sinks cannot flood the market so long as the 1 per cent rule explained above and the additionality rule remain in place (the requirement of 'additionality' provides that projects can generate credits if they produce climate benefits that would not have occurred in the absence of the

Integrating Climate Change and Forest Conservation Policies: the WWF Perspective (WWF-US, Washington, DC, 1999). <http://www.worldwildlife.org/news/pubs/forest_print.htm>.

[15] A de Roo, Draft Report on the Proposal for a European Parliament and Council Directive Amending the Directive Establishing a Scheme for Greenhouse Gas Emission Allowance Trading within the Community, in respect of the Kyoto Protocol's Project Mechanisms, European Parliament Committee on the Environment, Public Health and Consumer Policy (Brussels, 27 January 2004) 13, see <http://www.europarl.eu.int/meetdocs/committees/envi/20040315/518947en.pdf>.

[16] S Brown, 'Measuring, Monitoring, and Verification of Carbon Benefits for Forest-Based Projects', *Philosophical Transactions of the Royal Society* 360 (1797), (2002) 1669–83; Intergovernmental Panel on Climate Change (IPCC), *Land Use, Land-Use Change, and Forestry* (Cambridge University Press, 2000), see <www.grida.no/climate/ipcc/land_use/>.

[17] K Chomitz, *Evaluating Carbon Offsets from Forestry and Energy Projects: How Do They Compare?* (The World Bank, Washington, DC, 2000); J Sathaye *et al.*, 'Concerns about Climate Change Mitigation Projects: Summary of Findings from Case Studies in Brazil, India, Mexico and South Africa' (1999) 2(2) *Environmental Science and Policy* 187–98.

project activity, and more precisely without the incentive created by carbon finance). Moreover, affordable climate change mitigation is in line with the Kyoto Protocol's spirit.

As for the concern that cheap credits would reduce the incentive for modernization and efficiency gains, this depends on how the cost savings are used. For example, greater reliance on LULUCF projects in the early commitment periods could free up precious financial resources for energy research and development to produce the necessary emission reductions in the energy area in the longer term. Business could be given a long-term abatement schedule with gradual ratcheting down of the caps imposing significant energy efficiency improvements and alternative energy development, while LULUCF provided the necessary GHG concentration reductions in the meantime. This would call for more, not fewer, sinks in the early commitment periods.

With respect to the social and environmental impacts of LULUCF project activities, scientists have specified and illustrated the conditions in which LULUCF projects avoid negative outcomes and instead create the desired social benefits, such as additional food, cash income, new inputs and know-how, long-term capital, biodiversity protection, soil rehabilitation, and so on.[18] Additional environmental benefits can also be achieved, such as microclimatic improvements, preservation of hydrologic regimes, and preservation of natural habitat.[19] And the resources available under the UNFCCC can be put to good use to supplement the scarcer resources under the United Nations Conventions on Biological Diversity and to Combat Desertification. These benefits have led some analysts to argue that all categories of LULUCF projects, including prevented deforestation, ought to be included in trading regimes such as the CDM.[20] Excluding certain categories of sinks projects from the CDM imposes a non-tariff trade barrier preventing developing countries from exporting environmental services, which they can produce competitively.[21] Nevertheless, LULUCF projects continue to have a bad reputation with policy makers.

[18] Center for Environmental Leadership in Business (CELB), *Climate, Community and Biodiversity Alliance* (Washington, DC, 2004), see <http://www.celb.org/xp/CELB/programs/climate/ccba.xml> ; P May *et al.*, *Local Sustainable Development Effects of Forest Carbon Projects in Brazil and Bolivia: A View from the Field* (Instituto Pró-Natura and IIED, Brazil, 2003); J Smith and S Scherr, 'Capturing the Value of Forest Carbon for Local Livelihoods' (2003) 31(12) *World Development* 2143–60.

[19] J Niles, *Additional Benefits of Reducing Carbon Emissions from Tropical Deforestation*, Paper No 0084 (Center for Conservation Biology, Stanford University, Stanford, Calif, 2000).

[20] E Niesten *et al.*, 'Designing a Carbon Market that Protects Forests in Developing Countries', *Philosophical Transactions of the Royal Society* 360 (1797), (2002) 1875–88; M Totten, S Pandya, and T Janson-Smith, 'Biodiversity, Climate, and the Kyoto Protocol: Risks and Opportunities' (2003) 5(1) *Frontiers in Ecology and the Environment* 262–70; WBGU (German Advisory Council on Global Change), *Climate Protection Strategies for the 21st Century: Kyoto and Beyond* (WBGU, Berlin, 2003), see <http://www.wbgu.de/wbgu_sn2003_engl.pdf>.

[21] E Bettelheim and G d'Origny, 'Carbon Sinks and Emissions Trading under the Kyoto Protocol: A Legal Analysis', *Philosophical Transactions of the Royal Society* 360 (1797), (2002) 1827–51.

3 Specific Features of LULUCF Transactions

CDM or JI projects involving the generation of credits created by LULUCF projects distinguish themselves from energy projects in several respects.

3.1 People and the Landscape

LULUCF projects potentially involve many different people planting and tending trees or other types of vegetation and 'producing carbon'. Although some projects can be managed, and the credits corresponding to the emission removals sold by a single project entity from a single block of land, LULUCF projects are likely to be carried out by a multiplicity of people working independently of each other. The TIST Project in Tanzania involves over 20,000 village farmers organized in more than 2,000 independent groups in four large areas of Tanzania.[22] Such scattering over land and people makes it difficult for the tonnes of emission removals to reach the marketplace unless an entity aggregates them and deals with the buyer(s) on behalf of the sellers on the ground. LULUCF projects may also involve various plots of land with no connection with each other, yet all owned or managed by a single project entity. In this case aggregation is carried out by the project entity itself, which relies on its own staff or contracting agreements with the people on the ground. The PCF Moldova Soil Conservation Project, for instance, provides for the afforestation of around 14,500 sites over close to 1,900 different sites scattered across the country but all of them in the ownership or under the management of the national forest agency.[23] The aggregator will be the one signing the GHG removal purchase contract with the buyer and assuming the responsibility of delivering a minimum number of GHG removal credits over a certain time period to the buyer. There are no limits to the degree of aggregation as long as the project is carried out in accordance with full-scale project procedures. However, for a project to benefit from the streamlined modalities and procedures allowed for small-scale afforestation and reforestation projects under the CDM,[24] it must not exceed an overall level of 8,000 tonnes of GHG removals per year, as detailed by Ian Noble and Sebastian Scholz in chapter 14 above.

3.2 Non-Permanence

By contrast to energy projects, LULUCF projects face a specific risk, namely that of non-permanence. Carbon, sequestered in vegetation and soils, produces climate

[22] The International Small Group and Tree Planting Program (TIST), *The TIST Tanzania Project* (2004), <http://www.tist.org/tist/tanzania.php>.

[23] Prototype Carbon Fund (PCF), *The Moldova Soil Conservation Project* (2003), <http://www.prototypecarbonfund.org>.

[24] See <http://www.unfccc.int/resource/docs/cop9/06a02.pdf>.

benefits so long as the carbon remains sequestered. After the carbon is sunk, a number of disturbances may occur, which can reverse the sink, for example, fires, pests, or illegal logging. Such disturbances can cause the carbon previously sunk to be released into the atmosphere, thereby overturning the climate benefit. In comparison, emission reductions from an energy process are permanent, in other words the tonne avoided is avoided forever.

LULUCF transactions that take the COP9 draft decision into account will generally make provisions to mitigate the risk of non-permanence.

First, sinks accounting is designed to reflect this risk of non-permanence. As noted in chapter 14 above, COP9 has formalized the system of temporary crediting, which forces the re-verification and re-certification of credits generated by LULUCF CDM projects every five years. LULUCF JI projects are not subject to the same rule as the CDM since JI projects occur between two countries with inventories and caps on emissions. Provided that inventories are capable of precisely capturing changes in the stocks of carbon in vegetation, any reversals of initial removals of atmospheric carbon dioxide by sinks show up as debits in the host country's inventory. As a result, the country in which the debit is recorded loses AAUs in an amount equal to the loss, and is unable to sell these AAUs or has to make up for them in order to stay in compliance with its Kyoto target. If the JI inventory system works well, there is no need for temporary crediting. ERUs from LULUCF JI projects can therefore be considered quasi-permanent.

Secondly, parties to a contract for the sale and purchase of carbon credits from a LULUCF CDM project may decide to manage the risk of non-permanence through the contract itself. One way to do so is to provide for deducting from the number of tonnes sequestered in a given period, a proportion commensurate with the assessed risk of non-permanence and setting it aside in a reserve. This reserve can then be drawn upon to cover a loss of permanence affecting the tonnes embodied in the certificates already sold to the buyer. The reserve thus provides a type of self-insurance against loss of permanence.

This self-insurance system assumes that the tonnes of emission removal credits placed in the reserve do not lose value over time, however. In virtue of the rules adopted at COP9, credits attached to CDM sinks projects (tCERs and lCERs) need to be used in the commitment period in which they are issued. If they are not used in that five-year period, they lose all of their value. In a CDM context, the usefulness of the reserve is thus limited to five years maximum. A slight variant to the basic principle can be envisaged, however, which fulfils the same purpose. Instead of reserving a proportion of the tonnes sequestered in the previous period in a buffer and not selling them, the parties to the contract can agree that the seller will transfer, to the buyer, all of the credits corresponding to the tonnes sequestered for the same total contract price. If the loss of permanence materializes, the buyer starts losing some of the respective credits. Provided that the risk was assessed correctly in advance, the buyer ends up with the same number of credits at the end of the contract as with the reserve system. The major difference is, in the latter variant, the risk of non-permanence is fully borne by the buyer, as the seller loses the incentive of mitigating the risk during project implementation. The problem does not arise to the same extent in a JI project. Removal Units (RMUs) resulting from forestry activities undertaken in Annex I countries under Article 3(3) and (4) of the Kyoto Protocol, such as afforestation, reforestation, and forest management, cannot be banked from one commitment period to another. However, ERUs, into which these RMUs may be converted in the context of a JI project, can be banked. So RMUs put aside in a reserve do not lose their value over time.

tCERs and lCERs from a CDM project finally have to be replaced after a maximum of sixty years. If the seller and buyer believe that the sixty-year replacement rule from COP9 will still be applicable sixty years later, their contract has to provide which party to the contract will be responsible for the replacement. Ultimately, the responsibility rests with the buyer, since the buyer is the party using the tCERs and lCERs.

3.3 Monitoring

LULUCF projects differ from energy projects in regard to the monitoring system that has to be established in order to determine, at the end of each period, how many tonnes of removals of GHGs have been realized by the project (carbon payments are made according to actual achievements rather than plans). Most energy projects consist of one or several point sources where a fuel switch takes place, a new process or technology is introduced, or energy efficiency gains are made. In all such cases, monitoring the project's achievements in order to determine ex post the level of emission reductions is done at one or a few sites. Dozens of wind turbines may exist, but their output is centralized. There are exceptions, for example, a project involving a large number of bio-digesters or charcoal kilns that are not interconnected, in which case monitoring may involve data collection and centralization from many points. In the case of LULUCF projects, dispersion is the rule rather than the exception. A LULUCF project typically involves several thousands of hectares of non-contiguous land exhibiting various growth conditions and owned, managed or otherwise used by different people. The very stores of carbon, that is, the hundreds of thousands or even millions of trees or shrubs, are widely scattered. It may be impractical to measure the carbon sequestered in each tree or shrub and aggregate over the whole project. Instead, monitoring plots are established, which are representative of the whole project. The project lands are first categorized into a number of strata. Stratification ensures that the number of monitoring plots is not too high, so the cost of monitoring the project will be reasonable, yet sufficient to account for the diversity in project sites. In addition, the carbon pools to be measured are selected (live biomass including vegetation and roots, dead biomass and soil carbon). It may be, for instance, that accretion of soil carbon is so slow and variable from one site to the next that it is not worth the monitoring effort. Based on the number of strata and the variance of the carbon pools in each stratum, the monitoring plan sets the number of monitoring plots needed to achieve a given level of statistical confidence in the measurements.[25]

Another difference between the monitoring systems of energy and LULUCF projects concerns the monitoring periodicity. Whereas energy projects typically have monitoring and reporting periods of one year or less, LULUCF projects can have monitoring and reporting intervals as long as three to five years. The reason is the slow changes in the carbon stocks. Especially in the beginning of an afforestation or reforestation project, each seedling accumulates very little carbon. The changes that would be measured on an annual basis might fall within the margin of error. This applies even more to soil carbon, where build-up is slow throughout the project's life.

3.4 Social and Environmental Impacts

LULUCF project activities affect people and land to a much greater extent than energy projects. An energy project might radically transform a few dozen hectares of land, but LULUCF projects typically occupy several thousands of hectares. It is necessary, therefore, that an appropriate framework of rules and procedures be in place before project implementation so as to ensure that the project does not harm the local ecosystem. Many LULUCF projects seek to restore degraded land, in which case afforestation or reforestation bring much greater ecological benefits than costs to the land. However, even so, variables such as the choice of land preparation techniques (mechanized vs manual), exact location (abutting streams or not), and species to be planted (exotic vs endemic) may have various ecological impacts, for example, on biological diversity and water stream quality. Parties to the UNFCCC were aware of potentially perverse incentives in crediting reforestation activities and decided that reforestation projects must give evidence that the lands on which they are established were without forests as of 31 December 1989.[26] This rule prevents clearing natural forest for the purpose of claiming credits later under the Kyoto Protocol.

The lands on which LULUCF projects are established are generally owned or used by people, be they sedentary or nomadic. It would be unusual for a project to be located in an area that is not used by anybody. As for environmental impacts, a framework must exist to ensure that the project causes no harm, and even does good to the local people. Local people should be informed and consulted about the project's design, be given a way to express their opinions at an early stage, then have the opportunity to participate in project implementation, and share in the benefits.

Some organizations have proposed environmental and social requirements for LULUCF projects, in the hope that buyers would be able and willing to

[25] S Brown, 'Measuring, Monitoring, and Verification of Carbon Benefits for Forest-Based Projects', *Philosophical Transactions of the Royal Society* 360 (1797), (2002) 1669–83.

[26] See <http://unfccc.int/resource/docs/cop7/13a01.pdf>.

distinguish among superior and inferior projects and select only the superior projects.[27]

3.5 Late Delivery of Carbon

Afforestation and reforestation projects typically produce their carbon benefits later and for a longer time than energy projects. An energy project can start generating emission reduction credits up to its full capacity as early as the first year of implementation. This is rarely, if ever, the case for afforestation and reforestation projects, which take some years before the rapid growth in biomass occurs. In addition, LULUCF projects tend to last a longer time than energy projects, that is, beyond the end of the first commitment period of the Kyoto Protocol. As a result, carbon finance does not produce the same financial incentive for afforestation and reforestation as for energy projects since buyers are mostly interested in projects that deliver tonnes prior to 2012. Note that this late delivery of carbon by LULUCF projects does not invalidate their precious climate change mitigation role while durable solutions to climate change are found in the energy arena as noted earlier.

3.6 Lower Financial Profits

A large proportion of LULUCF projects produce benefits for the common good as opposed to private profits. Relatively fewer LULUCF projects attract private equity and debt finance, except in the commercial forestry area. However, at the current prices for carbon credits, commercial grade forestry projects do not benefit much from carbon finance. In particular, carbon payments are not sufficient to raise their rates of return much above the baseline scenario. It is therefore questionable whether many such projects are even additional in the sense of the Kyoto Protocol and Marrakesh Accords. In comparison, non-commercial grade forestry LULUCF projects tend to rely substantially on carbon sale revenues. In some cases, the sale of carbon is the sole source of hard currency revenues for these projects. These projects are clearly additional since carbon finance creates the financial incentive making these projects possible. However, these projects also find it much harder to reach financial closure. In other words, LULUCF projects face a sharper trade-off than energy projects between establishing additionality and finding financing. Consequently, LULUCF activities rely more heavily on sources of concessionary financing, including Official Development Assistance (ODA), and grants from the Global Environment Facility (GEF) or bilateral donors. This

[27] Center for Environmental Leadership in Business (CELB), *Climate, Community and Biodiversity Alliance* (Washington, DC, 2004), <http://www.celb.org/xp/CELB/programs/climate/ccba.xml>; The Nature Conservancy (TNC) *et al.*, *Letter to Heads of Delegation at CoP9* (TNC, Ballston, Va, 2003).

reliance begs the question of the acceptability to use such sources of concessionary finance to support climate change mitigation projects, which generate carbon credits. The Development Action Committee of the OECD has discussed this question and concluded that governments using ODA to buy carbon credits must choose ex post whether to classify the corresponding financing as ODA and lose the carbon credits or as carbon finance and deduct an equal amount from their ODA contributions.[28]

4 Conclusion

This chapter has shown that the market for LULUCF transactions is characterized by excess supply over demand and that GHG removal credits from LULUCF projects command lower prices than energy projects. This feature is due to a large extent to the complexity of LULUCF transactions, restrictions to their tradability, and the bad press given by certain NGOs.

LULUCF projects distinguish themselves from energy projects in some important respects. First, they tend to involve many people on the sellers' side, which makes it necessary to aggregate the supply at the contracting and delivery stages. Secondly, the credits generated by LULUCF projects are only permanent to the extent that the underlying carbon assets are permanent. The risk of non-permanence must therefore be managed in LULUCF transactions. Thirdly, the scattering of sites over the landscape calls for the monitoring system to be based on a statistically representative sampling of sites, rather than exhaustive data collection. Fourthly, land-use projects have particular social and environmental risks, which must be mitigated as a condition of the project's success. Fifthly, these projects take a longer time on average to deliver their carbon credits due to biomass growth patterns. Sixthly, they tend to be less financially profitable than energy projects and may require more concessionary financing.

A special effort must be made to order to alleviate the existing constraints on LULUCF projects. This effort entails expanding the number of projects available so as to show that the various risks faced by LULUCF projects can be managed in such a way as to achieve real climate, social, and environmental benefits. Facts will help dispel fears and misgivings among opinion leaders and policy makers and justifiably expand the role of LULUCF in climate change mitigation. The World Bank's BioCarbon Fund will contribute to this effort. The BioCarbon Fund will bring together capital from several private and public buyers, create a pool of high-quality carbon assets across various countries and technologies, blend various projects with different risk profiles, including non-permanence risk profiles, and apply novel non-permanence risk management measures.

[28] OECD, *ODA Eligibility Issues for Expenditures under the Clean Development Mechanism (CDM): Synthesis Note and Way Forward*, Report No DCD/DAC (2004) 17 (Paris, 16 March 2004).

PART VI

JI AND CDM: CARBON CONTRACTS

16

Carbon Contracts, Structuring Transactions: Practical Experiences

*Martijn Wilder, Monique Willis, and Mina Guli**

1 Introduction

The development of international, regional, and national regulatory frameworks to control the emission of greenhouse gases (GHGs) has resulted in the emergence of a new and distinct area of law dealing with the monitoring and mitigation of climate change through economic mechanisms such as emissions trading. While GHG emissions trading regimes seek to address environmental issues, the legal implications of emissions trading extend well beyond environmental law considerations and encompass and intersect with a number of other areas of law such as public international law, commercial law, finance law, project law, tax law and corporate and securities regulation.

As discussed in previous chapters, the UNFCCC and the Kyoto Protocol develop a set of public international law principles to address climate change on a global level. However, the Kyoto Protocol also provides that Parties (and entities which they have authorized to participate) may undertake approved emission reduction projects and trade Kyoto Protocol credits (AAUs, ERUs, CERs, RMUs) as a means of satisfying commitments under Annex B of the Protocol.

Undertaking projects and transacting credits must fit within existing legal and commercial regimes of the Parties. Early participants in the carbon market have begun to adopt innovative commercial principles and structures to utilize the Protocol's flexible mechanisms. Nowhere is this more evident than in the drafting and negotiation of carbon contracts and the structuring and financing of carbon transactions.

In practice, this requires the ability to overcome the 'disconnect' that often exists between international and national law and between private and public legal regimes, and to incorporate principles and structures, provided for in the Kyoto Protocol (such as CDM projects or CERs), into effective contracts which will bind the parties, comply with domestic law requirements, and also allow for

* Martijn Wilder and Monique Willis are members of Baker & McKenzie's Global Clean Energy and Climate Change Practice Group. Mina Guli is a legal counsel at the World Bank. The views expressed in this chapter are the views of the authors and do not necessarily represent the views of their employers.

enough flexibility to manage the constantly developing international legal framework.

This chapter focuses upon practical experience in the development of carbon contracts and the structuring of carbon transactions. It will focus specifically on:

(i) the definition of the carbon asset in various types of contracts (described generically in this chapter as an Emission Reduction (ER))[1] and how this has developed as the Kyoto Protocol rules have been refined;
(ii) the way in which the carbon asset can be transacted between parties;
(iii) the different types of contracts that are being adopted and the major considerations common to all carbon contracts; and
(iv) the different ways in which emission reduction projects can be structured.

This chapter is not designed to be a comprehensive guide to project finance or structures or to provide legal advice on the design of contracts to trade Kyoto Protocol credits and ERs. Rather, it gives a high level overview of some of the potential options for 'contracting for carbon', which includes the broad structuring of the project that will create the ERs and the transaction of the ERs through various sale and purchase arrangements. Derivative and secondary trading instruments have already entered the emissions trading marketplace. These instruments raise additional issues, which are not discussed within the scope of this chapter.

The initial challenge in contracting to sell carbon rights is defining the asset to be sold. Another of the major considerations is the mitigation and allocation of risks arising both from the operation of a particular project and from the development of the international regulations on climate change. The types of risks that should be considered when contracting for carbon are considered below.

2 Definition of the Carbon Asset

The first issue to address in contracting to sell ERs is the exact nature of the right that is to be transacted, and what the seller must do to deliver this right. This will largely depend on the purpose for which the buyer is purchasing ERs. Buyer motivations can include:

(i) immediate compliance with legal obligations such as those under the Kyoto Protocol (once it enters into force), the UK Emissions Trading Regime[2] and, as of 1 January 2005, the EU regime;[3]

[1] For example, in the World Bank CFB Emission Reduction Purchase Agreements, an Emission Reduction is defined as 'any right, interest, credit, entitlement, benefit or allowance to emit (present or future) arising from or in connection with any GHG Reduction by the Project and includes any right that may be created under any regulatory or legal regime as a result of the GHG Reductions whatsoever'.

[2] The UK Greenhouse Gas Emissions Trading Scheme 2002, rules available from <http://www.defra.gov.uk/environment/climatechange/trading/pdf/trading-rules_rev2.pdf> accessed 21 May 2004.

[3] EU Emissions Trading Directive 2003/87/EC as agreed on 22 July 2003.

(ii) an expected need to comply with a future legal obligation. In particular, significant transactions of carbon rights that comply with the Kyoto Protocol rules have occurred. Buyers here are seeking to acquire CERs or ERUs, which may then become offset against future Kyoto obligations. It is such transactions that the World Bank's Prototype Carbon Fund and the Dutch Government's CERUPT and ERUPT programmes are pursuing;

(iii) voluntary compliance or public relations reasons where buyers aim to use the ERs to meet part or some form of a voluntary target they have adopted;

(iv) retail schemes where specific organizations sell carbon to buyers to allow them to promote themselves as 'climate-neutral';

(v) responding to increasing pressure from shareholders insisting that companies minimize their climate impacts;

(vi) institutional pressure from stock analysts, investment institutions, and insurance companies re-evaluating and re-rating the risk that companies present in terms of climate change impacts and redirecting investments away from companies taking no action on climate change; and

(vii) the threat of 'climate litigation' whereby firms refusing to take action on reducing greenhouse emissions may be targeted in future legal proceedings for failing to reduce emissions.

In addition to such direct motivations, companies have also participated in the market to learn by doing, experiment with a broad range of different transactions and to influence policy.

The critical components of any contractual definition of any ERs should include the following:

(i) a reference to the physical reduction in levels of GHG emissions;

(ii) a clear definition of the GHGs to be included (generally being those under the Kyoto Protocol);

(iii) a clear unit of measurement usually being per tonne of carbon dioxide equivalent;

(iv) reference to the particular and clearly defined project creating the reductions;

(v) objective validation and assessment methodology; and

(vi) the basis against which such reductions are occurring (that is, against a pre-determined baseline).

A contract should ensure that the rights are of such a nature that they present the maximum opportunity for conversion into future rights. As with other kinds of assets, a carbon contract should ensure that the rights are unencumbered, freely transferable, and the party transferring or surrendering the carbon rights has the authority to do so, and will take no action(s) that affect the creation, generation, or transfer of emission reductions.

There are three main types of ERs, as described in Table 16.1 below.

The general tendency for early Emission Reduction Purchase Agreements (ERPAS) has been to define the asset to be transferred as including *all* rights and benefits associated with the reduction, removal, sequestration, or abatement of GHG emissions (that is, the generation of ERs) by the project. Purchasers who have intended to later use the ERs for compliance purposes, may also seek to negotiate an obligation upon the seller to work with the purchaser at a later stage to create compliance-based credits (for example, CERs or ERUs) from the ERs previously generated by the project.

Table 16.1: Main Types of Emission Reductions

Asset Type	Criteria	Example
Non-Compliance Contractually Based Right	Primarily for use in voluntary schemes, corporate programmes, PR, retail schemes which meet criteria agreed to between the parties.	• Chicago Climate Exchange • 500 PPM • Future Forests • Climate Care • APT—Cosmo Oil Deal (Australia)
Pre-compliance Contractually Based Right	Rights created in accordance with developing legal rules for future use in those developing legal regimes.	• Kyoto-compliant rights purchased by the World Bank, Dutch Government • EU Allowances (pre-2005)
Compliance Statutory Based Rights	Rights created in accordance with existing legal regimes for compliance purposes.	• EU Allowances (post-2005) • UK Emissions Trading scheme • New South Wales Abatement Certificates

2.1 Non-Compliance Contractually Based Rights

It is always possible for two parties to agree to transfer contractually based ERs from a project or an activity for non-compliance purposes, as there are no statutory requirements which must be complied with and no need for an existing emissions trading scheme. However, contractually based ERs outside of any international or domestic legal framework will generally only have value to purchasers who wish to use the rights for public relations purposes or for a voluntary emission reduction programme.

For example, some companies have developed carbon based 'retail products' which they use for public relations purposes to provide 'carbon neutral' products or services to customers (for example, 500 ppm, Future Forests, Climate Care).[4] In structuring such products and contracting to gain legal title to the relevant carbon rights, the carbon rights are contractually based and therefore may be less valuable in the marketplace than Kyoto Protocol compliant credits (for example, CERs or ERUs), which can be used for compliance purposes. Notwithstanding the 'mere' contractual nature of such rights and their absence of regulatory recognition, it is still important that the contractual right has actual physical action to reduce, remove, abate, or sequester GHGs to support it. In such cases an agreed approach to validation or audit of ERs, subject to the contract, is usually linked into the contract terms. Accordingly, it is not uncommon to refer to such tradable contractual rights as 'Verified Emission Reductions'.

Other companies have begun to 'bundle' their products with carbon offsets to make them more attractive to market purchasers (one example is the bundling of Australian coal with carbon sequestration rights from Australian forests to sell to a Japanese purchaser).[5] Again, this is structured by way of contractual rights and the credits are not necessarily Kyoto Protocol compliant (particularly in the case of Australia, which has not, at this stage, ratified the Kyoto Protocol).

Finally, some voluntary emissions trading schemes (for example, the Chicago Climate Exchange)[6] and corporate-based emissions trading schemes (for example, BP)[7] have arisen which will contract to buy and sell certain types of contractually based ERs.

2.2 Pre-Compliance Contractually Based Rights

Many of the transactions in the early carbon market, on an international level, have been through a 'learn by doing' approach purchasing contractual ERs in the hope

[4] See websites of these companies: <http://www.500ppm.com>,<http://www.futureforests.com/> and <http://www.climatecare.com/> accessed 21 May 2004.

[5] See, for example M Henley, 'RECs on the Railroad', *CFO* December 2003–January 2004 available at <http://www.cfoweb.com.au/stories/20031201/20962.asp> accessed 21 May 2004.

[6] See <http://chicagoclimateex.com> accessed 21 May 2004.

[7] See BP Global Choice at <http://www.bp.com.au/globalchoice/default.asp?menuid = ge> accessed 21 May 2004.

that these ERs will ultimately be able to be transformed into Kyoto Protocol credits (ERUs or CERs) to be used for compliance purposes. The transactions entered into by the World Bank carbon funds and the CERUPT and ERUPT funds are examples of this.

Early ERPAs for pre-compliance ERs generally provided that, although there was a preference to receive Kyoto Protocol credits (CERs or ERUs) from the seller, the purchaser would pay an identical unit price for each Verified Emission Reduction in the event that the CDM Executive Board was not yet issuing CERs. The purchaser therefore took the risk that the Kyoto Protocol would not enter into force, that the validation methodology was consistent with the methodologies approved by the institutional committees under the Kyoto Protocol and that the Verified Emission Reductions would not be worth anything to them for compliance purposes.

At the date of writing, the Executive Board is some way towards the 'prompt start' to the CDM and it is likely that projects will be registered and CERs issued within the foreseeable future. In addition, the policies of some Annex I Governments have become clearer in relation to their intentions to distribute AAUs and issue ERUs for certain project based activities. For example, the New Zealand Government has undertaken binding obligations to deliver AAUs to projects, which generate ERs. These future rights to AAUs (which took the form of a promissory note under New Zealand law) can be sold on the market and provide greater security for a purchaser than a generic contractual 'Emission Reduction'.[8]

As more certainty develops, it is therefore likely that purchasers will begin to insist on the delivery of electronic CERs or ERUs as they are described in the Kyoto Protocol and Marrakesh Accords rather than simply verified ERs. The risks arising in the interim period (for example, before the Executive Board can issue CERs or before the international transaction log is established) can be managed through the use of conditions precedent dealing with the entry into force of the Kyoto Protocol, or the establishment of functioning national registries.

However, in the uncertainty surrounding the future of the Kyoto Protocol regime, a purchaser may still wish to capture any benefits from the GHG ERs from the project in the event that the Kyoto Protocol does not enter into force or is replaced by another regime.

2.3 Compliance Statutory Based Rights

Some carbon contracts are designed to transfer ERs, which are recognized under an existing legal scheme. Generally, the definition of the carbon asset in such contracts is simpler, as there can be a reference to existing legislation or a treaty (for example, the New South Wales legislation which establishes the NSW Greenhouse Gas Abatement Scheme and creates an emissions trading regime), and the statutory framework generally provides verification requirements and an established means

[8] See n 1 above.

of transferring and recording the transfer of such rights. EU Allowances will be a statutory based right recognized under an existing legal scheme from 2005 and it is likely that there will be a proliferation of innovative contracts and standardized arrangements to transfer such rights to meet the compliance obligations of liable entities under the European Union Emissions Trading Scheme (EU ETS) once the registry system is operational.

3 Legal Ownership of Emission Reductions

Kyoto Protocol rights, such as CERs and ERUs, are created by way of a verification process of monitoring the reduction, abatement, or sequestration of GHGs by certain types of projects. In order for entities to be eligible to claim title to Kyoto Protocol credits they must also be able to claim legal title to the rights created from the abatement of GHGs by the project. This is not always a clear-cut issue. The general approach of most governments, early CDM and JI project developers and participants in the carbon market has been to assume that, in the absence of any law or contract to the contrary, the entity which owns or develops the project activity responsible for reducing GHGs is entitled to deal with any rights arising from this project (for example, CERs or ERUs).

However, it should not be an automatic assumption that private project developers are entitled to Kyoto Protocol credits created from a project activity. The Kyoto Protocol is an international agreement providing rights and obligations for national Party governments. The Marrakesh Accords do not deal with the ownership of project credits (CERs or ERUs), and it could be argued that the reduction of GHGs is, in effect, the management of a natural resource. In many countries, natural resources (such as air or water) are seen to be the responsibility and property of governments rather than private entities. Therefore, some governments may consider that ERUs or CERs are sovereign rights, which can only be owned and traded for profit by the government. For example, in its climate change strategy the New Zealand Government has announced that it will retain all rights and obligations arising from certain carbon sequestration activities. There have also been indications that China is also considering the extent to which CERs can be considered the property of the government.

Few domestic legal systems have laws in place that recognize ERs or determine the way in which legal ownership is to be allocated. If GHG ERs are considered to be a 'natural resource' (and therefore the property of the government), the government may claim ownership and require project developers to enter into concession arrangements to manage the resource.

A major concern of purchasers when entering into contracts to purchase ERs will be to ensure that the seller can establish legal title. A warranty to this effect should be included in any such contract with appropriate remedies if the seller cannot ultimately deliver title that is free of encumbrances.

Even where the government does not claim any rights over ERs, the title to underlying GHG reductions could be the subject of a dispute between different entities involved in a project, for example, the landowner, the lessee, project developer, contractor, manager, project owner, financiers, and purchaser. The legitimate entitlement to ERs should be clearly established before any contract to sell them can be entered into.

4 Contracting to Sell Carbon Rights

4.1 Key Issues in Carbon Contracting

In transacting to sell ERs, the key issues lie in negotiating and determining:

(i) the nature of the commodity which will be transacted (that is, contractually defined ERs, CERs, ERUs, EU Allowances);
(ii) the commercial terms, including price at which it will be sold; and
(iii) the risks and warranties that each party is prepared to accept and provide in the transaction.

The following sections set out issues which must be defined and managed in any type of contract to sell the rights or benefits associated with a reduction in GHG emissions, regardless of the type of project structure or contract, which is ultimately adopted.

4.1.1 Nature of the right

The contract needs to clearly specify what is being sold. Where a simple contractual right is being transacted (in the absence of any statutory or legal regime), it is likely that the purchaser will want provisions in the contract which require the seller to provide all necessary assistance for the purchaser to create whatever statutory right that may flow from the contractual right in the future that suits their purposes. Equally important is that the contract is clear as to what is not being sold. In this respect, the seller would not, if it merely sells a contractual emission reduction, want to provide any undertakings that this right will be capable of producing a Kyoto Protocol or European Union right unless that is pre-agreed. Where contractual rights are sold, the specific qualities of such rights will need to be outlined. For example, whether they are to be Kyoto compliant and how often they are to be independently verified.

4.1.2 Quantity of rights

The contract will need to clearly specify the total number of rights to be delivered. In order to minimize its risk and exposure for failing to produce sufficient numbers of rights, the seller may want to ensure that it retains some form of 'buffer' and therefore perhaps sells no more than 70 per cent of its total carbon stock. This will provide some protection in the event that the modelling of the carbon stock or the growth and yield forecasts proves to be incorrect, or against physical loss.

4.1.3 Legal title

The seller will need to demonstrate and warrant that they have clear legal title to the right and the ability to transfer the rights to the buyer without any third party consents or further government approvals.

4.1.4 Price and payment

Once a price has been agreed, the method of payment will be critical. In some cases, it may be possible for the project developer (seller) to acquire finance upfront, but this is unlikely. In most contracts, payment is not made until delivery of the rights is effected.

4.1.5 Transfer of legal title

Central to the contract will be the point in time to which legal title to the rights would be transferred. Where a contract right exists, transfer and payment usually occurs on receipt of the Verification Report. Where a statutory right exists, then it is most likely that a registry system will be in place to effect transfer of rights.

4.1.6 Failure to deliver

In the event that the seller fails to deliver the ERs or there is some kind of shortfall, then the contract will need to provide mechanisms to address this. Common mechanisms in carbon trading contracts for delivery shortfall range from obligations to find replacement rights from the seller's existing stocks or on-market or alternatively be far more onerous with the defaulting party required to pay liquidated damages.

4.2 Different Types of Carbon Contracts

The ERs arising from a project can be sold in a variety of ways and there is no standard formula for the way in which their creation and sale should be documented through contracts. However, the relatively small size of the carbon market (and the predominance of the World Bank's Carbon Finance Business and the Netherlands Government's CERUPT and ERUPT funds as purchasers) has meant that many project ERPAs to date have utilized similar definitions and concepts, refining these definitions as the international laws on emissions trading develop.

In the pre-Kyoto marketplace, contracts for the sale and purchase of ERs were generally individually negotiated for each project, taking into account the specific circumstances of individual projects, the domestic rules of the governing law for the contract and the legal and business constraints binding each of the parties. To source ERs, potential buyers and sellers have employed a variety of approaches. These include:

(i) implementing a competitive tender process for potential projects from which ERs will be purchased on the basis of a set of standard terms and conditions (this is the general approach taken by the Netherlands CERUPT and ERUPT funds);

(ii) issuing a request for particular types of ERs (for example, CERs or ERUs) from the market;

(iii) inviting the submission of potential projects through a project description and individually negotiating the purchase of ERs from such projects (this is the approach the World Bank's carbon funds have taken); or

(iv) sellers transacting with independent brokers to transact their ERs on the market (this to date has been a less common approach).

As the market for ERs, such as CERs, ERUs, and EU Allowances gains liquidity, it is likely that other approaches will become more popular, such as the use of trading platforms and brokers, as has occurred in some jurisdictions already with energy trading. Standardized procedures and regulations could be utilized to trade CERs, ERUs, or EU Allowances, as purchasers will be able to identify which project the ER came from by examining its electronic serial number.[9]

Approaches to contracting ERs have very much been influenced by levels of understanding of the Kyoto Protocol rules, an increasing number of sellers and buyers in the market, the gradual consolidation of the specific international rules surrounding the CDM and the 'early start' of the Executive Board. Approaches will continue to develop depending on whether the Kyoto Protocol comes into force and the extent to which the EU ETS will incorporate CDM or JI project credits. This will contribute to the 'immediacy' of compliance obligations for many Annex I Parties and will therefore lead to a growth in the market and a suite of contractual approaches being developed to suit particular buyer and seller circumstances.

There are a range of potential ways in which ERs can be sold into the market, including:

(i) a direct and immediate sale of a fixed amount of ERs which have been already generated by a project during a certain vintage (spot transaction);

(ii) the sale of a future stream of ERs from a project (forward transaction);

[9] Marrakesh Accords, Decision 17/CP.7 FCCC/CP/2001/13/Add.2, Appendix D, para 7 of and Decision 18/CP.7 FCCC/CP/2001/13/Add.2, para 46.

(iii) a sale of options to purchase agreed volumes of ERs at a later date at an agreed or market price, usually with a strike price to exercise the option itself; or

(iv) a combination of the above.

Early market transactions have generally been undertaken primarily as bilateral trades under ERPAs.

However, as the international and regional emissions trading schemes are developed and registry systems established it is likely that this will allow for some level of standardization among ERPAs. Attempts have already been made to develop standard trading contracts for the CDM and JI projects, such as by the International Emissions Trading Association (IETA).[10] It is possible that robust international emissions trading platforms may be established for various types of ERs with secondary markets in a similar manner to the way in which stocks and bonds are currently traded through registry systems. However, this is very much in the future and at present the negotiation of carbon contracts is occurring very much in a transition stage, with all parties attempting to hedge their risk exposure due to the developing nature of the international climate change regime.

5 Structuring and Financing CDM and JI Projects

The Kyoto Protocol establishes the general framework for CDM and JI projects and is elaborated on by the Marrakesh Accords. Article 12 envisages that a CDM project will benefit non-Annex I countries and that Annex I countries will be able to use CERs from a CDM project to meet their compliance obligations, while Article 6 envisages that two Annex I Parties can cooperate to undertake a JI project. However, the Protocol does not provide any details on how Annex I investment in a CDM or JI project is to be structured. CDM and JI projects can therefore be structured in a variety of ways, and the appropriate structure will depend on the extent to which the project requires external finance or is to be financed from within the host country or Annex I investor through the additional revenue stream of CERs or ERUs.

In the case of CDM projects, the discussions behind Article 12 of the Kyoto Protocol generally assumed that there would be a high level of Annex I Party involvement and technology transfer in CDM projects, with the Annex I Party providing underlying finance to the project or taking equity in the project.[11] However, in the early CDM market, CER purchasers have tended to limit their involvement in the project to the purchase of CERs, with discrete underlying project finance or equity

[10] See <http://www.ieta.org> accessed 21 May 2004.

[11] See, for example, UNFCCC's *A Guide to the Climate Change Convention and its Kyoto Protocol*, available at <http://unfccc.int/resource/guideconvkp-p.pdf> accessed 21 May 2004.

provided to the project participants from more traditional investors or financial institutions.[12] The major obstacle for many potential CDM projects has therefore been obtaining general finance to commission the project, before it even begins operation.

For JI projects, the creation and transfer of ERUs depends on the cooperation of the Annex I host country government to transform AAUs into ERUs and transfer them to the Annex I ERU purchaser. As JI projects will not be eligible to create ERUs until 2008, there is no international transaction log in place and Annex I countries have not yet been issued AAUs, there has generally been less progression with JI projects than with CDM, due to the 'prompt start' of the CDM Executive Board and the ability to issue CERs from the starting date of a registered CDM Project. However, some Annex I governments have nonetheless begun to develop infrastructure to support JI projects. For example, the Netherlands Government has developed an ERUPT fund which is designed to purchase AAUs or ERUs from JI projects and the New Zealand Government has launched a competitive tender process whereby it agrees to provide AAUs to emission reduction projects in return for verified ERs. The project developers are therefore able to on-sell these future AAUs to purchasers to make the project financially viable.[13]

Once an appropriate structure for a CDM or JI project has been chosen, a major issue for project participants, financiers and/or the purchaser of CERs or ERUs is to allocate the risks arising from the operation of the project and the developing nature of the Kyoto Protocol regime. This is generally managed by allocating risks through the network of different contractual arrangements, which make up a CDM or JI project. The price payable under each contract will reflect the allocation of risk between the parties to that contract. For example:

(i) The major financier to the project is generally crucial for the project to begin operation and will attempt to minimize its risk and protect its investment through means such as government or parent entity guarantees, binding deadlines for major steps in the project and penalties for non-compliance.

(ii) Offtakers from the project (for example, the electricity purchaser in a renewable energy project) will also attempt to ensure that the relevant commodities are delivered and will impose their own penalties for non-delivery.

(iii) The project participants should then allocate each of the potential project risks (for example, underperformance, delayed construction and so on) to the party who is best able to control those risks (for example, the supplier of fuel to the project, the construction contractor).

(iv) The various risks and uncertainty arising from the developing nature of the Kyoto Protocol legal regime should also be allocated as between the project

[12] See A Michaelowa, *Unilateral CDM: Chances and Pitfalls*, Division 44 Environment and Infrastructure.

[13] See discussion of 'Projects to Reduce Emissions' at New Zealand Climate Change Office website <http://www.climatechange.govt.nz/policy-initiatives/projects/index.html> accessed 21 May 2004.

participants and the purchaser of CERs or ERUs from the project. The allocation of these risks will generally reflect the price payable for the CERs or ERUs (that is, a buyer is likely to pay a higher price if it will only need to purchase CERs or ERUs once the Kyoto Protocol has entered into force).

Potential CDM or JI Project structures could involve a combination of the following options.

Host Country-led CDM or JI Projects

This is where a project is led and financed primarily by a host country with limited Annex I investor involvement. For example, a Unilateral CDM Project[14] or where an Annex I Party participant's role is limited to the purchase of CERs or ERUs. The contractual arrangements for a host country-led project will generally consist of (i) project agreements and financial arrangements governed by domestic host country law, and (ii) a separate ERPA with an Annex I purchaser of CERs or ERUs. Depending on the stage of the project in which the ERPA is entered into, the ERPA could be for a forward sale of CERs or ERUs or, alternatively (depending on factors such as how the registry rules develop and whether developing country entities will be eligible to trade CERs), a sale of existing CERs or ERUs from an operating project.

The ERPA will generally be negotiated directly with potential purchasers, who will be concerned to ensure that the seller has clear legal title to any CERs or ERUs arising from the project and that they have not been assigned to any third party. Payment will generally occur after the rights have been verified and certified by an independent third party Designated Operational Entity (DOE).

Carbon Finance

The term 'carbon finance' generally refers to the provision of finance by an Annex I country (or multilateral institution such as the World Bank's Carbon Finance Unit) to a potential CDM or JI Project in advance of the implementation of the project. This provides an additional revenue stream, which the project developer can rely on to obtain underlying project finance. The sale of CERs or ERUs occurs directly by an ERPA transferring the rights to CERs or ERUs from the project participants to a purchaser to raise finance for the project. This is the model that has sometimes been followed by the World Bank's Prototype Carbon Fund and the Netherlands Government's CERUPT and ERUPT funds through the provision of 'forward payments' for CERs or ERUs.

Debt Finance

There are also CDM or JI Projects where the Annex I Party investor has an involvement and interest in the operations of the underlying project, rather than merely

[14] See n 12 above.

an obligation to pay for CERs or ERUs as and when they are delivered. For example, the investing Annex I Party can loan money to finance the underlying project. The loan will generally be secured against the assets of the project (such as land, project infrastructure, payments under supply contracts, or other legal rights) and will attract interest, as with a financial lender for any type of project. However, for CDM or JI projects, innovative contracts can be entered into so that part of the repayment for the debt or the interest component can be made by way of CERs or ERUs. Alternatively, the transfer of CERs or ERUs could be viewed as a part down-payment for the loan provided.

Under these arrangements, the Annex I Party's interests in the project extend beyond a future interest in CERs or ERUs. It may therefore be more concerned to hedge various project and credit risks in the agreement as it has a financial interest in the success of the project.

Where debt finance is provided with party payment made in CERs or ERUs, the legal arrangements for the transfer of CERs or ERUs will be incorporated into the finance documentation as part of the repayment conditions. The financial contracts must also comply with the requirements of the laws of the host country with regards to foreign investment and project finance. It is worth considering whether the forward payment for CERs or ERUs in the form of a loan could be considered trading in derivatives or securities under domestic laws, and require various consents or licences.

Equity Investment

Alternatively, an Annex I investor may become even more involved in the project and purchase equity in the project through its investment. The return for the Annex I investment could be the assignment of CERs or ERUs arising from the project. Again, the contractual structure will be based in part on investment agreements in the host country of the CDM or JI project and must comply with all local investment law requirements, particularly with regards to foreign direct investment.

Non-recourse Project Finance

Project financing involves the construction of a new project and the servicing and repayment of finance dependent entirely upon the cashflows derived from the project. This structure means that the costs of developing a project are isolated from the ongoing business of the project developers, so that a failure to repay borrowed development funds would not impair the value of the project developers (or the various joint venture parties) developing the project.

CDM Technology Swap

This is an arrangement whereby an Annex I country provides or licenses technology to a project (that is, technology transfer as envisaged by Article 12), which is paid for with CERs from the project. This would involve a simple contract under which the consideration for the CERs is a technology swap. Appropriate warranties and indemnities would need to be included in the documentation.

Option arrangements

In addition to a general ERPA it is also possible to enter into an option arrangement whereby an independent third party or a project participant is provided with an option to purchase CERs or ERUs from the project at its discretion. This may be in addition to an existing ERPA, which is in place or may simply be an independent arrangement. In such cases options are usually granted at an option price with an agreed strike price for when that option is exercised.

Off-Take arrangements

In the case of CDM or JI Projects that produce other commodities in addition to CERs or ERUs, it is possible to structure off-take arrangements to bundle those products and the CERs or ERUs created on the basis of emission reductions from the project.

For example, the developer of a landfill gas energy project could sell CERs or ERUs bundled with either the electricity (under a power purchase agreement) or the fertilizer resulting from the project. If the purchaser of electricity or fertilizer had compliance obligations under their national legislation, the CERs or ERUs would assist to offset these obligations.

Establishment of funds/pooling carbon rights

Some entities (for example, the World Bank) have established 'carbon funds', whereby participants invest in the fund, which then purchases emission reductions on behalf of the participants. The participants are therefore entitled to a proportion of the emission reductions purchased by the fund, or a share of the proceeds from the sale of those emission reductions. Other potential fund structures for emission reduction projects could be broader than a pure 'carbon fund'. For example, some companies have looked at developing a sustainable forestry fund whereby investors are entitled to both the carbon sequestration rights from the timber and also the proceeds from the timber itself each time it is harvested.[15]

6 Managing Risk through Carbon Contracts

All projects entail risks, which must be managed through contractual arrangements. The constantly evolving regulations and political developments

[15] For example, the Pacific Forest Trust's Forests Forever Fund as described at <http://www.pacificforest.org/services/forever.html>, accessed 21 May 2004.

surrounding the Kyoto Protocol increases the risk profile for undertaking emission reduction projects in comparison to other types of projects (such as general construction or infrastructure projects). In addition, emission reduction projects will also face risks that are common to any project in developing or industrialized countries. Such risks include:

(i) the risk that an event beyond the parties' control (for example, a *force majeure* event) may impact the ability of the project to operate and produce Emission Reductions;
(ii) the risk that the market for the other products created by the project (that is, electricity) may rise or fall so that the project is no longer financially profitable; and
(iii) the risk of cost overrun from the project, such as due to delayed construction.

These types of risks should be mitigated through efficient management of the project and contractual clauses between the different parties to the project, including the project developer, the construction contractor, the project operator and the ER purchaser.

In the current developing Kyoto Protocol framework, CERs and ERUs are contractual rights based on an intangible reduction in GHGs, created under international law in a regime, which has not yet entered into force. There are therefore additional risks associated with the transaction of CERs or ERUs, including:

(i) the risk that the parties may not be able to establish legal title to the Emission Reductions on which the CERs/ERUs are based;
(ii) the risk that GHGs may not be properly monitored and accounted for so that the project will not generate as many CERs/ERUs as expected;
(iii) the risk that the Kyoto Protocol will not enter into force and CERs/ERUs cannot ultimately be created as envisioned by the Marrakesh Accords; and
(iv) the risk that the CDM or JI Project may face community or NGO opposition due to issues such as the development of a baseline or the nature of the project participants.

These risks are generally specific to the sale of Kyoto Protocol credits, rather than other aspects of the project, and should be dealt with through appropriate provisions in the contract to sell CERs or ERUs from the project.

As with any general commercial transaction, the greater the risk involved in purchasing ERs, the lower the price that a purchaser will be willing to pay and the stricter the conditions it will endeavour to place on the seller.

Generally, the parties will allocate the risks discussed above on the basis of which party has control over the occurrence of that particular risk. If neither party has control over a risk (such as the entry into force of the Kyoto Protocol), the assumption of this risk may be reflected in the price paid for the ERs. Alternatively, the parties may provide some flexibility in the contract in case the risk does occur (that is, the Kyoto Protocol does enter into force) and allow for the termination of the contract in that event or the acquisition of verified contractual ERs in place of CERs or ERUs.

Risk allocation will be dealt with through contracts between the various participants, financiers and purchasers of ERs through measures such as:

(i) conditions precedent to the entry into force of a contract (for example, a condition precedent to the effect that the contract will become operational upon the entry into force of the Kyoto Protocol);

(ii) guarantees from parent companies or governments involved as to the financial stability of the project developer;

(iii) force majeure clauses covering risk contingencies such as natural disasters or change in law or policy making the project unfeasible or leading to the seller failing to deliver the contracted amount of ERs;

(iv) laying off particular risks to other parties such as contractors or the verifier for the project;

(v) adequate warranties and indemnities in contracts;

(vi) procuring insurance for various project and credit risks; and

(vii) providing rights of termination in certain events.

The types of clauses appropriate will depend on the nature of the project, the type of ERs being transacted and the level of involvement of the ER purchaser in the underlying project (that is, whether it will be taking equity in the project or merely purchasing ERs). It is good practice to draft a risk allocation matrix at the outset of a project and prior to negotiation of term sheets for the sale and purchase of ERs.

7 Conclusion

Emission Reductions, including the various statutory manifestations that exist under national and international law in recognition of GHG abatement activities, are a new commodity. This commodity, and the demand for it, is the result of regulatory structures adopted by lawmakers in an attempt to harness market forces to drive lowest cost abatement of GHG emissions. Notwithstanding the statutory or contractual basis of ERs, they are ultimately commodities, and many existing contractual structures and risk mitigation techniques can be drawn upon in order for parties to agree and articulate commercial terms for their sale and purchase through 'contracting for carbon'. Having said this, the origin of ERs does require particular attention when defining the terms of their creation by emission reduction projects and sale and transfer between parties. Special consideration to risks peculiar to ER transactions should be at the forefront of parties' minds when negotiating and setting contractual terms.

17

CERUPT and ERUPT Contracts

*Daniel van der Weerd**

1 Introduction

This chapter describes how the Netherlands purchases carbon credits through Emission Reduction Unit Procurement Tenders (ERUPT) and Certified Emission Reduction Unit Procurement Tenders (CERUPT) and the contractual issues that arise. To provide a better understanding of why issues are addressed in a certain way, section 2 will start with an explanation of the background of the Netherlands' involvement in the carbon market and will set out the regulatory framework in which ERUPT and CERUPT contracts are located. Section 3 will describe how ERUPT and CERUPT contracts address the main obligations of the Project Company, such as delivery and reporting, and non-performance. Section 4 will set forth some conclusions.

2 Background and Regulatory Framework

2.1 Background

At the UN Conference on Environment and Development in Rio de Janeiro in 1992, the UN Framework Convention on Climate Change (UNFCCC) was created to stem the increase of the greenhouse effect by reducing the emissions of greenhouse gas (GHG). Greenhouse gas emission reduction targets were set for industrialized countries under the Kyoto Protocol adopted at the third session of the Conference of the Parties to the UNFCCC in December 1997. Accordingly, the Netherlands has to achieve a 6 per cent reduction in GHG emission levels over the first commitment period (from 2008 to 2012) relative to the emission levels in

* Daniel van der Weerd is legal counsel for Senter's Carboncredits.nl team. Senter is an agency of the Netherlands' Ministry of Economic Affairs. This chapter is not a policy paper and solely expresses the personal views of the author and his interpretations of the legal aspects of ERUPT and CERUPT contracts. The author would like to thank Messrs Ray Wijnands, Reginald Hernaus, and Adriaan Korthuis for their comments on this chapter.

the reference year 1990. This reduction is the Netherlands' contribution to the realization of the European Union's commitment under the Kyoto Protocol.

The Netherlands seeks to achieve a significant part of its reduction target through domestically implemented measures. The remaining part is to be realized through the flexible Kyoto mechanisms such as Joint Implementation (JI) and the Clean Development Mechanism (CDM). Within the Netherlands, the Ministry of Economic Affairs is responsible for the implementation of JI and the Ministry of Housing, Spatial Planning and the Environment for CDM. The Netherlands has allocated substantial amounts for purchasing JI and CDM credits. The initial overall budget was almost €1 billion.[1] The Netherlands also wishes to support sustainable development in host countries when purchasing carbon credits.

The Netherlands intends to meet its target under the Kyoto Protocol. Therefore it may be clear that, in any case initially, the Netherlands will be a buyer in the carbon market and not a trader. The Netherlands is first and foremost interested in actually receiving the credits it purchases and which it needs in order to achieve compliance with its target under the Kyoto Protocol. This has resulted in contracts that focus on incentives to deliver, deterrents against breaching the contract, and remedies that should provide the Netherlands with sufficient money to purchase alternative credits should projects not perform as expected.

2.2 ERUPT and CERUPT

The Ministry of Economic Affairs and the Ministry of Housing, Spatial Planning and the Environment have contracted Senter, a government agency, to purchase carbon credits through public procurement procedures. The procedure through which Emission Reduction Units (ERUs) are purchased is called ERUPT. To date, four calls for tenders have been published for ERUPT. The first call for tenders was published on 15 May 2000 and closed on 17 July 2000, the second was published on 1 December 2001 and closed on 4 March 2001, the third was published on 24 October 2002 and closed on 30 January 2003, and the last call for tenders was published on 22 October 2003 and closed on 15 January 2004. The procedure through which CERs are purchased is called CERUPT. So far one call for tenders was published on 1 November 2001, which closed on 31 January 2002.[2]

ERUPT and CERUPT contracts are by no means the only way in which the Netherlands purchases Emission Reductions. The Netherlands participates in the Prototype Carbon Fund (PCF) and has signed agreements with several financial institutions. The Ministry of Economic Affairs and the European Bank for Reconstruction and Development (EBRD) have recently set up a Carbon Fund.[3]

[1] Budget information is accessible at <http://www.ez.nl> and <http://www.minvrom.nl>.

[2] The Terms of Reference for these procedures are accessible at <http://www.carboncredits.nl>.

[3] The agreement between the Ministry of Economic Affairs and EBRD was signed in The Hague on 27 October 2003 and is available at <http://www.ebrd.com/country/sector/energyef/index. htm>.

The Ministry of Housing, Spatial Planning and the Environment has signed agreements with the Rabobank, the International Finance Corporation, the International Bank for Reconstruction and Development, and the Corporación Andina de Fomento.[4]

2.3 Procurement Procedures

ERUPT and CERUPT are procurement procedures through which the Netherlands purchases carbon credits. These procedures are set up and executed in accordance with the European Directives on the procurement of goods and services,[5] even though neither directive seems to apply to carbon credits. It is evident that at the time of adopting the directives and publishing the United Nations Common Procurement Classifications[6] and the Common Procurement Vocabulary (CPV),[7] carbon credits did not exist. The closest type of transaction to resemble contracts over the generation and transfer of a carbon credit would probably be to contracts for the issue, sale, purchase, or transfer of securities and other financial instruments. These services are not included within the scope of Directive 92/50 EEC.[8]

The European Council adopted several directives on the award of public contracts. These directives aim to increase the transparency of the market for public contracts by introducing uniform procedures for awarding public contracts. The directives also aim to create equality amongst suppliers by listing criteria a contracting authority may use for selecting suppliers and for awarding contracts. It is prohibited to discriminate amongst suppliers and the terms of reference for a procurement procedure shall not contain any disproportional requirements.[9]

Since the directives aim to create and increase transparency and equality amongst suppliers, the opportunities to change requirements after the closing date

[4] See the CDM homepage of the Ministry of Housing, Spatial Planning and the Environment, <http://www.cdminfo.nl>.

[5] Council Directive 92/50/EEC of 18 June 1992 relating to the co-ordination of procedures for the award of public service contracts [1992] OJ L209/1; and Council Directive 93/36/EEC of 14 June 1993 co-ordinating procedures for the award of public supply contracts [1993] OJ L199/1.

[6] The CPC can be accessed at <http://simap.eu.int/EN/pub/docs/CPV-standards/CPC.doc>.

[7] The CPV was created in 1996 as a tool for improving transparency and efficiency in the field of public procurement. Use of standard terms in the CPV makes it easier for potential suppliers to identify the procurement contracts in which they are interested. The CPV also facilitates fast and accurate translation of contract notices for publication in the EC *Official Journal*, and makes it easier to establish procurement statistics.

[8] See the preambles of Council Directive 92/50/EEC of 18 June 1992 relating to the co-ordination of procedures for the award of public service contracts [1992] OJ L 209/1.

[9] Requirements that discriminate against project companies that could equally well perform the demanded service or deliver the requested goods are disproportional, EH Pijnacker Hordijk and GW Van der Bend, *Aanbestedingsrecht: Handboek van het Europese en Nederlandse Aanbestedingsrecht* (2nd edn, Den Haag, Sdu Uitgevers, 1999).

of a call for tenders or to negotiate on the contract are very limited. Only the clarification of the terms of reference or the proposal are permitted.[10]

2.4 Civil Law

Three rounds of ERUPT and one round of CERUPT have shown that the terms of reference are not as clear on all issues as one would have liked or expected these to be. This is especially so for those project companies that are based in common law countries and are unfamiliar with civil law concepts. This realization has led to several changes and clarifications in both the terms of reference and the contracts themselves.

The Netherlands is a civil law country. When the Netherlands was conquered by Napoleon in the eighteenth century a Civil Code based on Roman law was introduced. Although the Civil Code has undergone quite some changes since then, the basic characteristics of the Dutch Civil Code resembles those of the Civil Codes of other civil law countries.[11]

The Civil Code sets out rules for contracts.[12] It describes, *inter alia*, when a contract is concluded. Pursuant to the Civil Code, a tender submitted under a procurement procedure constitutes an irrevocable offer. The expression of interest that is submitted in the first phase of a restricted procedure, the selection phase, is a revocable offer. The tender that is submitted by suppliers that are invited to participate in the second phase, the contract awarding phase, is an irrevocable offer. Effectively this means that no contract would be needed under ERUPT and CERUPT and that an explicit acceptance of an offer, the tender, would create a contractual relationship.[13] However, since not all details are addressed in the terms of reference and some issues may need clarification, the Netherlands does sign contracts with project companies.

The Dutch Civil Code contains mandatory rules and optional rules. Deviations from the mandatory rules are void. Deviations from the optional rules are, however, permitted. The optional rules will apply to contracts that do not specifically address the issues these optional rules address. Therefore it will not be necessary to address issues such as default and notices in a contract that is governed by Dutch law, unless it is desired to arrange these in a specific way that differs from the way the optional rules address these issues.

[10] Please note that only a very brief outline of some of the characteristics of the European procurement directives is provided in this section. This section in no way pretends to fully reflect all the rules, exceptions and nuances of EU procurement law.

[11] To get a better understanding of the difference between civil and common law systems: K Zweigert and H Kötz, *An Introduction to Comparative Law* (3rd edn, Oxford, Clarendon Press, 1998).

[12] The Raamwet EEG-voorschriften aanbestedingen (Law dated 31 March 1993, Stb. 212) obliges the use of the provisions of the European Directives as if the provisions of these directives were directly applicable. The texts of all directives in the EU languages can be obtained at <http://simap.eu.int>.

[13] Book 6 of the Dutch Civil Code contains the core of Dutch Contract Law. Under Dutch law, an offer and an acceptance can be oral instead of in writing and can create a contract.

2.5 Code of Conduct

Senter is an agency of the Ministry of Economic Affairs and as such part of the legal entity the State of the Netherlands. Netherlands administrative law sets standards for conduct of governmental institutions and administrative bodies. It limits the use of administrative powers and imposes certain obligations on administrative bodies. An administrative body should, *inter alia*, take due care in its operations and in making use of its powers, balance interests when making decisions, motivate its decisions, treat everyone equal in equal situations and honour acting upon reliance in good faith. Moreover, it is prohibited to use powers for purposes other than those for which these powers have been granted. These standards apply mainly in administrative law, in relations between the State and its citizens as well as anyone under Dutch jurisdiction. They do not apply amongst citizens in the same way. The Dutch Civil Code prescribes that any authority or power one derives from civil law may not be exercised contrary to rules of public law.[14] Therefore the standards for conduct that are set under administrative law will, to a certain extent, apply to the Netherlands in its activities under and pursuant to ERUPT and CERUPT. Whether, and to what extent, not living up to such a standard for conduct might impede the rights of the Netherlands to claim penalty payments or effect cancellation rights is hard to assess. It may be noted however that the Netherlands will, whether or not this would be fully enforceable, always aim to be a reliable contracting partner that takes due care in all its operations regarding a contract.

2.6 Reasonableness[15]

The Dutch Civil Code contains a doctrine on reasonableness and fairness. Contracting parties are obliged to behave in a reasonable manner towards each other.[16] As distinct from common law systems, a contract does not just have the legal consequences that are stated in the contract itself by the contracting parties, but also consequences resulting from legislation, custom or reasonableness. A provision in a contract may be non-applicable to the extent this would be unacceptable pursuant to standards of reasonableness and fairness.[17] Furthermore a judge can mitigate penalties and take reasonableness into account when setting amounts for damages.[18]

The remedies available to the Netherlands under ERUPT and CERUPT contracts have been criticized as harsh or even excessive by some potential suppliers. When exercising remedies, the Netherlands will however always assess to what

[14] Art 3(14) of the Dutch Civil Code. J Hijma and MM Olthof, *Compendium van het Nederlands vermogensrecht* (7th edn, Deventer, Kluwer, 1999).

[15] The full text for the Dutch Civil Code can be obtained at <http://www.overheid.nl> (wet en Regelgeving).

[16] Art 6: 2 of the Dutch Civil Code.

[17] Art 6: 248 of the Dutch Civil Code.

[18] Art 6: 94 and 6: 97 of the Dutch Civil Code.

extent it is reasonable to exercise a remedy, since it is required to do so by law. As stated above, the Netherlands and the project company are obliged to behave in a reasonable manner towards each other.

2.7 Activities Implemented Jointly

As a result of the approach the Netherlands' Government has taken towards Joint Implementation, the Netherlands was active in Activities Implemented Jointly (AIJ) at an early stage. The Joint Implementation Registration Centre was established in 1997. The Programme for Co-operation with Countries in Central and Eastern Europe (Programma Samenwerking Oost-Europa—PSO) has also supported projects that were aimed at reducing emissions. The general terms and conditions of the PSO have been the basis for the general terms and conditions of ERUPT and CERUPT. The PSO general terms and conditions focused on projects and contained several termination rights for the Netherlands, in order to allow the Netherlands to maintain some level of control and/or certainty about the performance and/or feasibility of a project. The terms of reference of ERUPT and CERUPT still contain provisions derived from PSO.

2.8 Early Players[19]

The Netherlands was an early player in the market for carbon credits. The first ERUPT call for tenders was published before the Conference of the Parties in The Hague (COP6). A call for tenders under CERUPT was published very shortly after COP7. It is therefore no surprise that the terms of reference for these calls for tenders are not as detailed as if they had been drafted at the end of 2003. Developments, such as the EU Emissions Trading Scheme, have not been taken into account in most calls for tenders. The terms of reference do however state that the contract and the general terms and conditions are subject to changes or supplementation by any international agreements between the parties listed in the Kyoto Protocol.

Some of the provisions in the ERUPT and CERUPT contracts may seem unbalanced and more favourable to the Netherlands. It should be noted however that, especially when drafting these first contracts, the Netherlands was, and still is, taking on most of the 'Kyoto risks'. If Kyoto does not enter into force, the Netherlands will pay for verified emission reductions. As a counterweight for taking on this biggest of all risks, the Netherlands has introduced some obligations for the project company, the most notable of which is the surplus provision.

[19] The Prototype Carbon Fund (PCF) and the International Emissions Trading Association (IETA) (and its members) should be mentioned as organizations that are playing a valuable role in developing the carbon market and carbon contracts; several of their ideas have found their way into ERUPT contracts.

This is a provision which obliges a project company to offer a right of first refusal to the Netherlands with respect to any carbon credits it generates in excess of the contracted volume.

3 Contracts

3.1 Introduction

The first ERUPT contract was a two page document. In reality, the contract is a lot more extensive than two pages. The terms of reference are incorporated by reference into the contract, as is the tender that is submitted by the project company. The contract also refers to the Letters of Approval (LoA).[20] The terms of reference contain the general terms and conditions pertaining to the contract. Incorporating all these documents into the contract makes it possible to limit the contract to a naming of the parties, price and volume. In practice provisions have been added that clarify the terms of reference, or that deal with lacunae in the terms of reference, or that accommodate changes in the requirements under the Kyoto Protocol.

The LoA performs an important role in ERUPT and CERUPT contracts. Apart from arranging the obvious, that is, approval of the project as a JI project by the host country, it deals with several other issues as well. The LoA also serves as the authorization of the Project Company to generate ERUs[21] and confirms the host country's commitment to the Kyoto Protocol and the requirements thereunder. Furthermore it states the total number or percentage of ERUs that will be transferred, it deals with taxation of ERUs,[22] transfer of AAUs if any,[23] transfer of the project, opens up Track One deliveries and arranges what happens if the Kyoto Protocol does not enter into force.

3.2 Delivery

The project company takes on several obligations under an ERUPT or CERUPT contract, the most important of which is delivery of the carbon credits. The Netherlands has chosen to adopt the Kyoto Protocol and Marrakesh Accords definitions for defining ERUs, AAUs, and CERs. By defining the deliverable as a Kyoto defined credit, any regulatory changes in the procedures through which such credit can be generated should be verified and will be issued and transferred, and are automatically part of the activities a project company should perform.

[20] At first the ERUPT contract was drafted in such a way that it also served as the Letter of Approval from the Netherlands. Currently separate Declarations of Approval are issued.

[21] See Art 6(4) Kyoto Protocol.

[22] Only in ERUPT 3 and 4. See sections 3.2.1 and 3.2.2 respectively.

[23] Not in ERUPT 1. See section 3.2.1.

The ultimate obligation of the project company is to deliver the Kyoto credit or, with respect to ERUs, to deliver such documents as will enable the Netherlands to successfully claim the transfer of ERUs to it by the host country. Deliveries are due on the dates indicated in the delivery schedule that are annexed to the contract. Delivery of carbon credits should be made on 1 April of the year following the year in which carbon credits have been generated.[24] Credits have to be delivered free of any levies or taxes.

3.2.1 ERUPT 1–3

Under ERUPT 1–3 the Netherlands will pay for delivered claims on ERUs.[25] Delivery of claims on ERUs occurs once the Netherlands receives verification reports that verify the generation of emission reductions and satisfy the standards of either Track One JI or Track Two JI.[26] The LoA, together with these verification reports will allow the Netherlands to claim the transfer of ERUs from the host country. Once emission reductions have been verified in accordance with the applicable standards and such verification has become final, the Project Company has fulfilled ITS obligations with respect to the delivery of claims on ERUs upon delivery of the verification report to the Netherlands.

Under Track Two JI, delivery of claims on ERUs occurs if a verification report is delivered that has become final in accordance with the verification procedure established under the Article 6 Supervisory Committee.[27] If the host country is eligible to verify, issue, and transfer ERUs under Track One JI[28] the Project Company will procure verification of emission reductions in a manner agreeable to the host country, but in a way comparable to that under Track Two JI.

The basis for Track One deliveries is not explicitly mentioned in the contract, but is mentioned in the LoA. Cooperation of the host country is pivotal for Track One deliveries; therefore reference to Track One procedures forms part of the standard format of an LoA that the Netherlands requires as part of the documentation that has to be submitted in the procurement procedure. Track One deliveries should create less transaction costs and less uncertainty for project companies and should therefore be a more attractive mode of delivery. It is however not within the scope of control of a Project Company to choose a mode of delivery.

[24] Except for AAUs which cannot be delivered before 1 April 2008.

[25] In ERUPT 4 a claim on an ERU has been defined as: 'means the documentary evidence and any other documents as required under and/or pursuant to the Kyoto Rules and/or as requested by the Host Country with respect to the generation and verification of an ERU for the effecting of the issuance and transfer of an ERU by the Host Country, such documentary evidence to be provided in the form of a verification report that meets the applicable requirements . . . .'

[26] For further detail, see ch 8 above.

[27] Decision 16/CP.7, Draft Decision -/CMP.1 (Article 6) Guidelines for the implementation of the Kyoto Protocol, Annex D, Participation Requirements, Art 24 and Annex E, Verification procedure under the Article 6 Supervisory Committee.

[28] ibid, Annex D, Participation Requirements, Arts 21 to 23 (inclusive).

Therefore the standard mode of ER delivery is the delivery in accordance with Track Two procedures.

As is clear from the above, any difficulties in obtaining the actual transfer of ERUs from the host country to the account of the Netherlands, is not a risk for the project company. The Netherlands has attempted to deal with this risk in the LoA. The standard format of an LoA clearly states, *inter alia*, that the host country authorizes the project company to generate ERUs and that it will accept the transfer of ERUs to the Netherlands.

Should the Kyoto Protocol not enter into force, the project company will still receive payments for the emission reductions that are generated by the project. The project company will procure the verification of the emission reductions by an independent entity. Delivery of those verification reports shall constitute delivery. The LoA states that, should the Kyoto Protocol not enter into force, the host country and the Netherlands shall consider the transfer of ERUs as a transfer of GHG emission reductions on a bilateral basis.[29]

The Netherlands does not assume any Kyoto-related risks with respect to so-called early credits[30] or claims on AAUs. The project company may offer the transfer of AAUs with respect to emission reductions generated by the project before 1 January 2008. In order to offer these AAUs, the LoA from the host country has to state that it will transfer AAUs under the Emission Trading mechanism of Article 17 of the Kyoto Protocol for such emission reductions. Delivery of these early credits occurs at the time the AAUs have been transferred to the account of the Netherlands. If a transfer of AAUs to the account of the Netherlands cannot be made, for whatever reason, the project company shall not receive payments for the verified emission reductions that have been generated by the project prior to 2008.

3.2.2 ERUPT 4

Under ERUPT 4, the developments with respect to the EU Emissions Trading Scheme (EU ETS) have been taken into account. This has resulted in a different definition of the deliverable. The deliverable is an emission reduction, which could be an ERU, an AAU, an EU Allowance, or a verified emission reduction. The terms of reference describe which specific type of emission reduction is the deliverable in a certain situation or period. Delivery of claims on ERUs and early credits still occurs in the same way as described above. Should the EU ETS however interfere with the project's ability to generate ERUs in such a way that the project, in whole or in part, becomes an obligated installation[31] under the EU ETS and receives an allocation of allowances, then EU allowances will become the deliverable instead of the claim on an ERU. This applies whether the Kyoto Protocol enters into force

[29] This was introduced in ERUPT 2; ERUPT 1 does not mention this.

[30] ERUPT 1 did not offer the possibility to offer early credits.

[31] An installation which falls within the scope of the EU ETS the operator of which requires a GHG Permit to continue operating.

or not. Delivery of EU Allowances occurs upon the transfer of such EU allowance to the account of the Netherlands. If the Kyoto Protocol does not enter into force and the project does not become an obligated installation, verified emission reductions will be delivered as described above in section 3.2.1.

3.2.3 CERUPT

Under CERUPT, delivery occurs upon transfer of CERs into the National Registry of the Netherlands or the Netherlands' temporary account at the CDM Registry. The project company is not obliged to deliver verification reports every year. To reduce the costs for the project company, verification and actual transfer of CERs will occur once every two years. In the intervening years, provisional payments will be made based on the number of emission reductions that are documented in the monitoring report. If the Kyoto Protocol does not enter into force, the Netherlands will pay for verified emission reductions.

3.2.4 Surplus

In consideration of the Netherlands' obligation to purchase and pay for the contracted carbon credits, and assuming the risk that the Kyoto Protocol does not enter into force, the project company is obliged to offer the Netherlands the first right of refusal, or an option right, with respect to any surplus carbon credits. Surplus carbon credits are those carbon credits that have been generated in excess of the contracted number of carbon credits.[32] Surplus carbon credits must be offered at the market price and the Netherlands shall notify the project company whether or not it accepts such carbon credits within two months of the date of the offer. Accordingly, it is a call option at the market price. The project company would not receive less money than in the situation where it could sell these credits to any third party.

3.2.5 First and following carbon credits

The project company is obliged to deliver the first and following carbon credits it generates to the Netherlands. The terms of reference of ERUPT and CERUPT have not stated this literally, but the surplus provision prevents the sale of carbon credits to third parties. In reality there are no opportunities not to deliver the first and following credits to the Netherlands or at least no benefits could be obtained from doing so. Theoretically it would be allowed to hold on to any credits that have been generated and which are not yet due for delivery. This would however be disadvantageous for the project company, since it would not receive any payments for these credits until

[32] ERUPT 3 stipulated that surplus is generated if more carbon credits are generated than originally estimated in the PDD, thus making it possible for the project company to sell part of the credits to a third party.

these are delivered. The project company will be obliged to deliver these credits to the Netherlands at a later stage or offer these as surplus credits to the Netherlands at a later stage, whichever is applicable. In ERUPT 4, it has been made explicit that the first and following credits that are generated have to be delivered.

3.2.6 Provisional arrangements

Under both ERUPT and CERUPT, provisional payments may be made on the basis of verification reports for as long as the Kyoto Protocol has not entered into force or for as long as the relevant Kyoto institutions are not fully operational. These payments do not become final until the verification report with respect to the verified ERUs has become final or until CERs have been transferred to the account of the Netherlands. Payments will correspond with the number of ERUs that are generated or the number of CERs that are transferred. Should that number be lower than the number of emission reductions generated according to the verification report on the basis of which provisional payments were made, the difference will either be deducted from future payments or will have to be repaid.

In addition, the terms of reference stipulate that, should the number of ERUs calculated under the guidelines as determined by the Supervisory Committee be less than 50 per cent of the number calculated under the guidelines that are part of the terms of reference, the contract may be cancelled. Should the number thus calculated be between 50 per cent and 100 per cent of the number as calculated by using the guidelines that are part of the terms of reference, payments will be based on the guidelines that are part of the terms of reference. The same applies to CERs.

3.3 Payments

Payments are made upon delivery of carbon credits. However both ERUPT and CERUPT offer the possibility of advance payments up to 50 per cent[33] of the contract value. This possibility has been introduced since most projects shall have a greater need for funding in the initial years than in later years. Advance payments shall be deducted from payments that are made upon delivery of verification reports. If 50 per cent of the contract value has been disbursed as advance payments, the Netherlands shall pay 50 per cent of the value of each delivery. The final contract value shall correspond to the amount of credits that has been delivered.

3.4 Project Operation

The project company is obliged to operate the project in accordance with the proposal it submitted under the procurement procedure. This proposal contains a Project Design Document (PDD) and a business plan, which describes the

[33] ERUPT 1 allows even up to 80 per cent. Under CERUPT no requests for advance payments have been made in the tenders submitted by the suppliers.

operation of the project in detail. This obligation limits the project company in its freedom to operate the project. It does not however prohibit any changes to the project, since approval may be obtained from the Netherlands for any proposed material changes to the project. The Netherlands wishes to maintain some level of control over proposed changes, since these changes might affect the ability of the project to generate emission reductions and/or might necessitate changes to the baseline or monitoring protocol.

3.5 OECD Guidelines for Multinational Enterprises

The Netherlands requires that project companies participating in ERUPT and CERUPT adhere to the OECD Guidelines for Multinational Enterprises. These Guidelines address several issues, such as employment and industrial relations, child labour, the environment, and combating bribery. If project companies refuse to declare their adherence to these Guidelines, they will be excluded from participation. Non-adherence to these Guidelines is also grounds for terminating the contract.

3.6 Reporting

The project company has several reporting obligations. It should report on the emission reductions it generated, by submitting monitoring reports to the Netherlands. Furthermore, it should submit quarterly reports to the Netherlands with information on progress, planned activities, proposed changes, and monitored emission reductions.

In addition to these periodical reports, the project company should report if anything occurs that might hinder the correct and timely delivery of carbon credits. Several contracts contain an extended version of this obligation. The rationale behind this incident reporting obligation is that both the Netherlands and the project company are fully aware of the problems an incident is creating or that it might create and the possible effect this might have on the delivery of carbon credits. Upon receiving such a report the Netherlands will request the project company to describe the possible effects and what activities the project company will undertake to prevent or minimize these effects and how it envisages to make up for any shortfalls in delivery.

The reporting obligation is an important obligation for several reasons. First, the Netherlands shall make no payments if reporting obligations are not fulfilled. Secondly, the Netherlands has a right to cancel the contract if, upon receipt of an incident report, it considers that the occurrence of the incident will have detrimental effects on the ability of the project to timely generate the required amount of emission reductions. Finally, upon receipt of an incident report, the Netherlands *may* make exceptions to any of the obligations of the project company if this would seem reasonable and not harm the interests of the Netherlands. The Netherlands could grant a grace period in which defaults should be cured, suggest or agree to (temporary) changes to obligations of the project company and could (temporarily) waive any of its rights, such as the right to receive penalty payments.

3.7 Non-performance

If, for any reason other than *force majeure*, a project company is unable to fulfil its obligations, it is in default.[34] Accordingly, only in circumstances of civil war, natural disasters, and nationalization of the project, is non-fulfilment of obligations excusable. Under Dutch law, any default *could* lead to termination of the contract, if such default would not be cured in an appropriate and timely manner. The contract lists several specific termination events in addition to this one.[35]

The contract contains a penalty provision that has given rise to a lot of discussion. If a delivery is not made on or before the date mentioned in the delivery schedule or the delivery falls short of the cumulative total number of credits that is due on the date stated in the delivery schedule, a penalty applies. In ERUPT 1 and 2 the penalty is 2.5 per cent of the total contract value every month that claims on ERUs which are due for delivery have not been delivered. This penalty is not linked to the number of claims on ERUs by which the delivery falls short. One claim on an ERU that falls short could trigger a penalty. The penalty is maximized on the total contract value. In ERUPT 3, the penalty is linked to the number of claims on ERUs by which a delivery falls short. A penalty of five times the price of a credit is payable for each non-delivered credit. If such credit is delivered within one year of the credit becoming due for delivery, 60 per cent of that penalty is reimbursed. In ERUPT 4, the penalty for a non-delivered credit is 120 per cent of the market price of such credit. In CERUPT, the same penalty as in ERUPT 1 and 2 applies, but there is a threshold. If 70 per cent or more of the credits that are due for delivery are delivered, the penalty is not imposed.

The penalty serves several purposes. It should trigger the project company to deliver carbon credits on time and to refrain from selling generated credits to others. Furthermore, the penalty could provide the Netherlands with money to buy alternative credits.

The possibility to deliver suitable alternative credits has been introduced in ERUPT and CERUPT contracts. It might be unreasonable if the Netherlands was to insist on receiving penalty payments in a situation in which the project company cannot deliver sufficient credits from the project, but offers to make up for that shortfall by delivering alternative credits. These alternative credits would have to meet certain standards in order to be acceptable to the Netherlands. Alternative credits should be credits that the Netherlands may use towards its Kyoto target. In addition, the alternative credits should be credits that arise from projects that fulfil the requirements for projects as mentioned in the terms of reference.[36] If the project company delivers

[34] Article 6: 74 and 6: 75 of the Dutch Civil Code.

[35] For example, in case of a bankruptcy of the project company.

[36] Under CERUPT alternative credits should arise from projects that use the same or a higher-rated technology than the technology the project itself is using, since the Netherlands used different price caps for different technologies.

suitable alternative credits to make up for a shortfall, no penalty shall be imposed and the suitable alternative credits shall take the place of the non-delivered contracted credits.

In addition to the penalty, damages could be claimed. This would in any case apply if the penalty payments that are received by the Netherlands with respect to non-delivered credits are insufficient to cover the damages caused to the Netherlands by the non-delivery of such credits. The Dutch Civil Code allows for claiming damages in addition to receiving penalty payments, but as stated in sections 2.4 and 2.6 above, reasonableness governs contracts, therefore it is extremely unlikely that excessive claims would ever be made or could ever be successful if contested in court. The paragraph on reporting, described in section 3.6 above, already mentions that the granting of grace periods and other exceptions may be made to the obligations of the project company.

3.8 Governing Law and Dispute Resolution

The ERUPT and CERUPT contracts are governed by Dutch law. Dispute resolution shall take place in The Hague before the District Court. Arbitration is a possible means of dispute resolution, if agreed to by both parties, and shall be under Dutch law.

3.9 Effectiveness and Assignment

The contract enters into force upon signature. So far all contracts have contained additional obligations for the project company with respect to outstanding issues. Should these issues not be addressed appropriately within the time period given, the Netherlands shall have the right to terminate the contract. The terms of reference provide examples of items or issues that could become additional obligations for the project company.

ERUPT and CERUPT contracts cannot be assigned without the prior written agreement of the Netherlands. As explained before, the Netherlands wants to maintain a certain level of comfort about the certainty of delivery of the contracted carbon credits. In the procurement procedure the financial and economic standing of suppliers is assessed, as well as their technical capacities. The Netherlands only selects those suppliers that meet a minimum standard on these criteria. It may be clear that proposed assignees will have to meet certain standards as well, before approval for assignment may be granted.

3.10 Confidentiality

The Netherlands is entitled to use and publish any data, reports, and publications, including tenders whether approved or rejected, compiled for or in the context of a Project, with the exception of confidential company information, and presented under the terms of the contract, without payment of any royalties provided the source is acknowledged. Parties can designate information as confidential.

4 Conclusion

The Netherlands is a reliable purchaser on the market for carbon credits and is committed to achieving its Kyoto target. Through the possibility of advance payments and the fact that payments do not depend upon the Kyoto Protocol entering into force, the Netherlands is offering some advantages. Some of the provisions in ERUPT and CERUPT are less advantageous for sellers of carbon credits. It should be borne in mind though that the Netherlands is bound by a code of conduct[37] in its actions and that reasonableness plays an important role in contracts that are governed by Dutch law.[38] By looking only at the contract, penalties and damages could theoretically turn out to take on draconian proportions. In reality, the Netherlands will not exercise any of its rights contrary to Dutch civil or administrative law and a reasonable exercising of rights is to be expected.

The contracts focus on delivery and contain obligations for the project company that should ensure that the Netherlands is confident of the full, correct, and timely delivery of contracted carbon credits. The Netherlands is focused on receiving credits that can be used towards its Kyoto target and seeks to limit the possibility that insufficient credits are delivered or can be bought.

The European Procurement Directives force the Netherlands to carefully assess the contractual package it wishes to present to potential suppliers, since changes and negotiations are only possible in a limited number of cases. Finding a balance between flexibility, clear and objective selection, and contract awarding criteria is a difficult exercise.

ERUPT contracts and terms of reference have been evolving over time to take into account the lessons learned from previous procurement procedures, Prototype Carbon Fund, International Emissions Trading Association, and others. Until such time that the Kyoto Protocol has entered into force, it remains a challenge to draft the perfect contract for purchasing Kyoto credits while maximizing the likelihood that something of value remains if the Kyoto Protocol does not enter into force, and at the same time maintaining balance in the contract.

[37] See section 2.5 above.

[38] See section 2.6 above.

18

Negotiating a JI Contract: A Project Developer's Perspective

*Anikó Pogány**

1 Introduction

The main purpose of an Emission Reductions Purchase Agreement (ERPA) is to lay out the obligations of the parties and to formulate the risk sharing agreed between the parties. This chapter's aim is to go through the key risk areas which sellers and purchasers of emission credits have to identify and to present how the subsequent risk allocation is reflected in an ERPA. Due to the author's own experience, the chapter follows mainly the viewpoint of the potential seller of emission rights.

The content and the analysis of this chapter is the result of the detailed study of several Joint Implementation (JI) projects as well as available transaction structures and purchase terms for those projects. The chapter could not capture all historic transactions, mainly due to the lack of publicly available information, and thus, its conclusions may not be applicable for all present and much less for future cases. Also, the author's empirical knowledge stems from the implementation of a renewable energy project as part of a combined heat-and-power plant, which represents only one of many possible JI project activities. Most of the conclusions of this chapter are applicable to JI project agreements in general, however, other project types and classes may also show important particularities.

2 The Basis for the Contractual Agreement: An Agreement on Risk Sharing

What are the concerns of a private sector company considering selling carbon credits from a project activity? Generally, before engaging in a project a careful assessment of the future cash flow and the risks that are embedded in the project is required: an

* Anikó Pogány is a senior adviser at the Hungarian Ministry of Environment. The views expressed in this chapter are the views of the author and do not necessarily represent the views of the Hungarian Ministry of Environment.

analysis of the risks and returns. In the case of carbon finance, due to the premature nature of the carbon market, the emphasis will always lie on the careful assessment of the risks of the project. Carbon credits are far from being a standardized commodity, instead where the parties rely or have to rely on the contractual definition of the carbon right, such right is an individually tailored product, newly created every time in every ERPA. Drafting a contract is a risk assessment, risk sharing and risk management game. Countless are the possibilities how a deal can go wrong, either internationally, at the macro level or at the project level. The agreement of risk sharing is finally reflected in contractual terms and conditions of the carbon contract.

3 The Long Path towards a Feeling of Comfort

Full understanding, if not full control, of the area in which a company engages and with respect to which it takes long-term commitments is a must in the successful conclusion of business. In general, decisions should be based on a feeling of comfort about the consequences of the decision. Today, the area of carbon finance for potential sellers is not prone to produce any feeling of comfort. Meeting the staff of institutional purchasers (government agencies or international organizations) with years of experience and comprehensive understanding can further raise the feeling of discomfort and of asymmetric knowledge between project participants.

Thus, the process of transaction structuring and contract drafting is an important internal learning process for project managers, a process of learning and teaching. Key decision makers in a company (for example, board members) have to understand and feel firm and comfortable about their decision when they confirm a decision to select a carbon purchaser, setting price limits, and approving key terms and conditions. Managers of many corporate areas (technical, financial, environmental, and public relations) also have to understand the basics of carbon finance and how it relates with their area of responsibility. Many considerations inherent to each carbon credit sale—setting the obligatory minimum annual amount of emission reductions to be sold, determining the range of flexibility within which delivery failure is not considered an event of default (if shortfall does not exceed that range buyer shall accept delivery), or setting the time frame within which a potential shortfall in the delivery of emission reductions should be recovered—require considerations of a mainly technical nature, which hugely impacts the overall risk profile and thus the overall value of the contract and thereby the value of the transaction.

4 Risks beyond the Control of the Project Developer

The first consideration, when structuring a carbon transaction, is to define the '*product*' which is to be sold and to define the transfer of the delivery risk from the seller to the buyer.

The preferred option for most carbon purchasers engaging in JI projects is to buy Emission Reduction Units (ERUs) as defined under Article 6 of the Kyoto Protocol and use them for compliance purposes. ERUs are issued by the host country to the extent verified emission reduction is generated by a project which was successfully approved and registered by the JI Supervisory Board (Track Two JI). The JI Supervisory Board will be formed after the Kyoto Protocol enters into force and will be responsible for the final approval of JI projects.

Other than ERUs, Emission Reductions (ERs), defined as rights to the greenhouse gas (GHG) emission reductions achieved by a project, are not defined by any regulation. ERs are rights which are only brought into existence by the implementation of a project in combination with a set of approved documents. Only if the baseline is solid, the Project Design Document (PDD) validated and the GHG emission reductions verified, can verified Emission Reductions be transferred and eventually be issued as an ERU. For the legal soundness of the whole transaction it is crucial that the documents supporting the creation of ERs are *final, unconditional, and unwithdrawable*. Unfortunately, final documents for JI projects cannot be produced before the Kyoto Protocol has entered into force and the JI Supervisory Board has been established. In the absence of final approval and registration of the project, underlying project documents are not final and are subject to possible change.

The fundamental question for the project participants is who bears the risk of this uncertainty, namely that the baseline changes or the methodology is considered flawed by the Accredited Independent Entity (AIE). When selling ERUs such risks stay with the seller. The key difference between ERUs and ERs is that ERUs are issued and transferred by the host country in the Kyoto Protocol's first commitment period after the Protocol has entered into force and the JI project has been successfully registered. ERs are created through successful verification of GHG reductions and are the basis for the issuance of ERUs. These transferable ERUs can be used for compliance under the Kyoto Protocol or the EU Emissions Trading Scheme (EU ETS).

A key question in the contracts negotiations, therefore, is which party takes the risk that ERs are not be converted into transferable ERUs, or the converted ERUs are not transferred.

This question covers:

- Kyoto Protocol ratification risk (ERUs are not generated due to absence of the instrument which creates the rights);
- Registration risk (ERUs cannot be transferred through failure of recognition of the project under the international rules);
- Country risk (ERUs are created but the country is unable to transfer them due to failure to comply with the JI eligibility criteria);
- Political risk (ERUs are created but the country refuses to transfer them, despite the commitment included in the Letter of Approval).

Most of these risks are already addressed in the definitions included in the ERPA. Selecting the legal nature of the product sold, selling ERs instead of ERUs means all the above risks stay with the purchaser. Additionally, it is up to the parties to carefully check that all other related terms and definitions in the ERPA are consistently used and reflect the agreed risk sharing.

Parties should also agree who would bear the risk of a potential change in the baseline study or in the monitoring and verification plan (which may be required by an accredited independent entity). The result of a revision of the baseline or PDD can be either that ERUs are not generated (in case the project is not any longer additional), or that the amount of ERUs generated may be negatively affected (in case the baseline case of the project is revised). Thus, the registration risk can affect the validity of the whole transaction as well as the amount of carbon credits sold (in case baseline is revised and the new baseline results in a lower amount of reductions).

5 Project Performance Risk

Generally, carbon sellers tend to be more willing to take project related than regulatory risks, because they have the means to control them. The two big areas are *project completion* risk, and *project performance* risk.

5.1 Project Completion Risk

Generally, the agreement between project developer and financier on the project completion and commissioning date is key, since with the completion of the project construction the overall risk of the project dramatically decreases and after that date the physical assets of the project provide an additional security in case of default.

In the case of carbon finance, the project completion date is essential for the carbon purchase as it helps to monitor the progress in the project and determines when (and if) the ERs are delivered. The carbon purchaser has an interest to terminate the agreement in case of delays, in order to limit the opportunity costs of the funds that are allocated to a specific project.

The project completion date is also relevant for a potential *upfront payment* since the project completion affects the *discount rate* the carbon purchaser applies on the upfront payment. According to the concept of JI, carbon finance is to bring additional capital to the project rather than just to strengthen the operational cash flow. For that reason, the project developer should aim to negotiate for a significant upfront payment of the total purchase price. When the parties decide on the date when the upfront payment is due (either on signature of the ERPA or on the date of the commissioning of the project) and also on a possible discount rate of the advance payment, they also assess the perceived risk of project completion. Also, since this creates an open position for the purchaser in the period of the highest risk, it may seek mitigation measures in the form of a security package (for example, bank guarantee, parent company guarantee) covering the amount of the advance payment.

5.2 Project Performance Risk

Unsatisfactory project performance results in lower achievement of ERs than anticipated. It is in most cases a production failure, a shortfall in the production of the project's primary product (for example, renewable electricity produced), which leads to the failure to transfer the agreed amount of ERs (delivery failure). Still, there are some cases when a delivery failure can happen without a failure in production. In the case of the generation of electricity from renewable electricity sources, the achievement of GHG emission reductions is often calculated against a baseline of electricity generated by fossil fuel combustion processes. In this case the baseline emissions can either be monitored and calculated by using standard emissions factors for an estimated generation of GHG emission reductions, or by calculating the off-set of emissions of the marginal electricity production at the time of the generation of electricity from the renewable source. In the second case, if the national electricity production changes towards unexpectedly carbon efficient technologies, it can result in a shortfall of generated ERs without a shortfall of power produced. A monitoring plan which relies on the actual calculation of the marginal off-set of electricity allows potentially higher emission reduction yields. However, such a monitoring method also creates a risk beyond the project developer's control.

The other performance risk type, which sellers may wish to share, is the regulatory risk in off-take prices. Renewable JI projects often operate in regulated market segments of the power sector where both the amount and the price of power are set in national legislation (for example, through preferential off-take tariffs). Since this green premium is the other leg of project finance beside the revenue from carbon credits sold, sellers are hugely exposed to any change of offtake prices or conditions and may wish to share that risk with carbon purchaser.

These specialties and all other factors (fuel supply problems, technical problems, and so on) which could result in a shortfall in the production of GHG emission reductions are reflected in the careful setting of the amount of ERs sold (per year and in total), and then are fine-tuned when defining the event of default and termination provisions.

5.3 Amount Sold

Usually the total amount of purchase is composed of two parts: an annual minimum amount which the project developer commits to deliver, and an additional amount which covers any excess amount generated over the minimum amount.

When setting the minimum amount, the project developer's key consideration is to maximize the minimum amount while seeking measures to limit the cases where a delivery failure can give rise to the termination of the agreement. A possible solution could be setting a flexibility range: if shortfall is within a certain percentage (5–15 per cent) of the minimum amount, the carbon purchaser agrees to accept the reduced amount and the shortfall in the transfer of ERs. Alternatively, seller and purchaser of ERs may agree on a period in which the seller can make up for the shortfall. The agreement on such a recovery period gives the project developer a grace period in which he can make up for the underdelivery without causing an event of default.

The availability of such options hugely affects the overall risk profile of the agreement, and can change the way termination causes, and provisions affect, the risk level of the whole agreement.

6 Additional Amount

Almost all agreements address the case when excess ERs are generated beyond the set and agreed minimum amount. Almost all purchasers wish to have granted a right of first refusal or option over these additional reductions. The seller may be asked to grant a call option to the purchaser over all additional ERs the project generated in a certain time frame. This purchaser generally does not wish to pay a premium for these additional ERs. This arrangement enables the purchaser to exercise the option at their own discretion and in view of the spot market price. Sellers may consider whether or not to give such an option, and whether they want to give it for free, since an option means that the rights of the seller are unilaterally bound or limited. The risk in granting such an option is risk of the lost opportunity, and the price of such an option should cover the reasonable opportunity cost of the seller, for example, in case of a price increase.

In all cases, when touching additional reductions and the possibility of the seller to sell ERs to third parties, the agreement should address the issue of the ownership of project documents. As mentioned, both ERUs and ERs are artificially created products which are linked to a number of documents, such as the PDD, or the Validation Report. If the preparation of such documentation is partly or fully financed by the carbon purchaser, he may retain the right of ownership and exploitation of these documents. However, the right over project documentation is essential for the seller to conclude a carbon contract with a third party, and the provisions on the ownership of project documentation should reflect it either by granting joint ownership or allowing the seller to exploit the project documents (with the consent of the carbon purchaser) for such purposes.

7 When Something Goes Wrong

The provisions describing default under and termination of the Agreement are essential for determining the value of a contract. In cases where the delivery of ERs does not only depend on the project entity's production of the primary product (for example, electricity), but on other external factors like the development of the power sector which may affect the baseline of a project, it can make sense to distinguish in the event of default between production failure (shortfall in generated power) and delivery failure (shortfall in generated reductions) and allow for a different treatment of these two kinds of failures in the event of default or termination causes, since one is within, the other beyond the control of the project developer.

Many carbon purchasers aim to enforce their interest in the rights to the emission reductions produced by the project through penalties for underdelivery. Carbon finance provides supplementary finance to a project—besides commercial loans and private equity—by creating a supplementary product additional to the primary product. More concretely, in case of a combined heat-and-power plant, for example, the plant becomes 'tri-generational' instead of co-generational: producing power, heat, and GHG emission reductions. For the first two, off-take agreements usually apply penalties for underdelivery reflecting standard commercial practice.

Penalties for ER underdelivery and provisions which oblige the supplier to deliver replacement ERs which the project developer has to purchase from other projects or on the spot market, increase the risk for the carbon seller considerably. Penalties can make the carbon cash flow, which is there to supplement the original cash flow of the project, negative and thus turn into a liability for the carbon seller. Penalties can thus endanger the project's viability in cases where the delivery failure is due to non-registration of the project or adverse changes in the project baseline. Thus, it may be sensible to negotiate penalties only for delivery failures which correspond to production failures. Such a penalty can provide an additional incentive for full delivery. On the other hand penalties, which apply where the underdelivery is related to the registration or host country risk should be avoided, since these risks are beyond the control of the project's owner and such penalties can put in danger an otherwise viable project.

8 Security Package

The providers of commercial project finance usually require a sophisticated security package to cover their open position. Similarly, carbon finance providers may require security to cover their open position wherever they provide an upfront payment to the project. The form of a security package can include a parent company guarantee or a letter of credit (bank guarantee). A parent company guarantee contradicts the principles of project finance, that is, every project has to be able to bear its own risk and the project failure does not allow recourse to the balance sheet (assets) of the parent company. In case of a bank guarantee, the financing institution will take into account the further increase of the letter of credit to the overall net open position (value-at-risk) of the bank. In order to limit its risk, the bank will aim to limit the trigger events which allow to call upon the bank guarantee. Also, to decrease the costs of the letter of credit for the carbon seller, it can be structured as a revolving bank guarantee under which the maximum amount decreases with the gradual amortization of the outstanding amount of the upfront payment. The ERPA should include provisions which limit the carbon purchaser's right to exercise its rights under the Letter of Credit, either by specifying the trigger events or the extent to which the bank guarantee can be drawn down.

9 Interaction with Financing Agreements

The ERPA and the financing agreement are linked. A direct link between the lenders and the carbon purchaser exists where the carbon purchaser agrees to make payments directly to the lending bank who receives the carbon revenues and thus the supporting cash flow. Also, financing institutions may require the right to approve all agreements which result in long-term contractual obligations for project entity, and as a result have a say in the conclusion of the ERPA as well.

It is therefore an important decision whether the ERPA should be part of the underlying financing documentation package or not; and accordingly, whether cross-default between the ERPA and the financing agreement is allowed or not. Where the ERPA is an integrated part of the financing documents (like the fuel supply agreement, off-take agreement, or the EPC contract) the financing bank has the right to approve the ERPA as well as any further amendments to the agreement. This implies that the financing bank has a direct influence on key carbon finance areas, and involves a third actor in the already difficult harmonization process of risk sharing between the carbon seller and purchaser. As for the cross-default, it is normally acceptable that the event of default or the termination of the financing agreement result in an event of default of the ERPA. Not acceptable is that a default under the ERPA triggers default under the financing agreements. The risk of the ERPA in the high uncertainty of the international regulatory environment is usually much higher than the risk of any other project financing documentation agreement.

A separate issue is the harmonization of the two security packages, which would create an indirect link between the ERPA and the underlying project agreements. Both finance providers (commercial loan and carbon) will seek alternative remedies in case the project defaults. It is therefore important to eliminate any possible overlap between the security package of the loan agreement and the security package of carbon finance, or set priority order (seniority or subordination) between the two. Luckily, both financiers usually seek different securities to hedge their exposure: while mortgages and claims over project assets are primary tools for creating securities for financing banks, carbon purchasers, where they make upfront payments, seek to mitigate risk through a parent company or commercial guarantee.

10 Counterparty Risk and Limitation of Liability

The most sophisticated underlying security package cannot replace trust and confidence in the general creditworthiness of the counterparty in a carbon transaction. Decisions, such as an upfront payment or the transfer of legal title to the ERs, require a careful judgment of the creditworthiness and soundness of the counterparty. The transfer of title to ERs under JI ERPAs usually takes place between 2009 and 2013, which leads to a significant performance risk taking into account the period between the conclusion of the contract and the delivery of the ERs.

Today, most carbon purchasers are still governments, government agencies, or international organizations. Therefore the purchaser can claim immunities, which may prevent the seller from having recourse against the purchaser in the event of breach of the purchaser's obligations. The careful consideration of these issues and the settlement to the extent possible is crucial for the required soundness of the agreement.

Additionally, the seller should seek to limit its liability. Generally, the seller's liability with respect to the rights to the emission reduction can be limited by specifying that all obligations of the seller end with the transfer of legal title to the defined ERs. Also, liability for losses resulting from a breach of the agreement, and, in particular, liabilities resulting in potential claims for consequential losses or liquidated damages, can be limited either through an agreed method as to how to calculate damages or through an agreement of an amount which limits the seller's liability.

11 Governing Law and Dispute Settlement

The Kyoto Protocol has not yet entered into force and few national legislations have integrated carbon transactions in their national legal framework. As a result, it is a challenge to decide which law might be the best suited to govern the ERPA. Parties can either agree that the EPRA be subject to a specified national law and take the risk associated with the future developments of that particular law, or they can leave the determination of the governing law to an arbitration tribunal. However, the lack of governing law provisions itself creates an uncertainty and a potential for disputes and interpretation which should be avoided. Also, the absence of a specified governing law increases the risk profile of the project and may negatively influence the appreciation of the income stream provided under the ERPA by any financing institution.

12 Risk of Unknown Financial and Accounting Treatment of ERs and ERUs

Currently, the institutional framework and the guidelines for the financial and accounting treatment of different ER rights are still missing. Additionally, a carbon transaction triggers a number of tax questions which are yet to be addressed under national law. Finally, the JI rules as set by the Marrakesh Accords do not yet specify the fees which will be levied for the registration of a JI project. All these factors add to the insecurities over the case flow the ERPA will bring to the project.

These risks cannot be mitigated, but need to be allocated between both parties to an ERPA. For the seller, it is usually more acceptable to bear host country VAT and corporate income tax than to expose itself to unknown international fees. On the other hand, where the purchaser is a national agency or inter-governmental institution it may be willing to bear the additional cost of possible international fees to be payable in relation to carbon transactions.

The international harmonization of the financial and accounting treatment of emission reduction credits and rights, as well as a common understanding on its legal nature (product/service/right, or possible securitization) across different countries is clearly desirable. Such agreements would facilitate the transparency and predictability of transactions and its financial results.

13 Conclusion

This chapter has tried to describe the challenges and risk which characterize the new and unknown market for JI projects. Whereas any carbon project requires a considerable amount of investment in time and learning effort, it opens the window to a new and fascinating financing instrument. A carefully analysed and negotiated ERPA can contribute to the robustness of a project while integrating the project in the international efforts to curb global warming.

Table 18.1: Major risks in relation to an Emission Reductions Purchase Agreement and their impact on and treatment in contractual terms

Risk	Consequence	Mitigation and place in the ERPA if Seller bears	Mitigation and place in the ERPA if Purchaser bears
Kyoto Protocol ratification risk Registration risk Country eligibility risk Political risk	No transfer of ERUs possible ERUs are not created ERUs are created but not transferred due to host country's failure to meet Kyoto Protocol eligibility requirements ERUs are created but not transferred due to host country's failure	*Definitions:* Baseline according to UNFCCC /Kyoto Protocol Rules (adapting to the regulatory changes) *Purchase and Sale:* Purchase of ERUs *Payment*: Payment is due on the transfer of ERUs from the national account	*Definitions:* baseline as set and agreed at the time of the signature of the ERPA *Purchase and Sale:* Sale of Emission Reductions instead of ERUs *Payment*: Payment is due on the receipt of the verification report by the Purchaser

Project performance risk	Shortfall in production (of power) leads to shortfall in delivery of Emission Reductions	Setting minimum annual amount Allow flexibility range in case of underperformance. Allow recovery period in case of underperformance	
Project performance risk: off-take regulation risk, changes in project baseline	Shortfall in delivery of Emission Reductions	No special provisions	Distinguish production and delivery failure
Project completion risk	Delayed delivery of emission reductions Upfront payment at risk	Possibility to terminate the ERPA in case of an extended delay Time of upfront payment: Due on commissioning of the Project	Time of upfront payment: Due on signature of the ERPA
Risk of interaction with financing agreement	The high uncertainty of international legal environment raises the overall risk profile of project	Distinguish production and delivery failure Termination provisions, cross default Security package: development and harmonization	
Counterparty risk	General creditworthiness Risk lies in the extended lifespan of project Risk in the nature and capacity of purchaser/seller Potential for consequential losses, liquidated damages	Limitation of liability Date of upfront payment, requirements on security package Background check on purchaser/seller	
Risk due to the absence of set financial and accounting treatment on ERs or ERUs	Host country's taxes and UNFCCC's fees may result in future change of net cash flow	Seller bears all applicable taxes in host country	Purchaser bears all applicable fees resulting from UNFCCC provisions and the Kyoto Protocol
Risk of the absence of a specified governing law	Creates legal uncertainty over the agreement; potential difficulties in having the ERPA accepted by financing institution	A business of trust and confidence. Arbitration clause Agreement on governing law	

19

Experiences with the Clean Development Mechanism in India

*Jaya Singhania**

The objective of this chapter is to present:

(i) an overview of the Clean Development Mechanism (CDM) as comprehended and presently under implementation in India;
(ii) a brief discussion on the regulatory framework governing various aspects of implementing the CDM;
(iii) the general implementation and documentation structure for infrastructure projects with particular reference to power projects and a discussion on the popular perception as to whether Emission Reduction Purchase Agreements (ERPAs) help promote renewable energy projects in India; and
(iv) comments and suggestions on certain provisions of the ERPAs in the context of Indian Law and interpretation.

1 Overview

The Government of India has taken serious cognizance of the requirement for sustainable development by environmentally friendly methods and technologies and preventive measures for greenhouse gas (GHG) reduction. Even before the popularity of CDM projects under the Kyoto Protocol, the Government of India has been promoting these methods and technologies. The focus is on renewable energy technologies, energy efficiency in the industry and transport sector and waste to energy projects, and the main activity thus far has been in the energy and transport sector. Renewable energy projects have been encouraged in as much as the availability of incentives under a wide range of programmes by the Government and vehicular emissions have been controlled by prescription of strict

* Jaya Singhania is a legal counsel at the Infrastructure Development Finance Company Limited in India. The views expressed in this chapter are the views of the author and do not necessarily represent the views of her employer.

standards and controls. The incentives include fiscal (in terms of tax, subsidy, soft loans, and specified tariff guidelines) and non-fiscal (in terms of simplified procedures for encouraging private, including foreign, investments) benefits. The spread of various renewable energy technologies, which include waste to energy as well as renewable energy projects, have been aided by policy and support measures by the Ministry of Non-Conventional Energy Sources (MNES), an independent ministry formed by the Government of India to encourage renewable energy projects in India. Besides, the Indian Renewable Energy Development Agency Limited (IREDA) has been formed to finance renewable energy projects and IREDA works directly under the MNES. The areas identified for renewable energy development are wind, small hydro, solar PV, solar thermal, biomass including co-gen waste to energy, energy efficiency, and biogas.

The IREDA has been providing interest rate and capital subsidies to these projects. In many instances where the projects are funded by debt funding from financial institutions, the interest rate subsidy is given to the financial institutions, which in turn pass on the benefits to the projects. To ensure sustainability of the project, the subsidies are rolled back in case the Project Entity fails to implement/operate the project in the manner envisaged.

The Ministry of Environment and Forests (MoEF) has acknowledged the purpose of the CDM, as defined in Article 12 of the Kyoto Protocol to the United Nations Framework Convention on Climate Change (UNFCCC), as twofold, namely: (a) to assist parties in the developing country in achieving sustainable development, thereby contributing to the ultimate objective of the Convention, and (b) to assist developed country Parties in achieving compliance with part of their quantified emission limitation and reduction commitments under Article 3 and set out the interim criteria for certifying projects as CDM projects keeping in view the same.[1]

Although the potential for carbon sequestration and reforestation are being acknowledged as important factors in sustainable development and legal framework for forest conservation, biomass developments are in place, forestry and agriculture are thus far largely unorganized. The issues in implementation and monitoring of CDM projects in these areas are still being debated and sorted out. This chapter thus focuses on CDM projects in the energy sector.

2 Indian Regulatory Framework

2.1 Legislative Framework

The Constitution of India is based on the federal concept and distributes sovereign as well as executive powers between coordinate constitutional entities, namely, the Union and the States. The Constitution delineates the rule making authority, lays

[1] See Art 12 Kyoto Protocol.

out the procedure for enacting laws, and promulgating regulations between the two. The basis of distribution is that, only those powers and authorities which are concerned with the regulation of local issues, are vested in the State, and the residue, especially those which tend to maintain the economic, industrial, and commercial unity of the nation are left with the Union.

The duty to secure protection and improvement of environment and safeguarding of forests is that of the State and every citizen. This is stipulated in Part IV of the Indian Constitution which sets out the Directive Principles of State Policy.[2] Primarily, the provisions of this part are by nature of directives and not enforceable by a court. Nevertheless, the principles laid down are fundamental to the governance of the country and when read with the Fundamental Rights enshrined in the Constitution, various environmental legislations promulgated by the Parliament and the State Legislatures in respect of environmental protection and pollution control, and precedents set by the Supreme Court of India pursuant to a spate of public interest litigation filed against polluting industries and practices, provide significant empowerment to the State and its instrumentalities to take punitive action against the non-compliant. In fact, the Supreme Court of India, through its judgments, has expressly laid down the applicability, in India, of the principles of strict and absolute liability with respect to environmental hazardous activities including with respect to injury from non-natural use of the land.[3]

Forests and the acquisition and requisition of property fall within concurrent jurisdiction, that is, a subject that comes under the jurisdiction of both the Central Government and the State Government, albeit the State Government may exercise its powers subject to the exercise of powers by the Central Government. Forest conservation vests completely with the Government as do the rights over forest land. The maintenance of land records, assessment and collection of revenues and records of rights, and alienation of revenues falls within the jurisdiction of the State Governments. India is an agri-society and the importance of local self-government bodies at the village level has been recognized by the Constitution of India. These government bodies (called panchayats) have been identified as being capable of being delegated with powers relating to land improvement,[4] minor irrigation, water management and watershed development,[5] rural electrification including distribution of electricity[6] and non-conventional energy sources[7] and could play a key role in the implementation of CDM Projects.

[2] Art 48A of the Indian Constitution states that the State shall endeavour to protect and improve the environment and to safeguard the forests and wild life of the country. See also: DD Basu, *Shorter Constitution of India* (12th edn, New Delhi, Prentice-Hall of India, 1996).

[3] *MC Mehta v Union of India* AIR 1987 SC 1086. See also P Joshi, *Law Relating to Infrastructure Projects* (2nd edn, New Delhi, Butterworths India, 2003) 165.

[4] Art 243A Constitution of India.

[5] Art 243B(1) Constitution of India.

[6] Entry 14, Eleventh Schedule, Constitution of India.

[7] Entry 15, Eleventh Schedule, Constitution of India.

Power is a concurrent subject. The policy, generation and transmission activities are undertaken by both the Central Government and the State Governments; distribution and supply of electricity to the final consumer and associated tariff setting are exclusively under the purview of the State Governments. The recent initiative of the legislature has brought the power sector under the umbrella of the Electricity Act 2003 (Act No 36 of 2003)[8] which came into effect as of 10 June 2003. Before this, the sector was governed by three main pieces of legislation and the rules and regulations thereunder. Significant to note is the Indian Government's policy initiative of 1991, which decided to open up the sector, which was until then a State monopoly, to the private sector. Pursuant to this policy decision, many State Governments promulgated State Reforms Acts to foster the ailing State Electricity Boards, and the sector, to health. Also promulgated was the Electricity Regulatory Commissions Act 1998 (Act No 14 of 1998)[9] for the establishment of the Central Electricity Regulatory Commission and the State Electricity Regulatory Commissions for, *inter alia*, the rationalization of electricity tariffs and promotion of efficient and environmentally friendly policies. The Parliament also enacted the Energy Conservation Act 2001 (Act No 52 of 2001)[10] to provide for efficient use of energy and its conservation and for matters connected therewith or incidental thereto. The Electricity Act 2003 has now been enacted to provide the entire legal framework governing the electricity sector. This enactment also seeks to promote renewable energy power including rural electrification. It directs the Central Government to prepare a national electricity and tariff policy including renewable energy power. It is also intended that the State Electricity Regulatory Commissions promote renewable energy-based projects by stipulating that a percentage of total consumption of electricity in the area of a distribution licence is that of renewable energy-based projects. Given the short period for which the legislation has been in force and the fact that many of the intended policies and rules under the legislation have yet to be formulated, it will take some time before the full potential of the initiatives on renewable energy projects can be realized.

2.2 Judicial Framework

The judicial power vests in a single set of courts and tribunals, civil, criminal, and revenue, to deal with disputes in respect of legislation, which is either State legislation or Central legislation. The jurisdiction of a court or tribunal is specified in the law governing the matter in dispute. Every High Court (in the States) has power of superintendence over all courts and tribunals throughout the territories in relation to which it exercises jurisdiction. High Courts are vested with certain powers to

[8] *Gazette of India, Extraordinary,* Pt II, Sec. 1, No 39, dated 2 June 2003.
[9] ibid, Pt II, Sec. 1, dated 2 July 1998 (w.e.f. 25 April 1998).
[10] ibid, Pt II, Sec. 1, dated 1 October 2001.

issue writs addressed to any person or authority, including in appropriate cases any government for enforcement of fundamental rights conferred by the Constitution. The Supreme Court is at the apex of the hierarchy of courts, civil, criminal, revenue, and of quasi-judicial tribunals.

For commercial disputes, the settlement procedures depend on the parties to the contract. In most commercial contracts, especially those involving foreign parties, including contracts for implementing energy projects, dispute resolution by conciliation and arbitration is preferred. After the enactment of the Arbitration and Conciliation Act 1996 (Act No 26 of 1996)[11] many commercial contracts provide for settlement of disputes by arbitration and conciliation under the same. This Act was specifically enacted to consolidate and amend the law relating to domestic arbitration, international commercial arbitration, and enforcement of foreign arbitral awards and also to define the law relating to conciliation, and is based on the model law of arbitration and conciliation mooted by the UNCITRAL.[12]

3 Process and General Documentation Framework

3.1 Process for Implementation

The process for implementing a CDM project would involve:

(i) project conceptualization by the project developers which includes the assessment of availability of resources (land, material, technology, and finances);
(ii) technical clearance by the authority for non-conventional energy development, a State Government Instrumentality;
(iii) general approval by different government departments such as the pollution control board, land revenue department, commission of industries and so on;
(iv) approval as a CDM project from the MoEF, a Central Government Ministry which acts as the enabling agency in respect of such projects; and
(v) tying up of resources for the projects from banks, financial institutions and in some cases government development agencies.

3.2 Project Documentation and Financing

To procure financing for a project, it is very important to have bankable project documents. If the project documents are not bankable, it is unlikely that the

[11] ibid, Pt II, Sec. 1, dated 19 August 1996, No 55, 1–36.
[12] UN Commission on International Trade Law.

project would procure financing from banks and financial institutions. The bankability of project documents depends on the commercial viability of the project, the legality of the transactions contemplated and the proper risk allocation and mitigation amongst various stakeholders. For merchant power generation projects the main project document is the power purchase agreement which sets out the technical parameters which the project should follow and the terms of bulk sale and purchase of electricity, in addition to setting out the rights and liabilities of the parties on contingencies of *force majeure*, defaults, and so on. The commercial viability of such a power project would depend on the ability of the project to perform during the tenure of the power purchase agreement, as also the continued revenue flows upon generation and sale of electricity. While the project proponent is often in a position to, and does control and mitigate the risks involved in implementation (construction and operations) by negotiating turnkey Equipment Procurement and Construction and Operations and Management contracts with proper milestones and penalties/liquidate damages for delays and laches, the provisions and clauses of the power purchase agreement per se may not be within its control. As discussed earlier, the policy and regulation with regards to the distribution of electricity is within the purview of the State Government. While most State Governments/State Electricity Boards involve technical and legal experts to negotiate and enter into bankable power purchase agreements for large and conventional power projects, it is observed that the documentation in respect of smaller projects, particularly the non-conventional energy projects with various uncertainties on account of environment conditions and irregular fuel supply suffer. Besides, an important factor that is considered while financing a renewable energy project is the tariff setting policy of the State Government within whose jurisdiction the power plant would fall. There is considerable debate on the kind of tariff policy required to promote non-conventional energy sources projects and what the State Governments should do to promote these projects given the high cost of implementation, albeit the low operations and maintenance cost of these projects. This is a different subject matter altogether and for the purpose of this chapter it will suffice to state that the Government of India by its various initiatives (including the Electricity Act 2003) is striving to strike a balance between the industry demand of low cost electricity and sustainable development.

Also pertinent to note for the purpose of this discussion is that while the conventional power projects are on a limited recourse or non-recourse basis and implemented by a special purpose vehicle, thus ensuring ring fencing and securing of the assets and revenues of the project, for renewable energy projects it is often not feasible to ring fence. This is mainly due to the fact that many such projects primarily consist of co-generation plants, which are also captive. Also, other renewable energy projects such as wind energy and solar energy projects are captive plants with the revenue stream being only from the industry for which it is set up. The financing of such plants therefore is often with recourse to the balance sheet of the company which is setting up the captive facility, thus depending on the commercial viability of the project as well as the other businesses of the project proponent. Even where the projects are not captive, banks and financial institutions, while financing small renewable energy projects, prefer to do so on a recourse basis, with recourse to other assets of the project proponent. While on the one hand such recourse gives additional comfort, it also exposes the project to insolvency risk of the project proponent (in most cases when recourse is sought on the other assets, the same has also to be extended on the project assets to the other creditors of the project proponent).

3.3 Emission Reductions Purchase Agreements (ERPAs) and their Contribution to the Bankability of a Project

Carbon purchasers such as the Dutch Certified Emission Reductions Procurement Tender (CERUPT) or the World Bank's Prototype Carbon Fund (PCF) and others have commenced activities for purchase of verified emission reductions[13] (ERs) from renewable energy projects implemented in accordance with CDM rules in India.

The contracts provide for purchase of ERs which are generated post the implementation period of the project and hence the ERPAs are structured in a way that money flow shall take place post facto. Given the uncertainties of realization of cash flows on account of various factors such as the international ratification of the Kyoto Protocol, continued performance against a baseline which could be adjusted during the crediting period due to technological innovation, fuel unavailability or technological underperformance and so on, the financiers often do not take into account the revenues expected from the ERPAs and usually take an exposure on the projects only if they are financially viable even without these revenues.[14] The revenues from the ERPAs therefore are at best incentive for equity investors who would gain incremental revenue if the Project performs as envisaged.

Also notable is the risk of uncertainty of the government policy with regard to CDM Projects (which would to a large extent be based on the Protocol and also the future treaties that the Indian Government would enter into with regard to emission reductions). There is a debate on potential risks of higher costs to the domestic industry due to a subsequent liability that may be taken on by the country with regard to emission reduction, on account of the ERPAs.

[13] ERs are generally defined as the rights accruing from a GHG reduction activity measured in tonnes of CO_2 equivalents.

[14] *See* for reference: V Deodhar, A Michaelowa, and M Krey, 'Discussion Paper on Financing Structures for CDM Projects in India and Capacity Building Options for EU-Indo Collaboration', HWWA Discussion Paper, 247, 1–31.

3.4 Approval Criteria for CDM Projects

The MoEF has indicated the following criteria to be adopted for classifying a project as a CDM Project.[15]

3.4.1 Eligibility

The project proposal should establish the following 'additionalities' in order to qualify for consideration as CDM project activity:

- Emission Additionality: The project should lead to real, measurable, and long-term GHG mitigation. The additional GHG reductions are to be calculated with reference to a baseline.
- Financial Additionality: The funding for CDM project activity should not lead to diversion of official development assistance. The project participants may demonstrate how this is being achieved.
- Technological Additionality: The CDM project activities should lead to transfer of environmentally safe and sound technologies and know-how.

3.4.2 Sustainable development indicators

Under the rules governing the CDM, it is the prerogative of the host Party to confirm whether a CDM project activity assists it in achieving sustainable development. The CDM should also be oriented towards improving the quality of life of the very poor from the environmental standpoint. The following aspects should be considered while designing CDM project activity:

- Social well-being: The CDM project activity should lead to alleviation of poverty by generating additional employment, removal of social disparities, and contribution to provision of basic amenities to people, leading to improvement in their quality of life.
- Economic well-being: The CDM project activity should bring in additional investment consistent with the needs of the people.
- Environmental well-being: This should include a discussion of the impact of the project activity on resource sustainability and resource degradation, if any, due to proposed activity, bio-diversity friendliness, impact on human health and reduction of levels of pollution in general.
- Technological well-being: The CDM project activity should lead to transfer of environmentally safe and sound technologies with a priority to the renewables sector or energy efficiency projects that are comparable to best practices in order to assist in upgrading of the technological base.

[15] See for example, Planning Commission, Government of India, 'Report for National Action Plan for Operationalising CDM in India' (September 2003).

3.4.3 Baselines

The project proposal must clearly and transparently describe the methodology of determination of the baseline. It should confirm the following:

- Baselines should be precise, transparent, comparable, and workable;
- Avoid overestimation;
- The methodology for determination of the baseline should be homogeneous and reliable;
- Potential errors should be indicated;
- System boundaries of baselines should be established;
- Intervals between updates of baselines should be clearly described;
- The role of externalities should be brought out (social, economic, and environmental);
- Baselines should include historic emission data-sets wherever available;
- The lifetime of the project cycle should be clearly mentioned.

The baseline should be on a project-by-project basis except for those categories that qualify for simplified procedures. The project proposal should indicate the formulae used for calculating GHG offsets in the project and baseline scenario. Leakage, if any, should be described. For the purpose of Project Idea Notes (PIN), default values may be used with justification. Determination of the base project which would have come up in the absence of the proposed project should be clearly described in the project proposal.

3.4.4 Financial indicators

The project participants should bring out the following aspects:

- Flow of additional investment;
- Cost-effectiveness of energy saving;
- Internal Rate of Return (IRR) without accounting for CERs;
- IRR with CERs;
- Liquidity, net present value, cost/benefit analysis, cash flow and so on, establishing that the project has good probability of eventually being implemented;
- Agreements reached with the stakeholders, if any, including power purchase agreements, Memoranda of Understanding, and so on;
- Inclusion of indicative costs related to validation, approval, registration, monitoring and verification, certification, share of proceeds;
- The proposal should indicate funding available, financing agency and also describe how financial closure is sought to be achieved.

3.4.5 Technological feasibility

The proposal should include the following elements:

- The proposed technology/process;
- Product/technology/material supply chain;
- Technical complexities, if any;
- Preliminary designs, schematics for all major equipment needed, design requirement, manufacturers name and details, capital cost estimate;
- Technological reliability;
- Organizational and management plan for implementation, including timetable, personnel requirements, staff training, project engineering, CPM/ PERT chart, and so on.

3.4.6 Risk analysis

The project proposal should clearly state risks associated with a project, including apportionment of risks and liabilities, insurance and guarantees, if any.

3.4.7 Credentials

The credentials of the project participants must be clearly described.

It can be noted from the aforesaid that the eligibility criteria set out by the MoEF are extensive. Ostensibly, it does appear that the project may be subjected to excessive scrutiny even at the host country approval stage, especially when compared with criteria stipulated by other countries which restrict the sustainable development objectives of the country by stipulating general parameters such as a positive and negative list of projects that may be considered for a CDM approval.[16] It can, however, be argued that the criteria only endeavour to establish beyond doubt that the project is sustainable and put it through some kind of preview of the project's conformity to other CDM validation requirements. Some Designated National Authorities (DNA) in other countries also contemplate this kind of review. In fact, the MoEF seems to have largely evolved the criteria from the CDM validation requirements. There are some additional requirements by way of financial indicators, which have apparently been stipulated to ensure that the transactions are genuine. It is pertinent to note the MoEF's commitment to the promotion of CDM projects in the country. The committee for granting approvals meets every month and as long as the application provides the required information, approvals are

[16] See for example, Republic of Colombia, Ministry of Environment, 'Criteria and Procedures for the Approval of CDM', Proposal under discussion—unpublished draft (Bogotá, December 2002).

granted expeditiously. Also pertinent to note is a typical approval from the MoEF, which provides that:

(i) Certified Emission Reductions (CERs) should not be sold to an entity which uses Official Development Assistance (ODA) to purchase the credits;
(ii) transaction details of CERs transferred should be disclosed to the DNA;
(iii) such information as may be required by the DNA from time to time should be furnished;
(iv) all statutory clearances required for the project should be obtained;
(v) transactions should be subject to the supervision of the Executive Board of the CDM;
(vi) the approval is not transferable.

It can therefore be noted that other than the requirement that CERs may not be sold to an entity which uses ODA funds (which in any case is a CDM requirement so as to ensure that developmental funds are not used for the purchase of ERs, but for implementation of the project), the MoEF clearance for a CDM project is quite routine. The approval process is hence rather complementary in establishing the bankability of the project, as both the financiers of the project and the Executive Board would also require information in line with that required by the MoEF while assessing the project. Also, given the fact that in its endeavour to promote the concept, the MoEF does not charge any fees for the approval process, it is quite possible that the authorities may desire a comprehensive review at the approval stage itself so as to ensure that only genuine cases are taken further. As with every new scheme of things, the ministry's approach towards the CDM also seems to be a cautious one and, given the Indian Government's focus on liberalization and globalization, one can be hopeful of evolution of a streamlined process with time.

3.5 Regulations applicable to Renewable Energy Projects

Compliance of regulation for implementing a power project depends on the scope of the project. For implementing small renewable energy projects, the licensing requirements and regulation are minimal. Under the Electricity Act 2003, the statute which now governs the power sector, the framework governing generating companies has been rationalized to a great degree. There is no licensing requirement for power generation as long as the generating station complies with technical standards relating to grid connectivity. In case of projects for captive generation using captive transmission lines, there are no licensing requirements whatsoever. However, a generating company intending to set up a hydro-electric generating station would require the appropriate authority's approval. These plants must be planned giving due attention to displacement of people, the use/misuse of water bodies, destruction of forests, and so on.

The appropriate electricity regulatory commission, the one that would have jurisdiction over the relevant power station of the generating company, will have the authority to specify the terms and conditions for determination of the tariff for the sale of electricity. While determining tariff policies, due regard will be given to the promotion of co-generation and generation for electricity from renewable sources of energy. All projects require a clearance from the relevant pollution control boards. These clearances are provided under the Water (Prevention & Control of Pollution) Act 1974 (Act No 6 of 1974)[17]as amended in 1988, and the Air (Prevention & Control of Pollution) Act 1981 (Act No 14 of 1981).[18] Larger projects and projects falling within particular areas also require to submit an Environment Impact Assessment before commencement of any activity. This includes a demarcation of the area which would be affected by development of the project, determination of the existing ecology of the project site and affected areas, social and economic impact of the project, providing for mitigation measures for adverse impact within an affected area, and so on.

4 Comments on the ERPAs and Conclusion

The ERPAs under negotiation by some of the CDM Projects cleared by the MoEF provide for sale and purchase of the ERs, delivery mechanism and transfer of title, exemptions from delivery for specified reasons, consequences for failure to buy and sell, and a dispute resolution mechanism.

As discussed in section 3.3 above, the ERPAs thus far do not make a substantial difference to the bankability of a project and are at best considered as additional revenues to the promoters/equity holders. Having said that, it would be desirable to have ERPAs which would:

(i) clearly define 'Emission Reduction' and the manner of measuring the same in the document itself. Also, since 'Emission Reduction' or 'GHG Reduction' or 'Certified Emission Reduction' are not properties or rights as recognized within the Indian legal system, it would be best if these contracts were governed by laws under which these rights were recognized;

(ii) clearly and certainly entitle the project to revenues upon establishment/ determination of ERs;

(iii) not burden the project with processing costs for the transaction. The buyer could initially bear the processing costs for the transaction which then could be set off against the payments made for CERs;

(iv) provide the project entity with flexibility in transferring the CERs derived. While the minimum quantities can be set out, the periodicity of the

[17] *Gazette of India, Extraordinary*, Pt II, Sec. I, dated 25 March 1974, 295–325.
[18] ibid, Pt II, Sec. I, dated 15 May 1981.

transfer and sale of additional CERs, if any should be at the option of the project entity so as to maximize the benefits for the project;

(v) not burden the project entity with onerous obligations/consequences on account of breach of provisions of the ERPAs due to reasons beyond its control. The buyers of CERs not being investors in the project per se should not expect to be mitigated for risks beyond the control of the project/project entity; and

(vi) provide for assignment of rights under the ERPA to Lenders to the Project, or their nominees.

To conclude, the CDM seems to have received due attention from the policy makers and regulators in India, as being a good initiative on the part of the international community towards sustainable development. There is hesitation however on the extent to which the Indian Projects should contribute to the same, particularly due to the fact that the Kyoto Protocol is yet to come into force. The policy initiatives seem to favour CDM, however the requisite legislative framework for giving legal sanctity to contracts for trade in ERs and CERs is yet to be put in place. Under Indian law therefore, as of now the ERPAs could be classified as contractual obligations resulting from the agreement of the Project Entity to implement and operate the Projects in a particular manner. It may also be observed that the time and costs involved in negotiating and entering into an ERPA does not seem commensurate with the returns a project entity may expect from the same, due to the uncertainties attached to the returns. The revenues under an ERPA will accrue to a project entity only if the transactions contemplated thereunder translate into benefits for the buyer, which benefits are dependent on the Kyoto Protocol, the host country stipulations, baselines, and so on. As observed earlier, these contracts are therefore being construed as being additional and incremental sources of revenues/profits to a Project Entity and not as contributions to the viability of a Project.

20

World Bank Carbon Finance Business: Contracts and Emission Reductions Purchase Transactions

*Charlotte Streck**

1 Introduction

The first World Bank[1]-administered carbon fund, the Prototype Carbon Fund (PCF), became operational in early 2000.[2] In that same year the International Bank for Reconstruction and Development (IBRD) acting as trustee for the PCF entered into its first Emission Reductions Purchase Agreement (ERPA) with the Republic of Latvia governing the first project to receive funding from the newly established PCF. Since then, the World Bank has executed twelve additional ERPAs, and launched a number of additional Carbon Funds, including the IBRD–Netherlands Clean Development Mechanism Facility, the Community Development Carbon Fund, the Italian Carbon Fund, and the Bio Carbon Fund, all designed to acquire Emission Reduction Credits[3] on behalf of the contributors to the funds (so-called Participants).[4] Developing and implementing these Carbon

* Charlotte Streck is a legal counsel at the World Bank. The views expressed in this chapter are the views of the author and do not necessarily represent the views of the World Bank.

[1] The World Bank Group is comprised of five associated institutions: the International Bank for Reconstruction and Development (IBRD), the International Development Association (IDA), and the International Finance Corporation (IFC), the Multilateral Investment Guarantee Agency (MIGA) and the International Centre for the Settlement of Investment Disputes (ICSID). The 'World Bank' as used in this chapter refers to the IBRD and IDA.

[2] Based on Resolution No 99-1 of the Board of the Executive Directors of the IBRD.

[3] For the purposes of this chapter the term 'Emission Reduction Credit' or just 'Credit' means any right to emit arising from or in connection with any removal or mitigation of greenhouse gases by a project and includes verified emission reductions as well as Certified Emission Reductions or Emission Reduction Units.

[4] Participants in World Bank Carbon Funds include industrialized country governments as well as private sector entities. The PCF, for example, holds contributions from six governments and seventeen private sector entities. Bilateral Funds, such as the Netherlands CDM Facility, were established through a bilateral agreement between a government and the IBRD. In this fund the World Bank purchases emission reduction credits exclusively for the benefit of the Government of the Netherlands.

Funds requires additional legal instruments necessary to support the final acquisition of Emission Reduction Credits.

By developing and implementing these Carbon Funds the World Bank has gained valuable experience in structuring carbon transactions, drafting ERPAs and in addressing contractually the unique features of carbon transactions. Cooperation between two or several parties in the implementation of a project needs to be given careful consideration to ensure that the credits generated will be eventually recognized under Article 6 (Joint Implementation (JI)) or Article 12 (Clean Development Mechanism (CDM)) of the Kyoto Protocol,[5] or under any other regulatory regime. Carbon transactions involve a long-term contractual relationship between the parties which, like any other long-term contract, requires continuous cooperation over several years or even decades.[6] At the same time, emission reduction transactions are still implemented in an environment characterized by regulatory insecurity. On the domestic level, governments have yet to integrate transactions involving the transfer of Emission Reduction Credits into their national legislation, and at the international level, the Kyoto Protocol has still not entered into force. Finally, carbon transactions are characterized by a special set of risks, arising from the regulatory system or the absence thereof, the need to secure continued host country support as well as the credibility of the monitoring and verification system, and from the transnational and cross-border nature of the cooperation.

It is therefore key that carbon transaction are based on solid contracts which clearly define the relationship between the partners, mitigate both regular project related risk and carbon finance specific risk, and remain flexible enough to adapt to a changing environment.

This chapter aims at sharing the lessons learned in drafting legal agreements for the World Bank Carbon Business.[7] It will therefore focus on carbon transactions involving CDM or JI project activities (together: Carbon Projects).[8] This chapter will be divided into two parts. The first, more general part, highlights the key features and challenges encountered in drafting carbon contracts and the second, more specialized part, discusses how these different challenges have been addressed in World Bank carbon finance contracts.

[5] Kyoto Protocol to the United Nations Framework Convention on Climate Change, Doc FCCC/CP/1997/L.7/Add.1 (not yet in force).

[6] Whereas JI contracts seldom cover contract periods longer than up to the end of 2012, most CDM-based contracts can run for up to twenty-one years, the maximum crediting period, and the lifetime of most CDM projects under current international rules. CDM contracts, which govern sequestration or 'Land Use Land Use Change and Forestry' (LULUCF) projects, can stretch for up to sixty years, which is the maximum lifetime of a LULUCF CDM project.

[7] Please see for further details on the World Bank Carbon Funds: <http://www.carbon finance.org> accessed 11 April 2004.

[8] Any agreements that are brokering or trading emission reduction units are not further discussed in this chapter.

2 Carbon Finance Contracts

A Carbon Contract foresees a payment against the achievement of an activity that leads to the removal or mitigation of greenhouse gases (GHGs). Once this has been verified by an independent auditor, the reduction or removal of GHGs may create a right under an existing or future regulatory regime. Carbon Projects need to be based on innovative and robust contracts which enable the parties to successfully implement the project long after the attention of initial project developers and consultants has moved on to other tasks.[9] Carbon contracts define the relationship between parties in an emerging market, which is characterized by a wide variety of uncertainties and risks. The contracts need to record the agreement between the parties, identify responsibilities, allocate risks, establish rights, and create clear and enforceable obligations. The following section will analyse the environment in which carbon contracts are currently being negotiated. It will identify the actors involved in such projects, describe the different type of carbon contracts and analyse the special challenges and features of carbon transactions.

2.1 Actors in Carbon Projects

Participants in Carbon Projects come from the public sector as well as the private profit and non-profit sector. Governments of countries included in Annex I of the UN Framework Convention on Climate Change (UNFCCC)[10] and the Kyoto Protocol have a prominent role as purchaser of Emission Reduction Units (ERUs) from JI projects or Certified Emission Reductions (CERs) from CDM projects. These units are purchased as a compliance tool to assist Annex I countries meet their emission reduction targets under the Kyoto Protocol. Another important role of Annex I country governments is to authorize private sector entities to participate in Carbon Projects, and to give the national private sector all necessary assistance to meet whatever requirements the Executive Board of the CDM or the JI Supervisory Committee may develop.[11]

CDM host countries support individual CDM projects through project approval where the country confirms that a project contributes to the host country's sustainable development objectives. In the case of JI projects, the assistance required of the host country goes beyond project approval and also involves the host country converting Assigned Amount Units (AAUs) and Removal Units (RMUs) into ERUs and transferring ERUs into the accounts of the acquiring Annex I Party. Additionally, a host country can also actively participate in a project

[9] Most Carbon Projects are still developed with extensive assistance of outside consultancy firms.
[10] Arts 6(3), 3(11), Annex B and UNFCCC Annex I.
[11] For more details on the governance of CDM and JI see ch 6 and ch 9 above.

as the project sponsor and seller of Emission Reduction Credits. Which role the host country assumes depends largely on its access to international finances and markets, its legislative framework, and the features of the individual project.

Early transactions under the pilot of project based emission transactions, the Activities Implemented Jointly (AIJ), typically involved governments only, namely an Annex I sovereign government as investor and a non-Annex I country as host.[12] Examples include transactions between Sweden and Estonia, or the Netherlands and Costa Rica. However, as AIJ matured, and even more prominently as the CDM and JI became operational, an increasing number of private sector entities have engaged in Carbon Projects. This has occurred either through the direct contracting of Emission Reduction Credits or through the participation in a fund following a portfolio model, such as through the PCF. Acquiring credits through a fund with a portfolio helps contain transaction costs, spreads the risk among the participants and maximizes the learning benefit by investing in a number of projects based on different technologies and located in different parts of the world.

Although Carbon Projects critically depend on government support from the parties to the Kyoto Protocol (Parties) involved in the Carbon Project, the development and implementation of Carbon Projects is eventually expected to be driven by private sector entities. CDM and JI were designed to attract the interest of the private sector in industrialized and developing countries alike and the mechanisms are expected to generate financing primarily from private rather than government sources.[13] Private sector entities from industrialized countries will focus on using the CDM and JI as a compliance tool to meet their own emission reductions targets, acting as broker in the trade with CERs and ERUs, or setting up and managing investment funds. Private sector entities from host countries, on the other hand, most likely see the possibility of generating and selling Emission Reduction Credits under CDM and JI projects as a means of making a difference in the internal rate of return of their projects and of pushing marginal projects into the realm of economic viability. The CERs or ERUs arising from a Carbon Project will have cash value that adds to the viability of a project from the sponsor's point of view.

2.2 Different Types of Contracts

Viable JI and CDM projects require resources and the specialized skills from a diverse group of people and institutions at several governance levels. Carefully structured and robust contracts contribute to the viability of projects by defining and regulating the relationship between the different actors required to implement a Carbon Project. The implementation of Carbon Projects leads to the

[12] K Danish, E Brenes, and J Rotter, 'Legal Dimensions of AIJ Project Development Activities', in RE Dixon, *The U.N. Framework Convention on Climate Change Activities Implemented Jointly (AIJ) Pilot: Experiences and Lessons Learned* (Dordrecht, Kluner International 1999) 281, 284.

[13] PH Sand, 'Trusts for the Earth', in PH Sand, *Transnational Environmental Law: Lessons in Global Change* (The Hague, Kluwer Law International, 1999) 291.

establishment of legal links between the different sovereign Parties and the private or public project participants involved.[14] Carbon contracts in the broader sense include all type of contracts that govern JI and CDM projects, including:

(a) Unilateral instruments of non-Annex I and Annex I Parties, such as Approval Letters for JI and CDM project activities;[15]
(b) Bilateral instruments between Annex I and non-Annex I Parties, such as Host Country Agreements or Memoranda of Understanding that govern the transfer of Assigned Amount Units (AAUs)[16] or ERUs under Article 6 or 17 of the Kyoto Protocol;
(c) Bilateral or multilateral instruments between project participants, which may include Annex I and non-Annex I Parties, private and public sector entities, and which govern the implementation of the Carbon Project, risk allocation, and the transfer of funds and Emission Reduction Credits.

The first two categories of agreements focus on the authorization of the project by the Parties involved and the transfer of the internationally defined Emission Reduction Credits to the extent that sovereign action is necessary. JI project activities for example, rely on the confirmation of the host Government to convert AAUs into ERUs once they have been verified.

For the purpose of this chapter, I will focus in my description and analysis on the last of the three categories, which includes the commercially structured contracts that govern the purchase and sale of rights to the GHG emission reductions (or removals) generated by the Carbon Projects. These agreements are referred to throughout this chapter as 'Carbon Contracts' (which is meant in the narrower sense) and are defined as all agreements that govern the relationship between project participants in a Carbon Project, namely the acquiescence, transfer, and distribution of all or parts of the carbon rights or Emission Reduction Credits generated by a CDM or JI (Carbon) project. Such Carbon Contracts may take various forms:

(a) the contract can be structured to purchase CERs or ERUs or options to these units, without specifying which projects have to generate the units;
(b) the contract can link the credits purchased to a specific project, either purchasing the emission reductions once they have been verified but before

[14] For the CDM, the Executive Board has clarified that a project participant is either a Party involved or a private and/or public entity authorized by a Party to participate, under the Party's responsibility, in CDM project activities which takes decisions on the allocation of CERs from the CDM project activity under consideration. See CDM-PDD Glossary of terms available under the documents section in <http://cdm.unfccc.int/> accessed 31 March 2004.

[15] Letters of Approval have an interesting double nature as administrative instruments under the domestic law of the host country and at the same time as unilateral statements under international public law generally directed towards an Annex I Party and authorizing private entities to participate in the project.

[16] See n 14 above.

they have been issued as units defined under the Kyoto Protocol (as verified emission reductions), or purchasing CERs or ERUs as units issued under the Kyoto Protocol. These contracts generally include project specific covenants and obligations;[17]

(c) the contract can be drafted as an option agreement, under which the project sponsor sells an option to verified emission reductions, CERs or ERUs generated by a specific project;

(d) the contract can require the transfer of a certain number of Emission Reduction Credits to investor(s) in a project as profits or dividends of an equity investment into the project;

(e) it can foresee the transfer of Emission Reduction Credits as down payment under a loan or a credit contract for a specific project, in which case the Carbon Contract would be included in or closely linked to the project finance agreements;

(f) finally the transfer of Emission Reduction Credits can be embedded in other project agreements such as Joint Venture Agreements, Risk Servicing Contracts, Concession Contracts or 'Build, Operate, Transfer' (BOT) Contracts.[18]

The choice of the right contract type depends on how risks will be shared between the parties, the involvement of the counterparties in the project, and how closely the purchaser of the Emission Reduction Credits wishes to monitor the progress of the project. The intended end-use of the Emission Reduction Credits is an important consideration for the acquiring party when deciding on which contract type is most appropriate. If the acquiring party wants to procure CERs and ERUs at minimal risk to meet international emission reduction targets, the purchase contract used will look different from a loan contract under which a bank accepts Emission Reduction Credits as part of the down payments of a loan, or the type of purchase contract a Emission Reduction Credit broker will need. Purchasers that wish to purchase Emission Reduction Credits but have neither the capacity nor an interest in supervising a Carbon Project may also choose to purchase CERs or ERUs leaving the risk of how to generate these rights with the selling party. However, leaving the risk of the project development and the issuance of CERs and ERUs fully on the selling party will likely result in a significantly higher price per credit purchased. Closer cooperation between the parties in

[17] Covenants mean formal agreements or promises, usually in a contract. Obligations imply the formal, binding agreement or acknowledgement of a liability to pay a certain amount or to do a certain thing for a particular person or set of persons, *Black's Law Dictionary* 7th edn, Bryan A. Garner (ed.) (West Group, St. Paul Minneapolis 2000).

[18] See IL Worika and T Walde, 'Contractual Architecture for the Kyoto Protocol: From Soft and Hard Laws to Concrete Commitments', (2000) 15 J. Land Use & Envtl. L. 489 for an analysis of different contract types. The authors' analysis is, however, still based on the earlier assumption that the interest in carbon credits could be the main driver between the implementation of a project. Reality has shown that carbon finance adds only 1–5 % to the rate of return of renewable energy projects.

creating the Emission Reduction Credits and a sharing of the project and regulatory risks is generally reflected in the negotiated price.

It is important for the entity developing and implementing the project to structure the contract so that it increases the financial viability of the project and maximizes the contribution the cash flow generated from the sale of Emission Reduction Credits makes to the overall financial package. As purchasers of Emission Reduction Credits tend to be located in industrialized countries, this cash flow is likely to come in a hard currency and can be assigned to the lending bank or the technology supplier. Because the security of the cash flow critically depends on which contractual party assumes the regulatory risks, the seller of Emission Reduction Credits has an interest to shift the regulatory risk to the buyer of the rights.[19]

Choosing the appropriate type of contract will also depend on legal constraints imposed or additional requirements by the Parties involved or the project participants. Many carbon transactions so far have been conducted in the context of international tender procedures[20] in order to comply with domestic procurement rules. Other carbon purchasers make it a condition of their participation that the project generates a number of ancillary social and environmental benefits emphasizing the contribution of the project to the sustainable development of the host country. These projects usually enjoy extensive trust and support from the local communities as well as from the entities involved in the project implementation. In this context the project components that generate direct local benefits have been included in some carbon contracts. If such additional environmental or social components form part of the project, additional provisions for creating and monitoring the ancillary benefits that are produced by the project, need to be included in the contract.[21]

Forward purchase contracts, so-called Emission Reductions Purchase Agreements or ERPAs, are currently the most common form of Carbon Contracts. These contracts do not regulate financial services, such as depth or equity financing, but govern the acquiescence of defined Emission Reduction Credits. As long as the purchaser of the contractually defined right or commodity does not make any upfront payment, his risk is limited to the preparation costs of becoming involved in the project and opportunity costs. Many Carbon Contracts, including the one used by the World Bank Carbon Finance Business, are structured more like long-term supply or off-take agreements[22] than financing agreements. ERPAs, which will be discussed in more detail below, are revenue-based contracts which guarantee a long-term revenue to a project that is equivalent to the volume of Emission Reduction Credits transferred to the purchaser.

[19] The exact definition of the legal nature of the emission reduction rights will depend on the law which governs the contract, as well as on local law.

[20] The Emission Reduction Unit Procurement Tender (ERUPT) is the most prominent example for such a tender. See for more detail ch 17 above.

[21] The World Bank's Community Development Carbon Fund will exclusively fund such projects. Similar arrangements are already in place for some of the PCF projects such as the Jepirachi Wind Park Project.

[22] Such as Power Purchase Agreements or Natural Gas Purchase Agreements.

2.3 Risk Allocation

As in any contract, sharing risk is a key element of each Carbon Contract. The perception of risk by either party is reflected in the assessment of the value of the contract and in the willingness to pay. Minimal risk will lead to a minimal discount of the value of the contract.[23] A condition of a realistic risk assessment by both parties is knowledge and expertise on both sides. However, whereas it is possible to identify risks specific to a Carbon Contract, given the term of many these contracts and the currently formative stage of the carbon market, it is next to impossible for either of the parties to monetize these risks. In the cases where the project depends on the revenues from the Carbon Project, the seller of Emission Reduction Credits is likely to limit exposure to any of these risks, even if this means selling at a lower price. This applies equally where the project sponsor seeks additional financing for the project and intends to use the Carbon Contract as collateral for a loan. Financing institutions that consider lending against contract revenue will carefully assess the value of the contract, taking into account the risk of contract failure. Only if the contract performance is predictable will lenders consider lending against the revenues of a Carbon Contract.

Contract risks can either involve a default by a party or events outside the control of either party. Risks related to a Carbon Contract can broadly be summarized under the following categories:

(a) Project related risks;
(b) Host country related risks;
(c) Baseline risks;
(d) Kyoto Protocol and other regulatory risks.

Effective risk management requires creating incentives to ensure performance of the contract, and assigning responsibility for performance to the party best able to mitigate against the risk of underperformance. Project implementation is the project sponsor's responsibility. The sponsor controls the project and is best equipped to mitigate any delays or other problems relating to the construction or operation of the project. The sponsor has sufficient incentive to minimize such problems independent from the Carbon Contract as in most instances the Emission Reduction Credit revenues are not the primary source of income for the project,[24] which would also be effected by failure in project implementation. The other risks are shared between the parties and reflected in the price to be paid for the Emission Reduction Credits. Whereas some early actors in the carbon market, such as the

[23] Maximal risk leads to maximal discount. Nobody would base income projects on a lottery ticket, for example.

[24] The case may be different for projects involving gases with a higher GHG potential than CO_2, such as Methane, HF_{23} or N_2O, where the revenues of Emission Reduction Credits represent a significant portion of the total revenue stream of the project.

World Bank's PCF or the Dutch Emission Reduction Units Procurement Tender (ERUPT), were willing to take large parts of the Kyoto Protocol and regulatory risks, it is unlikely that private sector players will readily assume such risks. With the increasing maturity of the market and standardized and reliable procedures for CDM and eventually also JI projects, it is, however, expected that the regulatory as well as baseline risk will be easier to manage for all parties.

Another important issue in Carbon Contract negotiations is whether the performance or delivery of obligations should be enforceable through penalty payments. Enforceability is a key consideration for the parties, which can be supported through both remedies and penalties. However, penalties should be used with care in Carbon Contracts.[25] Due to the emerging nature of the carbon market, it is very difficult for parties to accurately assess the risk of underperformance and failed transfer of the full amount of credits. Concern has also been expressed where the contract is based on fixed prices but penalties or damages are based on unknown, possibly higher future prices.[26] Where the contract provides the right to seek damages or penalties to the market price of CERs or ERUs, the costs of underperformance are even more difficult to assess. Many carbon sellers therefore consider penalties as 'unfair' and will be reluctant to accept any penalty clause. Penalty clauses are seen as an additional risk to prospective financiers, who inevitably offer higher lending rates associated with higher risks.

From the buyer's perspective, penalties may not necessarily contribute to the success of the project. Carbon Contracts rely largely on trust built and maintained between the parties. The insecure regulatory environment in which a Carbon Project will be implemented requires a closely maintained and managed relationship between the parties to the Carbon Contract. Penalties may create barriers in the communication where they discourage the seller from communicating problems to the buyer. The buyer may therefore only be notified about a problem after it is too late to take measures to mitigate the situation. Finally, carbon buyers should keep in mind that penalties may be hard to enforce in case of underperformance of the project. It is unlikely that that carbon buyer will have seniority over lenders or suppliers when it comes to serving debts or paying liabilities.

2.4 Key Elements of a Carbon Transaction

When implementing a Carbon Project, project participants must, as with any project-based foreign direct investments, contractually determine how they will arrange for project governance, allocation of risks, payment modalities, defaults and remedies, dispute resolution, insurance, and *force majeure* events.[27]

[25] It is also noted that penalty clauses are not recognized in all common law systems. However, these systems often have more flexibility when it comes to calculating 'liquidated damages' than many civil law systems.

[26] A Mehta, C Shekhar Sinha, and K Capoor, *Contracting for Carbon* (Environmental Finance, October 2003).

[27] See also for AIJ: Danish *et al.*, n 12 above.

Moreover, regardless of which contractual structure is chosen, Carbon Contracts, must address a number of specific issues peculiar to Carbon Finance. These issues include:

(a) A definition of the commodity or right purchased (verified emission reduction, CER, ERU);
(b) The transfer of legal title to the verified emission reductions and CERs or ERUs, as well as the transfer of the CERs and ERUs within the National Registries;[28]
(c) The schedule of generation and transfer of the Emission Reduction Credits;
(d) Management and allocation of carbon finance specific risks (including baseline and Kyoto Protocol risks);
(e) Validation of the Carbon Project and verification and certification of the GHG reductions or removals generated by the project;
(f) The implementation of monitoring requirements;
(g) Registration of the project for the purpose of the Kyoto Protocol; and
(h) The approval of the project by the host country as well as additional host country guarantees.

Only a careful consideration of all these issues, a project specific risk assessment and a fair allocation of the risks will result in robust Carbon Contracts. When accounting for these issues in a Carbon Contract, the following should be taken into consideration.

2.4.1 The insecure regulatory environment

Three years ago developing a Carbon Project was described as 'packing for a trip to a distant planet that is observable only through a telescope.'[29] Even though the Marrakesh Accords[30] have since laid out most of the international rules governing Carbon Projects, there is still a flurry of open questions with respect to Carbon Projects. The value of the Emission Reduction Credits acquired under a Carbon Contract depends on a regulatory system that bestows upon the owner or holder of the unit the right to use it under a defined regime for the purpose of compliance with either binding or voluntary emission reduction targets.[31] Such GHG

[28] See ch 7 above.

[29] KW Danish and JC Rotter, 'Drafting Contracts For Greenhouse Gas Offset Projects in Developing Countries', (2001) 15 WTR Nat Resources & Env't, 168.

[30] FCCC/CP/2001/13/Add.2, draft Decision -16/CP.7 and its Annex on 'The decision on Implementation Guidelines for JI Projects' and draft Decision -17/CP.7 and its Annex on 'Modalities and procedures for a clean development mechanism'.

[31] Emission Reductions are also purchased in limited amounts to meet voluntary commitments outside of any regulatory regime. Since these voluntary commitments are generally at least indirectly linked to regulatory regimes (and if only assumed to avoid the imposition of any binding targets) and will eventually only contribute a negligible share to the global carbon market.

emissions trading systems have emerged over the last decade. The Kyoto Protocol sets out the rules for an international trading system, the European Emissions Trading System (EU ETS)[32] establishes a regional trading system, and individual countries, such as the UK have established national Emissions Trading Systems.[33] While these schemes define different types of rights and allowances, the majority of all Emission Reduction Credits generated by specific emission reduction projects are attributable to projects that are implemented in accordance with the rules and modalities for CDM projects or the draft guidelines for JI projects. The value of these credits outside of the Kyoto Protocol depends on regulations that link the Kyoto Protocol to other schemes.[34] However, as domestic legislators are still in the process of integrating JI/CDM specific regulation into their national legislation, Carbon Contracts to date operate in a very insecure regulatory environment. This means the final use, and hence the value of the Emission Reduction Credits generated by the project is largely volatile. The Carbon Contract must be flexible enough to adapt to, and accommodate the possibility of operating under diverse regulatory environments. Alternatively, a contract could also terminate or be terminated if a certain trigger event occurs, such as the failure of the Kyoto Protocol to enter into force by a certain date, or the ineligibility of certain Emission Reduction Credits under the EU ETS.[35]

As each carbon transaction still pioneers a new legal domain and area, clear definitions in the agreement itself are key for it to be robust. At the same time the agreement needs to include the appropriate clauses to ensure it is adaptable to any changes to the future regulatory environment. Conditions, trigger events, sharing of responsibilities and risks as well as clauses that stipulate the general cooperation between the parties in order to ensure that the main objective of the contract, namely the generation and transfer of Emission Reduction Credits, are met, should therefore be included in all Carbon Contracts.

2.4.2 Undefined commodity

Today, many Carbon Contracts govern the forward purchase of CERs or ERUs whereby the seller meets its delivery and transfer obligation once it has transferred the agreed amount of CERs and ERUs to the buyer's account.[36] The receipt of the

[32] Directive 2003/87/EC of the European Parliament and of the Council, of 13 October 2003 establishing a scheme for greenhouse gas emission allowance trading within the Community and amending Council Directive 96/61/EC [2003] OJ L275/32.

[33] Under the UK Emissions Trading Scheme thirty-one organizations ('direct participants' in the scheme) have voluntarily taken on a legally binding obligation to reduce their emissions against 1998–2000 levels, <http://www.defra.gov.uk/environment/climatechange/trading/ukets.htm>, accessed 11 April 2004.

[34] See ch 29 below.

[35] See n 32 above.

[36] For example, the model contract of the International Emissions Trading Association, EU Allowances Emissions Trading Master Agreement V. 2.0 at <http://www.IETA.org> accessed 10 October 2004.

units in the registry account of the buyer usually triggers the payment of the purchase price. However, there is a long process between the GHG removal or mitigation activity and the issuance of CERs or the conversion of AAUs into ERUs. The rights associated with this activity go through a number of different 'developmental stages'[37] and contractually each of these stages can be defined as fulfilment of the seller's obligation shifting the risk of the eventual issuance of the unit eligible under the Kyoto Protocol or any other regime.

The seller of the Emission Reduction Credit achieves the reduction or removal of GHG against an agreed and validated baseline of emissions. The generation of the GHG reduction can be monitored and verified either by the host country (as in the case of Track One JI) or by an independent verifier (CDM and JI).[38] The project activity itself does not create a physical thing or commodity that can be delivered to the buyer. Outside of a regulatory system, it is the Carbon Contract that defines the rights associated with this physical action (the 'Emission Reduction (ER)').[39] These rights usually include the rights which may be associated with present or future GHG reductions, including rights bestowed under an existing or emerging regulatory regime (including the Kyoto Protocol). This means that the ER carries with it a right to be eventually issued as CER or ERU. However, it does not entail a guarantee for such issuance and the eligibility of the project under the Kyoto Protocol.

The legal nature of the Emission Reduction Credits will depend on the law governing the Carbon Contract. Where there is no legislative regime and the ER is a purely contractual right, the legal nature of this right does not need to be identical with any unit that is associated with this ER (CER, ERU, EU Allowance, among others). Units defined under the Kyoto Protocol are rights created under public international law and will rely upon sovereign government action (AAUs, ERUs) or approval (CERs) to come into existence. The contractually defined ER often embodies the claim to such a right, and even where the contract specifies the transfer of a CER or ERU as the fulfilment of the contract obligation, such transfer is usually preceded by the transfer of the underlying ER as a contractual right.

Therefore, one of the most important considerations of the Carbon Contract is for the Parties to agree on the nature of the product to be purchased, especially since such decision carries an important decision on the risk allocation between the Parties. A project sponsor or seller of ER is primarily seeking a reliable income stream. As such, they will be reluctant to take the risk of ensuring the issuance of the Emission Reduction Credits under a particular regime, especially when these regimes are still riddled with uncertainties. In these instances the seller may opt for selling verified ERs, thereby shifting the risk of the acknowledgement of the units

[37] The creation of a CER or ERU involves the validation and registration of the project, the implementation of the project, the monitoring, verification and certification of the GHG reductions or removals, and finally the issuance of CERs or conversion of AAUs into ERUs.

[38] See ch 8 above for an explanation of the characteristics of Track One JI.

[39] See also ch 16 above.

under a regulatory regime to the buyer. A seller, who has a better ability to manage the process of approval and issuance of the credits, may decide to sell units eligible under a regulatory regime reflecting this additional risk in a higher price.

2.4.3 Ownership of the emission reduction and right to transfer

AAUs, ERUs, and CERs are sovereign rights and can only be held and transferred following proper authorization by the government of an Annex I Party. However, private entities can define ERs contractually and regulate the transfer of the title of the defined rights. These rights emerge in the contractual context only; any link to a sovereign emission right is contingent upon it being bestowed by a State. In this context it is important to establish the owner of the ER and undertake the necessary due diligence to ensure that the alleged seller of the reductions is indeed authorized to do so. Where there are any doubts about the selling entity and in the absence of domestic legislation the right to the ERs must be established through the contractual relationship between the parties developing and implementing the project. The contractual agreement should be further confirmed by the approval of the host country.

The owner of the project that achieves the physical reduction or removal of GHGs is generally considered to be the owner and legitimate seller of ERs. Host country approvals of early CDM projects have confirmed this view. However, complex project financing arrangements may lead to situations where a multitude of legal entities, operators, financiers, and suppliers are implementing parts of the project or hold shares in the project company. In these cases as well as in cases where the buyer decides to enter into the Carbon Contract with the parent company of the entity achieving the GHG reductions, it is important to ensure that the right to sell the rights related to the ERs lie with the identified counterparty of the Carbon Contract.

Another, even more complicated problem, relates to the possibility of accounting twice for the same ER. What is commonly know as the 'double counting' problem is the concern that a Carbon Project, such as a grid-connected renewable energy project or a demand-side energy efficiency measure, will reduce emissions of GHG in the grid which could be sold by the energy supplier as well as the entity implementing the measure. The problem is even more prominent for regional grids or when a grid is linked to a non-Party to the Kyoto Protocol. In most cases, the entity that implements the GHG reducing activity is considered to be the legitimate owner of the Emission Reduction Credits generated by the project and will obtain authorization by the host country to sell the ERs. It is therefore also the appropriate partner for any Carbon Contract. However, the definition of the boundaries of the project is a crucial consideration in this context, and it is doubtful this entity can also claim the rights to emission reductions achieved outside of the territory of the host country. Any decision in this context should be discussed early with the regulating entities (for example, the Executive Board in the case of the CDM) and needs to be confirmed at the project level by the independent entity validating the project and verifying the ERs.

2.4.4 Long-term nature of the contract

Another feature of many Carbon Contracts is the long period over which Emission Reduction Credits are delivered and payments to the project are made. In the case of CDM projects involving carbon sinks, the crediting period for a project, and hence the possible term of the contract, can stretch up to sixty years.[40] CDM projects mitigating climate change through mitigation of GHG emissions have a maximum life of twenty-one years.[41] JI project activities are limited to generate ERUs during the first commitment period 2008–12. However, projects involving early or late crediting can involve contractual arrangement which determine a significantly longer contracting period.[42]

The potential for multi-decade agreements makes the establishment of a series of safeguards necessary to ensure successful long-term implementation of the project and the contract. The uncertainty surrounding the evolution of the carbon market raises additional issues, which require careful consideration when the contract is being designed. Issues like sudden increases or decreases in the price, backdrops against inflation, measures in case of underperformance or delays in the implementation of the project, change in ownership of the seller or the buyer as well as future additional requirements in the creating of the ER asset all need to be considered.

2.4.5 Extent of government involvement

Another issue peculiar to Carbon Contracts is the type of involvement and continuous support from governments as required under the Kyoto Protocol. To implement a Carbon Project under the Kyoto Protocol, all countries involved in the Project have to support the project through their approval. For JI projects, governments will also be involved in the creation of the ERUs through converting AAUs or RMUs into ERUs upon verification of the emission reduction for the purposes of the Kyoto Protocol.

[40] See draft Decision -/CMP.1, Modalities and procedures for afforestation and reforestation project activities under the clean development mechanism in the first commitment period of the Kyoto Protocol agreed at the 9th Conference of Parties to the UNFCCC in Milan.

[41] See the crediting periods for CDM projects as set out in the Marrakesh Accords CDM Modalities and Procedures FCCC/CP/2001/13/Add.2, para 49.

[42] The Marrakesh Accords have confirmed that ERUs can only be earned for the period between 2008 and 2012, when they provide '. . . projects starting as of the year 2000 may be eligible as Article 6 projects if they meet the requirements of the guidelines for the implementation of Article 6 of the Kyoto Protocol as set out in the annex below and that ERUs shall only be issued for a crediting period starting after the beginning of the year 2008.' FCCC/CP/2001/13/Add.2, Decision 16/CP.7, draft Decision -/CMP.1 See also chs 6, 7, and 10 above.

Governments are also involved in the international negotiations fora, where they decide, at the UNFCCC's Conference of the Parties (COP) and eventually at the Kyoto Protocol's COP to the UNFCCC serving as the Meeting of the Parties (COP/MOP), on the modalities, rules, and guidelines of Carbon Projects. Governments also decide on how to incorporate Emission Reduction Rights and carbon transaction into domestic legislation. The necessary and continuous support of the government of the project and the risk of losing such support needs to be considered at the time of drafting the contract and the negotiated risk allocation needs to be reflected in the document.

2.4.6 Cross-border nature of the transaction

Most Carbon Contracts involve the transfer of ERs across the border of two or several States. This cross-border nature of the contract triggers the issue of the governing law of the contract. Whatever law is chosen as the most appropriate to govern the agreement, the parties need to carefully examine the contract in the light of the chosen law—especially since there is no legislature clearly defining the nature of the rights to Emission Reduction Credits, and a body of jurisprudence has not yet been developed to guide arbitrators or local judges in common law jurisdictions.

Parties also need to decide on who will bear local and international taxes and fees (in case of JI and CDM projects). Such fees include the 'share of proceeds', a percentage of CERs to be deducted from the CERs issued for a CDM project, as well as an administrative fee levied by the UNFCCC Secretariat for the maintenance of the CDM registry.[43] Local taxes may include sales taxes, such as Value Added Tax (VAT) as well as fees related to the approval of the project (and the verification of the ERs in case of Track One JI) by the host country.

2.4.7 Baseline risk

Carbon Projects depend on the achievement of 'real' and 'measurable'[44] GHG reductions, which need to be monitored and verified against a baseline. Changes in the baseline scenario may affect the project by altering the projected GHG reduction scenarios. This risk becomes relevant for CDM projects at the end of the seven-year crediting period, in case the project participants seek a renewal of the crediting period. In the case of JI the rules governing the validation of the baseline and the verification of the GHG reductions depend on the host country as soon as the country qualifies for Track One JI. A change in the baseline needs to be considered when developing a project and risk of a change in baseline renders the seller of the Emission Reduction Credits unable to transfer the agreed amount of credits needs to be reflected in the contract. If the baseline changes, this may lead

43 FCCC/CP/2001/13/Add.2, Decision 17/CP.7, Draft Decision -/CMP.1, Appendix D, para 6(b).

44 Art 12(5) Kyoto Protocol.

to an adjustment of the purchased amount, or alternatively the parties may have agreed upfront to rely—for contractual purpose—exclusively on the baseline as agreed at the time of the signature of the contract.

3 World Bank Carbon Finance Contracts

Since the establishment of the PCF in 1999, the World Bank has been active in promoting the CMD and JI. The IBRD, as trustee of the PCF, purchases ERs from CDM and JI project sponsors and distributes them, on a pro-rata basis, to the participants in the fund. The other World Bank Carbon Funds developed and established since the creation of the PCF all follow the same basic model: the World Bank acts as trustee for the Annex I entities (public or private) and purchases Emission Reduction Credits for the benefit of the fund participants from entities developing Carbon Projects in CDM or JI host countries.

PCF ERPAs are individually negotiated for each project and each agreement contains provisions, which take into account the particularities of the project. Nevertheless, all ERPAs follow the same structure applying the same basic terms. The World Bank has developed standardized documentation including a model ERPA for all of its Carbon Funds. Standardization saves time and costs and makes the financing easier, provided that the financiers are familiar with contractual terms. On the other hand, it could defeat the premises for flexibility and dynamism in achieving contractual objectives.[45] Through the ideal of an 'individually negotiated standardized ERPA' the World Bank has tried to standardize as much as possible, while at the same time to negotiate and accommodate the specific case as much as necessary.

The projects that receive financing from the PCF follow the rules and procedures developed by the decisions of the UNFCCC COP, which set out the rules for CDM and JI project activities.[46] The implementation of the PCF is governed by the 'Instrument establishing the Prototype Carbon Fund' adopted by the World Bank Board Executive Directors.[47] Under the terms of the PCF Instrument, the Trustee undertakes to provide for as much flexibility as possible within the terms of its contractual arrangements for the purchase of Emission Reduction Credits. All this in the acknowledgement that the regulatory framework of the UNFCCC and the Kyoto Protocol relating to the ownership, holding, and transfer of ERs is still under development and that flexibility is necessary to maximize the likelihood that the PCF will

[45] Worika and Walde, n 18 above, 497.

[46] Marrakesh Accords, FCCC/CP/2001/13/Add.2. Draft Decision 16/CP.7 and its Annex on 'The decision on Implementation Guidelines for JI Projects' and draft Decision 17/CP.7 and its Annex on 'Modalities and procedures for a clean development mechanism', ibid.

[47] International Bank for Reconstruction and Development, Resolution No 99-1 Authorizing Establishment of the Prototype Carbon Fund as amended by Resolutions Nos 2000-1 and 2003-3.

achieve its stated objectives.[48] The PCF[49] purchases ERs from projects of the nature of Article 6 or Article 12 projects. The PCF anticipates that such projects will be able to generate CERs or ERUs once the project has been registered as CDM Project or once the Protocol has entered into force and after the beginning of 2008 (for JI Projects).[50]

3.1 Main Features of the PCF ERPA: Carbon Specific Clauses

Definitions. In this context, the definitions included in Article I of the PCF model ERPA are among the cornerstones of the model, and are crucial to ensure consistency in the operations of the fund. Definitions are even more important where there is no legislative framework that could be used as a reference point. The only reference point so far for the implementation of the CDM and JI is the Kyoto Protocol itself together with the decisions of the COP adopted in relation to these mechanisms. These definitions, included in the PCF ERPA, remain unchanged in all ERPAs and define the subject of the contract as well as the main parameters of the creation of the ERs.

Emission reductions. In accordance with the PCF Instrument, the IBRD acting as the trustee of the PCF purchases 'Emission Reductions (ERs)' after they have been verified and certified by an independent third party. The PCF trustee takes the risk that the Kyoto Protocol will not enter into force as well as that the ERs purchased will eventually be issued as CERs or transferred as ERUs. ERs are therefore defined in the PCF ERPAs as:

> any right, interest, credit, entitlement, benefit or allowance of the Host Country to emit (present or future) arising from or in connection with any GHG Reduction by the Project and includes any right that may be created under any regulatory or legal regime as a result of the GHG Reductions whatsoever.[51]

GHG reductions refer to the result of the mitigation activity, namely the:

> removal, limitation, reduction, avoidance, sequestration or mitigation of GHGs by the Project below the Baseline and measured in tCO_2e as monitored in accordance with the Monitoring Plan, and which have been generated in a manner consistent with the International UNFCCC/Kyoto Protocol Rules and any applicable specific environmental and social mitigation requirements identified in the Monitoring Plan.[52]

These definitions are crucial elements of the agreement as they establish the rights that are sold and purchased. They ensure that the PCF will receive all the rights that

[48] See, PCF Instrument, Art XIII, Sec. 13(1).

[49] Legally it is the IBRD acting as trustee of the PCF which acts as a contractual party, for the sake of the simplicity the text refers also to the 'PCF' as purchaser of emission reductions.

[50] To date the World Bank has entered into twelve PCF ERPAs, eight of which are for CDM projects with the remaining four JI projects. Another fifteen agreements are currently under negotiation with project sponsors and are expected to be signed before the end of 2004.

[51] Section 1.01 of the PCF Model CDM ERPA.

[52] ibid. All terms starting with capital letters are defined in the PCF ERPA.

are, or will be associated with a tonne of GHG reduced or sequestered by the project, while at the same time giving the seller the confidence that once they have generated and verified the ER, they cannot be held responsible for any further use of the reduction under a regulatory framework. The focus of the PCF ERPA therefore lies in ensuring that the project is implemented in accordance with the emerging JI/CDM framework and that the GHG reductions are monitored and verified applying the emerging rules.

Full Title. An important element of the ERPA is the assurance from the seller that it has full title to all of the GHG Reductions generated by the project free of any encumbrance and that it has not sold, transferred, assigned, licensed, disposed of, granted, or otherwise created any interest in the GHG reductions or contracted ERs to any third party.

Assignment of Claims to CERs/ERUs. The ERPA also foresees the conversion of the ERs into CERs or ERUs and ensures that the seller assigns to the PCF all of its rights to any CERs or ERUs arising from the contracted ERs.

Monitoring. For the purposes of the Carbon Project the seller commits to establish and maintain data measurement and collection systems for all indicators listed in the monitoring plan and to 'install, operate and maintain the facilities and equipment, and employ and train staff, necessary for gathering all such data as may be required by the Monitoring Plan.'[53]

Reporting. Based on the collected data, the seller submits an 'Annual ER Report'[54] to the PCF that includes the amount of the monitored GHG Reductions generated in the previous year (or agreed monitoring period). The Annual ER Report is defined as:

> a report provided by the Project Entity setting out the number of GHG Reductions generated by the Project during the previous Year as monitored in accordance with the Monitoring Plan and which includes all other data as may be required to be collected and recorded by the Monitoring Plan and which upon entry into force of the Kyoto Protocol shall serve as the monitoring report required to be provided under the International UNFCCC/Kyoto Protocol Rules.[55]

Verification. This Annual ER Report provides the basis for the verification of the GHG reductions achieved by the project. The PCF contracts the verifier in

[53] Section 7.01 of the PCF Model CDM ERPA.

[54] The Annual ER Report 'means a report provided by the Project Entity setting out the number of GHG Reductions generated by the Project during the previous Year as monitored in accordance with the Monitoring Plan and which includes all other data as may be required to be collected and recorded by the Monitoring Plan and which upon entry into force of the Kyoto Protocol shall serve as the monitoring report required to be provided under the International UNFCCC/Kyoto Protocol Rules'.

[55] Section 7.02 of the PCF Model CDM ERPA.

a competitive process from the list of the UNFCCC accredited verifiers.[56] The verifier conducts the verification applying the rules and procedures for CDM projects (awaiting further guidance for JI projects).

Other Kyoto Protocol Related Clauses. The ERPA includes the agreement on the crediting period for the project for CDM projects, agreement that the PCF shall serve as the focal point for all communications with regard to the project with the Supervisory Committee or the CDM Executive Board,[57] and that the project entity will cooperate in the issuance of CERs or conversion of ERUs. PCF ERPAs for JI projects also contain an article dealing with a possible conflict between the EU ETS and the JI project activity.

3.2 Main Features of the PCF ERPA: Commercial Clauses

Delivery and Payment. Under the PCF ERPAs the World Bank buys an agreed total amount of ERs at a fixed price per ER for the lifetime of the contract. The PCF pays upon 'delivery' of the ERs, such delivery to be understood as the submission of the verification report to the PCF. Upon receipt of the verification report the PCF makes the payment for the ERs due to it and the legal title to the ERs passes from the seller to the World Bank.

Price and Delivery Risk. The agreement does not provide for future renegotiations of the price should the market price for comparable Emission Reduction Credits increase or decrease. Presented with this, the seller needs to weigh the advantage of an early contract and a secure cash flow against the possibility of taking the risk that ERs will be issued as more valuable CERs or ERUs some time in the future. The best way to manage these diverting interests and at the same time create a robust contract is for the PCF to purchase only a portion of the GHG reductions projected to be generated by a project. Such so-called 'over-collareralization' allows the seller to take advantage of future up-sides of the carbon market by giving the seller the possibility of selling surplus ERs to future buyers at higher prices. The PCF benefits from such a structure as it gives the seller incentive to implement the project and over-perform on its promise to generate GHG reductions. In this arrangement the PCF reserves the right to a certain annual minimum ERs to be generated and transferred. It generally also claims the right to all reductions generated above and beyond the annual minimum amount until the total purchase amount is met, plus any agreed annual amount which can be sold to a third party.

[56] Before such accreditation had been finalized, the PCF contracted the validators and verifiers from the list of the applicant entities. For both lists see the UNFCCC website: <http://cdm.unfccc.int/> accessed 11 April 2004.

[57] Generally, the PCF takes care of methodological issues and assumes the responsibility to submit the project for registration with the EB.

Cost Recovery. The PCF bears large parts of the upfront development costs of a Carbon Project. Additional to its own staff costs, it pays for the establishment of the baseline, the development of the monitoring plan and the project design document, and the validation of the project and the aforementioned documents. This relieves the project developer from advancing these significant up-front costs. The PCF recovers these costs from the annual payments of the purchase price following a recovery schedule agree with the seller.

Option. In some cases the seller has granted to the PCF an additional first-right of refusal, or a simple call option, to purchase additional ERs beyond the contracted amount. These rights help the World Bank manage the PCF portfolio, ensuring that at the total amount of the PCF Trust Fund (US$180 million) will be used efficiently to purchase ERs. With simple option arrangements, the PCF hedges its risk of project failure across its portfolio.

Production Failure. In case a project default results in a failure to generate and/or transfer the agreed minimum annual amount of ERs (serving as default amount), the PCF reserves the right to require the seller to propose a plan of action to remedy the shortfall, which includes appropriate adjustments to the minimum amounts. If there is persistent under-delivery, as well as in a number of other default events[58] that make it impossible or highly unlikely for the project to operate and achieve GHG reductions, the PCF has the right to terminate the agreement. The seller has the reciprocal right of terminate the agreement if the PCF fails to make timely payments.

Additionally, PCF ERPAs contain a clause that ensures the PCF has the right to recover damages from the seller in case of fraud, gross negligence or wilful misconduct by the seller. This clause is particularly important to dissuade the seller breaching the contract and selling the Emission Reduction Credits to a third party, if, for example there has been a raise in the market price.

Implementation of the Project. Under the PCF ERPA, the seller of the ERs undertakes to implement the project 'with due diligence and efficiency and in conformity with appropriate administrative, financial, engineering and environmental practices and all other relevant requirements of this Agreement.'[59]

Miscellaneous. Finally, the ERPA contains clauses covering the amendment of the agreement, confidentiality, *force majeure,* notices, IBRD immunities, arbitration, and assignment of the contract. The clause dealing with IBRD immunities appears in all World Bank Carbon Contracts, and states the IBRD reserves its privileges and

[58] Such as dissolution of the project company, failure to conclude construction in time or failure to obtain the permits necessary for project operation.

[59] Section 9.01 of the PCF Model CDM ERPA.

immunities, granted by its Member States and established in its Articles of Agreement.[60] Naturally, this clause will not appear in other Carbon Contracts. Arbitration clauses, on the other hand, are common in most contracts. When drafting the arbitration clause, project participants need to consider the governing law, the arbitration procedures, the selection of the arbitration tribunal as well as the place of arbitration. The PCF refers to arbitration rules developed by the UN Committee on International Trade Law (UNCITRAL).

4 Conclusion

The objectives of this chapter were to give an introduction to the main issues that need to be considered when drafting a CDM or JI project agreement. As an introduction to the topic, this chapter claims neither to be comprehensive nor to have treated those points mentioned in depth. It has not treated the content of early project agreements such as Confidentiality Agreements, Letters of Intent or Mandate Letters, or Term Sheets. It has also left aside Letters of Approval or Host Country Agreements. However, it hopes to have provided a checklist of the most important issues a Carbon Contract must deal with.

In the absence of a legislative national framework the work of a lawyer working on Carbon Projects is fraught with uncertainties. The experience gathered in the implementation of the Carbon Finance Business may help any lawyer faced with the challenge of drafting a Carbon Contract to access helpful sources and understand the basic principles.

In summary, it can be said that the most important step in structuring a carbon transaction is the development of a risk matrix. The allocation of risk will provide the backbone of the contract, supported by clear, unambiguous definitions and clauses that ensure the flexibility of the agreement. Preciseness is required in order to ensure robustness, and flexibility should not be confused with vagueness. In a final analysis, it must be said that most Carbon Contracts will require amendments during their lifetime; they will grow and develop with the project. It is the challenge of the lawyers working on the project to provide the framework and the basis for the successful implementation of the project. It is the challenge of seller and buyer to develop and maintain the trust and spirit of cooperation to ensure the success of the project.

[60] Art VII, IBRD Articles of Agreement as amended 19 February 1989.

21

Arbitration in 'Flexible-Mechanism' Contracts

*Dane Ratliff**

To give a satisfactory decision as to the truth it is necessary to be rather an arbitrator, than a party to the dispute.

Aristotle (384 BC–322 BC)

1 Introduction

In the above quote, Aristotle recognizes the nature of third-party dispute settlement, and its ability to produce a 'satisfactory' decision. A party to a dispute is necessarily affected by its role in the dispute; is necessarily biased toward its own position and the truth of that position. An arbitrator, on the other hand, must be impartial and independent of the dispute, weighing the statements of both parties regarding the truth (the subject matter of the dispute) against his[1] own experience and knowledge. Only in so doing is he able to arrive at a 'satisfactory decision as to the truth'; 'satisfactory' meaning that both parties must accept his decision, for in having given their consent to appoint a third-party arbitrator to resolve their dispute, they each admit their own respective inability to agree with the truth of the other's position.

Arbitration is a form of dispute settlement that has changed little in its general principles since the nineteenth century, that is:

- Parties consent to settle their dispute through arbitration;
- If there is more than one arbitrator each party may choose its own arbitrator and the two so chosen appoint a president;
- Arbitrators should be neutral and impartial;

* Dane Ratliff is legal counsel at the Permanent Court of Arbitration. The views expressed in this chapter are the views of the author and do not necessarily represent the views of the Court.

[1] The use of the masculine implies the feminine, and vice-versa throughout this chapter.

- Parties are free to designate the applicable law and rules of procedure the tribunal is to follow;
- The arbitral award is final and binding.[2]

All of the basic tenets of arbitration lend themselves to resolving international disputes. In international disputes, rarely would a party prefer to submit to the courts of the opposing party, for such fears as lack of neutrality, unfamiliarity with the foreign legal system, and language barriers. Arbitration is the only legally binding dispute settlement option capable of meeting such concerns.[3]

Being able to rely on international law and designate the applicable law, is a particular advantage of arbitration in contrast with resolving disputes in domestic courts, especially in countries where there are ill-developed legal systems regarding a particular issue. Until today there are few legal systems that explicitly classify and define emissions rights, or deal with the ownership, trading, or taxation, of such rights. Also, arbitral awards are binding and final and could in most cases be widely enforced by relying on recognition and enforcement laws and/or broadly accepted international conventions, whereas there may be difficulties in getting the judgment of a domestic court enforced abroad.[4]

The philosophy of dispute settlement and arbitration aside, this chapter is concerned with the dispute settlement potential and reality in international flexible mechanism (flex-mech) related contract disputes: specifically, transactions under the Clean Development Mechanism (CDM) and Joint Implementation (JI) under the Kyoto Protocol (Protocol) to the United Nations Framework Convention on Climate Change (UNFCCC or Convention),[5] and generally with the International Emissions Trading (IET) regime, as these appear to be the most likely areas in which disputes will arise, and where arbitration clauses will be needed.[6]

The uncertainties in drafting emissions trading and other flex-mech related contracts are great—as a corollary, so is the risk of a poorly drafted contract. As one author in this book has pointed out elsewhere regarding IET:

> Because the Kyoto Protocol is not yet in force and, in many cases, neither are the domestic rules that create emissions reduction credits or allowances, these often only exist as contractual rights to credits or allowances that may come into existence at some future time. Therefore, for the time being and even after the Kyoto Protocol comes into force, clear

[2] See Redfern and Hunter, n 9 below, 2–10.

[3] See section 3.3 below on 'Form and Content of the Arbitration Agreement /Clause'.

[4] See A Parra, 'Provisions on the Settlement of Investment Disputes in Modern Investment Laws, Bilateral Investment Treaties and Multilateral Instruments on Investment' (1997) 12 *ICSID Review—Foreign Investment Law Journal*, 288–9.

[5] Arts 6 and 12 Kyoto Protocol to the United Nations Framework Convention on Climate Change (Protocol), (1998) 31 ILM 22. As previous chapters have dealt with these instruments in detail, it is not considered necessary to enter into a detailed analysis of them here.

[6] The potential for disputes in these mechanisms is high because of the complexity of the transactions, and multiplicity of actors of mixed origin involved in them.

and well-drafted contractual obligations will be essential to this emerging market in 'carbon'.[7]

The fact that arbitration clauses are separable and may survive the underlying contract[8] even if its provisions are struck down, is one of many reasons making arbitration clauses in such contracts an indispensable necessity for mitigating risk.

There are two sections in this chapter, the first giving an overview of the climate change dispute settlement regime, and the second dealing with arbitration in flex-mech related contracts (also: Carbon Contracts). In treating the latter topic, the potential parties to a dispute, the form of the arbitration clause, and the legal framework in which the arbitration clause exists, will all be examined. This chapter is not however, a general exegesis of the principles of international arbitration, as there are exhaustive encomiums thereon.[9] It is hoped however, that the reader will benefit from application of those general principles of arbitration to the particulars of flex-mech related contracts.

A brief overview of the UNFCCC and the Protocol[10] dispute settlement provisions is set out in following.[11]

2 Dispute Settlement in the UNFCCC and Kyoto Protocol

The dispute settlement procedures under the UNFCCC and the Protocol are only open to States Party to the Convention and, as the case may be, the Protocol, in which case Article 19 of the Protocol states that the provisions of Article 14 of the UNFCCC apply *mutatis mutandis* to the Protocol. Article 14(1) of the UNFCCC on dispute settlement reads as follows:

> In the event of a dispute between any two or more Parties concerning the interpretation or application of the Convention, the Parties concerned shall seek a settlement of the dispute through negotiation or any other peaceful means of their own choice.

[7] A Hobley, 'Is Kyoto Dead? Climate Change After Bush' (2002) 5 *Environmental Liability* 167–83, at 179.

[8] Redfern and Hunter analyse the separability of the arbitration agreement from the underlying contract by describing the contract as actually having two parts. The first part deals with the commercial relationship, and the second part contains the arbitration clause. See, Redfern and Hunter, n 9 below, 54.

[9] Arguably the best known examples of such works are: E Gaillard and J Savage (eds), *Fouchard, Gaillard, Goldman, on International Commercial Arbitration* (The Hague, Kluwer Law International, 1999); GB Born, *International Commercial Arbitration* (2nd edn, The Hague, Kluwer Law International, 2001); L Craig, W Park, and J Paulsson, *International Chamber of Commerce Arbitration* (New York, Oceana, 2000); A Redfern and M Hunter, *Law and Practice of International Commercial Arbitration* (3rd edn, London, Sweet and Maxwell, 1999).

[10] United Nations Framework Convention on Climate Change (UNFCCC), 1771 UNTS 165 (1992), and the Protocol, n 5 above.

[11] It is argued below that formal inter-State dispute settlement is now more unlikely to obtain due to the elaborate provisions on compliance in the Marrakesh Accords. See Decision 24/CP.7, UN Doc FCCC/CP/2001/13/Add.3, 21 January 2001. For history and analysis of the compliance regime, see R Lefeber, 'From the Hague to Bonn to Marrakesh and Beyond: A Negotiating History of the Compliance Regime Under the Kyoto Protocol', in *Hague Yearbook of International Law* (The Hague, Kluwer Law International, 2001) 25–54.

Article 14(1) of the UNFCCC imposes a duty on the Parties to a dispute to first attempt to negotiate or use 'other peaceful means of their choice'[12] to resolve it. This multi-tiered approach is not uncommon in international law, as Sir Robert Jennings has remarked, 'many if not most jurisdiction clauses of treaties require that negotiation shall be tried first, and only if that fails can there be a valid resort to a court or tribunal'.[13] Multi-tiered approaches are also not uncommon in private emissions trading contracts as will be seen below.

Although it is not clear from Article 14(1) how the dispute will be made known since there is no duty of notification, Article 14(5) states that (subject to Article 14(2)) conciliation proceedings may be initiated 'at the request of any of the parties to the dispute' after twelve months following prior 'notification by one Party to another', that a dispute exists between them. Clearly from the negotiating history, unilaterally triggerable compulsory arbitration was a contentious issue,[14] and the emphasis was placed on using negotiation and the Conference of the Parties (COP) to resolve disputes before resorting to formal proceedings. It could be argued that because of that, even the official notification of one Party to the other that a dispute exists was not made mandatory, as such notification sets a formal procedure in motion, leading to the ability of one Party to unilaterally submit the dispute to conciliation.[15]

Article 14(2) is drafted so as to allow Parties to accept, by means of a declaration, compulsory submission of any dispute 'concerning the interpretation or application of the Convention [or Protocol as the case may be]' to either the International Court of Justice (ICJ), or arbitration in accordance with procedures to be adopted by the COP.[16] So far, only the Solomon Islands has made such a declaration. As mentioned in footnote 11 above, the likelihood of inter-State arbitration under Article 14, much less adjudication before the ICJ, has decreased given the elaborate compliance procedures in the decisions adopted by the seventh session of the Conference of the Parties to the UNFCCC, the so-called Marrakesh Accords,[17] and the focus therein on measures designed to enable compliance. The facilitative arm of the compliance branch can provide advice to facilitate compliance. Taken

[12] These could be for example, any of the peaceful means noted in Art 33(1) of the UN Charter on which Art 14 UNFCCC is modelled: negotiation, enquiry, mediation, conciliation, arbitration, judicial settlement, or resort to regional agencies or arrangements. Parties would be free to choose the procedures pertaining to these other peaceful means.

[13] R Jennings, 'Keynote Address', in *Resolution of International Water Disputes*, Permanent Court of Arbitration (ed.), (The Hague, Kluwer Law International, 2002) 27.

[14] H Ott and S Oberthür, *The Kyoto Protocol: International Climate Policy for the 21st Century* (Berlin, Springer, 1999) 223, n 53.

[15] Again, this is predicated on the condition that after twelve months from notification of the dispute, negotiations have failed to settle it.

[16] These procedures have not yet been adopted, however, the Permanent Court of Arbitration has been promoting the use of its 'Optional Rules for Arbitrating Disputes Relating to Natural Resources and/or the Environment' to serve as the annex of arbitration procedures contemplated under Art 14(2)(b) UNFCCC, and the corresponding environmental conciliation rules to serve as the annex of conciliation procedures contemplated under Art 14(7) UNFCCC. See <http://www.pca-cpa.org/ENGLISH/EDR> for more information.

[17] See n 11 above.

together, both types of compliance procedures may ultimately serve to prevent disputes from arising, by encouraging Parties to keep their house in order on the issues that might cause disputes, that is, meet their targets, ensure proper accounting, and so on.[18]

It is worth noting that section XVI of Decision 24/CP.7 of the Marrakesh Accords on compliance, sets out that the procedures and mechanisms relating to compliance operate 'without prejudice' to dispute settlement under Article 19 of the Protocol. This means that compliance branch decisions will not affect the outcome of dispute settlement proceedings, and that dispute settlement proceedings and compliance branch proceedings can operate concurrently. The notion of concurrent compliance branch and dispute settlement proceedings dealing with the same or related subject matter, yet resulting in conflicting decisions, raises some interesting questions about how such conflicting decisions of the two bodies could co-exist. Reference to Lefeber and Goote's distinctions between compliance proceedings and dispute settlement helps clarify that the objectives of the two proceedings will however, probably not produce much overlap:

- A dispute settlement procedure provides a state with an instrument to primarily protect its individual legal interests, whereas a compliance procedure is an instrument to protect a common treaty interest;
- As opposed to dispute settlement procedures, compliance procedures are non-adversarial;
- Compliance procedures are future-oriented and proactive . . . Dispute settlement procedures, however, are past oriented and reactive. They provide an avenue for the injured state to seek redress. Responses under dispute settlement procedures focus on the consequences of internationally wrongful acts, namely cessation, non-repetition, and reparation, such as restoration, compensation and satisfaction.[19]

Finally, arbitral proceedings are sometimes stayed at the request of a party when there are concurrent proceedings (*lis pendens*) on the same or related subject matter in another judicial forum.[20] Yet the compliance branch, although it acts in a quasi-judicial manner potentially imposing sanctions for non-compliance with legal obligations under the Convention and Protocol, is not a judicial body. An arbitral tribunal would nonetheless likely take account of any concurrent

[18] See Lefeber, n 11 above. Obviously parties engaging in emissions trading are convinced of the benefit of doing so, and would have an incentive to resolve any disputes that may put them in non-compliance, so as to be able to continue to transfer.

[19] See M Goote and R Lefeber, *Compliance Building Under the International Treaty on Plant Genetic Resources for Food and Agriculture* (Food and Agriculture Organization of the UN, Background Study Paper No 20, December 2003) 5 and 6, available at <http://www.fao.org/ag/cgrfa/compliance.htm>.

[20] The Ireland/UK 'MOX Plant Tribunal' recently suspended its proceedings pending the outcome of related proceedings at the European Court of Justice. See 'Permanent Court of Arbitration: The MOX Plant Case, Text of Decision', reprinted in (2003) 42 ILM 5 at 1187, and for more information, see the PCA's website at <http://www.pca-cpa.org> under 'Recent and Pending Cases'.

and related proceedings in the compliance branch, and the danger of producing conflicting 'judgments'. Furthermore, arbitral awards are inter partes, and cannot bind third parties without their consent.[21] The compliance branch's enforcement arm could however, by suspending the future eligibility of a Party to the Protocol to transfer emissions reductions, theoretically affect Parties beyond those involved in a specific dispute, for example, private entities involved in aspects of the transfer. Interesting as it may be to delve into the details of the foregoing hypothetical, such detailed analysis would be beyond the scope of this chapter, and is merely raised here to survey the context in which the dispute settlement provisions under the UNFCCC/the Protocol exist.

3 Arbitration in Carbon Contracts

There is potential for a dispute to arise at any stage in a transaction where rights and duties are ill-defined and open to interpretation. Although the rules for the flex-mechs are nearly fully elaborated and transactions are already taking place, it could be argued that the integrity of the system will not be fully tested until disputes arise and are litigated.

3.1 Potentials for Dispute

Disputes may arise at any stage of a flexible-mechanism transaction and could be contractual or under requirements of the Protocol, involving combinations of States and private parties. Examples of the main areas where disputes might arise under Carbon Contracts could include the following:

- Auditing, Monitoring, and Verification;
- Brokering;
- Ownership/Title to Emissions Reduction Units (ERUs) or Certified Emissions Reductions (CERs) and the underlying verified Emission Reductions;[22]
- Delivery of ERUs or CERs;
- Performance in terms of agreed reductions not being met;
- Validation;

[21] Although consent to arbitrate may in some circumstances be considered implied, and in some jurisdictions, consolidation of proceedings with the same or related subject matter have been ordered. See B Hanotiau, 'Complex Multi-Contract-Multiparty-Arbitration' (1998) 14 *Arbitration International,* 369–94. See also JG O'Connor, 'Maritime Arbitration Without Consent, Vouching, Consolidation, and Self-Execution—Will New York Practice Migrate to Canada?' (1993) 10 *Journal of International Arbitration (JOIA),* 161–83.

[22] Most contracts define the Emission Reductions as a claim to the CERs/ERUs or as the bundle of rights arising from a tonne of greenhouse gas emission reductions achieved by a project activity.

- Expropriation of a project or its benefits;[23]
- Interpretation of obligations in project design and baselines.

3.2 Parties

Before looking at the individual flex-mech contracts that may require dispute settlement clauses, it is important to first identify who the potential parties to disputes might be, as that may affect the available mechanisms to enforce an arbitral award. In general terms there will be three possible combinations of parties involved in a dispute:

- State–State;
- State–Private party;
- Private party–Private party.

More specifically, certain flex-mech related contracts and agreements may include the following parties as shown in Table 21.1.

3.2.1 CDM

In the CDM, disputes could arise between the States undertaking to engage in a project, between the private and/or public project-participant(s), or between

Table 21.1: Flex-Mex Related Contracts

Flex-mech	Contract/Agreement Type	Parties
JI	Host Country Agreement	Carbon purchaser (State or intergovernmental organization) and Annex I State
JI and CDM	Emissions Reduction Purchase Agreement (ERPA)	Carbon purchaser (public or private) and Project Owner (public or private)
CDM	Letter of Endorsement (LoE)	Host country sends to (public) Project Owner
CDM	Letter of Approval (LoA)	Host country sends to (public or private) Project Owner
Emissions Trading	Emissions Trading Contract	State–State, State–Private, Private–Private

[23] The CERs themselves that a project generates are not meant here, as it is doubtful that they could be directly expropriated. Only the CDM Executive Board can issue CERs, but the host country could affect the means of producing them and exert influence on the verifier and DOE, resulting in indirect or 'creeping' expropriation. In land use and land use change forest (LULUCF) projects there is a particular danger of land expropriation. See Werksman *et al.*, n 41 below, 11–12.

private and/or public project-participant(s) and a fund such as the World Bank Prototype Carbon Fund (PCF).

3.2.2 JI

In JI projects, disputes could arise between States, or a combination of private and/or public project participant(s).

3.2.3 IET

In IET, disputes will most likely occur between two private parties engaged in emissions trading where they have been authorized to do so by the respective Annex I country, but could also arise between the two Annex I countries engaged in trading.[24]

For all of the above, where two State Parties are involved in a dispute relating to the interpretation or application of the Convention or the Protocol, they must follow the provisions of Article 14, unless they agree on 'other peaceful means' pursuant to Article 14(1). Where private actors are involved in disputes between or among other private, or public actors, they must rely on contractual arbitration clauses or bilateral investment treaties (BITs) as the case may be.

3.3 Form and Content of the Arbitration Agreement/Clause

An international arbitration agreement is:

> An agreement in which two or more parties agree that a dispute which has arisen or which may arise between them, and which has an international character, shall be resolved by one or more arbitrators.[25]

This definition covers both an arbitration clause whereby the parties to a contract agree to resolve any dispute arising in relation to that contract through arbitration, and a submission agreement, whereby parties to an existing dispute agree ad hoc to submit that dispute to arbitration.

In international arbitration, the arbitration clause generally contains specifications as to:

- the intent of the parties to resolve any and all disputes between or among them in connection with the contract by arbitration;

[24] Arts 6(3) and 12(9) Protocol afford non-State actors a role in JI and CDM projects respectively. FCCC/CP/2001/13/Add.2, Decision 18/CP.7 sets out the requirements for involvement of non-State actors under Art 17 of the Protocol. Generally on the role of non-State actors in the Kyoto Protocol see LB Campbell, 'The Role of the Private Sector and Other Non-state Actors in the Implementation of the Kyoto Protocol', in WB Chambers (ed.), *Inter-linkages: The Kyoto Protocol and the International Trade and Investment Regimes* (Tokyo, UNU Press, 2001) 17–41.

[25] See Fouchard *et al.*, n 9 above, para 385.

- the number of arbitrators;
- place of arbitration;
- procedural rules;
- method of appointing arbitrators and/or appointing authority;
- language;
- applicable law.[26]

By not including the foregoing elements in a clear and unambiguous manner in the arbitration clause, parties may risk its being declared invalid in potential future litigation.[27] Parties may also specify the amount of time within which the arbitration should take place and other procedural matters relating to the conduct of the arbitration, such as whether to hold hearings or have a 'papers only' proceeding where no hearings are held, specification of a registrar or administrative-secretary to assist the tribunal, rules of evidence, and the amount of confidentiality attaching to the proceedings.[28] However, procedural matters are often determined by the tribunal and parties after the arbitration has commenced, and as circumstances warrant.

In applying the foregoing to Carbon Contracts, where the amount in dispute is relatively low, parties may wish to consider the use of one arbitrator only, as doing so could help keep costs to a minimum, and potentially speed up the proceedings. When the amount in dispute is relatively high, three arbitrators may be considered. Designating the place of arbitration matters as the *lex loci arbitrii* may affect the arbitral proceedings and procedural matters, and more importantly, issues such as the arbitrability of the dispute, which in turn may affect the enforcement of the award.[29] Parties should thus exercise caution in choosing the place of arbitration, having regard to the requirements of domestic law. As some practitioners point out, the more requirements domestic law imposes on arbitration, 'the greater the risk of the award being . . . set aside.'[30]

[26] See Redfern and Hunter, n 9 above, 163–9. See also Born, n 9 above, 5.

[27] Born states thereto, 'A poorly-drafted arbitration agreement can plant the seeds for disputes over jurisdiction and other matters, impose unworkable or undesirable procedures, and may be unenforceable' ibid.

[28] It is commonly held that arbitration is inherently confidential unless the parties specify otherwise. This is not necessarily the case, as arbitral tribunals and courts have ruled that if parties do not specify that confidentiality applies to arbitration proceedings, those proceedings cannot be considered confidential ipso facto. Parties would therefore be well-advised to err on the side of caution and specify the amount of confidentiality applying to their proceedings either in the arbitration agreement, or in a separate agreement (that is, arbitrators' terms of reference) after the commencement of the arbitration. See LY Fortier, 'The Occasionally Unwarranted Assumption of Confidentiality' (1999) 15 *Arbitration International,* 131–40.

[29] Enforceability will be discussed below and depends on for example, whether the State where the arbitration takes place has an arbitration act or law, and if that State is Party to an agreement on mutual recognition of arbitral awards such as the 1958 Convention on the Recognition and Enforcement of Foreign Arbitral Awards (New York Convention or NYC), 330 UNTS 38 (1959). See Redfern and Hunter, n 9 above, 79–92.

[30] See Pelloonpää and Caron, n 33 below, 59.

The applicable law could be specified as international law, and any particular rules of law such as CDM or JI requirements, or even national law relating to the transaction or project. Often, pre-existing procedural rules contain clauses stating that the tribunal shall determine the applicable law where parties do not designate any, or fail to agree thereon.[31] Parties should as a rule try to agree on the applicable substantive law to avoid any surprises in the arbitral award, but where they cannot agree or are unsure, they may even be better off leaving the decision on applicable law to the tribunal. In any case, parties will likely be afforded an opportunity to present their arguments on applicable law and jurisdiction during the proceedings.

Using a pre-existing set of procedural rules such as the 1976 United Nations Commission on International Trade Law (UNCITRAL) Arbitration Rules, or the 2001 Permanent Court of Arbitration (PCA) Optional Rules for Arbitration of Disputes Relating to Natural Resources and/or the Environment (PCA Environmental Rules),[32] will ensure that the selected procedures are reliable and widely accepted. The UNCITRAL Arbitration Rules were created primarily with resolution of international commercial disputes in mind, and although there are no official figures, it is estimated that many thousands of such disputes have been resolved using them.[33] Parties using the UNCITRAL Arbitration Rules ought to specify an appointing authority.[34] Should the parties fail to agree on, or fail to designate their arbitrator, the appointing authority may be requested by any party to make an appointment, so that a recalcitrant party cannot frustrate the constitution of the tribunal. The appointing authority also decides challenges to arbitrators, which may be brought when a party considers that an arbitrator is not independent and impartial.

[31] Art 33(1) of the Arbitration Rules (UNCITRAL) provides: 'The arbitral tribunal shall apply the law designated by the parties as applicable to the substance of the dispute. Failing such designation by the parties, the arbitral tribunal shall apply the law determined by the conflict of laws rules which it considers to be applicable.' The PCA Optional Rules for Arbitration of Disputes Relating to Natural Resources and/or the Environment (Environmental Rules) are based on the UNCITRAL Rules, but give the tribunal a wider range of choices in determining the applicable law in Art 33(1): 'In resolving the dispute, the arbitral tribunal shall apply the law or rules of law designated by the parties as applicable to the substance of the dispute. Failing such designation by the parties, the arbitral tribunal shall apply the national and/or international law and rules of law it determines to be appropriate.' For copies of the UNCITRAL Arbitration Rules, see <http://www.uncitral.org>. For copies of the PCA Environmental Rules, see <http://www.pca-cpa.org/ENGLISH/EDR>.

[32] These rules were drafted with international environmental or natural resource related disputes in mind, including climate change issues, and are based on the UNCITRAL Arbitration Rules in an attempt to ensure applicability to public and private international law disputes. They specify that the PCA Secretary-General will act as default appointing authority, and make lists of government nominated arbitrators and scientific experts available to parties and the PCA Secretary-General for constituting the tribunal. They are referred to in a number of emissions trading contracts as the rules of choice.

[33] See M Pelloonpää and D Caron, *The UNCITRAL Arbitration Rules as Interpreted and Applied—Selected Problems in Light of the Practice of the Iran-United States Claims Tribunal* (Helsinki, Finnish Lawyers' Publishing, 1994) 2.

[34] Specifying the appointing authority may serve to reduce the time it takes for the tribunal to be constituted. For a general description of the role of the appointing authority, see Redfern and Hunter, n 9 above, 202–3. The PCF Legal Instrument arbitration clause, naming the PCA Secretary-General as default appointing authority, will be examined in the following section.

The UNCITRAL Arbitration Rules are non-institutional, yet specify the PCA Secretary-General as the person to *designate* an appointing authority.[35] The PCA Secretary-General acts as the appointing authority in the PCA Environmental Rules, a number of emissions trading contracts, the PCF Legal Instrument,[36] as well as hundreds of arbitration clauses specifying the UNCITRAL Arbitration Rules as the procedural rules for dispute settlement.

In the context of flex-mechs, parties referring to the UNCITRAL Arbitration Rules as the dispute settlement procedures should consider specifying the PCA Secretary-General as the appointing authority, as he can rely on lists of government appointed arbitrators to help constitute the tribunal, many of whom have climate change expertise. Parties and/or the tribunal may also wish to engage the services of an arbitral institution such as the PCA or International Centre for Settlement of Investment Dispute (ICSID) to: administer the proceedings, take charge of archives, coordinate communications among the parties and tribunal, arrange for and supervise hearings, support the tribunal and ensure procedural matters are properly addressed.[37]

After an agreement to carry out a CDM or JI project is entered into, a variety of contracts involving private actors to implement the resultant project could be concluded (each contract containing an arbitration clause), and could include concession contracts, build operate transfer (BOT) contracts, and joint venture and service contracts.[38] Emission Reductions Trading Agreements (ERPAs) (and

[35] The drafters of the UNCITRAL Arbitration Rules sought an institution with permanence, experience, and neutrality to anchor the appointment procedure to. See P Sanders, 'Commentary on the UNCITRAL Arbitration Rules', in *Yearbook Commercial Arbitration*, Vol. 2 (The Hague, Kluwer Law International, 1977) 172–224. See also H Holtzmann and J Neuhaus, *A Guide to the UNCITRAL Model Law on International Commercial Arbitration—Legislative History and Commentary* (Deventer, Kluwer, 1989) 37, 60, which states in relevant part ' . . . if a party needed assistance in connection with the appointment of, or challenge to arbitrators, or with their failure to act, the party could request the Secretary-General of the Permanent Court of Arbitration at The Hague . . . The Secretary-General [of the PCA] is an internationally-recognized and experienced neutral office.'

[36] The 'International Bank for Reconstruction and Development Resolution No. 99–1 Authorizing Establishment of the Prototype Carbon Fund—As Amended by Resolutions No. 2000–1 and No. 2001–3', hereinafter referred to as the 'PCF Legal Instrument', is available from <http://www.carbonfinance.org>.

[37] On the various tasks that an institution can perform in supporting arbitration proceedings and the arbitral tribunal see R Jennings, 'The Differences Between Conducting a Case in the ICJ and in an *ad hoc* Arbitration Tribunal—An Inside View', in N Ando *et al.* (eds.), *Liber Amicorum Judge Shigeru Oda* (The Hague, Kluwer Law International, 2002) 893–909, esp. 905–9. See also A Watts, 'Enhancing the Effectiveness of Procedures for International Dispute Settlement', in Frowein and Wolfrum (eds.), *Max Planck Yearbook of United Nations Law*, Vol. 5 (The Hague, Kluwer Law International, 2001) 21–35. For a practitioners guide on supporting arbitral tribunals see 'UNCITRAL Notes on Organizing Arbitral Proceedings' 1996, available from < http://www.uncitral.org>. Other arbitral institutions such as the International Chamber of Commerce (ICC) Court of Arbitration and London Court of International Arbitration (LCIA) provide similar services. For more information on the functions of the ICC Court of Arbitration which is given a scrutiny and review function in the ICC Arbitration Rules see Craig *et al.*, n 9 above.

[38] For a more detailed discussion of the various possible contract forms, see IL Worika and M Brown, 'Contractual Aspects of Implementing the CDM and other Flexibility Mechanisms Under the Kyoto Protocol', in WB Chambers (ed.), *Inter-linkages: The Kyoto Protocol and the International Trade and Investment Regimes* (Tokyo, UNU Press, 2001) 215–59.

emissions trading contracts) may subsequently be concluded for the transfer of project resultant emissions reduction credits. It is to the use of arbitration in that project cycle that we will now turn.

3.3.1 CDM/JI private project participants and host countries—bilateral and multilateral investment treaties

Where private actors are involved in disputes with States pertaining to a CDM or JI project investment, they should note whether any bilateral (or multilateral) investment treaties (BIT or MIT respectively) exist between the host country and the investing Annex I country. BITs often contain arbitration clauses that afford the investing private party a right to arbitrate claims against the host country where the investment is made.[39] Blackaby and Paradell's recent scholarly survey of investment treaty arbitration reveals that BITs offer a number of protections to foreign investors against host country interference with the investment, and a list of these is reproduced here:

- National treatment (foreign investments must receive equal treatment to [or treatment better than] domestic investments);
- 'Most favoured nation' treatment (investments from one country may not be treated worse than the best treatment afforded to the investment of any third country);
- General standards of 'fair and equitable' treatment;
- 'Full protection and security', no investments by arbitrary and discriminatory measures, and observance of specific investment undertakings;
- No expropriation without prompt, adequate and effective compensation; and
- The right to transfer investment-related funds into and out of the host country.[40]

However, caution may be warranted in relying on investment treaty protections to attempt to initiate arbitration in the context of flex-mech related contracts. BITs usually contain a definition of investments, and although these are usually broadly formulated, it should not automatically be assumed that the project amounts to an investment under the treaty. Moreover, according to some scholars, there is a danger that, for example, the functioning of the CDM could be viewed as

[39] See N Blackaby and L Paradell, 'Investment Treaty Arbitration', in *Bernstein's Handbook of Arbitration and Dispute Resolution Practice*, Vol. 1 (London, Sweet & Maxwell, 2003) 709–39. The authors cite the rationale for the right of the investor to arbitrate against the host country as follows: 'By recognizing that effective investor protection has to be firmly within the hands of the investor and independent of state action, BITs incorporate a means of settling disputes between investors and their Host Countries . . . in a binding manner. Whether or not a contract or agreement has been concluded with the Host Country, after an initial cooling off period for negotiation, the investor may commence an arbitration directly against the Host Country of the investment.' It should be noted that older BITs often contain clauses that require parties seeking arbitration to first exhaust local remedies, and/or clauses requiring the consent of the host country to arbitrate.

[40] ibid at 712.

requiring a type of discrimination by the host country between foreign and domestic investors that would violate, *inter alia*, the 'national treatment' standard at the heart of most BITs.[41] Werksman *et al.* state in that connection:

> In the Kyoto Protocol context, goals of equity are actually pursued by treating countries differently on the basis of their varying national circumstances. A potential for conflict with an IIA [International Investment Agreement] may arise if a Party hosting a CDM project is encouraged or required by the Protocol to expressly discriminate between investors on the basis of their nationality and the status of their home country. CDM Rules could directly discriminate between investors on the basis of their nationality in at least the following . . . ways.
>
> i. Party v. Non-Party . . . ;
> ii. Complying v. Non-Complying Party . . . ;
> iii. Foreign investors from Annex I versus Foreign Investors from Non-Annex I Parties . . . ;
> iv. Foreign v. Domestic Investor . . . .[42]

Werksman *et al.* go on to argue that the investment agreement being the governing law in a dispute 'could have the unwelcome effect of trumping emission reduction and sustainable development policies taken pursuant to the CDM'.[43] Case law reveals that arbitral tribunals in disputes between investors and host countries under BITs most often find the BIT to be the 'primary source of law'.[44] However, although the BIT may be the 'primary source' of law in an investment dispute, 'to construe and supplement BIT provisions, tribunals have resorted to international law as the most naturally applicable source of law . . . .'[45] In that connection, two facts may serve to avoid the undesirable scenario of reliance on investment treaty arbitration leading to a trumping of Kyoto Protocol and CDM policies. First, an arbitral tribunal would probably be loath to ignore the provisions of the Protocol and the specific CDM requirements as sources of international law governing the relationship between the parties. Secondly, the arbitral tribunal might look to a Letter of Approval (LoA), Memoranda of Understanding, and/or other instruments of consent for further indication as to what the intentions of the host country were with regard to the specific CDM project the foreign investor was involved in, and apply the provisions of the investment treaty in light of that.

From an academic perspective, and since there is no relevant case law, the authors seem right to suggest that caution should be exercised in relying on

[41] See J Werksman, KA Baumert and DK Navroz, 'Will International Investment Rules Obstruct Climate Protection Policies', Climate Notes (World Resources Institute, April 2001) 8–10, 14. See also J Werksman and C Santoro, 'Investing in Sustainable Development: The Potential Interaction between the Kyoto Protocol and the Multilateral Agreement on Investment', in Chambers (ed.), n 38 above, 191–213.

[42] ibid at 9.

[43] See Werksman *et al.*, n 41 above, 12.

[44] See Blackaby and Paradell, n 39 above, 720.

[45] ibid.

investment treaty arbitration clauses for CDM disputes, and (in favour of the host country's right to determine what will contribute to sustainable development) to suggest that Protocol Parties should agree that the rules of the Protocol will prevail in case of a conflict between any BITs/MITs in force between them, and the Protocol.[46]

A private flex-mech related project investor could also directly negotiate a contract allowing for arbitration with a host country for CDM or JI investments. In that connection, the involvement of actors such as the World Bank through its PCF, 'multilateralize' CDM and JI transactions, and may level the playing field, '. . . to encourage the participation of small hosts and investors, to promote balanced bargaining positions between investors and hosts, and to spread risk . . .'[47] Since private parties engaging in flex-mech contracts do not have recourse to the dispute settlement provisions of the Protocol and Convention, and may not enjoy investment treaty protections, they should mitigate risks by including arbitration clauses in their contracts.

3.3.2 The CDM LoA

In the beginning, there were two countries, an Annex I country seeking to meet its commitments under the Kyoto Protocol, and a non-Annex I host country seeking to benefit from CDM investment leading to sustainable development. Article 40(a) of the Marrakesh Accords requires that each Party (to the Protocol) wishing to engage in a CDM project, give 'written approval of voluntary participation' in the CDM through their Designated National Authority (DNA), and that the host country confirm in that written approval that the project activity 'assists it in achieving sustainable development'.[48] That written approval has come to be known as the LoA, and is for all intents and purposes, the instrument of consent of each country wishing to participate in a CDM project. The Designated Operational Entity (DOE) responsible for the project must submit these LoAs when applying to the CDM Executive Board for validation of emissions reductions achieved by a project, additional to any that would have occurred in the project's absence.

LoAs have been executed in varying forms. Some emissions reduction purchasers, such as the PCF, require a host country to issue a 'Letter of Project Endorsement' or 'Letter of No Objection' before the host country LoA is issued.[49] Why require these additional instruments? It could be argued that the additional requirements are a way of building a case (which could be argued before a court or tribunal) showing the clear intent and consent of the issuing country from an early stage to commit itself to carrying out a particular project in good faith. In public

[46] See Werksman *et al.*, n 41 above, 12, 14.

[47] See Werksman and Santoro, n 41 above, 194.

[48] See Annex: Modalities and Procedures for a Clean Development Mechanism, Decision 17/CP.7, FCCC/CP/2001/13/Add.2. See also chs 11 and 12 above.

[49] See 'PCF Operational Cycle', available at <http://www.carbonfinance.org>.

international law (PIL) terms, expecting a State to honour its agreement, even in the form of a unilateral declaration, follows the doctrine of *pacta sunt servanda*: 'agreements are made to be kept and enacted in good faith.'[50]

At present, there are no dispute settlement clauses or agreements covering the Letters of Endorsement and Approval. Some public and private CDM investors have however expressed interest in the potential use of arbitration before the project is registered, to settle any disputes resulting after the LoA is issued, but before any contracts containing arbitration clauses (such as ERPAs) have been concluded.[51] Annex I country representatives interviewed argue in that connection, that there is a legal vacuum in terms of not having a right of recourse to any court or tribunal before the Protocol enters into force (after which time they would have recourse to Article 19 of the Protocol), should they suffer losses resulting from activities undertaken pursuant to the LoA; and for example, the project is not completed, or even expropriated. They seem to be correct that, aside from attempting to go before the ICJ which could result in a lengthy high-profile proceeding, or obtain the consent of the host country to arbitrate the dispute, there is no express right of legal recourse pursuant to the LoA (although disputes regarding the LoA would likely be subject to administrative law in the host country).

Even if there were a unilaterally triggerable arbitration clause covering the LoA, could reliance on the LoA itself, leading to the detriment of a party, give rise to a cause of action?[52] Recall that the Marrakesh Accords simply require the LoA to state that each Party is participating voluntarily, and that the host country considers that the project will assist it in achieving sustainable development. A definitive answer would depend on how the LoA was formulated, and of course, if there were a forum in which to bring such a claim.

Some model LoAs, such as the World Bank's, state that the host country 'is committing itself to render such assistance as may be necessary in the future registration, certification, issuance and transfer for the purposes of the UNFCCC and the Kyoto Protocol, of the emission reductions generated by the [project]'.[53] The issuance of an LoA on the latter terms of the PCF Draft LoA could be viewed as a unilateral act of State amounting to 'promise'. Brownlie notes, 'unilateral declarations involve, in principle at least, concessions which are intentional, public, coherent, and conclusive of the issues'.[54] There is however, a distinction

[50] Art 26 of the 1969 Vienna Convention on the Law of Treaties (1155 UNTS 331) states 'Every treaty in force is binding upon the parties to it, and must be performed by them in good faith.' This is, for better or worse as some argue, considered a general principle of international law. On the pros and cons of *pacta sunt servanda*, see R Jennings and A Watts (eds.), *Oppenheim's International Law* (9th edn, Vol. I, Essex, Longman, 1996) 1206, n 4.

[51] This information is taken from conversations of the author with Annex I representatives from June and December 2003, who were willing to discuss this matter on the basis of anonymity.

[52] On the basis of estoppel, which in PIL can be described as 'the element of conduct which causes the other party, in reliance on such conduct, detrimentally to change its position or to suffer some prejudice', see I Brownlie, *Principles of Public International Law* (6th edn, Oxford, Oxford University Press, 2003) 615–16.

[53] Draft LoA available from <www.carbonfinance.org>.

[54] Brownlie, n 52 above, 614.

between *acta jure imperii* (Government acts which are expressions of sovereignty), and *acta jure gestionis* (the Government doing business), and the unilateral act here may be viewed as the latter. That distinction matters when assessing the intent of the State's action, and could be used to determine applicable law in a dispute.[55] The use of the words 'commits itself to render such assistance', show intentional concession, the document is public,[56] and the issues, that is, 'future registration, certification, issuance and transfer', are coherent and conclusive.

Taken together then, an Annex I and host country LoA relating to the same project could be viewed by a tribunal as an intergovernmental agreement[57] to freely undertake[58] the particular project described in the LoA, especially when there have been other instruments indicating intention to enter a legally binding relationship, such as Memoranda of Understanding or LoEs prior to the issuance of the LoA.[59] Both States may also take action according to reasonable expectations derived from the LoA, on the basis of estoppel.[60] It therefore seems reasonable to wish to have a contractual right of recourse to arbitration if a dispute arises over one Party's activities in the other Party's territory, undertaken pursuant to the terms of the that Party's LoA. Yet what could the form of such an arbitration clause or agreement be, and more importantly, is it necessary?

In order to arbitrate such disputes as described above in this section, a separate agreement between the parties to the CDM project would have to be concluded, as without such consent, an arbitral tribunal could not be constituted and would have no jurisdiction, unless such jurisdiction were provided for in some other agreement such as a BIT or MIT. Such a submission agreement would include standard elements of arbitration clauses, specifying the parties, scope, applicable

[55] See M Rubino-Sammartano, 'Developing Countries vis-à-vis International Arbitration' (1996) 13 JOIA 21–36.

[56] See n 48 above. According to Art 40(a) Marrakesh Accords, the LoA goes to the DOE and later CDM Executive Board before certification can be effected, and is thereby made public.

[57] Prof Sir Robert Jennings and Sir Arthur Watts point out that 'International agreements may be given a variety of designations, such as treaties, agreements, Acts, conventions, declarations and protocols. These various designations are used in no consistent manner. Although the designations may *indicate the level of importance* which the parties attach to an instrument, *designation alone does not in international law affect the binding force* of the instrument or its characterization as a treaty.' (Emphasis added.) This is said assuming expectations of *quid pro quo* in terms of a unilateral declaration is not necessary for a unilateral act to have binding effect. See *Oppenheim*, n 50 above, 1208–9.

[58] The notion of entering an agreement freely corresponds with the notion in Arts 51 and 52 of the Vienna Convention on the Law of Treaties, that use of coercion in making a treaty may invalidate that treaty, ibid.

[59] The Netherlands in its Certified Emission Reduction Unit Procurement Tender (CERUPT) Terms of Reference 'Draft Letter of Approval' states that if the Kyoto Protocol does not enter into force, the host country and The Netherlands 'consider the transfer to The Netherlands as a transfer of greenhouse gas emissions reductions on a bilateral basis'. This clause supports the view that the relationship created by the two LoAs for a particular CDM project is a bilateral agreement, even if not declared as such. Yet of course a court or tribunal would have to determine this on a case-by-case basis, and the terms of the LoA. See 'CERUPT Terms of Reference', <http://www.carboncredits.nl> under 'Publications'.

[60] See n 52 above.

law, procedural rules, number of arbitrators, language and place of arbitration. An example or model[61] could read as follows:

> Any dispute, controversy, or claim between [*insert Party to the Protocol*] and [*insert Party to the Protocol*] arising out of, or relating to the interpretation or application of [*insert title and description of project*], shall be settled by final and binding arbitration in accordance with [*insert procedural rules*] as in effect [*insert date of agreement or time of commencement of arbitration*]. The tribunal shall apply international law and the provisions of the Kyoto Protocol. Where there is a conflict between international law and the provisions of the Kyoto Protocol, the latter shall prevail. The number of arbitrators shall be [*insert one, three, or five*]. The place of arbitration shall be [*insert place*]. The language of the arbitration shall be [*insert one or more languages*].

However, the above clause has never been and may never be used, and is listed here for purposes of illustration only. It also in no way advocates that an arbitration agreement covering the LoA between two Kyoto Protocol Parties is desirable. In discussions with practitioners, some have noted that although the LoA commits both Parties to a certain project in a legally binding manner, negotiating an additional arbitration clause or agreement could prove politically difficult, and jeopardize issuance of an LoA for a project. They also argue that Article 40(a) of the Marrakesh Accords is very simple in its requirements, and LoAs need not be formulated in contractual language.

Since States could rely on either the dispute settlement provisions of the Protocol, or attempt to use the non-compliance procedures of the Protocol once it enters into force should a dispute arise of the type described above, there may be some merit to the practitioners' view. Yet the procedures for arbitration and conciliation in the Convention have not yet been adopted, and until they are (if they are), State Parties wishing to have recourse to dispute settlement before a project is registered will have to agree such clauses on a case-by-case basis, which may prove difficult and onerous for the Party seeking redress. Yamin has commented that, 'the legal framework for dispute settlement created by Article 14 UNFCCC is not well suited to creating legal certainty, even for Parties', which may explain their hesitance in finishing the annexes on arbitration and conciliation to the UNFCCC.[62]

Another question regarding the use of arbitration for claims regarding the LoA is where two States were to agree to an arbitration clause similar to the one above, whether that clause could also be structured to give private parties a right of recourse to arbitration—an especially salient issue since in most cases the addressee of the LoA is a private party. States could agree to insert a provision in the arbitration agreement similar to that in a BIT, affording private parties from one Party

[61] This model clause is based on the UNCITRAL model clause contained in the footnote to Art 1(1) UNCITRAL Arbitration Rules, see <http://www.uncitral.org> for a copy of the UNCITRAL Arbitration Rules.

[62] See F Yamin, 'Operational and Institutional Challenges', in J Goldemberg (ed.), *Issues and Options—The Clean Development Mechanism* (New York, United Nations Development Program, 1998) 53–79, at 79.

a right to arbitrate claims arising from the project described in the LoA. At that point, the agreement would start to resemble a BIT, but exclude the express investor protections in BITs. Assuming there were already a BIT in force between the two States Parties, inserting a clause giving private parties a right to arbitrate disputes arising 'out of or in connection with' the LoA, the terms of the latter might be seen as lex specialis, and the BIT protections could perhaps be relied upon in a supplementary manner.

It has also been argued that the UNFCCC Conference of the Parties (COP) acting as the Kyoto Protocol Meeting of the Parties (MOP) should adopt a decision allowing investing private parties a direct right of recourse to arbitrate against either the COP/MOP and Executive Board, the foreign State in CDM projects, or to use the non-compliance procedures of the Protocol.[63] It is doubtful that such a decision could ever be agreed by the COP/MOP, given the historically contentious nature of the negotiations regarding dispute settlement,[64] and given the broad scope of the issues such procedure might open to arbitration (and potential abuse thereof). A politically more workable solution, as it would only bind the parties agreeing to it and not the whole COP, might be to limit the scope of the private party's right to arbitrate with the foreign State to a specific CDM project—a separate arbitration agreement would be a natural place for this to occur.

3.3.3 JI agreements

The foregoing ideas to conclude a separate arbitration agreement to cover disputes arising between Kyto Protocol Parties could equally apply to JI projects, as there are no dispute settlement clauses in instruments concluded before the ERPAs or Carbon Contract. The Netherlands Government through its Emissions Reduction Unit Procurement Tender (ERUPT 4), asks the host country to sign a Memorandum of Understanding, and to issue a Letter of Project Endorsement, and an LoA before entering an agreement to purchase ERUs, although any disputes between the public or private entity developing and/or implementing a project and the Netherlands will be 'dealt with exclusively before the District Court of The Hague' under Dutch law.[65]

The World Bank PCF also requires that a number of letters indicating consent are issued (LoE, LoA) before conclusion of an ERPA for a JI project, and moreover that an agreement between the bank as trustee, and the host country be concluded, obliging the host country to transfer or authorize the transfer of ERUs. Additionally, a Letter of Intent is signed with the private or public Project Entity which secures the exclusivity between the Parties, and contains an arbitration clause specifying that disputes will be resolved under the UNCITRAL Arbitration Rules with the PCA Secretary-General acting as appointing authority.

[63] ibid. [64] See Ott and Oberthür, n 14 above.
[65] See 'ERUPT Terms of Reference 4', <http://www.carboncredits.nl>.

3.3.4 Private contracts and ERPAs

As noted above, private parties/entities concluding a contract relating to a CDM or JI project[66] or an International Emissions Trade, should always insert arbitration clauses in their contracts to mitigate risks, as they do not have recourse to the Protocol's dispute settlement procedure, and may not be able to rely on BIT arbitration clauses should disputes arise. Being able to arbitrate in a neutral forum under international law instead of having recourse to domestic courts only is especially important when contracting with States who could change a law affecting the contract to the detriment of the private party. Moreover, once a State agrees to arbitration, it cannot invoke its own law as a defence to resist arbitration.[67] An arbitration clause should also contain 'standard' elements of arbitration clauses, and could have the following form:

> Any dispute, controversy, or claim arising under, out of, or relating to the interpretation, application, or performance of this agreement, including its existence, validity, or termination, shall be settled by final and binding arbitration in accordance with [*insert procedural rules*] as in effect [*insert date of agreement or time of commencement of arbitration*]. The number of arbitrators shall be [*insert one, three, or five*]. The place of arbitration shall be [*insert place*]. The language of the arbitration shall be [*insert one or more languages*].[68]

The above clause is, as the previous model clause above, included for illustrative purposes only, and parties should of course rely on legal counsel on a case-to-case basis to ensure that the arbitration clause they select is valid, enforceable, and advisable in the circumstances.

ERPAs are concluded between the carbon purchaser and the private or public entities transferring the rights arising from the emissions reductions at an agreed price—in effect forward selling the emissions reductions. ERPAs have been concluded for both JI and CDM projects. The ERPAs the author has seen on a confidential basis all contain binding arbitration clauses, some tiered so that arbitration may commence after negotiation has been tried and fails, allowing any party to initiate arbitration pursuant to either the UNCITRAL Arbitration Rules, naming the Secretary-General as the appointing authority, or to the PCA Environmental Arbitration Rules.

Disputes surrounding the ERPA could be between a combination of public and private entities (including funds purchasing emissions reductions) entering into the ERPA, and arise regarding any number of issues pertaining to the contract including failure of the project, monitoring, verification, validation, or general contractual issues such as penalties, and so on.

[66] As noted in the text above, these may include concession contracts, build operate transfer (BOT) contracts, joint venture and service contracts, or ERPAs and emissions trading contracts.

[67] See R Briner, 'The Iran United States Claims Tribunal and Disputes Involving Sovereigns' (2002) 18 *Arbitration International*, 299–303, at 302.

[68] See n 61 above.

3.3.5 PCF legal instrument

The 'International Bank for Reconstruction and Development Resolution No. 99–1 Authorizing Establishment of the Prototype Carbon Fund—As Amended by Resolutions No. 2000–1 and No. 2001–3'[69] contains the 'Instrument Establishing the Prototype Carbon Fund' in Annex I, which in turn contains provisions for the functioning of the fund, and for disputes to be settled by arbitration. Section 18(2) on Arbitration sets out:

> Any dispute between the Trustee and a Participant arising out of or relating to this instrument or such Participant's Participation Agreement shall be settled by arbitration in accordance with the UNCITRAL Arbitration Rules as at present in force. The number of arbitrators shall be three. The appointing authority shall be the Secretary-General of the Permanent Court of Arbitration at The Hague. In the event of a conflict between the UNCITRAL Arbitration Rules and the terms of this Instrument or of the Participation Agreement, the terms of the Instrument and Participation Agreement shall prevail.

Again, this clause contains standard elements of arbitration clauses, specifying the parties (who could be States or non-States), the intention of parties to arbitrate all disputes arising out of or relating to the underlying contract, procedural rules, number of arbitrators, appointing authority, and a conflict of laws clause.

Although the International Centre for the Settlement of Investment Disputes provides arbitration services and is part of the World Bank Group, the UNCITRAL Arbitration Rules specifying the PCA Secretary-General as appointing authority were chosen in order to detach the arbitration procedure from the trustee.[70]

3.4 Enforcement of the Award

It is a maxim of international arbitration that, 'in principle, the eventual result of every arbitration should be a final award that is enforceable at law'.[71] One of the most often cited advantages of arbitration is that the arbitral award may, in contrast to judgments of domestic courts, be recognized and enforced throughout the world. Most procedural rules state that the arbitral award is 'final and binding', but if one party does not perform the judgment ordered in the award, the other will probably seek to have the award enforced either at the place of arbitration, or in another country where the opposing party has assets. Enforcement will normally be sought in the national courts, as after the tribunal has rendered its award and

[69] Available from <http://www.carbonfinance.org>. The PCF arbitration clause has also been adopted for the Community Development Carbon Fund and the BioCarbon Fund and has been included in the respective instruments governing the operations of the funds.

[70] Comments made by Mr Antonio Parra, Deputy Secretary-General of ICSID, at a Workshop held on 23–24 January 2003 at the World Bank on 'Legal Implications of the CDM on Host Countries—Legal Nature and Ownership of Emission Reductions.'

[71] See Redfern and Hunter, n 9 above, 363.

such becomes res judicata, the tribunal is considered to be *functus officio*, no longer having jurisdiction.[72]

Parties may have other curial remedies such as challenging enforcement of the award, depending on the law of the place where such measures are sought. Perhaps a testament to the tenacity of the arbitral award is that a few awards set aside in one country have nevertheless been deemed enforceable by the courts of another country.[73]

Throughout this chapter, several factors related to enforceability have been mentioned, including that the arbitration agreement is valid, and the subject matter of the dispute is arbitrable as a matter of public policy under the law of the place of arbitration. Parties should also inform themselves of the particular procedural and legal requirements of the place where enforcement is sought, as the tribunal will not be able to foresee every legal or other situation that might affect enforceability of the award.[74]

Arbitrating with States may pose difficulties for enforcement and execution of the award, although a full treatment of that topic is beyond the scope of this chapter.[75] States have come to accept, albeit grudgingly in some cases, arbitration with private parties in investment and commercial disputes, with an ever-growing number of investor-State arbitrations over the last ten years under multilateral investment treaties such as the North American Free Trade Agreement (NAFTA)[76] and the ICSID Convention[77] taking place. Some treaties, such as the ICSID Convention, may be seen to afford additional security regarding the enforceability of an award. Although recent case law may reveal a trend toward courts not recognizing claims of sovereign immunity from jurisdiction and execution, parties ought to require explicit waivers where possible.[78] It is also recommended that the private

[72] Many procedural rules provide for the arbitral tribunal that rendered the award to interpret or correct it, or issue an additional award on matters raised in the arbitration but not covered in the award, upon application of a party within a certain time period after the award has been issued, usually thirty days as in Arts 35–37 UNCITRAL Arbitration Rules. An exception to seeking enforcement in national courts would be where a treaty-based mechanism such as the 1965 Washington Convention has its own enforcement mechanism. See n 77 below. On *functus officio* see JG Bauer, 'Once a Catchy Phrase, Always Immutable Law—The Origins and Destiny of Three Famous Mantras: Functus Officio, Once on Demurrage Always on Demurrage, Manifest Disregard of the Law' (1994) 11 JOIA, 41–56.

[73] See P Ramaswamy, 'Enforcement of Annulled Awards—An Indian Perspective' (2002) 19 JOIA 461–72. See also, GH Sampliner, 'Enforcement of Nullified Foreign Arbitral Awards—Chromalloy Revisited' (1997) 14 JOIA 141–66.

[74] See Redfern and Hunter, n 9 above, 363.

[75] ibid, 477–481. See also, WM Reisman, 'International Arbitration and Sovereignty', in (2002) *Arbitration International* 18 231–239.

[76] (1993) 32 ILM 289. See <http://www.naftalaw.org> for recent NAFTA investor-State case law and scholarly articles on NAFTA investor-State dispute settlement. See also L Trakman, 'Arbitrating Investment Disputes Under NAFTA' (2001) 18 JOIA 385–416.

[77] Convention on the Settlement of Disputes Between States and Nationals of Other States, UNTS 575, 1966. In general, see C Schreuer, *The ICSID Convention—A Commentary* (Cambridge, CUP and ICSID, 2001). See also Blackaby and Paradell, n 39 above.

[78] See Albert Jan van den Berg's treatment of the *Letco v Liberia* case in this connection: 'Recent Enforcement Problems under the New York and ICSID Conventions' (1989) 5 *Arbitration International* 2–20. For a more recent treatment of the enforcement and execution issues with sovereigns, see

party examine whether the sovereign it is contracting with is a party to the New York Convention (NYC) or other treaty providing for recognition and enforcement of foreign arbitral awards.[79]

The NYC is the primary international instrument used for seeking recognition and enforcement of foreign arbitral awards, although domestic laws and bilateral or regional agreements on recognition and enforcement may be relied upon as well (some being more favourable than the NYC), and the latter may be the only available instruments where the State in question is not a party to the NYC.[80] Paulsson notes regarding the NYC, 'Its very existence without doubt makes it unnecessary . . . to proceed to enforcement; parties voluntarily carry out the award, knowing that in any event they must comply since there is an effective enforcement mechanism'.[81]

The NYC applies generally to awards made outside the State where enforcement is being sought.[82] A State may make reservations that it will only recognize and enforce awards made in another contracting State, and that it will only apply the NYC to awards considered commercial under the laws of the State making the reservation.[83] The latter 'commercial reservation' has on a few occasions led to problems of application of the NYC in certain jurisdictions as there are naturally differences among States as to what can be considered 'commercial'.[84] The latter problem may apply to Carbon Contracts, although it is improbable. In the US, courts have ruled that emissions trading is essentially a commercial activity,[85] but

NB Turck, 'French and US Courts Define Limits of Sovereign Immunity and Execution and Enforcement of Arbitral Awards' (2001) 17 *Arbitration International* 327–42. Turck points out that the approach of domestic courts to attaching State property varies widely. She deals with, *inter alia*, *Creighton Ltd v Qatar*, whereby a French court ruled that acceptance of an ICC (or similar) arbitration clause constitutes a waiver of execution, in that the State accepts the award to be final and binding, and undertakes to execute it without delay (ICC Rules, Art 24). US courts were favourable to this view as well, although limiting it through application of the Foreign Sovereign Immunities Act. Note that some procedural rules such as the 'PCA Optional Rules for Arbitrating Disputes Between Two Parties of Which Only One is a State' provide that the use of the rules constitutes a waiver of immunity from jurisdiction, but state that a waiver of immunity from execution must be explicit. See <http://www.pca-cpa.org/ENGLISH/BD/>.

[79] See H Fox, 'State Immunity and Enforcement of Arbitral Awards—Do We Need an UNCITRAL Model Law Mark II for Execution Against State Property?' (1996) 12 *Arbitration International* 89–94. The author argues that for a state to rely on sovereign immunity as a bar to enforcement, it must follow international law, good faith, and issues of estoppel arising from its contracting with a private party. The author notes in that connection, that for commercial matters, states tend to cooperate in the arbitral process, and not impede enforcement.

[80] Generally on the New York Convention, see AJ Berg, v.d., *The New York Convention of 1958* (Deventer, Kluwer, 1994). In some cases, local laws and these instruments may provide for an even more favourable enforcement scheme. The New York Convention contains a non-prejudice provision in Art VII(1).

[81] See J Paulsson, 'Cross-enrichment of Public and Private law Dispute Resolution Mechanisms in the International Arena' (1992) 9 *JOIA* 59–68, at 70.

[82] Art I(1) NYC. In some cases, countries recognize awards made on their territory as foreign.

[83] Art I(3) NYC. [84] See Redfern and Hunter, n 9 above, 457–8.

[85] The Court held in *Ormet Corp v Ohio Power Co*, 98 F. 3d 799, 65 USLW 2295 (1996) that 'in establishing a system of marketable [emissions] allowances, Congress intended for disputes among allowance holders to be resolved in the same manner *as are other private commercial disputes*' (emphasis added).

there may be other countries where emissions trading is viewed differently, and which have, like the US, made a commercial reservation to the NYC.

The NYC allows an opposing party to resist a request for enforcement on five grounds set out in ArticleV(1). These grounds for resisting enforcement are in brief: invalidity of the agreement, lack of due process, jurisdictional issues, improper constitution of the arbitral tribunal, and that the award has been set aside or is not binding in the place where it was rendered. The bar is set high in most jurisdictions for setting aside arbitral awards. Therefore, parties may place confidence in the fact that their contracts contain an arbitration clause, which can be enforced using a variety of domestic and international mechanisms.

4 Conclusions

There is a risk that the Kyoto Protocol may not enter into force, yet despite that risk, emissions reduction investment and trading, all in the name of sustainable development, is taking place and will continue to take place. Private parties and States or State agencies will be involved in these transactions, and will seek to mitigate their risk exposure through carefully drafted contracts. In order to ensure good faith in the performance of these contracts, parties are relying on arbitration clauses, given the international nature of most of the trades taking place and the general advantages of arbitration, including party autonomy to structure the process.

The involvement of diverse private and public actors in flex-mech activities will bring together equally diverse legal systems and cultures, and the convergence of these systems could lead to different interpretations of contractual obligations—precipitating a dispute, and leading one party to commence an arbitration where the contract allows for it.

It used to be the case that developing countries, Latin America in particular, were averse to international arbitration, the latter's objections being based on notions of inequity derived from protectionist 'Calvo clauses' and mistrust of the power of western investors.[86] Today however, international arbitration is being viewed by these same countries as an essential mechanism for doing business and encouraging investment.

International arbitration is best viewed, not as a tool of the powerful to protect the powerful, but as a bridge between legal systems and cultures, and one that can be travelled with equal alacrity by all parties agreeing to use it.

Most flex-mech contracts if well drafted, especially ERPAs and emissions trading contracts, will limit disputes to discrete matters, and such disputes could conceivably be settled more quickly and thus at less expense than the same proceeding in a domestic court. Costs are something parties can keep to a minimum by ensuring that the underlying contract is unambiguous in its definitions and

[86] See n 55 above.

performance requirements, and also by limiting procedural matters such as the length of the arbitration where appropriate, having sole arbitrators where the amount in dispute is low, and through careful selection of arbitrators. Using tested and reliable arbitration procedures backed up by institutional support, as mentioned above, will help ensure that any arbitration once commenced runs smoothly and mitigates the effect of dilatory tactics.

Perhaps the best argument for arbitration in flex-mech contracts is similar to the one that Aristotle pointed out so long ago. Where a contract clearly has such an international nature, and no party will wish to submit to the opposing party's courts, only a neutral third party, freely agreed upon by the parties to the dispute, can give them a satisfactory decision as to the 'truth'—satisfactory, because in the ideal case, it would necessarily resolve the dispute, and be in line with the objectives of the Convention and Protocol.

PART VII

ARTICLE 17 OF THE KYOTO PROTOCOL: EMISSIONS TRADING

22

Emissions Trading under Article 17 of the Kyoto Protocol

*Rutger de Witt Wijnen**

1 Introduction

Article 17 of the Kyoto Protocol introduces a relatively new concept of approaching environmental issues: the trading of emission rights. Emissions trading is not entirely new. The first programmes were introduced in the US in the 1970s under the Clean Air Act. In the agricultural sector, some European countries developed trading systems in the 1980s for manure rights, an emission avant la lettre.

Emissions trading has proven to be a cost-effective way to achieve environmental goals. Not only are environmental concerns addressed, but consideration is also given to economical sustainability. Environmental issues are presented as an economic opportunity rather than as just a cost factor.

The first part of this chapter explains the basics of emissions trading. Various systems of emissions trading will be described. The second part describes the emissions trading system of Article 17 of the Kyoto Protocol.

2 Emissions Trading

2.1 Emission Rights

The phrase emissions trading is not entirely correct. Emissions are not traded, emission rights are. An emission right can be defined as the right to emit a certain quantity of a specified substance during a defined period of time. This definition has four elements:

(1) the right to emit;
(2) a specified substance;

* Rutger de Witt Wijnen is a partner with the law firm De Brauw Blackstone Westbroek.

(3) of a certain quantity;
(4) over a defined period of time.

2.1.1 Right to emit

The definition assumes that a right is needed to emit. Without legally binding constraints one could argue that, at least from a legal perspective, one could emit whatever one wants. For hazardous substances, most, if not all, jurisdictions have legislation in place limiting the right to emit such substances above a certain threshold quantity. The most common requirement is to obtain a permit for emitting certain pollutants. The operator of an oil refinery will need to have an environmental permit to emit certain quantities of pollutants. Should the oil refinery emit more than allowed under the environmental permit, the competent authorities will generally have a variety of compliance tools available, such as the imposition of penalties and, as an ultimate remedy, the closing down of the facility.

In the establishment of the rights to emit, the identity of the emitter may play a role. Competent authorities may decide for a variety of reasons to exempt certain industries or classes of installations from emission restrictions. Companies which are not obligated to participate, because their emissions are below a certain threshold, could then be given the option to participate on a voluntary basis. A recent example can be found in the draft National Allocation Plan of the Netherlands. Under the European Emissions Trading Directive[1] all EU Member States have to develop a National Allocation Plan in which they allocate emission rights to operators of installations for activities described in Annex I to this Directive. The Dutch Government proposes to exclude installations that emit less than 25 kilo tonnes of carbon dioxide (CO_2) per year from the scheme, as the administrative burden of including those installations would be disproportionate to the environmental benefits. These excluded installations account for less than 1.5 per cent of the CO_2 emissions. However, the excluded installations have the option to participate in the trading scheme voluntarily.

The right to emit, as used in the definition, refers to the regulated right to emit.

2.1.2 Specified substance

The right must relate to a certain substance or a certain group of related substances. Emission rights, under the EU Emissions Trading Scheme (EU ETS), relate to CO_2 emissions only, emission rights under the Kyoto Protocol relate to certain specified greenhouse gases (see section 3.2.2 below). If emission rights are granted in respect of various related substances, a base unit has to be set which can be used to convert the other substances so as to maintain uniformity and avoid any confusion in calculations across the board.[2]

[1] Directive 2003/87/EC [2003] OJ L275/32.

[2] In the case of the Kyoto Protocol, such base unit are CO_2 equivalents. See Section 3.2.2 below for further discussion.

2.1.3 Certain quantity

The right has to be in respect of a certain quantity of emissions. This quantity can be absolute or relative. An absolute quantity is a certain fixed, maximum, amount that an installation may emit whatever the circumstances. A relative quantity is a flexible number relating to another variable such as energy consumption or product output. The installation may generate so much of a certain emission per kW energy consumption or per kilo tonne product output. If energy consumption or production goes up, the emissions may go up as well. This is the so-called performance based rate or standard.

2.1.4 Defined period of time

The period of time during which the emissions may be made has to be restricted. The emission right is only valid during this period of time, often called the compliance or commitment period. Provisions can be made to extend the validity to a following period. Emission rights not used during the relevant compliance period may then be used in a subsequent compliance period. This concept is called banking. The opposite of banking is borrowing. Emission rights of a future compliance period may be used in the current period.

2.2 Trading

Trading can only take place if the underlying asset is transferable. Therefore, all emissions trading schemes will have to provide that the emission rights are transferable. This transferability can be limited in scope. Some schemes provide that transfers may only be made among the participants, namely the entities having an emission reduction obligation. Such schemes are called closed schemes. To enlarge the market, trading can also be made available to parties who do not have a reduction obligation themselves, such as brokers and financial institutions.

Trading can be limited to transactions within a certain scheme. However, trading can also be linked to other schemes. Linking is only possible if a mechanism is put in place to convert units from one scheme into units of the other scheme. As a result of linking, a net import of emission reductions can occur so that the total available quantity of emission reductions in a certain scheme increases.

A distinction should be made between emission rights trading and emission rights transfers (see section 3.6 below).

2.3 Forms of Emissions Trading

2.3.1 Cap and trade

The most common form of emissions trading is the cap-and-trade system. In this system, a competent authority establishes an aggregate maximum of emissions of a certain substance during a specified period of time. This maximum is allocated over the participants in the scheme, the industries and installations regulated by the emissions trading programme. The initial allocation to the relevant companies can be made for free or pursuant to an auction. After the initial allocation, the emission rights can be used for compliance by the receiver of the rights or for trading. Each participant may as a general rule only emit the quantity of emissions for which it has been allocated emission rights. Typically the participant will have to submit an amount of emission rights to the competent authority periodically equal to the amount of emissions generated during such period. It is important that a reliable monitoring system is put in place so that the regulators can verify the actual emissions during the relevant period. Should more emissions have been generated than for which emission rights had been allocated, the operator of the installation concerned has the right to purchase additional emissions rights on the free market to avoid penalties. Should fewer emissions have been generated, the excess may be sold. This mechanism provides a cost-effective way to reduce emissions. A company will assess the costs of taking emission reduction measures itself in comparison to purchasing emission rights on the market. If a company can implement reduction measures internally for a lower price than the current market price, it will have an incentive to take additional measures so as to be able to sell the excess emission rights.

2.3.2 Baseline and credit

In this system emission rights are not awarded by a competent authority, but have to be earned. A participant in the scheme establishes a baseline of the emissions, which would occur in the business-as-usual scenario. This baseline and the assumptions and data used therein have to be validated by a competent authority or unrelated, neutral, institution. The actual emissions will be monitored and verified from time to time. If the participant has been successful in making emission reductions, the difference between the baseline and the actual emissions will be converted into emission credits which can be traded. Should actual emissions be higher than the baseline, additional credits up to the excess need to be purchased in order to comply.

3 The Kyoto Protocol

3.1 Introduction

The Kyoto Protocol provides the framework for the first ever global scheme of emissions trading. Sovereign states[3] have accepted binding and enforceable

[3] A regional economic integration organization can also become a Party. The European Community has made use of this option. For the purpose of this contribution, I will restrict myself to the role of States.

commitments to reduce greenhouse gas (GHG) emissions. This commitment can be achieved not only by taking domestic measures, but also by making use of the so-called flexible mechanisms of the Kyoto Protocol: Emissions Trading (ET), Joint Implementation (JI), and Clean Development Mechanism (CDM). Another unique feature of the Kyoto Protocol is the combination of a cap and trade system with a baseline and credit system. On the basis of the general characteristics of emissions trading system described in the first part of this chapter, the way the trading systems of the Kyoto Protocol have been set up will be explained. Finally some legal issues will be discussed regarding emissions trading under the Kyoto Protocol as further developed in the Marrakesh Accords.[4]

3.2 The Kyoto Emission Rights

3.2.1 Right to emit

Under the Kyoto Protocol, Parties included in Annex I to the United Nations Framework Convention on Climate Change (UNFCCC) with a commitment inscribed in Annex B to the Kyoto Protocol (Annex B Party) have the legal obligation to reduce their emissions of certain greenhouse gases (GHGs). Those parties are generally sovereign states. Annex B Parties will be assigned an amount of GHGs (the Assigned Amount). The emissions in such party's territory may not exceed the Assigned Amounts, unless additional emission rights have been acquired through the use of the flexible mechanisms.

3.2.2 Specified substance

The Kyoto Protocol is about the reduction of GHG emissions. Six of those gases fall within the scope of the protocol: carbon dioxide (CO_2), methane (CH_4), nitrous oxide (N_2O), hydrofluorocarbons (HFCs), perfluorocarbons (PFCs), and sulphur hexafluoride (SF_6). In order to make these six gases compatible, the concept of global warming potential has been introduced. Each gas has a certain index due to the varying power to accelerate global warming. The index has been provided by the Intergovernmental Panel on Climate Change in its second assessment report of 1995.[5] On the basis of this index each gas is translated into CO_2 equivalents, which is the base unit under the Kyoto Protocol.

3.2.3 Certain quantity

The Assigned Amount shall be equal to the percentage inscribed in Annex B to the Kyoto Protocol of a country's aggregate anthropogenic CO_2 equivalent emissions

[4] Decisions 1–39/CP.7.

[5] For a complete overview of the Global Warming Potentials, see Table 1 in the UNFCCC guidelines on reporting and review, FCCC/CP/2002/8.

Table 22.1: Global Warming Potential (GWP)

Greenhouse gas	Global warming potential (GWP)
Carbon dioxide (CO_2)	1
Methane (CH_4)	21
Nitrous oxide (N_2O)	310
Hydrofluorocarbons (HFCs)	140–11,700
Perfluorocarbons (PFCs)	6,500–9,200
Sulphur hexafluoride (SF_6)	23,900

of the GHGs listed in Annex A in 1990, multiplied by five.[6] Annex B Parties may choose to use 1995 as the reference year for HFCs, PFCs, and SF_6.[7] The Assigned Amount shall be divided into Assigned Amount Units (AAUs), each of which is equal to one metric tonne of CO_2 equivalent.

3.2.4 Defined period of time

Under the Kyoto Protocol there are distinct commitment periods of five years each. The first commitment period shall run from 1 January 2008 up to and including 31 December 2012. For each commitment period an Assigned Amount will be established.

3.3 The Kyoto Trading Systems

3.3.1 Baseline and credit: JI and CDM

The Kyoto Protocol provides for two baseline and credit systems: JI and the CDM. Under both systems credits can be earned which may be used by Annex B Parties for the purpose of meeting their reduction commitments. Under JI the credits to be earned are called Emission Reduction Units (ERUs), under the CDM they are called Certified Emission Reductions (CERs). Once an ERU or CER has been issued to an Annex B Party, it cannot only be used by such party for compliance purposes, but also for trading with other Annex B Parties. The JI and CDM are described in detail in the previous sections of this book.

3.3.2 Cap and trade

Article 17 of the Kyoto Protocol forms the basis for a global emissions trading system among Annex B Parties, which may be used by Annex B Parties for compliance purposes.

[6] Art 3(7) Kyoto Protocol. For parties with an economy in transition, another base year may be used, according to Art 3(5) Kyoto Protocol.

[7] Art 3(8) of the Kyoto Protocol.

3.4 Emissions Trading under Article 17

Article 17 of the Kyoto Protocol

The Conference of the Parties shall define the relevant principles, modalities, rules and guidelines, in particular for verification, reporting and accountability for emissions trading. The Parties included in Annex B may participate in emissions trading for the purposes of fulfilling their commitments under Article 3. Any such trading shall be supplemental to domestic actions for the purpose of meeting quantified emission limitation and reduction commitments under that Article.

The framework provision of Article 17 of the Kyoto Protocol forms the basis for a trading system, which is generally, but erroneously, called 'emissions trading'. As mentioned above, emissions as such cannot be traded. Only rights in respect of emissions, the right to emit, can be traded. Additionally, as will be explained in more detail below, Article 17 is not about trading, but rather about transferring and acquiring of emission rights.

The material rules for emissions trading will have to be set by the Conference of the Parties serving as the Meeting of the Parties to the Kyoto Protocol (COP/MOP).[8] Decision 18/CP.7 of the Marrakesh Accords[9] makes a recommendation to COP/MOP to adopt at its first session the rules annexed to this Decision. Emissions trading, to stick to this terminology, in accordance with these rules will be discussed in this chapter.

3.5 What can be Traded?

Although Article 17 is mostly referred to in the context of trading of AAUs, the restriction to AAUs can not be found in Article 17. Paragraph 2 of Annex to Decision 18/CP.7 (Decision 18/CP.7) states that also ERUs, CERs, and RMUs[10] can be transferred and acquired. This solves one of the major hurdles in international emissions trading which is the definition of the asset to be traded. AAUs, ERUs, CERs, and RMUs (jointly: Kyoto Units) are all defined assets.[11] The common denominator of each of these units is that each is a unit issued pursuant to the relevant rules of the Kyoto Protocol or rules promulgated pursuant thereto and each represents one metric tonne of CO_2 equivalent.

[8] Art 17 refers to the Conference of the Parties. This must have been an oversight as only COP/MOP can make decisions in respect of the Kyoto Protocol.

[9] FCCC/CP/2001/13/Add.2.

[10] RMUs or Removal Units, are units which can be issued to reflect a removal of GHGs by sinks on the basis of Art 3(3) Kyoto Protocol.

[11] See Decision 18/CP.7, FCCC/CP/2001/13/Add.2, para 1.

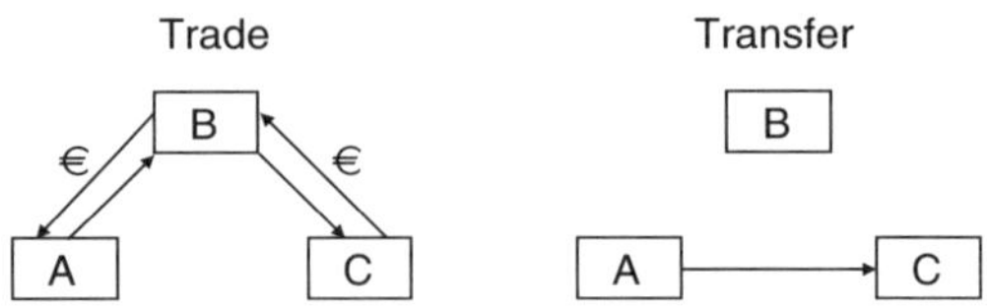

Figure 22.1. Trade vs Transfer

3.6 Trade versus Transfer

Paragraph 2 of Decision 18/CP.7 also corrects an imperfection in Article 17 as it refers to 'transfer and/or acquire' rather than to 'trading'. Strictly speaking Article 17—and the modalities, rules and guidelines for emissions trading set forth in Decision 18/CP.7—are not about trading but rather about transferring and/or acquiring. Trading is buying and selling, transferring is delivering/accepting the sold/bought asset. A transfer and acquisition may be the final step of a trade. However, not every trade results in a transfer. If a buyer (B) buys an asset, (B) generally has the right to require the seller (A) to transfer the asset to (B). However, if (B) resells the asset to a third party (C) before it has been transferred to (B), (B) may instruct (A) to transfer the asset to (C) directly.

In this way it is conceivable that trading systems of Kyoto Units develop which do not lead to transfers of such units or which lead to transfers only at certain, pre-agreed, points in time.

Article 17 deals with transferring and/or acquiring of Kyoto Units and does not impose conditions on trading of such units. Therefore, trading in Kyoto Units could begin at any time. Transfer of Kyoto Units can only take place as from 1 January 2008 assuming the Kyoto Protocol has entered into force.

Trading prior to 1 January 2008 will have to be in the form of forward trades. In the case of AAUs, ERUs, and RMUs, as these assets can only be created after 1 January 2008. Although CERs may be issued prior to this date, the transfer and acquisition of CERs after their initial issuance can only be made pursuant to Article 17 and can therefore only occur after 1 January 2008.

3.7 Who can Transfer?

3.7.1 Eligibility requirements

The Kyoto Protocol is an agreement among the parties thereto (each a Party). Generally those Parties are, as described above, sovereign States.[12]

As follows from the previous paragraph, Article 17 does not regulate trading of Kyoto Units. Everybody, whether a State or not, whether a Party to the Kyoto Protocol or not, can trade Kyoto Units. However, if a trade leads to a transfer, such transfer can, as a general rule, only be made by Parties to the Kyoto Protocol.

[12] See n 3 above.

But just being a Party to the Kyoto Protocol is not sufficient to be able to transfer or acquire Kyoto Units. First of all, only an Annex B Party is eligible to transfer or acquire Kyoto Units. This is a logical consequence of the requirement in Article 17 that emissions trading may be participated in for the purpose of fulfilling emission reduction commitments. Only Annex B Parties have such commitments.

In addition, paragraph 2 of Decision 18/CP.7 lists the following requirements for eligibility:

(a) it has to be a Party to the Kyoto Protocol (indeed a redundant requirement, but listed in paragraph 2 nevertheless);
(b) its assigned amount has been calculated and recorded;
(c) it has in place a national system for the estimation of anthropogenic emissions by sources;
(d) it has in place a national registry;
(e) it has submitted annually the most recent required inventory;
(f) it submits the supplementary information on assigned amounts;

all in accordance with the relevant provisions pursuant to the Kyoto Protocol.

Annex B Parties shall be considered to meet the eligibility requirements after sixteen months have elapsed since the submission to the UNFCCC Secretariat of its report to facilitate the calculation of its Assigned Amount, unless the enforcement branch of the compliance committee finds that the Annex B Party does not meet the requirements. To make life easier for persons wanting to verify the ability of a Party to participate in emissions trading, the UNFCCC Secretariat will maintain a publicly accessible list of Parties that meet the eligibility requirements, according to paragraph 4.

3.7.2 Legal entities

Unlike Article 6 for JI and Article 12 for CDM, Article 17 makes no reference to participation of legal entities in emissions trading. Once again, most likely an oversight which has been remedied by the Marrakesh Accords. Paragraph 5 of Decision 18/CP.7 provides that a Party that authorizes legal entities to transfer and/or acquire emission rights under Article 17 shall remain responsible for the fulfilment of its obligations under the Kyoto Protocol and shall ensure that such participation is consistent with the applicable rules. Participation of legal entities may have seemed so obvious that a general provision allowing Parties to authorize legal entities to participate in emissions trading has not been included in the Marrakesh Accords.

A legal entity may, after due authorization, only transfer and/or acquire Kyoto Units if the authorizing Party is eligible to do so. A Party shall maintain an up-to-date list of the legal entities it has authorized to participate and make such list publicly available.[13]

[13] See para 5 of Decision 18/CP.7.

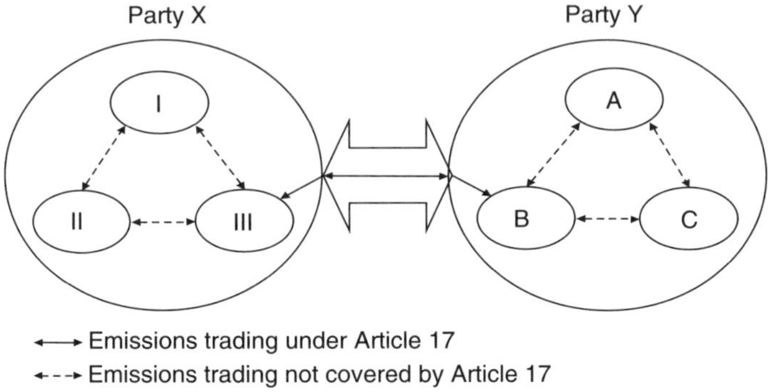

Figure 22.2. Transfer of Kyoto Units

Paragraph 5 of Decision 18/CP.7 makes no distinction as to the country of incorporation or domicile of the legal entity. The question has come up whether an Annex B Party could authorize a legal entity not incorporated or having its domicile in such party's jurisdiction to participate in emissions trading. In my view there is no reason why this is not possible, even if the legal entity would be incorporated or have its seat in the jurisdiction of a non-Annex B Party. All transactions will be channelled through the national registry of the authorizing Annex B Party (see Figure 22.2 above). This implies that each participating legal entity will need to have an account in the national registry of the authorizing Annex B Party. The 'nationality' of the legal entity is irrelevant. All Kyoto Units credited to its account will be part of the overall national account of the authorizing Annex B Party. Therefore, also legal entities from non-Annex B Parties can participate in emissions trading under Article 17, provided they are authorized to do so by an Annex B Party.

Some Annex B Parties argue that only legal entities, which have a GHG reduction commitment themselves, are eligible for authorization to participate in the flexible mechanisms of the Kyoto Protocol. In this view, only European companies, which participate in the trading scheme established pursuant to the European Emissions Trading Directive, could get an authorization to buy CERs and other Kyoto Units. This is a very narrow interpretation of the Kyoto Protocol for which there seems to be no legal basis in the texts of the Kyoto Protocol or the Marrakesh Accords.

3.8 Transfer Procedures

Paragraph 5 of Decision 18/CP.7 provides that transfers and acquisitions between national registries shall be made under the responsibility of the Parties concerned in accordance with the provisions in draft Decision -/CMP.1 (Modalities for the accounting of assigned amounts).[14] National registries shall be established by each

[14] Decision 19/CP.7.

Annex B Party to ensure the accurate accounting of the issuance, holding, transfer, acquisition, cancellation, and retirement of Kyoto Units, according to paragraph 17 of this Decision. The working of registries is described in Chapter 7 above. It is important to realize that all transfers, including transfers initiated by legal entities, which have been authorized to participate in emissions trading, will be made through the national registries of the authorizing Annex B Party. A transfer of Kyoto Units by a legal entity in Party X to a legal entity in Party Y is, at the same time, a transfer of such Kyoto Units by Party X to Party Y. The total amount of Kyoto Units in the national account of Party X will be decreased by the number of Kyoto Units transferred by the legal entity.

As indicated in this figure, transfers of emission rights can also take place within a registry. Such transfers are not covered by Article 17, as these transfers are purely a domestic matter.

3.9 The Commitment Period Reserve

Even if all eligibility requirements are met, a transfer cannot proceed if, as a result thereof, the aggregate amount of Kyoto Units in the national register of the transferring Party would fall below the so-called Commitment Period Reserve. Paragraph 6 of Decision 18/CP.7 provides that each Party included in Annex I[15] (to the UNFCCC) shall maintain, in its national registry, a commitment period reserve which should not drop below 90 per cent of the Party's Assigned Amount calculated in accordance with Article 3 of the Kyoto Protocol, or 100 per cent of five times its most recently reviewed inventory, whichever is lowest.

During the negotiations on the rules for emissions trading some delegates expressed concerns that an Annex B Party could be tempted to sell too much of its Kyoto Units which might jeopardize its ability to comply with its reduction obligations under the Kyoto Protocol. In order to protect these overzealous sellers of Kyoto Units, the Commitment Period Reserve was introduced. A Party shall not make a transfer, which would result in these holdings being below the required level of the commitment period reserve, according to paragraph 8 of Decision 18/CP.7.

The provision of paragraph 6 on calculating the level of the Commitment Period Reserve is hard to digest. The following example may clarify the meaning. Assume that the emissions of GHGs in Country A equalled 60 Mtonnes in 1990. The reduction commitment of Country A inscribed in Annex B to the Kyoto Protocol is 95. The assigned amount for Country A for the first commitment period will be 5 times 95 per cent of 60 Mtonnes, which equals 285 Mtonnes.[16]

[15] Obviously this obligation only applies for Parties included in Annex I with a commitment inscribed in Annex B to the Kyoto Protocol.

[16] See para 5 of the Annex to Decision 19/CP.7. I have applied the general rule for calculating the Assigned Amount only.

In this fictional example, the Commitment Period Reserve will be 256.5 Mtonnes, which is 90 per cent of 285 Mtonnes. Assume that the most recent inventory reviewed of Country A shows that in a year 50 Mtonnes of GHG have been emitted, the Commitment Period Reserve will be reduced to 100 per cent of five times 50 Mtonnes, which equals 250 Mtonnes. Should a subsequent inventory show an increase of emissions, the Commitment Period Reserve will go up again, provided that it will not exceed 256.5 Mtonnes in this example.

For legal entities participating in emissions trading under Article 17, the Commitment Period Reserve is an obstacle as a transfer could be frustrated for reasons out of the control of such legal entities. If a seller of Kyoto Units has sufficient Kyoto Units in its account to honour its obligations to transfer a certain quantity to a buyer in another jurisdiction, but such transfer would at a national level decrease the Kyoto Units in the national account below the level of the Commitment Period Reserve, such transfer cannot be completed. Buyers and sellers would be well advised to agree in advance whether such inability to complete the delivery of Kyoto Units qualifies as *force majeure* or not.

3.10 Applicable Law

The Kyoto Protocol and the Marrakesh Accords are silent on the law applicable to emissions trading and to the transfer and acquisition of Kyoto Units. It is a thorny issue, which may, especially in this international setting, lead to confusion. Various legal systems are involved which may have a different concept of trading and transferring.

Once again, a distinction should be made between trading on the one hand and transferring on the other.

3.10.1 Trading

Trading is the contractual agreement to sell Kyoto Units. This is not covered by Article 17. Parties to an international agreement typically add a choice of law clause to their contract in order to avoid uncertainty as to the applicable law. If no choice of law is made, the applicable law will be determined by conflict of laws rules, which the competent court or arbitral tribunal considers applicable. The conflict of laws rule may be based on an applicable treaty, such as the EC Convention on the law applicable to contracts of 1980[17] or on the national law of the forum adjudicating the matter.

Contrary to the private sector, sovereign States sometimes prefer not to add a choice of law clause to a commercial contract when contracting with other States. This preference seems to be based more on political considerations than anything else. If no reference to an applicable law is included in the commercial contract at all, the applicable law will have to be determined in the same way as described

[17] Rome Convention of 19 June 1980.

above for private entities. States sometimes refer to principles of public international law, *lex mercatoria* or similar concepts. If an agreement among states is governed by international law, the Vienna Convention on the Law of Treaties of 1969[18] will apply.

3.10.2 Transfers

Transfers of Kyoto Units are governed by Article 17 of the Kyoto Protocol and the provisions of Decision 18/CP.7. Any disputes as to transfers of Kyoto Units will have to be adjudicated on the basis of those provisions. As the Parties remain responsible for the transfers and acquisitions between the national registries, it is hard to imagine a dispute in respect of a transfer between (authorized) legal entities. Their role in the transfer process is limited, giving instructions to the relevant national register. On the other hand a dispute between a legal entity and the authorizing Annex B Party are very well conceivable. Such disputes are most likely governed by the laws of the jurisdiction of the relevant Annex B Party, as such jurisdiction will have enacted national legislation to allow legal entities to apply for an authorization to participate in emissions trading.

4 Summary

This chapter describes the mechanism of emissions trading under Article 17 of the Kyoto Protocol. As an introduction, the concept of emissions trading is described in general terms. Various basic terms such as emission rights, trading and transfers and various trading regimes as explained. Article 17 of the Kyoto Protocol is discussed against the background of the general description of emission trading. Finally, some specific issues such as the participation of legal entities, transfer procedures, the commitment period reserve and applicable law are described.

[18] 1155, UNTS 331.

23

The Emissions Trading Scheme of the European Union

*Robert Dornau**

1 Introduction

Europe is taking the lead in greenhouse gas (GHG) emissions trading. On 22 July 2003 the European Council approved a Directive[1] setting the framework for an EU-wide system of GHG emission allowance trading at company level. This European Emission Trading Scheme (EU ETS) will start 1 January 2005. The Directive requires each Member State to impose binding caps on emissions of CO_2 from installations involved in energy activities, the production and processing of ferrous metals, the mineral industry (for example, cement, glass, or ceramic production), and pulp, paper or board production. Other GHGs and sectors will be added in later periods.

The EU ETS is seen as the essential measure for the European Community (EC) and its Members to fulfil its commitments under the Kyoto Protocol to reduce their aggregate anthropogenic emissions of GHGs listed in Annex A to the Kyoto Protocol by 8 per cent compared to 1990 levels in the period 2008 to 2012.

With the adoption of the Directive, Europe takes the lead in GHG emissions trading when it put in place a regime that will cover around 13,000 installations[2] through the enlarged EU25 Community, representing about 46 per cent of the EU's total CO_2 emissions in 2010.[3] The European Commission, Council, and Parliament are convinced that emissions trading delivers cost-effectiveness

* Robert Dornau is an independent climate change consultant. The views expressed in this article are the views of the author.

[1] Directive 2003/87/EC of the European Parliament and of the Council of 13 October 2003 establishing a scheme for greenhouse gas emissions allowance trading within the EC and amending Council Directive 96/61/EC [2003] OJ L275/32.

[2] See a summary table of national allocation plans of the EU25 (No data available for Spain, Lithuania, and Malta) as of 22 January 2004 as published by the European Commission at <http://europa.eu.int/comm/environment/climat/emission_plans.htm>accessed 26 January 2004.

[3] For this and the following figures see for example, speeches by Margot Wallstroem, EU Environment Commissioner, at <http://www.europa.eu.int/comm/environment/climat/emission.htm> accessed 26 January 2004.

by reducing the cost of meeting EU Kyoto commitments by 35 per cent, representing €1.3 billion benefit per year to 2012. Analysts expect the total size of the trading market at €5–10 billion per year.

As the scheme allows for links with other Kyoto Parties it can be regarded as the nucleus of a possible global trading scheme.

This chapter describes the EU ETS and provides a background on its general and unique features. It will also analyse the impact and relevance of the scheme and certain provisions in the scheme.

2 The Birth of the Carbon Economy—Legislative History

The Directive was drafted by the EU Commission and amended in Parliament and Council over a period of roughly three years. The consultation process for an EU-wide emissions trading scheme was started with a Green Book published by the Commission in March 2000, the final Directive was adopted in July 2003.

The submissions in the Green Book[4] and later lobbying showed that many views had to be taken into account in the negotiations. While governments, environmental non-governmental organizations (NGOs) as well as industry were, in general, in favour of using emissions trading as an instrument to combat climate change, they had different concerns, which they lobbied for in the negotiations: environmental NGOs, argued in favour of robust environmental effectiveness, requiring strict targets, monitoring, verification, and compliance functions (that should be exercised at Community level). Industry was united in not wanting an emissions trading system to place European businesses at a competitive disadvantage, with most industries wanting as wide a coverage as possible, both in terms of market participants/sectors and in terms of GHGs.[5] They were also concerned that industry should not bear an unreasonable burden in relation to the transport and household sectors. Some industries, and their associations, favoured voluntary agreements to reduce GHGs and opposed a cap-and-trade system. Depending on the type of industry, different systems for allocation of allowances was favoured. The formulae that received the most support were for a 'framework' to be established at European level, while still allowing some flexibility for Member States to respond to national circumstances.[6]

The Commission in October 2001 published a first proposal of the Directive. Against the opposition of some industries and also governments it aimed

[4] European Commission, 2000, Green Paper on greenhouse gas emissions trading within the European Union COM (2000) 87, <http://europa.eu.int/comm/environment/docum/0087_en.htm> accessed on 14 March 2004.

[5] See for example, IETA Position on the EC emissions trading proposal, <http://www.ieta.org/Documents/New_Documents/IETA_Position_on_the_EC_Proposal_May_2002.doc>accessed on 26 January 2004.

[6] See for example, Peter Vis, Climate Change Unit European Commission, Summary of Submissions on the Greenhouse Gas Emissions Trading within the European Union, 14 May 2001.

at a scheme at company level in order to obtain a sufficiently large, competitive, and cost-effective market. As stated in the explanatory Memorandum to the Proposal: 'A Community scheme would minimise distortions of competition and potential barriers to the internal market that might otherwise arise as a result of disparate numbers of trading schemes being established in the European Union.'[7]

The Council of Environment Ministers provided its seal of approval in December 2002 following a first reading of the proposal in the European Parliament in September. While the Parliament would still aim for amendments to the Directive the main political hurdle was cleared. A common position of Parliament and Council was developed in March 2003, incorporating 23 of the 73 amendments proposed by the European Parliament in the first reading. The main amendments negotiated at this stage and later were: the scope of the Directive was broadened, the possibility of sectoral opt-outs was removed and the percentage of emission allowances to be granted free of charge was fixed. The Commission, Council, and Parliament then worked closely together in agreeing on language to allow for approval of the Directive by Parliament in a second reading on 2 July 2003, making the final approval by the Council on 22 July 2003[8] just a procedural step without the need of further debate. This early finalization of the process provided business with certainty and allowed Member States to start preparing for the implementation of the Directive in national law in order to facilitate a prompt start in 2005. As Jorge Moreira Da Silva, Rapporteur of the European Parliament for the Emissions Trading Directive, stated at the press conference on 2 July 2003: 'The carbon economy is born!'[9]

3 The Trading Scheme

This section provides a description and analysis of the scheme's features.

3.1 Scope—Sectors and Gases

The Directive requires each Member State to impose binding caps on emissions of CO_2 from installations above certain production capacity or output thresholds (see Table 23.1) in the following sectors: (1) energy activities; (2) the production and processing of ferrous metals; (3) the mineral industry (for example, cement, glass, or ceramic production); and (4) pulp, paper or board production.

[7] See p 2, Proposal for a Directive of the European Parliament and of the Council establishing a framework for greenhouse gas emissions trading within the European Community and amending Council Directive 96/61/EC, Brussels, COM(2001)581 final 2001/0245 (COD).

[8] The final text of the Directive was published on 13 October 2003.

[9] Jorge Moreira Da Silva, MEP, Rapporteur for the Emissions Trading Directive, at the press conference in Strasbourg after the adoption of the Emissions Trading Directive by the European Parliament on 2 July 2003.

An 'installation' as defined in the Directive is 'a stationary technical unit where one or more activities listed in Annex I[10] are carried out and any other directly associated activities which have a technical connection with the activities carried out on that site and which could have an effect on emissions and pollution'.[11]

In the first period only CO_2 emissions will be included. As from 2008, Member States may unilaterally include additional installations/sectors and GHGs, subject to approval by the Commission. In addition a general review of sectors and gases to be included is foreseen. The chemicals, aluminium, and transport sectors are specifically mentioned in the Directive[12] as likely to be included at that time with a view to further improving the economic efficiency of the scheme. The regulations for a unilateral opt-in of gases and installations from 2008 on can be seen as a compromise. The Parliament initially wished to include all six GHGs and the aluminium and chemical sectors from 2005.

Many companies only realized at a late stage that they would be covered by the Directive because they have one or more on-site combustion activities (for

Table 23.1. Sectors and threshold values in the EU ETS[13]

Energy activities
Combustion installations with a rated thermal input exceeding 20 MW (except hazardous or municipal waste installations).
Mineral oil refineries.
Coke ovens.
Production and processing of ferrous metals
Metal ore (including sulphide ore) roasting or sintering installations.
Installations for the production of pig iron or steel (primary or secondary fusion) including continuous casting, with a capacity exceeding 2.5 tonnes per hour.
Mineral industry
Installations for the production of cement clinker in rotary kilns with a production capacity exceeding 500 tonnes per day or lime in rotary kilns with a production capacity exceeding 50 tonnes per day or in other furnaces with a production capacity exceeding 50 tonnes per day.
Installations for the manufacture of glass including glass fibre with a melting capacity exceeding 20 tonnes per day.
Installations for the manufacture of ceramic products by firing, in particular roofing tiles, bricks, refractory bricks, tiles, stoneware or porcelain, with a production capacity exceeding 75 tonnes per day, and/or with a kiln capacity exceeding 4 m^3 and with a setting density per kiln exceeding 300 kg/m^3.
Other activities
Industrial plants for the production of: (a) pulp from timber or other fibrous materials; (b) paper and board with a production capacity exceeding 20 tonnes per day.

[10] See Table 23.1 Sectors and threshold values in the EU ETS. [11] See Art 3(e).
[12] Arts 24 and 30(a). [13] Source: Annex 1 of the Directive.

example, generators, furnaces, boilers, and so on) which alone, or in aggregate, have a 20 MW or greater thermal input. This has the effect that also many chemical and aluminium sites will be regulated with their emissions from those activities, even though these sectors as such are not yet targeted.

3.2 Commitment Periods

The Directive provides for an initial three-year commitment period that runs from 2005 to 2007. This period is regarded as a learning-by-doing period for Member States as well as operators of installations covered by the Directive. Subsequent periods will be of five years. The first five-year period starts in 2008, concurrently with the first commitment period under the Kyoto Protocol.

The EU ETS will however operate independent of entry into force of the Kyoto Protocol. It is important to understand the difference between the two approaches: the Kyoto Protocol is negotiated between governments imposing reduction targets on a country's total emissions (including industry, household, transport, as well as land use, land use change and forestry), whereas the EU ETS is a scheme at company level, based on the EU Directive and national legislation, regulating emissions of industry installations only. The overall targets of the Kyoto Regime are used to derive national allocations for the industry but the entry into force is not necessary for the EU/national regulation to be binding for companies.

3.3 Permits and Allowances

The Directive requires Member States to ensure that from 1 January 2005 the operators of installations that fall under the scope of the EU ETS (see Table 23.1) hold a GHG emissions permit.[14] The permit specifies monitoring and reporting requirements and states an obligation to surrender a number of allowances equal to the total verified emissions of the installation in the previous calendar year. The Directive defines operator as 'any person who operates or controls an installation or, where this is provided for in national legislation, to whom decisive economic power over the technical functioning of the installation has been delegated'.[15]

Relevant departments in each Member State (the competent authorities)[16] should be contacting installations that will be included in the scheme. However, it is the operator's responsibility to apply for the correct permit.[17] Any installation not in possession of a valid permit by the time the National Allocation Plan (NAP) is submitted to the EU Commission (deadline was 31 March 2004 for the first period, however the majority of EU Member States missed the deadline) will be regarded as a new entrant.[18]

[14] Art 4. [15] Art 3(f).
[16] See Art 18 of the Directive for a definition of competent authority.
[17] According to the rules set out in Art 5.
[18] Consequences of being a new entrant are outlined in section 3.6 below.

Table 23.2: Permits and Allowances

Permits	Allowances
Site-specific	Issued by Member States
Set monitoring and reporting obligations (EU Guideline)	Entitlement to emit a tonne of CO_2e
Set obligation to hold allowances matching emissions each calendar year	Held in the national registry system
Non-transferable	Tradable

It is important to note that, where one operator carries out several activities falling under the same subheading of Table 23.1 in the same installation or on the same site, the capacities of such activities are added together. If together they exceed the respective threshold, a permit is required. Permit holders will be issued allowances into their holding accounts in their national registry according to the NAPs.

Allowance[19] means an allowance to emit one tonne of carbon dioxide equivalent (tCO_2e)[20] during a specified period. These allowances are transferable, thereby creating the possibility of emission allowance trading. Starting in 2008, transfers of allowances to another Member State will involve corresponding adjustments of assigned amount units under the Kyoto Protocol (assuming it has entered into force).

The characteristics of permits and allowances are summarized in Table 23.2 above.

3.4 Compliance

The number of allowances an installation receives (its Allocation) is decided upon in advance for the full commitment periods of three and five years respectively. However, compliance with actual emissions is required each calendar year. Member States have to ensure that, by 30 April each year, at the latest, the operator of each installation surrenders a number of allowances equal to the total verified emissions of the installation in the previous calendar year.[21]

In order to provide Operators with some flexibility, the Directive allows for banking and borrowing between years of one commitment period. 'Banking' means that allowances of, for example, year 2005 that are not used for compliance of the year 2005 can be used for compliance in consecutive years, at least in the first commitment period 2005–7. Whether banking of allowances from the first into the second

[19] Art 3(a).

[20] CO_2 equivalent: the universal unit of measurement used to indicate the global warming potential (GWP) of each of the six GHGs. It is used to evaluate the impacts of releasing (or avoiding the release of) different GHGs. The six gases and corresponding GWPs are: carbon dioxide (1), methane (21), nitrous oxide (310), halocarbons (HFC) (140 to 11,700), sulphur hexafluoride (23,900).

[21] Art 12(2).

commitment period 2008–12 will be allowed will be decided at national level when the Directive is implemented in national law and the allocation plans are finalized. 'Borrowing' means that allowances of for example, year 2006 can be used for compliance of the directly preceding year, in that case year 2005. This is made possible as allowances for the current year are issued at the latest by 28 February whereas the deadline for compliance for the previous year is 30 April.[22]

Flexibility is of course also achieved through the possibility to buy or sell allowances in the allowance market (Emissions Trading).

Failure to meet the emissions caps will result in penalties[23] of €40 for the period 2005–7 and €100 from 2008 on for each tonne of carbon dioxide by which installations exceed their caps. It is important to note that in addition to paying the penalty, companies will also have to surrender a compensating amount of allowances in the subsequent year. This has the effect that the penalty is not a possibility to buy out of the reduction commitment and hence does not represent a price cap.

3.5 Monitoring, Reporting, and Verification

The principles governing emission monitoring, reporting and verification are outlined in Annexes IV and V to the Directive. Two options for monitoring are provided: calculation or measurement using standardized and accepted methods. A Commission Decision establishing guidelines for the monitoring and reporting of GHG emissions will provide additional guidance. Reporting requirements for the EU ETS are planned to be coordinated with existing reporting requirements to minimize the burden to businesses.

The Directive requires verification of emissions within the EU ETS to ensure that the reported emissions are credible, reliable and accurate. The consistent testing of emissions data by the application of these principles is expected to be provided through the accreditation of the organizations, which perform the verification.

3.6 Allocation

One of the core tasks in the run-up to the implementation of the EU-wide GHG allowance trading scheme is the elaboration of NAPs[24] by Member States. Each NAP will contain a list of all installations covered and, more importantly, the quantities of allowances intended to be allocated to each installation in the relevant commitment period. This procedure provides industry with certainty in planning the emission reductions required during each commitment period.

The allocation processes will have a large impact on supply and demand within the market. Receiving a fair allocation is of utmost importance for every company. Allowances are allocated to a large degree for free but as they are market instruments, will gain a market value. Receiving too little allowances will force companies

[22] Art 11(4). [23] Art 16. [24] Art 9.

to buy allowances in the market, whereas a favourable allocation will leave them as sellers in market. Hence, sectors have been lobbying hard to try and get the most favourable allocation possible.

For the period starting 2005, Member States had to submit these plans by 31 March 2004 at the latest for review and approval by the European Commission. Installations that are not listed in the NAP by that date will be regarded as new entrants[25] in the current period.[26] This procedure puts new installations on equal footing with existing installations' expanding capacity.

For later periods NAPs will have to be submitted to the Commission at least eighteen months before the beginning of the relevant period. The Commission has six months to review and approve the plan or reject it within three months after submission if it is incompatible with criteria outlined in the relevant allocation guidance.

The Directive itself provides relatively little guidance on allocation criteria.[27] Additional guidance to help Member States establishing their allocation plans was given by the Commission in the form of a non-paper in April 2003[28] and a Commission Communication Paper in January 2004.[29] The criteria for Allocation are for example:[30]

- Method of Allocation. At least 95 per cent (90 per cent) of the allowances have to be allocated free of charge in the first (subsequent) periods. The difference to 100 per cent may be allocated by the Member State in an auction.
- Kyoto Commitments. The total quantity of allowances to be allocated shall be consistent with assessments of actual and projected progress towards fulfilling the Member States' contributions to the Community's commitments under the Kyoto Protocol. Prior to 2008 the total quantity shall be consistent with a path towards achieving or overachieving a Member State's target under the burden sharing agreement[31] and the Kyoto Protocol.
- Potential to reduce. Quantities of allowances to be allocated shall be consistent with the potential, including the technological potential, of activities covered by this scheme to reduce emissions. Member States may base their distribution

[25] As defined in Art 3 of the Directive.

[26] New entrant status is a temporary issue: an installation is defined, as new entrant in one period will no longer be a new entrant in subsequent periods.

[27] See Annex III of the Directive.

[28] 'The EU Emissions Trading Scheme: How to develop a National Allocation Plan', Non-Paper, second meeting of Working 3, Monitoring Mechanism Committee, 1 April 2003, <http://www.europa.eu.int/comm/environment/climat/pdf/030401nonpaper.pdf>accessed on 27 January 2004.

[29] Communication from the Commission on guidance to assist Member States in the implementation of the criteria listed in Annex III to Directive 2003/87/EC establishing a scheme for greenhouse gas emission allowance trading within the Community and amending Council Directive 96/61/EC, and on the circumstances under which *force majeure* is demonstrated, COM(2003)830 final, Brussels, 7 January 2004.

[30] Other criteria listed in Annex III are: Assessments of emissions developments, consistency with other legislation, clean technology, involvement of the public, list of installations, competition from outside the Union.

[31] Decision 2002/358/EC.

of allowances on average emissions of GHGs by product in each activity and achievable progress in each activity.

- Non-discrimination between companies or sectors. The plan shall not unduly favour certain activities or sectors. Member States shall not allocate more than is likely to be needed. Allowances allocated under the NAPs will constitute 'state aid' and will therefore have to be notified to the Commission for assessment under state aid rules. Competition policy rules will also apply. Allocations to individual installations or sectors must not constitute incompatible state aid, which would distort or even threaten to distort competition.
- New Entrants. The plan must contain information on how new entrants will be able to participate in the EU ETS in the Member State concerned. The guidance outlines three options to implement the Member State's obligation to provide new entrants with access to allowances: allow new entrants to buy allowances in the market; provide access to allowances on the basis of periodic auctions; or provide access to allowances free of charge out of a reserve. In the case of a reserve, the Member State would need to adjust the allocations to existing installations accordingly—meaning existing installations might receive fewer allowances in the allocation—and also decide on what would happen to unallocated allowances should the number of new entrants turn out to be fewer than anticipated.
- Early Action. The plan may accommodate early action and shall contain information as to the manner in which early action is taken into account. This is important from a fairness point of view. Those installations that have already reduced GHG emissions in absence of or beyond legal mandates should not be disadvantaged vis-à-vis other installations that have not undertaken such efforts.

Member States must distribute allowances to installations by 28 February each year. The percentages of the total allocation of allowances issued each year are fixed in the NAPs. Market participants argue that the majority of allowances should already be issued in the first year of each Commitment Period to help the development of a liquid market.[32] Such a procedure would enable especially smaller companies to hedge their risks through engaging in the market by trading allowances rather than more complex forward or option contracts. In addition trading partners might not want to accept the risks of default associated with such contracts.

3.7 Market Organization

Allowances can be transferred (a) between natural or legal persons within the European Community and, (b) between persons within the European Community and persons in countries listed in Annex B to the Kyoto Protocol

[32] As the scheme allows for borrowing such a procedure would not cause distortions.

which have ratified the Protocol, provided, however, that such Annex B countries have entered into an agreement with the Community on the mutual recognition of their allowance trading schemes.[33] The regulators expect this to increase the cost-effectiveness of achieving the Community emission reductions target.

Allowing persons and not only operators to engage in the allowance market can help enhance market liquidity as speculative traders can act as market makers. Entry of such liquidity providers however, also depends on the organization of the market. The Commission stated already in its memorandum to the first proposal of the Directive that it 'wishes the organisation of the market in allowances to be left open to solutions driven by the private sector.'[34] The Directive therefore does not touch upon issues such as the legal classification of an allowance (for example, financial instrument or commodity) and taxation. While these points are to be solved on a national basis by governments, other elements essential for an efficient market like standard trading contracts were developed by the private sector.[35] Standardization of regulatory treatment of allowance is essential to create a truly European Market. Differences between Member States will lead to differing allowance prices and market fragmentation.

Trading in EU Allowances is already taking place, most of it on a bilateral basis directly between companies or through brokers. While trading volume is still low, the price for allowances rose from around €6 to around €12 per Allowance in 2003. Different institutions currently explore the possibility for an electronic trading platform in the allowance market. However, a liquid transparent market will not evolve unless national governments provide legal certainty in relation to an allowance (the product being traded) and how the trading of allowances fits within the existing legal regime.

3.8 System of National Registries[36]

An allowance will be an electronic unit clearly identifiable by a serial number. Member States are requested by the Directive to establish and maintain a registry in order to ensure the accurate accounting of the issue, holding, transfer, and cancellation of allowances. In addition, the Commission will establish a Community Independent Transaction Log (CITL). The CITL will check all transactions of allowances for irregularities in the issue transfer and cancellation of allowances. If irregularities are identified the transfer of allowances will not be allowed to be completed. The public will have limited access to account information held within the Registries System, to ensure transparency of its environmental integrity.

[33] See Arts 12 and 25.

[34] Proposal for a Directive of the European Parliament and of the Council establishing a scheme for greenhouse gas emission allowance trading within the Community and amending Council Directive 96/61/EC, Brussels 23 October 2001.

[35] See the IETA Emissions Trading Master Agreement, available on <http://www.ieta.org> accessed on 27 January 2004.

[36] Art 19.

The Directive foresees that Member States may maintain their registries in a consolidated system, together with one or more other Member State. Currently only France and the UK have developed or are in the process of developing national registries for the EU ETS. Their experience shows that the development of a secure and efficient registry incurs costs of €2–3 million. It is therefore likely that not all Member States will develop registries but make use of the possibility of maintaining legally independent registries on a consolidated system, realizing economies of scale.

The Commission published a substantive first draft of a Commission Regulation[37] elaborating on standards for national registries and the CITL in October 2003. The objective of the Regulation is to establish a standardized and secured system of registries including the CITL in the form of standardized electronic databases containing common data elements, to operate together as the Registry system. It sets out functional and technical specifications required of each Registry and the CITL and prescribes how they will operate in the registry system.

3.9 Other Provisions

3.9.1 Temporary exclusion (opt-out)

The current draft Directive foresees that individual installations or economic activities can temporarily be exempted from the emission trading scheme during the three-year period 2005–7.[38] However, such opt-outs are subject to approval by the Commission on strict conditions: the excluded installations or economic activities have to limit their emissions as much as would be the case if they were subject to the Directive. Furthermore, the rules with regard to monitoring, reporting and verification of emissions apply to the exempted installations or economic activities.

The opt-out provision is of particular importance in the UK and Germany. This is the case in Germany because industry sectors as a whole are currently reducing emissions according to voluntary agreements with the Government. The UK on the other hand insisted on an 'opt-out' possibility for sectors in order to enable it to continue its national emissions trading system and climate change agreement. This goal was not completely achieved, as Member States may not opt out whole industry sectors but individual installations only. The regulators in Brussels feared that opting out of whole sectors would limit market liquidity and cause distortions in the internal market.

3.9.2 Pooling

The draft Directive provides for the possibility for operators of installations to pool their emission[39] allowances in the first and/or second commitment period

[37] Draft COMMISSION REGULATION No.../2004 of xx/xx/2004 for a standardized and secured system of registries pursuant to Article 19(3) of the Directive 2003/87/EC and Article 6(1) of Decision 2003/xx/EC, COM, 2004 .

[38] Art 27.

[39] Art 28.

(not subsequent periods). The installations forming the pool will nominate a trustee who will receive the total allowances of the pool and will be responsible/liable to surrender sufficient amounts of allowances for compliance. If the trustee fails to comply, individual installations in the pool are responsible for their own installations.

Pooling is subject to approval by the Commission. The Commission may within three months reject a Member State's application for pooling. No examples triggering a rejection are given in the Directive but the following points were mentioned in discussions and presentations: the pooling systems may not be too large, as large groups will limit the liquidity of the market and the pooling system should be transparent. No Member State will be able to force companies to join trading pools. The pooling provision was mainly negotiated by Germany, in light of the voluntary agreements with industry.

3.9.3 Force majeure

Member States may apply to the Commission for certain installations to be issued with additional allowances in case of *force majeure*[40] to be demonstrated on an installation and case-by-case basis. The Commission has provided further guidance on this in a Commission Communication Paper in January 2004.[41] In principle allocation decisions are made by the Member State before the beginning of the relevant trading period, thereby avoiding uncertainty in the allowance market. The Commission will therefore carefully consider the justification and potential effects of any application of such allowances. Circumstances that could have been insured will not be considered *force majeure*.

3.9.4 Review and further development

Article 30 of the Directive states that the Commission may propose to the Parliament and Council by 31 December 2004 to include further GHGs or activities in the EU ETS on the basis of progress being achieved in the monitoring of relevant emissions. Such an amendment to the Directive would influence the NAPs in the next commitment period.

In addition the Commission is requested to provide a report on the application of the Directive to the Parliament and Council by 30 June 2006. This includes an analysis of the possibility to link the Kyoto Project Mechanisms in the EU ETS.

[40] Art 29.

[41] Communication from the Commission on guidance to assist Member States in the implementation of the criteria listed in Annex III to Directive 2003/87/EC establishing a scheme for greenhouse gas emission allowance trading within the Community and amending Council Directive 96/61/EC, and on the circumstances under which *force majeure* is demonstrated, COM(2003)830 final, Brussels, 7 January 2004.

4 Linking the EU ETS with Kyoto Project Mechanisms

In the context of the EU ETS, 'linking' means that operators can use credits from Kyoto Project Mechanisms (JI and CDM) to fulfil their obligations under the EU ETS. It implies the recognition of JI/CDM credits as equivalent to allowances from an environmental and economic point of view.

Regulation on the inclusion of Kyoto Project Mechanisms (JI and CDM) in the EU ETS was not included in the Directive. The Commission feared that this would stall a quick approval process. A first draft of a Directive amending the Directive establishing the EU ETS, in respect of the Kyoto Protocol's Project Mechanisms, was published on 23 July 2003.[42] The compromise version was agreed and finalized on 7 April 2004.[43] This Directive is commonly referred to as the 'Linking Directive'.

The Council held an in-depth debate on the above proposal for a Directive in December 2003 discussing mainly two issues: the timing of conversion of credits and the possibility for a quantitative limit on the total amount of credits to be used for compliance in the EU ETS. Several delegations supported the possibility to use Certified Emission Reductions (CERs) of the CDM for compliance in the EU ETS in the period 2005–7, opposing the draft of the Commission that suggested allowing the use of credits for compliance only from 2008 onwards.[44]

Early recognition of those credits in the EU ETS will provide market participants with more flexibility and certainty, increasing liquidity of the EU emissions trading market and reducing allowance prices and thus compliance costs. It will also be an important signal that the EU will contribute to the sustainable development in CDM and JI host countries and promotes the transfer of environmentally sound technologies.

Accepting the CDM within the framework set by the UN process without adding stricter provisions or even exclusions (as suggested in the discussions, for example, credits from forestry) will show that the EU is committed to the Kyoto Protocol and trusts the international community in the development of sound rules to tackle climate change.

5 Conclusions

The EU ETS in its current form could clearly be criticized as a second-best solutions as it might be too complex and was maybe set up too fast. But, repeating

[42] Proposal for a Directive of the European Parliament and Council amending the Directive establishing a scheme for greenhouse gas emission allowance trading within the Community, in respect of the Kyoto Protocols project mechanisms, Brussels, 23 July 2003 COM(2003)403 final, 2003/0173 (COD).

[43] Directive of the European Parliament and of the Council amending Directive 2003/87/EC establishing a scheme for greenhouse gas emission allowance trading within the Community, in respect of the Kyoto Protocol's project mechanisms.

[44] See EC Council press release 16140/03 (Presse 376) 7, provisional version 22 December 2003.

a sentence often heard in the negotiation process: 'The best will be the enemy of the good.' Devising a scheme ex ante that fits perfectly in all twenty-five EU Member Countries, and is in place in time to help address the Kyoto challenge can be considered an impossible task. The first period of 2005–7 is clearly seen as a learning by doing period leading to revisions and extensions of the Directive.

It is important to understand that the EU ETS is an essential means for the Community to address its Kyoto commitment. The need to reduce GHG emissions does not stem from the EU ETS but from the commitment made by ratifying the Kyoto Protocol. Governments could have chosen to achieve the needed reductions through means of taxation or regulation only. But, by choosing a market mechanism regulators acknowledge that the market itself will be in a much better position to decide where the most reduction is achieved with each Euro invested. And studies have shown that trading delivers cost-effectiveness by reducing the cost of meeting EU Kyoto commitments by 35 per cent, representing €1.3 billion benefit per year to 2012.

The open architecture of the EU ETS with provisions to link to other Kyoto Parties and the Kyoto project mechanisms of CDM and JI shows the importance European regulators assign to the Kyoto Protocol as an instrument to tackle climate change that is accepted by the clear majority of the world's population.

This approach will also allow the EU ETS eventually to become the nucleus of a global carbon market allowing the assumption that GHGs will one day be globally traded in euros rather than US dollars.

24

Implementing the European Emissions Trading Directive in Germany

*Simon Marr**

1 Introduction

On 17 December 2003 the German Cabinet adopted the Greenhouse Gas Emissions Trading Act (GETA) and the Ordinance implementing the European Emissions Trading Directive for Installations of the Federal Immissions Control Act (ETO). Both legal instruments transpose the European Directive establishing a scheme for greenhouse gas emission allowance trading within the Community and amending Council Directive 96/61/EC[1] (ET Directive) in national law and thus prepare the legal conditions for an emissions trading system in Germany.[2]

Subsequently, the German Parliament (*Bundestag*) on 12 March 2004, in its first reading, adopted a revised draft law of the coalition of the Socialdemocrats and the Green Party, integrating the ETO in the GETA thereby creating one single legal instrument and transferring most of the competencies to the Federal German Emission Allowance Trading Authority (Authority), which will be in charge of supervising the German emissions trading system.[3] In the following, the core provisions of the draft of the GETA as adopted by the *Bundestag* on 12 March 2004 will be set out in its context, while taking into account, in particular, legal questions that arise when developing an emissions trading system in Germany.[4]

* Simon Marr is a counsel at the Federal Environmental Agency. The views expressed in this chapter are the views of the author and do not necessarily represent the view of the Federal Environmental Agency.

[1] Directive 2003/87/EC [2003] OJ L275/32.

[2] Treibhausgas-Emissionshandelsgesetz and Verordnung zur Umsetzung der Emissionshandels-Richtlinie (Bundestagsdrucksache 15/2540), both available at: <http://www.bundestag.de> or <http://bmu.bund.de>. For further reference, see S Marr, *Emissionshandel in Deutschland* (Zeitschrift für Europäisches Umweltrecht, 2004) 10–21.

[3] Beschluss des Gesetzentwurf der Fraktionen SPD und Bündnis 90/Die Grünen—Drucksache 15/2681, available at: <http://www.bundestag.de>.

[4] The draft Law of parliament of 12 March 2004 in Art 1 contains the GETA, and in Art 2 contains an amendment of the Federal Immissions Control Act.

2 Transposition of the European Emissions Trading Directive

2.1 Background

On 25 October 2003 the ET Directive entered into force, thereby establishing a cap-and-trade emissions trading system for greenhouse gases (GHGs) in the European Union, including the newly acceded Eastern European Member States.[5] Pursuant to Article 249(3) of the European Community Treaty (ECT) 'A directive shall be binding, as to the result to be achieved, upon each Member State to which it is addressed, but shall leave to the national authorities the choice of form and methods.' Thus, Member States are obliged to transpose the provisions of a Directive into national law, and they have the discretion as to how they do this in comparison to a Regulation pursuant to Article 249(2) ECT which is 'binding in its entirety and directly applicable in all Member States.'

In general the ET Directive provides that as of 1 January 2005, in each Member State, operators of installations exceeding 20 MW can only emit CO_2 emissions, if they have a permit. In the first period (2005–7) only operators of the energy and industry sector emitting CO_2 fall within the scope of application of the ET Directive. However, the ET Directive provides that as of 2008 other GHGs and sectors could be included by a Member State provided that it obtains the consent of the European Commission. As a consequence, as of 1 January 2005 the European Union Emissions Trading Scheme (EU ETS) will consist of twenty-five linked emissions trading systems.[6]

As has been mentioned above, the ET Directive will most likely be implemented in Germany through one single instrument—the GETA—despite the fact that the German Government adopted a slightly different Greenhouse Gas Emissions Trading Act and the ETO in December 2003. However, the adoption of these two instruments started the legislative process of setting out an emissions trading system in Germany which, according to German constitutional law, finally involves the adoption of the laws by the *Bundestag* and the chamber representing the *Länder* (*Bundesrat*).[7] It can be assumed that in the course of the ongoing legislative process, the GETA will finally be adopted similar to its current draft, although it is likely that it will undergo some further changes most likely relating

[5] For a general background of the ET Emissions Allowance Trading System in the EU, see J Lefevre, 'Greenhouse Gas Emission Allowance Trading: A Background' in (2003) 3 Yearbook of European Environmental Law 148–93.

[6] All ten acceding countries will need to transpose the ET Directive, including Cyprus and Malta which, however, as non-Annex I countries, do not have to commit themselves to a greenhouse gas emission reduction pursuant to the Kyoto Protocol.

[7] The *Bundesrat* is the body that represents the interests of the *Länder*, or federal States, within this federal structure. It participates in the Federation's policy-making and thus acts as a counterweight to the *Bundestag* and the Federal Government, whilst also serving as a link between the Federation and the federal States.

to the division of competencies between the *Länder* and the Federal Government. As of 15 May 2004 negotiations of the GETA in the conciliatory committee of the *Bundestag* were still pending. Presumably the Law will enter into force in June or July 2004.

The GETA sets out a framework for the German emissions trading system and provides for the necessary procedures and the administrative structure to make emissions trading operational in Germany. For example, it regulates which installations and GHGs fall within its scope of application,[8] the permit procedure for the emission of GHGs, that is, CO_2 emissions,[9] the procedure of issuing allowances,[10] the further design of a national emissions trading registry,[11] as well as provisions on the transfer of allowances[12] and compliance provisions.[13]

The core provisions for the allocation, for example, the determination of the overall emissions budget of allocated allowances, and the different rules and methods of the allocation will be set out in a separate law—the National Allocation Plan Act (NAPA). Thus the NAPA is of prominent importance for the German Emissions Trading System. It will be based on the National Allocation Plan (NAP), and simultaneously with the latter's notification to the European Commission it will run through the German legislative process. Theoretically, any possible amendments of the NAP required by the European Commission would need to be reflected also in the NAPA. This also holds true for amendments to the NAPA required by the German legislator. In practice, close coordination between the European Commission and the German legislator will be necessary. Most likely, NAPA will enter into force in the beginning of August 2004. According to German administrative law allocation of allowances is only possible if NAPA has legally entered into force. Thus, the German Emission Allowance Trading Authority can only allocate allowances when NAPA and its respective regulations have finally entered into force. It remains to be seen whether this will happen in a timely manner.

2.2 The Greenhouse Gas Emissions Trading Act

2.2.1 Scope of application

Annex 1 and § 2 GETA determine the scope of application of the ETS in Germany. Similar to the ET Directive the scope of GETA's application is broad. GETA already provides for a broad definition of a number of terms.[14] Thus, in case that after 2008 other sectors and GHGs will be included in the EU ETS, it would be relatively easy to adopt the legislation amending the GETA. Its overall systemic approach would not need to be amended. Pursuant to the results of the concluded data collection for notification of the NAP to the European Commission,

[8] The scope of application will be determined by reference to Annex 1 and § 2 GETA.
[9] § 4 GETA. [10] § 9(2) GETA. [11] § 14 GETA. [12] §§ 15–16 GETA.
[13] §§ 17–19 GETA. [14] For example 'responsible person' and 'activities'.

approximately 2,400 installations will be covered by the GETA[15] with an overall emissions budget in the energy and industry sectors of 503 million tonnes CO_2 per annum for the period 2005–7.

Annex 1 GETA only lists installations of groups of certain economic sectors pursuant to the ET Directive. However, uncertainties still remain as to what kind of installations exactly fall within the scope of application of the ET Directive.[16] The problem is related to some ambiguities in the interpretation of the term 'combustion installation' pursuant to Annex 1 of the ET Directive.

The European Commission is of the opinion that pursuant to the first subheading of Annex 1 'Energy activities', *all* combustion installations of a size exceeding 20 MW fall within the scope of application, regardless whether its particular sector is listed in Annex 1 of the ET Directive or not. This finding would be particularly relevant for the chemical industry, in particular its process-related emissions, which would hereby indirectly be included in Annex 1 of the ET Directive. Some Member States, however, are of the opinion that only combustion installations of the respective sectors as listed in Annex 1 are covered by the ET Directive; others do not go that far but argue that the sector 'energy activities' does not include any installations in which the products of combustion are used directly as an integral part of a production process. However, lastly it is only the European Court of Justice or an agreement between the Member States and the Commission themselves that can resolve this issue conclusively.

2.2.2 *Permit*

Pursuant to the ET Directive the emission of GHGs by covered installations requires permission in Member States.[17] This will on the one hand secure the participation of operators in the emissions trading system and due to the permit requirements on the other hand guarantee that operators will be able to monitor and report its emissions.[18] Without holding a GHG permit, operators of covered installations will not be allowed to emit CO_2 from the beginning of January 2005. Pursuant to the ET Directive and the 'Guidelines for the Monitoring and Reporting of Greenhouse Gas Emissions pursuant to Directive 2003/87/EC of the European Parliament and of the Council'[19] a GHG emissions permit may also cover one or more installations on the same site operated by the same operator.[20]

Pursuant to § 4(1) of the GETA, installations covered by it will require a GHG emission permit, which will be issued by the Federal Environmental Agency. Thus, in its current draft, GETA provides for a *separate* GHG emissions permit, whereas

[15] For details of the data gathering process, see <http://www.bmu.bund.de>.

[16] For reference of the installations covered by the German ETS, see ibid.

[17] Arts 4–6 ET Directive.

[18] See also Arts 5 and 6 ET Directive.

[19] Decision of 29 January 2004 European Commission establishing guidelines for the monitoring and reporting of greenhouse gas emissions pursuant to Directive 2003/87/EC of the European Parliament and of the Council, COM (2004) 130 final.

[20] See Art 6(1) ET Directive; Annex IV Decision of 29 January 2004, COM (2004) 130 final, 60.

its precursor—the governmental draft of 17 December 2003 of the Greenhouse Gas Emissions Trading Act—integrated the GHG emissions trading permit in the permit required under the Federal Immissions Control Act. This also had the consequence that the Immissions Control Authorities of the *Länder* were the competent authorities for issuing the respective GHG emission. It is likely that in the ongoing conciliation procedure the GHG emissions permit will be reintegrated in the permit procedure required pursuant to the Federal Immissions Control Act. Article 8 of the ET Directive expressly provides for this option in its second sentence. The integration of the two permit procedures is based on the consideration that the permit pursuant to the Federal Immissions Control Act already contains a permission of the emission of GHGs.[21] In addition, both permit procedures are similar to each other in that they both formulate precautionary requirements in relation to global warming and serve the protection of the atmosphere and climate. For new and already existing installations this means that the GHG emissions permit, pursuant to the ET Directive, is issued pursuant to the Federal Immission Control Act. The unequivocal advantage of this integrated permit procedure is its efficiency, because it would spare the issuance of approximately 2,400 new permits.[22] In addition, it is hard to imagine how effective monitoring of installations could be guaranteed without the competent immission control authorities of the *Länder*, which are specialized in monitoring installations. On the contrary, a divide of competencies between a monitoring of emissions by federal authorities on the one hand and partial implementation by the *Länder* authorities on the other, could cause possible competence problems. The German Emissions Allowance Trading Authority will generally supervise the ETS in Germany, that is, allocate and issue allowances and ensure compliance, whereas the *Länder* are generally in charge for executing the Integrated Pollution Prevention Control (IPPC) Directive and thereby the Federal Immissions Control Act and would thus have the final competence to eventually shut down an installation.

2.2.3 Allocation of allowances

2.2.3.1 Emission budget

The allocation of allowances is of key importance in any ETS, since it ensures the effectiveness of emissions trading as an environmental policy instrument. Depending on the respective political decision, allowances can be auctioned or allocated free of any costs.[23] Also by determining the overall budget of allowances

[21] For further reference, see C Bail, S Marr, and S Oberthür, in Rengeling (ed.), *Handbuch zum europäischen und deutschen Umweltrecht* (Köln, Carl Heymann, 2003) 254–304 (299).

[22] Currently approximately 2,400 installations are covered by the German ETS, see <http://www.bmu.bund.de>.

[23] Art 10 of the ET Directive provides for the possibility of auctioning up to 5 per cent of all allowances in the period 2005–7 and up to 10 per cent in the period 2008–12.

allocated to the participants in the scheme, the allocation of allowances can guarantee the cap of each ETS. Operators of installations who exceed their allowance budget will either need to reduce emissions themselves or buy allowances on the market. In the German ETS the allocation procedure is threefold: first, the German Ministry for Environment sets up the NAP. Basically, it contains the methods and general rules for the allocation of allowances, for example, the Macro and Micro Plan, compliance factors and provisions on new entrants, closure of installations, preferential treatment for process emissions as well as 'early action', and banking. The NAP does not fall under any categories of planning instruments defined under German law, but is a plan *sui generis*. Secondly, the European Commission will examine the NAP and can require amendments and eventually reject the NAP. Pursuant to the ET Directive due account will be made of comments on the NAP from the public.[24] The NAP will provide the basis for the NAPA which will eventually provide for legally binding allocation methods and rules in Germany. Public comments on the NAPA will be taken into account prior to its entering into force. Thirdly, after the entry into force of the NAPA, the Authority will make a decision on the individual allocation of EU Allowances to each operator.

Operators will need to apply for the respective individual allocation of allowances presumably until 15 August 2004 the latest for the allocation in the period 2005–7; for the period 2008–12 operators are required to submit allocation applications at 1 April 2007 at the latest.[25] Thus, the overall budget of allocated allowances cannot be determined before the allocation procedure has been finally concluded, which according to the ET Directive must happen by 30 September 2004. To this end, deviations from the overall allocation budget originally notified to the European Commission will be expected. Pursuant to the draft registry regulation, these deviations, would need to be notified to the European Commission.[26] Prior to submission each allocation application must be verified by verifiers that will be admitted by the Authority.[27] Operators can take legal action against allocation decisions pursuant to ordinary administrative procedural law, that is, before regular administrative courts.

The overall budget of allocated emission rights will be determined by the Macro and Micro Plan.[28] The overall emissions budget will be determined by the international and European obligations for the reduction of GHGs (Macro Plan).

[24] § 8(1) GETA. In Germany, the list of installations sent to the European Commission was published on the internet.

[25] § 10(3) GETA. Meeting the first deadline of 15 August 2004 could be difficult for operators, since it is not entirely clear as yet, if the ETA will enter into force by then, in that case operators would be granted the right to submit the applications for allocation later.

[26] See Arts 15(2) and 16(2) of the Draft COMMISSION REGULATION (EC) No . . . /2004 of xx/xx/2004 for a standardized and secured system of registries pursuant to Article 19(3) of Directive 2003/87/EC and Article 6(1) of Decision 2003/xx/EC, COM, 2004.

[27] § 10 (1) ETA.

[28] See NAP (English version), pp 7ff, available at <http://www.bmu.de>.

Pursuant to the EC burden sharing agreement, Germany is obligated to reduce its aggregate emissions of a 'basket' of six GHGs[29] by 21 per cent over the period 2008–12 compared to its 1990 emissions.[30] Thus, an overall reduction from emissions of 1,014.4 million tonnes of CO_2 in the base year (1990) to 846 million tonnes per annum in the period 2008–12 and 859 million tonnes of CO_2 per annum as derived emission target for the period 2005–7 is mandated.[31] Originally the emission reduction targets for the sectors energy and industry, had been determined pursuant to the voluntary commitments of the German industries which had obligated itself to a reduction of 45 million tonnes CO_2 per annum until 2010 and 20 million tonnes CO_2 per annum until 2005 compared to its 1998 baseline.[32] However, due to strong political pressure from industries during the NAP negotiations reduction targets merely equivalent to the figures in the self commitment declaration were finally watered down to a 10 million tonnes per annum reduction between 2008–12 which in effect means that industry has given up on the targets pursuant to its voluntary commitment.

The above mentioned Macro Plan will set the cap for the Micro Plan which is determined by the overall number of allowances allocated to operators measured on the basis of their historical emissions for the base year period 2000–2. The German Ministry for Environment has gathered preliminary data information of the CO_2 emissions of installations covered by the ETS in Germany for the years 2001–2. It based its collection of information on data from the immissions control authorities of the *Länder* and the operators itself.[33] The compliance factor serves to reconcile the Macro Plan with the Micro Plan, taking into account the reserve and other special allocations.[34] This has already been done in the NAP on the basis of the preliminary data information. Accordingly, the compliance factor notified to the European Commission in Germany is 0.9765.[35] In addition, the NAP now provides for a fixed reserve for new entrants and corrections of 3 million tonnes per annum. Due to the political decision in Germany to provide for a fixed compliance factor and reserve for new entrants in the NAPA, deviations in the allocation process, for example, for more early action allocations as calculated could have the consequence that the budget will in fact be exceeded. According to the current wording of the NAPA the German reserve is fixed and will be managed on a first-come-first-served basis if it is exceeded. It seems, however, quite likely that this provision will be amended and some instrument of state liability will be

[29] Carbon dioxide (CO_2), methane (CH_4), nitrous oxide (N_2O), hydrofluorocarbons (HFCs), perfluorocarbons (PFCs), and sulphur hexafluoride (SF_6).

[30] See Bail *et al.*, n 21 above, 287.

[31] See NAP, pp 12ff.

[32] See draft NAP of 29 January 2004, pp 27ff; for the declaration of industries' voluntary reduction commitment, see <http://www.bewag.de/db_files/d455b3_f2f609fbb7_-6f28192616861660/Vereinbarung_110501.pdf>.

[33] For the data gathering process, see NAP, pp 8ff.

[34] See NAP, pp 44ff.

[35] The overall compliance factor was already subsequently amended by § 5 NAPA to 0.9755; it is calculated as follows: 0.9755 (compliance factor) = (reduction factor) − (reserve) − (process-related emissions) − (early action) − (special allocation); for further reference, see NAP, pp 44.

introduced during the ongoing legislative proceedings of NAPA to ensure more security for investors.[36] In case a Member State exceeds its overall notified allowance budget, it will need to notify the European Commission.[37] It will be interesting to see whether the allocation rules as set out in the NAPA and the calculated extra emissions for early action or process emissions will ensure that Germany stays within its notified budget.

2.2.3.2 Key Allocation Rules and Criteria in Germany

The overall emissions budget and allocation rules in Germany are set out in the NAP, although for Germany, they will be legally binding through the NAPA, which was adopted by the German Cabinet on 21 April 2004 and will presumably enter into force by the beginning of August 2004.[38] In Germany allowances will be allocated free of charge in the period 2005–7, and most likely also in the subsequent period from 2008–12. For existing installations allowances will in principle be allocated on a grandfathering method where allocation is based on the average absolute emissions of the respective individual installation in the period 2000–2. The total budget for CO_2 emissions in the sectors energy and industry in the period 2005–7 is 503 million tonnes per annum; for the period 2008–12 it is 495 million tonnes per annum.[39] The total amount of CO_2 emissions for installations falling within the sectors energy and industry and actually participating in emissions trading is 499 million tonnes per annum for the period 2005–7.[40] The reserve for new entrants installations for the period 2005–7 is strictly limited to 3 million tonnes per annum.[41] The NAPA accommodates *early action* of a total of 110 million tonnes per annum CO_2 emissions and 61 million tonnes CO_2 emissions per annum for process emissions.[42] It takes also into account 1.5 million tonnes estimated CO_2 emissions per annum due to the closure of two nuclear power plants.[43] Closed installations can keep the amount of allowances issued before the closure up until the year of the closure. New installations can receive allowances on the basis of the allocation of a closed installation for a maximum of four years. Thereafter no compliance factor will be applied for another fourteen years.[44] This means in effect that in those cases new installations will not need to reduce emissions for up to eighteen years. On the one hand this creates a strong incentive and legal security for investors to modernize old

[36] Despite the wording of the current NAP with its fixed reserve, this issue is still under negotiation, because the German Government during its NAP negotiations finally could not agree on how to handle this issue properly. Thus it can be expected that some sort of state liability will be introduced during the further ongoing legislative process of NAPA.

[37] See Art 15 of the Draft Commission Regulation (EC) No . . . /2004 of xx/xx/2004 for a standardised and secured system of registries pursuant to Article 19(3) of ET Directive and Article 6(1) of Decision 2003/xx/EC, COM.

[38] NAPA is available at <http://www.bmu.bund.de>; or as *Bundestagsdrucksache* 15/2966 at <http://www.bundestag.de>.

[39] § 4 NAPA.

[40] NAP, p 23.

[41] § 6(1) NAPA.

[42] See reasoning of the NAPA, pp 21 and 23, available at <http://www.bmu.bund.de>.

[43] § 15 NAPA.

[44] § 10 NAPA.

installations and build new, efficient ones. On the other hand with this provision, subsequent more ambitious reduction of GHGs will need to be rendered by other sectors, such as transport and households, which do not even have the possibility of using the flexible mechanisms.

New entrants that do not claim such a transfer rule of allocations of a closed installation will receive a cost free allocation on the basis of a benchmark of 750 g/kWh, minimum 365 g/kWh, however, never more than actually needed.[45] This is ensured by a strict ex post correction of the respective factual emissions.[46] In case the yearly emission reports show that factual emissions of any installation covered by the German ETS drop below 60 per cent an ex post correction is applied and allowances will be withdrawn.[47] In Germany no banking of allowances will be possible between the periods 2005–7 and 2008–12.[48]

2.2.4 *Administration of allowances*

The Emission Allowance Trading Authority will also be the competent authority for the issuance and administration of allowances in a German ETS. In addition, the Authority will be hosting the emissions trading registry pursuant to the draft registry regulation of the European Commission which provides for a purely electronic emission trading system in which allowances solely exist electronically.[49] Thus, the GETA provides for the installation and operation of a registry as an electronic database. Electronic communication between the individual and the public administration is permissible pursuant to German public administration law.[50] Pursuant to the draft registry regulation of the Commission trading with emission rights generated by the other flexible mechanisms of the Kyoto Protocol, such as Certified Emission Reductions (CERs) and Emission Reduction Units (ERUs) (together 'Kyoto Units') prior to its entry into force might also be possible.[51] The registry will be designed pursuant to the provisions of the Commission's Registry Regulation once it enters into force.[52] In addition, in Germany the registry might also display restrictions of the transferability of allowances, for example, by a lien to ensure its protection. Any person will receive an account which is to his or her disposal. Operators will be obligated to install an operator holding account. Any account holder will have guaranteed access only to information relating to his own account. However, access to account information relating to third persons is limited and will only be granted pursuant to the

[45] § 11 NAPA. [46] § 11(6) NAPA. [47] § 7(9) NAPA. [48] § 20 NAPA.

[49] See Draft Commission Regulation, n 37 above.

[50] See German Public Administrative Procedural Law Act § 3a (*Verwaltungsverfahrensgesetz*).

[51] Pursuant to the Proposal for a Directive amending the Directive establishing a scheme for greenhouse gas emission allowance trading within the Community, in respect of the Kyoto Protocol's project mechanisms (hereinafter referred to as Linking Directive) trading with CERs and ERUs will be feasible, even without the Kyoto Protocol entering into force, for reference, see COM(2003)403 final, available at <http://www.europa.eu.int/comm/environment/climat/emission.htm>.

[52] § 14(1) GETA.

German Access to Environmental Information Act (*Umweltinformationsgesetz* (UIG)).[53] Thus, operators can deny access to its account information to third persons on the grounds of confidentiality. Nonetheless, once CERs and ERUs will be included in an EU ETS, it would make sense to grant broader access rights to identify these units in the respective operator's accounts. This way transparency of how operators comply with its respective reduction obligations could be ensured, in particular, whether they hold CERs or ERUs, which in some cases might be generated from projects considered as not sufficiently complying with sustainable development standards.

2.2.5 Emissions allowance trading

2.2.5.1 Buying Allowances

The contract governing the transfer of allowances will be determined by Member States law, due to a lack of a Europe-wide regulation. In Germany, trading with allowances will constitute the purchase of a legal right pursuant to § 453 Civil Law Code (*Bürgerliches Gesetzbuch* (BGB)). The fact that the allowances are presumably of a public legal nature is of no avail to this, since public legal rights can also be subject of purchase by way of § 453 BGB.[54]

2.2.5.2 Transfer of Allowances

To understand the German civil law system one needs to take note of the so called 'abstraction principle', which basically differentiates two legal acts occurring in a simple transaction of goods: the mere closing of a contract as a first step and the subsequent substantial transactions of the respective goods or money involved. The transfer of allowances, as regulated in the GETA, refers only to the 'second' substantial transaction under German civil law.

The transfer of allowances regulated in the GETA follows the draft registry regulation of the Commission. Thereafter the transfer of allowances is 'complete' after its transaction has been successfully executed and the allowance has been registered in the 'receiving' holding account.[55] Following the original draft of the registry regulation such transaction should generally be irreversible to underpin the legal certainty of the whole emission trading system. Thus the legal ownership of allowances is transferred by the mere registration of the respective transaction. This 'minimalist'[56] solution has the undisputed advantage that it does not burden the emissions trading system with the validity of the respective underlying settlement of each transaction. According to the draft registry regulation buyer and seller will exchange a number of message sequences similar to offer and

[53] § 14(3) GETA.

[54] For further reference, see G Wagner, 'Handel mit Emissionsrechten' (2003) 15 Zeitschrift für Bankrecht und Bankwirtschaft 409–24.

[55] See, Art 13(7) of the Draft Commission Regulation, n 37 above.

[56] See Wagner, n 54 above.

acceptance pursuant to German contract law. A valid settlement is a condition for the registration of each transaction. However, any complete transaction assumes a preceding settlement.[57] Thus, for the transaction of the legal ownership of allowances in Germany the registration of the transaction is decisive. After successful registration of the transaction the transaction is irreversible.[58] This provision provides for broad legal certainty, in particular the buyer when he is in good faith of the legal ownership of the seller. This provision, however, also protects the buyer who is in bad faith of the legal ownership. Only the receiver of allowances who is in bad faith of his legal ownership when the allowances are allocated to him is not protected by this broad provision and cannot claim irreversibility.[59] This quite broad protection even of the buyer who is in bad faith is nevertheless justified because allowances are exchangeable and do not embody an individual value. Thus the mere possibility to gain redress by way of compensation instead of restoration of the legal ownership of the transferred allowances seems justified.[60]

2.3 Financial Law Aspects of Allowances

Pursuant to § 15 of the GETA, allowances are not securities as provided for in the German Financial Credit Act (*Kreditwesen Gesetz* (KWG)), and thus trading with such allowances does not fall under the supervision of the Federal Agency for financial services (*Bundesamt für Finanzdienstleistungen* (BAFin)). Transactions of allowances for mere compliance purpose would not fall under the supervision of BAFin in any event. The purpose and objective of this regulation is that small and medium sized companies can save transaction costs by allowing that allowances can be managed by an external portfolio management without the need of complying with the requirements pursuant to the KWG.[61] Nonetheless, § 15, second sentence provides that trading with derivatives which have underlying allowances, are subject to the supervision of BAFin. This supervision 'light' by BAFin is necessary, because trading with derivatives is more complex and not as transparent as trading with allowances, and thus poses a bigger risk for investors.

2.4 Monitoring and Sanctions

Section 6(1) of the GETA sets out the key obligation of operators in a future emissions trading system in Germany. Thereafter operators are obligated to surrender allowances in an equivalent amount of their emissions of the preceding year.

[57] § 16(2) GETA. [58] Art 13(7) of the Draft Commission Regulation, n 37 above.

[59] § 16(2) GETA.

[60] See Wagner, n 54 above, who argues that restoration of the legal ownership should be possible in such situations.

[61] This possibility to administer allowances more cost efficiently is to differentiate from pooling pursuant to § 24 GETA, where the trustee will be held liable for the compliance obligations of participating operators.

In addition, operators are obligated to submit their verified emission reports to the Authority, until 1 March of each year[62] at the latest, that carries out plausibility and spot checks.[63] Emissions of the preceding year of each operator will need to be verified by a verifier and reported to the Authority. In case the verified reports will not be submitted properly or not within the deadline of 1 March of each year (beginning in 2006) the operator's account in the registry will be blocked for any transfers of allowances out of the account. In case operators do not comply with their key obligation pursuant to § 6(1) of the GETA to surrender enough allowances, operators will be fined by the Authority €40 in the period 2005–7 and €100 in the period 2008–12.[64] The Authority is also empowered to estimate the respective number of emissions of the preceding year in case operators fail to submit any emission reports or report incorrectly.[65] However, operators remain obligated to surrender the number of allowances pursuant to the estimation until 30 April of each subsequent year. To this end, the GETA provides for an inherent compliance system, in that the respective amount of allowances lacking will be deducted from the amount of allowances that will be issued for the following year.[66] In addition, the GETA provides for the 'naming and shaming' of operators who did not submit an equivalent amount of allowances pursuant to § 6(1) of the GETA.[67]

3 Emissions Trading and Environmental Law in Germany

The draft Law of the *Bundestag* of 12 March 2004 also amends the Federal Immission Protection Act.[68] Hereby it transposes Article 26 of the ET Directive, which amends the IPPC Directive to the extent that the GHG emissions permit may not include emission limit values for direct emissions, unless it is necessary to ensure that no significant local pollution is caused. In addition, it leaves the discretion to the Member States to provide for an exemption of the energy efficiency principle as set out in Article 3(d) of the IPPC Directive for direct emissions of installations.

With this amendment of the Federal Immission Protection Act, precautionary GHG emissions limits are not permissible any more, except to avoid environmental hot spots of local environmental pollution. With respect to the energy efficiency principle pursuant to Article 3(d) of the IPPC Directive operators of installations are, however, still obliged to use their energy efficiently, for example, in case operators receive external energy or when using heat.[69]

[62] § 17(1) GETA. [63] § 5(4) GETA.

[64] § 18(1) GETA and reasoning (*Begründung*), p 21. [65] § 18(2) GETA.

[66] § 18(3) GETA. [67] § 18(4) GETA.

[68] Art 2 of the draft Law transposing Directive 2003/87/EC establishing a scheme for greenhouse gas emission allowance trading within the Community and amending Council Directive 96/61/EC.

[69] See Bail *et al.*, n 21 above, 253; P Beyer, *Die integrierte Anlagenzulassung* (Frankfurt am Main, Peter Lang, 2001) 96ff.

Despite Article 26 of the ET Directive it remains nonetheless unclear how the underlying conflict between the precautionary principle as set out in Article 3(a) of the IPPC Directive and a cap-and-trade emissions trading system can be resolved, in particular, as regards technique-related precaution by way of best available techniques.[70] This problem exists in particular in German environmental law and has not been resolved by the above mentioned amendment of the Federal Imissions Control Act either. This problem might imply further amendment of the IPPC Directive.

In German environmental law the precautionary principle plays a prominent role in the Federal Immisions Control Act.[71] Pursuant to § 5(1) No. 2, operators are obliged to use the best available technique (BAT) standards for their installations (technique-related precaution), which could eventually lead to the consequence that due to stringent mitigation of emissions there simply might not be enough emissions available for operating a cap-and-trade emissions trading system. This poses the question how the underlying conflict between the precautionary principle as set out in Article 3(a) of the IPPC Directive and a cap-and-trade emissions trading system can be resolved in German environmental law.[72] According to Koch and Wienecke the precautionary principle and an emissions trading system would be applied in parallel and trading with allowances would only be limited to the share of GHGs, which could not be avoided even by applying BAT standards. This approach, however, does not seem to ensure a sufficiently large market for allowances. It obliges operators to curb emissions with the consequence that selling emissions on the market will simply be restricted. This conflict between the precautionary principle as set out in the German Immissions Control Act pursuant to the IPPC Directive and a cap-and-trade emissions trading system can thus only be resolved properly by taking the approach that a cap-and-trade emissions trading system simply has become an application of the precautionary principle pursuant to the IPPC Directive.[73]

4 Conclusion

The ET Directive establishes a legally binding cap-and-trade emissions trading system in Germany from the beginning of January 2005. In addition, the EC

[70] For further information, see HJ Koch and A Wieneke, 'Klimaschutz durch Emissionshandel' (2001) Deutsches Verwaltungsblatt 1093; E Rehbinder and M Schmalholz, 'Handel mit Emissionsrechten für Treibhausgase in der Europäischen Union' (2002) Zeitschrift für Umwelt und Planungsrecht 6–7; M Burgi, 'Die Rechtsstellung der Unternehmen im Emissionshandelssystem' (2003) *Neue Juristische Wochenzeitschrift* 2488–9.

[71] See S Marr and A Schwemer, 'The precautionary principle in German environmental law' (2003) 3 Yearbook of European Environmental Law 125–48 (137).

[72] For further information, see Koch and Wieneke, n 70 above, 1093; E Rehbinder and M Schmalholz, 'Handel mit Emissionsrechten für Treibhausgase in der Europäischen Union' (2002) Zeitschrift für Umwelt und Planungsrecht 6–7; Burgi, n 70 above.

[73] See also Burgi, n 70 above.

burden sharing agreement sets out the obligation for Germany of a 21 per cent reduction of GHGs between 2008–12 compared to 1990. The entering into force of the Kyoto Protocol would surely help give Member States' reduction obligations more political legitimacy. In addition, its entering into force is an important condition for further drastic measures to curb and stabilize GHGs in the atmosphere.[74]

Nonetheless, the entering into force of the Kyoto Protocol is not a legal requirement for the already existing legal obligations pursuant to EC law which are clearly separable from the Kyoto Protocol. Thus with the adoption of the GETA and NAPA by the *Bundestag* the most prominent framework for a legally binding reduction obligation in Germany has been established. In the face of vehement resistance against the establishment of an emissions trading system by some German industries it is difficult to say at this stage how the emissions trading system will finally look in Germany. The struggle for the biggest share of the cake of allowances has fully erupted within industry and will continue throughout the whole legislative and allocation process. So far it has watered down a quite ambitious first draft of the NAP with the result that the NAP now even falls behind what industry a few years ago had committed itself to voluntarily. It will be interesting to see whether the timely transposition of the ET Directive in Germany will be further jeopardized by industries' intervention.[75] The further design of Germany's ETS will certainly also depend on the finding of the European Commission on Germany's NAP and lastly on the Parliament's decision. With the entering into force of the Linking Directive and the Commission's Registry Regulation in the course of 2004, the legal field emissions trading will become even more complex. Thus, the numerous legal questions raised in this chapter and also generated in general by the implementation of the Kyoto Protocol within Europe will only be the beginning of the development of a new legal field.

[74] See Bail *et al.*, n 21 above, 253.

[75] See *Handelsblatt*, 23 January 2004, p 4; *Financial Times (Deutschland)*, 23 January 2004, p 26; 30 January 2004, p 14; 2 February 2004, p 1, 12; 31; *Frankfurter Allgemeine Zeitung*, 2 February 2004, p 13; *Handelsblatt*, 2 February 2004, p 6; *Spiegel*, 2 February 2004, p 19.

25

UK Emissions Trading Schemes

*Chris Dodwell**

1 Introduction

The UK plays a leading role in the fight against climate change and continues to press for a strong global response to climate change during international negotiations. In a key speech in September 2004, the UK Prime Minister, Tony Blair, called climate change 'The World's greatest environmental challenge', stating that it would be a top priority for the UK's presidency of the G8 in 2005 and citing emissions trading as a key mechanism for reducing emissions (see further section 6 below).

Over the last thirty years, the gross domestic product (GDP) of the UK has doubled and, while energy use has remained static, carbon dioxide emissions have fallen by more than 20 per cent, meaning that carbon dioxide emissions per unit of GDP have halved. This effect has been particularly marked in the last ten years where GDP rose by 25 per cent at the same time as CO_2 emissions were falling by 8 per cent, reflecting a drop in carbon intensity of more than 25 per cent. Action taken throughout the 1990s as a result of domestic and industrial energy efficiency improvements and the 'dash to gas' (fuel switching from coal to natural gas) in the electricity generation sector has significantly reduced greenhouse gas (GHG) emissions.

2 Effects of Climate Change in the UK

The UK's climate has followed the global trend towards higher temperatures, with Central England temperatures rising by almost 1°C over the last century. Since 1990, the UK has experienced its warmest decade (the 1990s) and two of the warmest years (1990 and 1999, joint with 1949) in the last 230 years.

* Chris Dodwell heads the team at the UK's Department for Environment, Food and Rural Affairs which is responsible for the implementation of the EU ETS. The views expressed in this chapter are the views of the author and do not necessarily represent the views for the Department for Environment, Food and Rural Affairs. The author would like to thank Chris Leigh and Jill Duggan for their contributions to this chapter.

In 2002, the UK Climate Impacts Programme, commissioned by the Department for Environment, Food and Rural Affairs (DEFRA), published climate change scenarios for the UK.[1] The scenarios, based on outputs of the Hadley Centre's regional climate model, show that:

- average annual temperatures across the UK may rise by between 2° and 3.5 °C by the 2080s;
- high summer temperatures will become more frequent and very cold winters will become increasingly rare;
- winters will also become wetter and summers may become drier across all of the UK;
- heavy winter precipitation will become more frequent, while the amount of snow could decline by 60–90 per cent by the 2080s; and
- extreme high water levels, which currently have a 2 per cent annual probability of occurring, could become 10 to 20 times more frequent at some east coast locations by the 2080s.

Through the UK Climate Impacts Programme, we are building a picture of what such changes could mean for the UK. Some key impacts could include:

- increased flooding risk in many lowland areas, due to more frequent river flooding and more severe storm surges;
- increased water demand and more droughts could create worsening water supply problem;
- increased weather variability could affect UK businesses (including the insurance industry), with the possibility of transport disruption and damage to buildings;
- warmer, drier, and sunnier summers could benefit domestic summer tourism;
- possibly 20,000 fewer cases of winter cold-related mortality, but 2,000 more cases of heat-related summer mortality; and
- a mixture of effects on agriculture and biodiversity, with some species/varieties benefiting and others losing out.

3 International and Domestic Emission Reduction Targets

At Kyoto in 1997, the European Commission agreed to reduce its GHG emission by 8 per cent below 1990 levels by 2008–12. Following discussions with the EU and other Member States on sharing out the EC's target, the UK's legally binding target under the Burden Sharing Agreement[2] is to reduce its emissions by 12.5 per cent.

[1] See <http://www.ukcip.org.uk/scenarios>.

[2] Council Decision 2002/358/EC.

In 1997 the Labour Government was elected with a manifesto aim of reducing CO_2 emissions by 20 per cent from 1990 levels by 2010, well beyond the UK's Burden Sharing Agreement target. In November 2000, the Government drew up the ambitious UK Climate Change Programme, which sought to achieve the target by a mixture of policies and measures targeting in particular further energy efficiency measures and new renewable energy sources.

In February 2003, the Department of Trade and Industry published its Energy White Paper '*Our energy future—creating a low carbon economy*'. The White Paper set the overall goals for UK energy policy in the long term and in it, the Government accepted the recommendation of the Royal Commission on Environmental Pollution that the UK should put itself on a path towards a reduction in carbon dioxide emissions of some 60 per cent from current levels by about 2050.[3] This recommendation—with a commitment to demonstrate real progress by 2020—forms one of the four goals for UK energy policy set out in the White Paper.[4]

4 UK Climate Change Programme

The UK Climate Change Programme published in November 2000 sets out how the UK intends to meet its Kyoto target and move towards its domestic goal. The programme is designed to:

- reflect the importance of tackling climate change by the UK's ambitious domestic goal;
- take a balanced approach, with all sectors and all parts of the UK playing their part;
- safeguard and, where possible, enhance the UK's competitiveness, promotes social inclusion and reduces harm to health;
- focus on flexible and cost effective policy options that will work together to form an integrated package; and
- take a long-term view, looking at targets beyond the Kyoto commitment period and considering the need for the UK to adapt to the impacts of climate change.

The Programme sets out the approach that the UK is taking to reduce GHG emissions across six broad sectors: energy supply, business, transport, domestic, agriculture, forestry and land use, and public sector. The strategies for addressing the business sector reflected the recognition by many firms that action to reduce emissions could bring wide-ranging benefits including lower costs, increased

[3] Available at <http://www.dti.gov.uk/energy/whitepaper/ourenergyfuture.pdf>.

[4] The other three goals are: maintaining the reliability of energy supplies; promoting competitive markets in the UK and beyond; and ensuring that every home is adequately and affordably heated. ibid 11.

competitiveness, and new market opportunities. At the same time, Government recognized that businesses needed a clear, flexible, and stable framework, which would allow for long-term planning and investment decisions in reducing emissions. In particular, this framework included the introduction of economic instruments such as the UK Emissions Trading Scheme (UK ETS) and the climate change levy and Climate Change Agreements (CCAs) for energy intensive industries (see section 5).

The Programme will be reviewed later in 2004. The review will consider the extent to which the existing policy measures are delivering the expected emission reductions and whether future measures are needed to keep the UK on track to its Kyoto commitment and national goal.

5 UK Emissions Trading Scheme

A key part of the Climate Change Programme is the UK ETS, which is managed by DEFRA. The UK ETS was the first GHG emissions trading scheme in the world, which allowed companies from several sectors of the economy to participate.[5]

One of the many novel features of the UK ETS is that it demonstrates that it is possible to incorporate participation on both a cap-and trade basis in relation to absolute targets (that is, by direct participants—see section 5.2 below); and on a baseline-and-credit basis in relation to relative or unit-based targets (that is, by participants with CCAs—see section 5.3 below).

The details of the UK ETS are set out below under the following headings:

- Development of the UK ETS;
- Direct participants;
- Climate Change Agreement participants;
- Results and conclusions.

5.1 Development of the UK ETS

In March 1998, the Chancellor of the Exchequer asked Lord Marshall of Knightsbridge, Chairman of British Airways, to consider whether economic instruments could be used to improve energy efficiency in business and reduce GHG emissions from that sector, in order to help the UK meet its emissions reductions target under the Kyoto Protocol. Lord Marshall, assisted by a task force of officials from the Department of Trade and Industry (DTI), the Department of the Environment, Transport and the Regions (DETR) and HM Treasury, submitted his

[5] The earliest trading schemes established in the US were designed to address other pollutants. The only schemes prior to the UK ETS addressing GHGs were: the Danish scheme which was only open to eight companies in the electricity generation industry; and in-house schemes such as those operated by Shell and BP.

report in November 1998[6] in which he concluded that:

- all sectors of the economy need to contribute to the achievement of the UK's Kyoto target, which was likely to be only the first of the commitments required to secure adequate climate change;
- a range of economic instruments would be necessary to ensure that the Kyoto target was met;
- an emissions trading system would be required, but only a pilot scheme was practicable for the present; and
- there was a role for a tax if businesses of all sizes and sectors are to contribute to improved energy efficiency and to help meet the UK's emissions targets.

The design of the UK ETS was developed jointly by Government and industry through the UK Emissions Trading Group. This Group had been set up in June 1999 by the Confederation of British Industry and the Advisory Committee on Business and the Environment (ACBE). The Group presented its proposals to Government for a UK ETS in March 2000. The detailed policy framework was published in August 2001[7] and the rules of the scheme itself were published in March 2002.[8]

The Government's principle aims in launching the UK ETS were: to secure a significant quantity of emission reductions at reasonable cost; to give organizations early practical experience of emissions trading; to ensure that UK business and finance would be ready to exploit the opportunities available from an international market, allowing the City of London and the UK to become established as a centre for emissions trading; and to influence the development of European Union action to meet Kyoto targets, by demonstrating that emissions trading was practical and demonstrating the UK's commitment to it.[9]

5.2 Direct Participants

The core of the UK ETS is a voluntary 'cap-and-trade' allowance trading scheme for companies who have made commitments to reduce their emissions in return for financial incentives from the Government. These companies are referred to as 'direct participants'. The reductions to be made by direct participants and the incentives to be paid for them were set in an auction held in March 2002.

[6] 'Economic Instruments and the Business Use of Energy,' Government Task Force on the Industrial Use of Energy, October 1998.

[7] See 'Framework for the UK Emissions Trading Scheme' <http://www.defra.gov.uk/environment/climatechange/trading/pdf/trading-full.pdf>.

[8] See 'Rules for the UK Emissions Trading Scheme' <http://www.defra.gov.uk/environment/climatechange/trading/pdf/trading-rules_rev2.pdf>.

[9] See *inter alia*, 'The UK Emissions Trading Scheme: A New Way to Combat Climate Change', National Audit Office (2004). The full report is available at <http://www.nao.org.uk/publications/nao_reports/03–04/0304517.pdf>.

The UK ETS also allows companies with 'Climate Change Agreements' to participate on baseline and credit basis. Details of CCAs and their participation are discussed in section 5.3 below.

The design of the direct participant element of the UK ETS is discussed below under the following headings:

- coverage—regulated sources and gases;
- the overall cap and the distribution of allowances to participants;
- trading rules;
- monitoring and compliance;
- legal basis and dispute resolution.

5.2.1 Coverage—regulated sources and gases

Direct participation in the UK ETS was potentially open to all persons in the UK with emissions of GHGs from 'eligible sources'. The main categories of non-eligible sources were: electricity generators (except where energy is generated and used on-site); land and water transport; domestic households; and facilities covered by CCAs (able to participate via another route—see section 5.3 below).

Thirty-four companies successfully bid for targets in the March 2002 auction, including companies from the chemicals, oil and gas, minerals, air transport, retail and financial services sectors, as well as universities, one local council and the National Natural History Museum in London—living up to Government hopes that participation would be economy-wide.

Participants were required to choose whether they would participate solely in relation to CO_2 emissions or emissions of all six GHGs. An unusual feature of the UK ETS in comparison with other schemes is that participants participate in respect of the *indirect emissions* arising from their electricity usage.[10] On environmental grounds, there are merits in this approach as it will increase the speed of change to a lower carbon economy (as opposed to simple price signals on fuel—likely to be the result of including electricity generators in trading schemes—which have proved relatively slow in influencing behaviour). Opening the scheme up to companies without direct emissions of CO_2 is certainly one of the reasons for the wide variety of direct participants in the UK ETS.

In order to prevent leakage of emissions by the transfer of production to other plants not covered by the scheme, participants are required to enter all sources operating in any sector and the scheme had detailed rules governing change in operation and source substitution.[11]

[10] The emissions resulting from electricity usage are calculated by reference to average emissions factors, except where the participant can demonstrate that electricity is wholly from renewable sources.

[11] See Schedule 3 to the Scheme at <http://www.defra.gov.uk/environment/climatechange/trading/pdf/trading-rules_rev2.pdf>.

5.2.2 Overall cap and distribution of allowances

The emission reductions, which direct participants agreed to deliver below their baseline emissions, were set in the March 2002 auction.

The design chosen for the auction is known as a 'descending clock'. Potential participants were asked to bid the quantities of emission reductions they were prepared to make at an opening price per tonne of reduction below baseline. If the total of the reductions bid multiplied by the offer price exceeds the total budget available (£215 million in total the five years of the scheme), the auction moved on to the next round in which the offer price was reduced and bids were requested at the lower price. After nine such rounds, the participants agreed to reduce their annual emissions by just over 4 million tonnes of carbon dioxide equivalent in 2006 in return for £215 million. This is equivalent to reductions of 12.1 million tonnes in total over the duration of the scheme at a price of £17.79 per tonne.

Participants were required to agree to overall reductions to be delivered in 2006, with delivery of interim targets on a straight line to those overall reductions. The allocation to a direct participant in any year is the participant's baseline less its target for that year. Targets increase pro-rata on a cumulative basis. Thus, the target in Year 3 is three times the Year 1 target.

Baselines are calculated on the basis of average measured emissions from all sources from 1998 to 2000,[12] unless the participant already has a regulatory limit imposed on emissions from any source. If so, for baseline years where measured emissions exceed the regulatory limit, the baseline is calculated by reference to the regulatory limit rather than measured emissions.[13] This was necessary first in order to satisfy the European Commission in relation to possible state aids issues[14] and secondly to ensure that Government funds were not used to pay for what was already required by regulation.

The total cap for the participants is the total of all baselines less the total of all annual targets bid in the auction. In total, the emission reductions bid in the auction represented a 13 per cent reduction in emissions by the end of the scheme.

The relationship between the reductions bid in the auction and baseline emissions can be seen from the following Figure 25.1.

5.2.3 Trading rules

Accounts for direct participants are automatically opened in the Emissions Trading Registry, which operates via secure internet access in real time. In addition, anyone may apply for a trading account in the Registry. The electronic registry minimizes

12 A shorter time period may be used if the participant does not have adequate information.

13 See 'Guidance on Treatment of Regulatory Requirements in the UK ETS' <http://www.defra.gov.uk/environment/climatechange/trading/pdf/trading-reg_require.pdf>.

14 See European Commission Decision on State Aid No N416/2001 (21 November 2001).

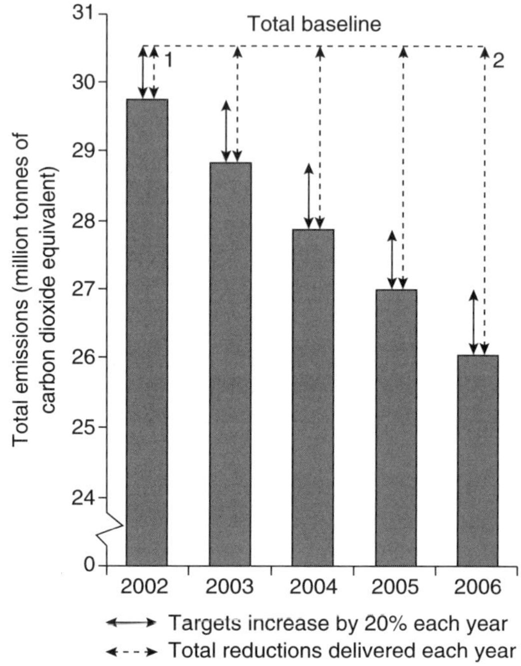

Figure 25.1: Direct Participants' total targets
Direct Participants need to deliver both their target for 2006, and targets for 2002 to 2005

1 2002 total target, 0.79 million tonnes (20% of 2006 target, 3.96 million tonnes).

2 Promised total reductions, 3.96 million tonnes.

Source: National Audit Office.

transaction costs and is likely to serve as the model for future registries (for example, for the EU Emissions Trading Scheme discussed in Chapter 23 above).

The registry is split into two sectors, the absolute sector and the relative sector (for those participating through CCAs with relative targets—see further section 5.3). Trading within sectors is unlimited but trades from the relative sector into the absolute sector must pass through a 'gateway' which prohibits a net import of allowances from the relative sector.

Banking between years is allowed for the duration of the scheme but borrowing between years is prohibited.

5.2.4 Monitoring and compliance

Direct participants are required to monitor and report emissions in accordance with approved protocols. A number of protocols covering indirect CO_2 combustion emissions and an extensive range of CO_2 process protocols were prepared by the Government. If direct participants wished to enter sources not covered by these protocols or sources emitting other GHGs, they were required to prepare them and submit them for approval. This is an innovative step, which put the onus on participants to demonstrate that reporting mechanisms are accurate to remove doubts about uncertainties.

Direct participants' compliance with the monitoring and reporting requirements must be verified by independent accredited verifiers before emissions are reported. These verifiers are required to certify the accuracy of both the initial 'baselines' and reported annual emissions in each year of the scheme. They are also required to certify compliance with special rules designed to ensure that the closure or sale of parts of a business by a direct participant does not lead to a windfall of allowances.[15]

Trading is based on seller liability, which will promote liquidity. Penalties for non-compliance with the requirement to surrender allowances equal to annual emissions were initially limited to non-payment of incentive money, the docking of allowances for the subsequent year at a ratio of 1:1.3 for each tonne of excess emissions and the 'naming and shaming' of the participant concerned. In 2003, the Government introduced legislation providing for financial penalties of £30 per tonne of excess emissions.[16] These will apply in relation to compliance with emissions targets for 2004 and beyond and at the same time the docking element of the penalty will reduce to 1:1.

5.2.5 Legal basis and dispute resolution

The participation of direct participants in the UK ETS is governed by contracts between the Secretary of State and the direct participant, known as a Direct Participant Agreements.[17] The detailed rules of the UK ETS[18] are incorporated by reference into the Direct Participant Agreement.

In particular, under the terms of the Direct Participant Agreement, the Secretary of State agrees to:

- open and maintain accounts at the Registry in the name of the direct participant and to allocate allowances to an account in the name of the Direct Participant;
- pay to the direct participant incentive payments;[19]
- apply the rules of the scheme to other participants as necessary to promote the purposes of the scheme; and
- comply with the other obligations imposed upon her[20] by the rules of the scheme so far as the same directly concern the Direct Participant.[21]

[15] See Schedule 3 to the Scheme (n 11 above).

[16] Waste and Emissions Trading Act 2003, s 39, see <http://www.legislation.hmso.gov.uk/acts/acts2003/20030033.htm>.

[17] <http://www.defra.gov.uk/environment/climatechange/trading/pdf/trading-dp_agreement.pdf>.

[18] See n 9 above.

[19] The statutory authority for these payments derives from s 153(4) of the Environmental Protection Act 1990.

[20] The 'her' refers to Margaret Beckett, the current Secretary of State. The UK ETS rules refer to the Secretary of State as female throughout.

[21] See clause 4.1 of the Direct Participant Agreement (see n 17).

In return, the direct participant agrees to abide by the rules of the scheme including: limiting its emissions to the agreed levels; monitoring, calculating, reporting, and submitting emissions statements and verification opinions in respect of its baseline and its annual emissions; and repaying incentive monies in certain circumstances. The Direct Participant Agreement also sets out detailed procedures for dealing with the withdrawal or insolvency of the direct participant.

Schedule 4 of the rules of the Scheme sets out the procedure for resolving disputes between a direct participant and the Secretary of State. The parties are required to use mediation to discuss in good faith and settle any dispute prior to referring the matter to an expert adjudicator. To the extent that any dispute involves a challenge to any decision, act or omission of the Secretary of State, Schedule 4 requires the Secretary of State to reconsider her position in the light of the adjudicator's report and to notify any participant affected of the action she proposes to take after reconsideration and of her reasons. Otherwise, each party will be free to accept or reject the adjudicator's report or recommendation.

5.3 Climate Change Agreement Participants

An important feature of the UK ETS is its link with the Climate Change Levy, an energy tax on non-domestic energy users announced by the UK Government in 1999, partway through the development of the UK ETS. The levy has been payable on all business use of electricity, gas and coal since April 2001. Energy-intensive sectors are eligible for an 80 per cent discount from the levy if they commit to negotiated reductions in emissions or energy use under CCAs.[22]

The government signed ten-year CCAs in 2000 with forty-four sectors and more than 5,000 companies operating more than 12,000 individual facilities, including the UK's most energy-intensive manufacturers such as steel, aluminium, cement, chemicals, paper, and food and drink producers.[23] The majority of the targets in CCAs are efficiency targets (referred to as 'relative' targets) expressed in emissions/energy use per unit of production, as opposed to absolute targets.

The 5,000 or so companies with CCAs are able to participate under the UK ETS on a 'baseline-and-credit' basis. In other words, Climate Change Agreement Participants (CCA participants) can use the UK ETS to buy allowances—either

[22] The Climate Change Levy and the CCAs are provided for in Sch 6 to the Finance Act 2000, available at <http://www.hmso.gov.uk/acts/acts2000/20000017.htm>.

[23] Eligibility to join a CCA is currently confined to those facilities carrying out an activity listed under Pt A1 or A2 headings in the Pollution Prevention and Control (PPC) (England and Wales) Regulations 2000 as amended. The Government is currently consulting on the potential to extend CCA eligibility beyond the PPC regulations and is considering the use of a Specific Energy Intensity Threshold. Budget 2004 announced that 'businesses that exceed a 12 per cent energy threshold will be allowed to apply for a CCA without first having to meet a test of international competition. Businesses currently eligible for CCAs will remain eligible but, in addition, the Government will extend CCAs to businesses that pass an energy intensity threshold of 3 per cent and can in some cases demonstrate the existence of international competition issues.'

from other CCA participants or direct participant—required to meet their targets or to sell any over-achievement of their targets. In some cases, sector associations trade as a trading group on behalf of those of their members that wish to participate in the group.

The detailed rules for CCA participants are set out in Part D of the rules of the UK ETS.[24] These rules are incorporated by reference into the terms of both Umbrella and Underlying Agreements. The key provisions of Part D can be summarized as follows:

- A CCA operator or sector who wishes to participate in the scheme must register. Compliance accounts will be opened for each target unit or trading group registered.
- A CCA overachievement may be converted into allowances on application to the Secretary of State. Overachievement must be verified in accordance with the rules of the scheme before any allocation is made.
- If a CCA operator has not met the target for a target unit, or a sector association for a trading group, the target may be increased by operators or sector associations placing sufficient allowances to cover the deficit in their compliance account from where they will be retired.

5.4 Results to Date

5.4.1 Emissions reductions

In terms of emissions reductions from direct participants, the first two years of results from the UK ETS are certainly impressive, with reductions against baseline of 4.64 million tonnes of CO_2e in 2002 and 5.2 million tonnes of CO_2e in 2003, overachieving the annual targets by 3.6 and 3.5 million tonnes of CO_2e respectively. These reductions are certainly impressive and will contribute significantly to the UK Government's climate change ambitions.

In relation to CCA Participants, a total of 580,000 allowances were bought and used to meet more than 1,000 targets for 2002 while a further 4.0 million tonnes of overachievement were either allocated or ring-fenced for future use.

5.4.2 Level of trading and market development

The market in allowances has been variable, with allowance prices peaking at £12.50 per tonne in autumn 2002, before dropping to around £5.00 per tonne.[25] The large reductions made by direct participants in excess of their targets has led to an increase in supply in allowances which has yet to be matched by demand from CCA participants, although the review of CCA targets in 2004 may stimulate

[24] See n 9 above.

[25] Figures quoted from NAO study, see n 9 above.

further demand. The price of allowances rose in the weeks before the reconciliation deadline in 2004, mirroring the increased activity.

The level of trading in the UK ETS has been good, despite the number of allowances on the market, and in comparison to other trading schemes. In the year leading up to the first reconciliation, over 2,000 transfers took place of over 7 million allowances. In the year to the second reconciliation, the number of transfers had fallen, reflecting the absence of milestones for CCA holders, but at 325 was higher than might be expected.

Despite the low price, the impressive reductions from the scheme indicate that the UK ETS has provided an incentive for companies to make additional reductions beyond their targets. Moreover, the experience of the UK ETS appears to have left the UK well placed to participate fully in international emissions trading schemes, in particular in relation to verifiers and advisory services.[26] The National Audit Office (NAO) concluded that it was still too early to say whether the UK ETS had contributed to the City of London becoming a global centre for emissions trading but that London was perceived as being more developed as a centre for trading than its likely rivals, Frankfurt and Paris.

5.5 Conclusions: National Audit Office Report

In late 2003, the NAO[27] carried out a 'value for money' investigation into the UK ETS and published its report[28] in April 2004.

In particular, the NAO wanted to look into the question of whether the emissions reductions offered by participants were sufficiently stringent and in particular whether the emissions baselines of some participants were set at too high a level. Given the UK ETS's voluntary nature, there was always a risk that this might be the case although by incorporating a financial incentive which is linked to the quantum of reductions offered, the UK ETS attempted to address this problem. However, certain risks remained: whether the approach taken could overcome participants' self-interest in setting non-demanding targets and whether in a novel market, such as this, bidding would be low rather than high risk, with the result that the market will be populated by sellers rather than buyers. In addition, it was alleged that this issue is exacerbated for the UK ETS by the consequences of the approach towards 'regulatory additionality' (discussed at section 5.2.2 above). The Government's rules on this issue have been criticized by those who claim that they have failed to prevent a large amount of 'hot air' (that is, allowances which are surplus to requirements) entering the scheme.

[26] The scheme has enabled the UK to 'definitely offer[ed] leadership: we have schooled a number of the essential advisory communities, law, insurance and banking to prepare us for these bigger markets and so we have done ourselves a big favour by starting off on a UK Scheme' (James Cameron in evidence to The Select Committee on the European Union (Sub Committee D) on 24 March 2004).

[27] The NAO is an independent body which scrutinizes public spending on behalf of the UK Parliament, see <http://www.nao.org.uk/> for more details.

[28] See n 9 above.

The NAO's research into the four companies which had reduced their emissions by the greatest extent beyond their targets established that the companies had, in practice, made significant additional efforts to cut emissions and that 66 per cent of the reductions reported by the companies was attributable to the scheme. The report also acknowledged that the Department felt unable to set more demanding targets as the scheme needed to be based on even-handed application of general principles. However, it recommended that the Department should consider further ways of improving the value obtained from the payment of incentives, for example by agreeing further emissions reductions and/or voluntary limits on the sale of surplus allowances. Following the second year reconciliation the Department undertook a review of the scheme, both within the current agreement period, and to consider options beyond 2006 when Direct Participant Agreements end. The results of this review will be available later in 2004.

The NAO made a number of additional recommendations about future trading schemes:

- in implementing the European scheme, the Department should continue to press for UK companies to retain as much benefit as possible from their experience in the UK scheme and for elements of the UK scheme to be adopted by other Member States;
- in its plans for developing other trading schemes (notably for waste and sulphur dioxide trading), the Department should:
    - take early advice from technical experts on the industries concerned, when designing the rules for new schemes;
    - develop the way it uses any other regulatory requirements on scheme participants, based on experience from the UK scheme;
    - fully inform participants about plans to introduce trading and provide enough time to prepare;
    - consider carefully the size of pilots to test scheme design, especially where knowledge of the market is limited;
    - develop explicit plans to make the best possible use of the scheme to improve its information on, for example, the cost of measures to reduce emissions;
    - continue to share the good practices developed by the Department's emissions trading team with other parts of the Department and other government departments; and
    - ensure that risk management procedures provide for sufficient challenge.

Overall, the NAO's assessment of the UK ETS was positive. Its first key finding was as follows:

> The Department has successfully set up a novel and functioning emissions trading scheme, which has the potential to benefit the UK economy. Companies participating in the

Scheme told us that they have gained greater understanding of how they can reduce emissions and practical experience of using the emissions market. Companies providing emissions trading services, such as brokerage and verification, have established themselves in the UK market and gained experience that places them in a strong position to gain further business as European and international emissions trading develop.[29]

The findings continued:

The UK Scheme has encouraged the development of the European Scheme and influenced its design in some aspects. The experience gained in establishing the Scheme is helping both the Department and industry prepare for the launch of the European scheme in 2005. In addition, the Department is adapting the registry system for the UK Scheme, which records the numbers of allowances held by participants, for use in the European scheme and other trading schemes. The Department is collaborating with a number of European Union member states who have expressed interest in adopting the UK system for their own registries. . . .[30]

Finally, in conclusion, the NAO stated:

The Scheme is a pioneering initiative. Innovation in policy-making carries risks, and the issues identified above have resulted, at least in part, from the development of policy in a completely new area. And these issues must also be put against the significant achievements of the Scheme, not least in setting up a well-functioning emissions trading system and encouraging participants to identify emissions reductions and make them available to others. A key aim of the Scheme has been 'learning-by-doing' and it is important that the Department makes the most of its opportunity to learn from the experience of the Scheme in further developing this Scheme, in continuing to influence the European emissions trading scheme and in designing other trading schemes planned in the environmental area.[31]

6 UK Policy on EU Emissions Trading Scheme

As a footnote to the UK's experiences with the UK ETS it is worthwhile considering briefly its policy on the implementation of the EU Emissions Trading Scheme (EU ETS) (see Chapter 23 above).

Although the UK ETS adopted a different approach on some key design aspects from the EU ETS (in particular the treatment of emissions relating to electricity generation and emissions from GHGs other than carbon dioxide), the UK has been an active supporter of the introduction of wider trading across the European Union. The 2003 Energy White paper stated that the EU ETS should be 'a central plank of [the UK's] future emission reduction policies, through which the traded carbon market can set a signal for the value of carbon reductions in the economy'.[32]

[29] See n 9 above, 2. [30] See n 9 above, 2. [31] See n 9 above, 5.
[32] See n 3 above, 29.

The UK published its National Allocation Plan for the first phase of the EU ETS in early May 2004. In the plan, it was stated that the UK intended to use the first phase of the EU ETS to move beyond its Kyoto Protocol target towards its domestic goal. The EU ETS would be used to deliver additional carbon dioxide savings which it was envisaged in the UK Climate Change Programme would be delivered by emissions trading but for which the UK had not yet introduced firm policies. The UK has decided that these additional savings (which equate to a total of 5.5 million tonnes of CO_2 between 2005 and 2007) will be delivered by the electricity supply industry. On the basis of the latest emissions projections, were this policy to be extended into the second phase of the EU ETS (2008–12), it would mean that the projected reduction in the UK's total emissions of CO_2 by 2010 would be increased from a reduction of 14.3 per cent[33] on 1990 levels to a reduction of 15.2 per cent.

Moreover, the UK Government has used the National Allocation Plan to reaffirm its commitment to the domestic goal of moving towards a 20 per cent reduction in CO_2 emissions below 1990 levels by 2010. It has stated that its policy for the second phase of the EU ETS from 2008 to 2012 will be to ensure that the total quantity of allowances allocated to the installations covered by the scheme is consistent with the achievement of that goal. The details of policy for the second phase will be considered further in the review of the UK Climate Change Programme, which will take place later in 2004.

Commenting on the EU ETS in September 2004, Tony Blair, the UK Prime Minister, stated:

> trading is the most cost-effective way to reduce emissions. The [EU ETS] is of great importance to our goals and to those of Europe. The establishment of a carbon trading market throughout the world's most important economic area next year will be an enormous achievement and will change the way thousands of businesses think about their energy use. Cutting carbon emissions is the way the future will be and . . . there are advantages to British industry from early action.

[33] This reduction already takes into account the wider policies and measures set out in the Climate Change Programme.

26

Pollution Permit Trading in Chile

Marcela Main S. *

1 Background

Santiago de Chile is situated in a valley surrounded by mountains as high as 3,000 metres. It encompasses an area of approximately 15,500 km^2 in which nearly 6 million people live, 40 per cent of Chile's total population. Santiago accounts for around 50 per cent of the gross domestic product (GDP) of the country. The manufacturing sector is the second largest economic activity of the region.

This combination of geographic and economic characteristics, given the weather conditions present in the area, contributes to an unfavourable dispersion of pollutants in the autumn and winter seasons. In fact, the valley significantly restricts the circulation of wind and, therefore, the renewal of air inside the basin. The area also has a heat inversion layer that produces a ceiling that keeps the pollutants enclosed. Furthermore, the growing economic activity in the region in recent decades and the natural growth in population have caused a horizontal expansion of the city that forces the population to travel significant distances on a daily basis. In conjunction with structural problems in the city and a poor public transportation system, there are externalities, such as vehicle congestion, that in turn help increase air pollution indicators.[1]

This situation led the Chilean Government to implement an Air Decontamination and Prevention Plan for the Metropolitan Region in 1998.[2]

* Marcela Main S. is a legal counsel with the Chilean Comisión Nacional del Medio Ambiente. The views expressed in this chapter are the views of the author and do not necessarily represent the views of the employer.

[1] In fact, the transportation sector is the most contaminating sector in the region, responsible for the emission of 48 per cent of breathable particle matter (PM_{10}), for 84 per cent of nitrogen oxides (NO_x) and for 91 per cent of carbon monoxide (CO). Added to the foregoing is the significant share of 30 per cent of the emissions of volatile organic compounds (VOC) and 34 per cent of sulphur oxides (SO_x). Stationary sources also contribute to PM_{10} (21 per cent) and SO_x (64 per cent). International references indicate that the share of this sector in VOC emissions would also be significant. Lastly, the most significant residential emissions are VOCs, with 38 per cent. Home use of firewood generates 13 per cent of the PM_{10}.

[2] Decreto 16 of 1998, which establishes the 'Plan de Prevención y Descontaminación Atmosférica para la Región Metropolitana'.

Subsequent to the adoption of the Plan, emissions have decreased significantly. At the same time, however, the cost of further reducing the pollution levels is becoming a burden for competitiveness of the regional economy and the efficiency of the measures is lowering compared with the benefits that come from the emissions reductions. Taking into account these considerations, the Chilean Environmental Agency—CONAMA (Comisión Nacional del Medio Ambiente)—has recently developed the basis for an emissions trading scheme within the Pollution Control Plan.

The different experiences in market based approaches for pollution control that have been implemented in Santiago will be presented in this chapter. CONAMA's studies of market based approaches for pollution control date back to the early 1990s. As early as 1992, an Offset Programme for breathable particle matter (PM_{10})[3] was established. Later, the Pollution Control Plan set targets for existing sources and established an Offset System for new projects. And finally, a Bill of Law for an Emissions Trading Scheme was presented on July 2003.

2 Air Pollution Control Regulations

The first attempt to introduce market based regulations for pollution control in Chile was made in 1992. This system was based on the Californian RECLAIM programme[4] and established emissions standards of PM_{10} for existing stationary sources and an Offset System for new ones. However, national regulation constraints and lack of experience within the competent institutions made the outcome of this measure not as successful as expected. Nevertheless, it set the basics for inventories and registries, which became one of the key elements while designing the Pollution Control Plan.

2.1 Executive Decree No 4

Given the particular conditions of air pollution in Santiago indicated above, even before an environmental framework was set up, the health authority established several regulations to control air emissions, in particular from stationary sources. Among these regulations is Executive Decree No 4 of 1992 of the Ministry of Health (ED No 4), that set down PM_{10} emission standards for stationary sources and created the first emission offset system.

Under ED No 4, existing sources—industrial boilers and furnaces registered by the date the Decree entered into force—received a perpetual daily emissions permit proportional to a predetermined level of emissions. Due to problems in monitoring and control, in practice the authority had to determine the emission

[3] 10-micron Particulates (PM_{10}).

[4] REgional CLean Air Incentives Market (RECLAIM). The programme was approved by the California Air Resources Board in October 1993. It covered initially over 400 sources which accounted for 10 per cent of the SO_x and NO_x emissions of stationary sources in the South Coast Air Management District (Los Angeles), <http://www.aqmd.gov/reclaim/reclaim.html>.

capacity of the source each year instead of the emission level and grant it a permit based on that capacity and not based on the emission. As a result, the source was obligated to maintain a permit at all times that would cover its maximum emission capacity. New sources, on the other hand, were not granted any permit. This meant the recognition of the historic emission and froze those emissions at a level equal to the aggregate of all permits granted.

Although it was nominally an offset system, in practice ED No 4 created a Permit Trading System. The offset was defined as an agreement among source owners so that one of the parties would decrease its particle matter emissions by at least the amount by which the other would increase them. The offset could take effect between the existing sources in order to create a mechanism for flexibility in complying with new standards or regarding new sources that had to offset all of their PM_{10} emissions. The amount of the trade-off for existing sources was set at 25 per cent for 1993 to a 100 per cent for 1996, with an increment of 25 per cent annually. ED No 4 also established an Emissions Report Register and a Trade-Off Register.

At the same the system suffered from a number of imperfections. Its enforcement system was insufficient as it was governed by the general rules on sanitation, which were not adequate to the requirements of an offset system and its implications. Furthermore, the subsequent Decree of the Ministry of Health (ED No 812/95) complemented the regulations and created a procedure of prior approval of trade-offs. This provoked uncertainty among the potential permit suppliers since the future purchase of permits, if they were needed, was subject to these restrictions, which limited the creation of a permit market. Additionally, monitoring restrictions made the permit permanent in nature instead of daily, which made the market even more illiquid. By July 1999, although 401 sources were participating in the scheme—with 3,981.3 kg/day authorized—only thirty-two trade-offs had been approved of which only two were trade-offs between unrelated sources. However, at the same time, around 200 new sources needed to offset their emissions but had difficulty finding a supplier. So the approval process became a bottleneck for the installation of new industries. Lack of enforcement made it easier to overpass these requirements.

It must be emphasized, however, that the application of ED No 4 helped generate significant information through records and inventories that would later contribute to the establishment of new regulations, in particular the Air Decontamination and Prevention Plan for the Metropolitan Region.

2.2 The Air Decontamination and Prevention Plan for the Metropolitan Region

The framework of reference for Chilean environmental laws is *Ley de Bases Generales del Medio Ambiente No 19300* (LBMA).[5] This Law states that the authority must

[5] Environmental Framework Law No 19300, Ley 19.300: Bases Generales del Medio Ambiente, promulgated 1 March 1994 by the President of the Republic Patricio Aylwin Azocar and published in the *Diario Oficial* on 9 March 1994.

establish Decontamination Plans when the environmental quality standards are systematically exceeded in a location (non-attainment area). When there is a danger than these standards would be exceeded (attainment zone), *Plans de Prevención y Descontaminación Atmosférica* (PPDA)[6] are issued.

The standards in Santiago were exceeded to such a point that in 1998, by means of Executive Decree No 16 of the Office of the Presidential Chief of Staff, a PPDA for Santiago was established for PM_{10}, total suspended particles (TSP), carbon monoxide (CO), ozone (O_3), and nitrous oxide (NO_2). The PPDA contains information relative to the effects of the regulated pollutants on health, information on the region as a basin, information on the activities whose emissions have an impact on the area, an emissions inventory and a calculation of the proportional contribution of the activities to emissions, the baseline for emissions and their projection, and the emissions reduction target globally and by activity.

The principal focus of the PPDA is a reduction of the smaller part of breathable particle matter ($PM_{2.5}$) because it is the greatest aggressor against the health of the population. To achieve that reduction, the PPDA proposes a set of strategies, courses of action, and measures to attain the emission reduction goals in a period of fourteen years. Among the most relevant are the improvement in the quality of fuels and an industrial conversion to natural gas, the rational use of cars, the permanent restriction of vehicles during winter, a strict cleaning and washing programme and the reforestation of urban green areas and the slopes of hills.

The PPDA also contemplates the perfection of the PM_{10} trading system for stationary sources (derived from ED No 4) and adds the exigency for gas emissions trade-offs. In order to make the system once again reliable, the PPDA defined an immediate stage and another medium-term stage. The first stage began with the effectiveness of the PPDA. It requires the offset of 120 per cent[7] of the emissions of all new projects and activities in Santiago de Chile that have to conduct an Environmental Impact Assessment (EIA). The specifications of the offset, for example if emissions have to be compensated through forestry projects or fuel switching of an existing source or the replacement of old buses for new ones and so on, is determined on a case-by-case basis through the EIA. The second stage has been elaborated in the updated plan (see below).

The achievements of the plan thus far are a reduction of nearly 30 per cent in the concentration of pollutants in the city, with a particular effect in the reduction of PM_{10}.[8] By 2002, all major property development projects (a total number of fourteen) had offset their emissions through forestry projects, all old taxis have been taken off the streets, and nearly 2,000 buses have switched from diesel to

[6] Decontamination and Prevention Plans.

[7] The trading requirement was raised to 150 per cent in 2000.

[8] $PM_{2.5}$ has also dropped by nearly 50 per cent in a scenario of economic growth in which the GDP of the country doubled. As of 2000, the industrial sector had progressed 66 per cent in reducing PM_{10} and 33 per cent in NO_x. However, the transportation sector had increased its emissions of PM_{10} by 17 per cent and of NO_x by 15 per cent.

natural gas. All new projects that have to offset their emissions cannot be implemented unless the condition established trough the EIA has been fulfilled.

2.3 Updating and Reformulation of the PPDA for Santiago

As of 2001, the PPDA was reformulated[9] in order to establish measures, principally of technological impact, to continue going forward with the progressive improvement in the quality of air in Santiago, and reach the goal of compliance with established air quality standards by 2010. The most important measure is having set an overall goal for emission reduction in the transportation sector of 75 per cent for PM_{10} and 40 per cent for NO_x. This goal should be met by 2005.

The updated plan defines a framework for an emissions trading system covering the five pollutants identified in the plan. Details with respect to the trading programme will be determined based on the physical-chemical characteristics of the pollutants and the spatial location of the activities or projects. As part of the identification of economic tools to control pollution, the new PPDA establishes a cap on transportation and industry emissions, applies an offset system to all type of sources, and orders a Bill of Law to be presented on Tradable Emission Permits. Such a Bill of Law corresponds to the Tradable Emission Permit Bill of Law[10] that is analysed below.

3 Tradable Emission Permit Bill of Law

3.1 Command-control Measures versus Emissions Trading Schemes

As indicated above, the enforcement of the PPDA has brought about a significant improvement in the air quality of the basin. Paradoxically, the cost/benefit analysis of command-control measures as needed to keep the lowering of emissions, reveal that these kinds of tools will become more and more inefficient. In fact, a preliminary model revealed that the cost of meeting the goals in the PPDA through an emissions trading scheme versus the application of command-control measures would significantly lower the cost of compliance. The net social benefit would amount to US$178 million, with a cost of compliance of US$276 million for command-control versus US$98 million for emissions trading.[11]

[9] The Final Project on Reformulation of the PPDA was approved by the Steering Committee of CONAMA by resolution No 204–2002 dated 29 May 2002. At the time of writing this chapter, it was being processed with the Office of the Controller General of the Republic.

[10] The SPET being debated in Chile was named in Spanish the Decontamination Permit System (*Bonos de Descontaminación*) as one way to reinforce the positive idea that underlying the trading of a permit there is a reduction in emissions by a source.

[11] National Environmental Commission, Tradable Emission Permit Law, draft of 2 July 2003 (CONAMA, Chile, 2003).

3.2 Political Setting

The elements of efficiency associated with an emissions trading scheme (ETS) and the flexibility it gives, made the Chilean Government and the private sector agree, in January 2002, through a Pro-Growth Agenda Agreement, that the debate of an ETS Bill of Law would be given priority in the discussion of further reducing local pollution in Chilean cities.[12] Therefore, under the public–private agreement, given the experience gained by CONAMA in the control of air pollution in Santiago, on 5 June 2003, the President of the Republic, Mr Ricardo Lagos E., sent an ETS Bill of Law to Congress for debate. At this time it has been approved unanimously in its first constitutional round and is expected to become effective in the course of 2004.

It is of note that in the Bill elaboration process, CONAMA received the collaboration of representatives from different public services, including the Ministries of Finance, of Economy, of Health, of Public Works, of Transportation, and the Office of the Presidential Chief Staff as well as the Superintendency of Securities and Insurance. It also developed different instances of stakeholders' participation. Of particular importance in this latter regard were academics, businessmen, and non-governmental organizations (NGOs). It also took into account the opinions of the Santiago Stock Exchange, the Electronic Exchange, and the Central Securities Deposit.

3.3 Framework Approach

The ETS Bill of Law aims at establishing a legislative framework for a Trade-Off System of emissions trading. Article 4 of the Bill of Law states that specific schemes shall be applied by means of an executive decree and may refer to air, water, or soil in any predetermined basin or geographic area and be related to any pollutant. Thus, in order to apply the emissions trading system to Santiago, a specific executive decree would be required.

The Bill of Law establishes all general regulations applicable to the ETS and offers design alternatives for a better adaptation to each particular case. In this regard, in order to allow the authority to adopt a technical/political decision that best suits the circumstances of the basin or geographic area where the ETS will be applied, Article 10 of the Bill gives the alternative to apply cap-and-trade, which is called 'pure and simple' for these cases, or credit-based, called 'prior certification', schemes.

The articles in the Bill set out the stages of implementation of the ETS. The provisions contain rules to establish the schemes, standards for the initial

[12] Although the idea of legislating on this matter dates back to 1994 and Arts 47 and 48 LBMA discuss the use of emissions trading to limit pollution, several prior efforts to draft such a proposal were fruitless.

allocation, regulations on the market and provisions for certification and reporting of emission reductions.

3.4 Contents of the Bill of Law

The contents of the Bill of Law is organized under the following sections:

Title I—General provisions and definitions
Title II—Tradable Emission Permit System

- Establishment of Tradable Emission Permit System
- Allocation of emission quotas
- Emission quotas
- Tradable permits
- Characteristics common to emission quotas and tradable permits
- Limitations and restrictions on the use of emission quotas and the transfer of tradable permits

Title III—New sources, voluntary sources, sinks and unregulated participants

- New sources
- Voluntary sources and sinks
- Unregulated participants

Title IV—The tradable emission permit market
Title V—Registers
Title VI—Emission certification and control

- Methodologies
- Emission certification agencies
- Control

Title VII—Sanctions
Title VIII—Appeals
Title IX—System financing
Final Title—Final title

3.5 The ETS in Relation to Decontamination and/or Prevention Plans

The proposed Chilean ETS is conceived as one of the tools that Decontamination and/or Prevention Plans can use to achieve the necessary emission reductions in order to meet the plan goals. The ETS is one more tool in a battery of environmental management tools and as such, it does not seek to replace emission or quality standards but rather complement their application and provide a more efficient mechanism for compliance. However, it is not obligatory for a PPDA to contemplate the establishment of an ETS in its design. The decision to apply an ETS is left to the environmental authority, which may choose this mechanism if circumstances warrant.

Additionally, by applying the preventive principle inspired by the LBMA, the Bill of Law also includes the alternative of establishing an ETS outside of plans in order to prevent situations of default on environmental quality standards. In fact, Article 8 of the proposed Law stipulates that 'in those zones for which there is no Decontamination or Prevention Plan, the Regional Environmental Commissions may request that the Steering Committee of CONAMA apply a System in order to avoid reaching levels of non-attainment when the environmental assessment of a project concludes that execution thereof will run the risk of reaching levels of attainment or non-attainment.'[13]

3.6 Main Characteristics of the ETS

Competent authority. Since Article 70 of the LBMA provides that CONAMA must act as the consultant, analyst, and coordinator in environmental matters, Article 2 of the Bill of Law stipulates that CONAMA is the authority responsible for ensuring appropriate application of the scheme.

Basic definitions. Article 3 of the Bill of Law contains the basic definitions. Among others are the concepts relative to source, emission, basin capacity, and methods of design such as banking, reconciliation period, and phases defined in Article 3 of the Bill of Law. Although some of the definitions established in the Project were taken from the regulations on air pollution and stationary sources, many were rewritten given the difficulties in their interpretation or to make room for the application of the ETS in cases of water or soil pollution. Article 3(u) defines the scheme as the 'set of provisions . . . that are intended to control the total emissions of one or more pollutants and that establish the mechanisms and procedures relative to the allocation of emission quotas, the creation of emission permits and the characteristics of the trading scheme in a given geographic zone'. This definition provides, in turn, two key concepts that must be clarified for a better understanding of the Chilean ETS: (1) emission quotas; and (2) emission permits.

Emission quotas. The emission quota is the authorization for a source to emit a certain quantity of pollutant in a given period and will be established through each specific Decree that applies an ETS to a given area. Article 3(e) provides that the emission quota is a 'certain magnitude of emission of one or more pollutants authorized for an existing or regulated source for a certain period of time and phase.' In other words, the legal nature of the emission quota is essentially that of an authorization (or permit). In this same regard, Article 17 provides that

[13] In these cases, the load capacity of the basin shall be determined first.

'regulated sources may only emit the pollutants in a System if they have a valid and sufficient quota.' Following the logic that it is a framework Bill, Article 14 provides that the 'allocation should be based on historic emissions and/or benchmark emission'. Article 14 also specifies that the 'allocation of emission quotas shall be at no cost.' The proposed Chilean ETS therefore follows the example of most international experiences, such as the allocation of allowances under the EU ETS.[14]

The Bill does not consider the figure of emission quota reservations, which could be auctioned to new entrants. In fact, the decision was made not to include such quota reserves given the complexities involved in recording the income for the Government from the sale of those quotas. As a result of this decision, new sources can only set up business in the zone where the ETS is applied if they first acquire sufficient quotas from an existing source to back their emissions.

One of the elements that is needed to grant certainty to those regulated by the scheme is the permanent nature of the initial allocation which cannot retroactively be adjusted. However, despite the permanent condition of the initial quota allocation, Article 11 further provides that the permanence of the allocation does not withstand 'the periodic adjustments to which the quota is subject in each phase of the System.'[15] In fact, since the ceiling on global emissions must be reduced over time to achieve the expected reduction in emissions, the quotas are subject to the condition that they will be adjusted from time to time, according to the emission reduction target in the plan. These periodic adjustments should be notified sufficiently in advance so as to allow the quota holders to have the necessary information to allow businesses to take relevant investment decisions. The Bill of Law states that adjustments need to be notified at least one allocation phase in advance, that is with one period in advance for which an emissions reduction target was established. The adjustment factor cannot be statically pre-defined in the decree applying a scheme because of the dynamics in the management of pollution. It could be overestimated or underestimated and thus impair the stability of the system. International experience shows that revisions and/or modifications of the scheme must be considered, which are, in this case, associated with these moments of adjustment.

Emission permits. The emission permit is, intrinsically, the tradable instrument that represents a reduction in emissions by a source that will allow another source, by purchasing such instrument, to set up a business or expand without increasing the overall pollution in the basin. Article 3(b) of the Bill of Law provides that the emission permit is 'the tradable permit or instrument in the System that represents all or part of an emission quota assigned by the authority and includes a term of effectiveness and magnitude of emission.'

[14] For further detail see ch 23 above.

[15] Each 'phase' refers to each allocation period.

In order to avoid any appearance of 'ownership' associated with quotas and permits, the Bill is careful not to create material instruments containing the quotas or permits. The incorporeality of the instrument reveals that what is traded is a conditional authorization that does not create a right of property. The quota is therefore defined as an authorization and the credit 'will be created by mere registration of the respective emission quota in the Permits Register. The respective emission permit will be deemed in effect as from the time of registration by mere operation of the law and available for sale' (Article 19). In other words, the credit is an incorporeal instrument by which an authorization to emit a certain quantity of a specific pollutant can be traded. Along these lines, it is important to note that the Bill of Law foresees a Registries System that plays a fundamental role in the integrity of the ETS. It is through the Register System that the information on permits, supply, and transactions is centralized.

As indicated earlier, one of the weaknesses in the Off Set System created by ED No 4 is the rigidity of the object traded. Indeed, instead of being a daily emission permit proportional to a predetermined level of emissions, it ended up being a yearly permit based on capacity which forced sources to maintain permits at all time equal to their maximum capacity. In order to avoid this effect in the ETS, the Bill of Law provides that the permit 'may be divisible both in terms of magnitude as well as term of effectiveness' (Article 3(b)). Article 21 further states that 'emission permits may be divided in terms of magnitude as well as time so their transfer can be partial.'

Ensuring a Liquid Market. In an approach that went from very conservative to more pragmatic, following the logic of the economic tool and international experience, the Bill of Law lays down rules to ensure transparency, security, liquidity, and continuity of prices on the market, but it does not regulate the specific workings of the market. The regulation of intermediaries was eliminated, all methods of sale were recognized and the direct sale of permits was expressly authorized—although the involvement of intermediaries is not precluded—and transactions must be registered for purposes of information and of keeping the registers current.

In fact, in order to ensure sufficient authentic information both for interested parties as well as regulators, Article 30 of the Bill of Law provides that 'emission permits will be transferred by registration of the transaction in the Permits, Prohibitions and Forfeitures Register.' This means that there is no transfer without a corresponding new registration and, therefore, the buyer of the permit will be the one most interested in updating the registrations and the updated register is what will allow the regulator to know, with certainty, the authorized emission of any covered source. The registration procedure itself follows the model of RECLAIM. Similar to the Californian model it contains two levels of review, a single form for buyer and seller (to avoid matters of interpretation of the respective contract) and price information.

Limitations and Restrictions on the use of Quotas and Permits. Throughout the debate of the drafting of the Bill of Law, different concerns were analysed regarding the indirect effects of the application of the scheme as well as the local increases in emissions or the behaviour of sources during critical episodes. This, together with the experience of RECLAIM during the electricity crisis of California a few years back, led to the conclusion that the regulation of the ETS must allow certain limitations and/or restrictions to be established for exceptional cases where there may actually be some distortion in the system.

It was finally ordered that the decree applying an ETS may contain limitations and/or restrictions on the use of quotas and the transfer of permits between different categories of sources, between different types of pollutants, between sources located in different zones and during critical episodes, all with the goal of avoiding increases in the emissions of other pollutants or local situations of contamination.

Voluntary Sources, Sinks, and Unregulated Participants. Although an ETS is applied to a predetermined group of sources—basically those sources which are most responsible for the total emissions in an area—the Bill of Law does consider the possibility that this involvement can be extended to other categories of sources, which may be interested in voluntarily participating in the scheme and in taking advantage of the economic tool to co-finance technology improvements and which may simultaneously become important permit suppliers. Article 23 of the Bill of Law provides that voluntary sources and sinks can participate provided their baseline and their emissions, capture, or abatement can be measured or estimated and certified, monitored, and controlled according to procedures approved by the authority. Otherwise, these projects cannot participate because there is no way to determine their quota or permit nor could their emissions or capture be controlled.

The Bill of Law also contemplates the figure of the unregulated participant, that is, 'an individual or body corporate who acquires an emission permit but whose emissions, if any, are not regulated by the Permit System' (Article 3(q)). Ultimately, what is allowed is, for example, the participation of NGOs who want to help decontaminate by acquiring credits.

Registries System. The administration of the ETS is based on a registries system centralized at CONAMA that will provide the information necessary for the system to function correctly especially in regard to the permits market, emission control, and regulation. Article 31 of the Bill of Law considers the creation of the following Registers:

(a) A Sources, Sinks and Unregulated Participants Register that should include a Signatures and Authorities Registry;
(b) Emission Quotas Register;
(c) Emission Permits, Prohibitions and Forfeitures Register; and
(d) Emission Certification Agency Register.

Since the correct identification of existing sources is fundamental to quota allocation, Article 33 provides that 'they will not be allocated an emission quota, if pertinent, unless the requested information has been satisfactorily provided.'

Emissions Certification and Enforcement. From the viewpoint of enforcement, the Bill of Law relies on an emissions report system where specialized private accredited entities conduct certification according to methodologies defined by CONAMA. This allows the government's regulatory work to be focused, more effective and efficient.

However, the Bill does not change the regulatory authority of the enforcement agency but rather, as explained above, establishes mechanisms that neutralize the deficiencies now existing in this matter. Moreover, since the Health Code contemplates very low sanctions, which would debilitate the system, the Bill establishes a range of sanctions coherent with the infringements committed and adds a set of standards to determine the specific sanction applicable based on the seriousness of the infringement.

3.7 Implementation of an ETS in Santiago

The first case where an ETS will be applied in Chile will be Santiago, as part of the Reformulation of the PPDA. The system is expected to enter into effect around 2005 and involves initially only the large-scale emitters, meaning stationary sources that issue above a certain quantity, and public transportation. The pollutants in the scheme will be PM_{10} and NO_x. At this time, CONAMA is in the process of designing the scheme for the municipality of Santiago.

3.8 Relationship between the Chilean ETS and the Kyoto Protocol

The framework nature of the Bill of Law allows its application to any environmental element (air, water, soil) and to any pollutant. Therefore, theoretically, it could be applicable to greenhouse gases (GHGs), which would require a solution compatible with the Kyoto Protocol. However, since Chile does not have GHG emission reduction goals under the Kyoto Protocol at this time, there has been no need to agree upon the way to relate the Systems.

Nonetheless, that same fact makes the relationship of the Chilean ETS and the Kyoto Protocol more interesting, especially when considering the status of projects implemented under the Clean Development Mechanism (CDM) because the same project could participate in both the CDM and the ETS as different pollutants would be regulated. Therefore, a source located in Santiago could reduce its PM_{10} emissions and create credits while reducing GHG emissions and creating Certified Emission Reductions under the Kyoto Protocol. There would be synergy between both mechanisms.

However, the Chilean ETS will be flexible enough to allow for additional requirements added to the scheme, such as the emission reductions certification and registration, if the linking of other emissions trading schemes to the Kyoto Protocol, such as the EU ETS, makes such amendments necessary.

4 Conclusion

The last decade of experience of CONAMA regarding air pollution control has shown that when dealing with severe pollution problems, market based approaches may become a very efficient solution. Nevertheless, it is extremely important to take the time and produce enough information so as to understand the pollution problem, identify the relevant sources, set realistic targets and involve all stakeholders.

Chile may be the first developing country introducing these kinds of policies but it took policy makers in the country a decade to set the basis for this discussion. The attempt through ED No 4 and the step-by-step approach of Santiago's PPDA made it possible to be prepared to present the Bill to Parliament. At this time, controversial topics, such as allocation, have been already solved through the PPDA.

27

Emissions Trading Schemes in Australia

*Josh Carmody and Monique Willis**

1 Introduction

Australia is a country rich in natural resources with one of the fastest growing economies amongst the members of the OECD.[1] Its relatively small national population of 20 million people, combined with the high emissions intensity of many Australian export commodities (such as coal, steel, and aluminium) and the country's reliance on coal fired electricity generation to supply most of its energy needs, has provided Australia with the legacy of one of the highest emissions profiles, on a per capita basis, of any nation in the world. This goes some way to explain Australia's emission reduction commitment target in Annex B to the Kyoto Protocol[2] (the Protocol), which is set at 108 per cent of 1990 levels and is the second lowest emission reduction target, in percentage terms.[3]

Following the announcement by the US in 2001 that it would not ratify the Kyoto Protocol, the Australian National Government adopted a similar policy position and stated its rejection of the Kyoto Protocol in its current form.

Notwithstanding that national policy position of non-ratification (which remains at time of writing this chapter), the Australian Federal and State Governments have implemented a number of policies and measures seeking to address climate change and Australia's high emissions profile. Australia was the first country to introduce a market based renewable energy trading scheme. Nearly all Australian States recognize the carbon sequestration properties of forest plantations and have created tradable statute based property interests arising from this carbon sequestration process. In January 2003 a state-based greenhouse gas (GHG) trading scheme was launched, targeting the energy sector in Australia's most populous State

* Josh Carmody and Monique Willis are members of Baker & McKenzie's Global Clean Energy and Climate Change practice group. The views expressed in this chapter are the views of the authors and do not necessarily represent the views of Baker & McKenzie.

[1] Organization of Economic Cooperation and Development.

[2] Kyoto Protocol to the United Nations Framework Convention on Climate Change 1997, UN Doc FCCC/CP/1997/L.7/Add.1 reprinted in (1998) 37 I.L.M. 22 (not yet in force).

[3] The country with lowest emission reduction commitment in percentage terms listed in Annex B to the Protocol is Iceland.

of New South Wales (NSW). Indications are that this scheme, or something like it, will soon extend into other Australian State jurisdictions. These and other Australian initiatives to address climate change outside of the Protocol framework are discussed below.

2 Australia and Climate Change

During 2001 and 2002 Australia endured its most severe drought in recorded history. The Australian Greenhouse Office (AGO) estimates that Australian average temperatures have risen by 0.7°C over the last century. Water supply levels reached near-record lows throughout much of the country in 2002 and 2003. About 80 per cent of Australia's population live within 50 kilometres of the coast, with increased vulnerability to tropical cyclones and storm surges. Bush fires of unprecedented intensity have devastated several Australian cities in recent years. A guide to Australian climate change released by the AGO in 2003 states:

> Australia is vulnerable to changes in temperature and precipitation projected for the next 50 to 100 years, because it already has extensive arid and semi-arid areas, relatively high rainfall variability from year to year, and existing pressures on water supply in many areas. In addition, vulnerability arises due to high fire risk, Australian ecosystems sensitive to climate change, and invasion by exotic animal and plant species introduced by human activity. Australia also has a high concentration of population in coastal areas, an economy strongly dependent on world commodity prices, tourism dependent on the health of the Great Barrier Reef and other fragile ecosystems, and economically and socially disadvantaged groups of people.[4]

3 Australia and the Kyoto Protocol

The Australian Federal Government holds the '*foreign affairs*' power to sign international treaties under the Australian Federal Constitution. The Federal Government ratified the United Nations Framework Convention on Climate Change (UNFCCC) on 30 December 1992.[5] It was one of the first national governments to do so. By the time the Kyoto Protocol was negotiated and agreed in Kyoto in December 1997, a change of Federal Government had taken place and a new conservative Government was in power. This conservative coalition of the Australian Liberal Party and the Australian National Party continues to govern Australia at the federal level nationally.

[4] B Pittock (ed.), *Climate Change: An Australian Guide to the Science and Potential Impacts* (Australian Greenhouse Office, 2003), <http://www.greenhouse.gov.au/science/guide/> accessed 3 May 2004.

[5] Australian Treaty Series 1994 No 2, Department of Foreign Affairs and Trade, Canberra, United Nations Framework Convention on Climate Change (New York, 9 May 1992), entry into force generally and for Australia 21 March 1994.

It is current Federal Government policy that Australia will not ratify the Protocol, but it is Government policy to meet Australia's commitment under the Protocol to reduce national emissions of GHGs to a level which is equivalent to 108 per cent of Australia's level of emissions in 1990.

A summary of the Federal Government's position in relation to the Kyoto Protocol is provided by the following quotations from the Federal Minister of Environment, Dr David Kemp:

> Australia has not ratified the Kyoto Protocol and at present has no plan to do so, but we are still actively engaged in promoting truly global action. The Protocol is not in our national interest and will remain that way without a clear pathway for the involvement of developing countries, and without the United States.
>
> July 2003 Speech
> Minister for the Environment[6]

> We are committed to our Kyoto target on emissions of greenhouse gases, and we are developing a national climate change forward strategy that will help position Australia as a strong, competitive economy with a lower greenhouse signature.
>
> December 2003 COP9 Media Release
> Minister for the Environment[7]

The Australian policy position is heavily influenced by:

(i) the Federal Government's view as to the implications of Protocol ratification on participants in certain export exposed energy intensive industries which comprise a large and important part of Australia's economy;

(ii) the decision of the world's largest emitter of GHGs, the United States not to ratify the Kyoto Protocol; and

(iii) the failure of developing countries (such as China and India) to accept binding commitments to reduce their national GHG emissions during the Protocol's first commitment period.

The Government is of the view that Australia will meet its Protocol target without the assistance of the Protocol flexibility mechanisms.

Federal Government policy and legislative initiatives that address climate change at a national level include:

(i) The introduction in 1997 of Electricity Generator Efficiency Standards to reduce GHG intensity from thermal coal fired power stations;[8]

[6] Federal Minister for the Environment and Heritage The Hon Dr David Kemp, MP, Australian Resources and Energy 2003 National Conference, Sydney, 16 July 2003, <http://www.deh.gov.au/minister/env/2003/sp16jul03.html> accessed 3 May 2004.

[7] Federal Minister for the Environment and Heritage The Hon Dr David Kemp, MP, Speech to the Renewable and Sustainable Energy Roundtable Side-event, COP 9, Milan, 9 December 2003, <http://www.deh.gov.au/minister/env/2003/sp09dec03.html> accessed 3 May 2004.

[8] See *Program Guidelines: Generator Efficiency Standards* (Australian Greenhouse Office, 2001), <http://www.greenhouse.gov.au/ges/program_guidelines.html> accessed 3 May 2004.

(ii) The 1998 launch of a National Greenhouse Strategy, being a national strategic framework combining Federal and State initiatives for advancing Australia's greenhouse response;[9]

(iii) The introduction in 2000 of a national legislative scheme creating a market based Renewable Energy Certificate trading market, designed to increase generation capacity from approved renewable sources by 9,500 gigawatt hours (GWh) by 2010. A government-appointed independent panel recently recommended that this target should be modified to include a long term target of 20,000 GWh of additional renewable energy capacity by 2020 to stimulate further investment in new renewable energy infrastructure in Australia;[10] and

(iv) The introduction of a suite of targeted financial assistance and funding packages to encourage GHG abatement projects as well as voluntary agreement initiatives between Government and participants in key industry sectors to bring about voluntary long-term GHG emission reductions.[11]

4 Australian Policy Responses

This chapter sets out the emergence of renewable energy and emissions trading regimes in Australia at a State-wide and national level. These schemes create existing, in most cases statutory, commodities of value. They can be traded through the economy and be used as compliance instruments to comply with Australian-imposed regulations and compliance requirements. A domestic market for these Australian 'credits' currently exists, notwithstanding Australia's current decision not to ratify the Protocol. These regulatory commodities and trading regimes are driven by compliance requirements of Australian law, they create additional market value to the generation of renewable energy and activities that sequester carbon or create a reduction in GHG emissions.

Australian and overseas corporations have begun to prospect for opportunities where some additional commercial returns can be realized from these new regulatory markets. Subject to future legislative and policy developments in Australia, it is possible that additional value may, in future, be available to owners of some of

[9] National Greenhouse Strategy, 'Strategic Framework for Advancing Australia's Greenhouse Response' (release date: 26 November 1998), <http://www.greenhouse.gov.au/pubs/ngs/ngs.html>.

[10] Renewable Energy (Electricity) Act 2000 (Cth) (No 174, 2000) and Renewable Energy (Electricity) (Charge) Act 2000 (Cth) (No 129, 2000). See also, *Overview of the Mandatory Renewable Energy Target* (Office of the Renewable Energy Regulator), <http://www.orer.gov.au/overview.html> accessed 3 May 2004.

[11] For example, the Greenhouse Gas Abatement Program introduced as part of the Measures for a Better Environment Package in May 1999, which provides government funding to eligible emission reduction projects (see <http://www.greenhouse.gov.au/ggap/index.html>) and the Greenhouse Challenge, launched in 1995, whereby companies or industries can enter into voluntary emission reduction agreements with the government (see <http://www.greenhouse.gov.au/challenge/about/index.html>).

these projects and the rights they create under the Joint Implementation (JI) mechanism if Australia ratifies the Protocol in the future. In order for this to occur, Australia must not only ratify the Protocol but also ensure compliance with its terms. It would also be necessary for the Australian Government to pass laws permitting private entities to participate in international emission trading activities.

4.1 The National Renewable Energy Trading Scheme

The Renewable Energy (Electricity) Act 2000 (Commonwealth),[12] implemented by the Australian Federal Government, creates a mandatory obligation for Australian electricity retailers to source increasing amounts of their required electricity from approved renewable sources (the Mandatory Renewable Energy Target (MRET)). The aim of the scheme is to increase the contribution of renewable energy sources in Australia's electricity mix to 9,500 GWh per year by 2010. Under the scheme, accredited renewable energy generators are granted the ability to create 'Renewable Energy Certificates' (RECs) for each megawatt hour (MWh) of electricity generated from accredited generation facilities. RECs can be sold to liable entities independently of the electricity to which they relate.

In order to discharge their obligations under the scheme, liable entities (predominantly electricity retailers and large users of electricity) must surrender a certain number of RECs each year to the Office of the Renewable Energy Regulator (ORER) based on the proportion of electricity they consume from the national grid. A liable entity may create RECs itself, if it complies with the criteria to do so, or it can purchase RECs from a third party which is eligible and accredited under the renewable energy legislation to produce them. Where a liable entity does not have enough certificates to surrender, it is required to pay a renewable energy certificate shortfall charge. Currently, the rate of charge is AUD$40 per MWh or AUD$57 tax adjusted.

Each REC is created in an electronic form approved by the Regulator, containing a unique ID code and other identifying data. A registry has been established by the Regulator to record transfers and creation of RECs. After creating and registering RECs, the renewable energy generator may sell the RECs to third parties.

An independent review in 2003 of the MRET Scheme recommended that the Federal Government maintain the existing target of an additional 9,500 GWh generated from renewable sources by 2010.[13] However, the review also recommended that the target should be increased beyond 2010 at a rate equal to the rate before 2010, to eventually stabilize at 20,000 GWh by 2020. The panel estimated that by the time the Protocol compliance period of 2008 to 2012 was reached, the

[12] Accessible at <http://scaleplus.law.gov.au/html/pasteact/3/3431/pdf/174of2000.pdf> accessed 3 May 2004.

[13] *Renewable Opportunities, A Review of the Operation of the Renewable Energy (Electricity) Act 2000* (Australian Greenhouse Office, 2003), accessible at <http://www.mretreview.gov.au/report/pubs/mret-review.pdf> accessed 3 May 2004.

MRET would have contributed about 10 per cent of Australia's current projected GHG abatement. This new target was seen as a means to further stimulate investment in new renewable energy infrastructure in Australia. At the time of writing, the Federal Government has not yet formally adopted this recommendation.

The independent panel found that renewable energy generation to date had primarily come from the hydroelectric power generation and the installation of domestic solar hot water appliances. There was also strong growth in the Australian wind sector coming from a small base. Generation from biomass had not been as significant as expected prior to the introduction of the MRET, and the measure had only had a marginal influence on the installation of new generation from solar power photovoltaics.

4.2 National Green Power Programme

In addition to the legislation-based emissions trading and renewable energy trading scheme discussed above there also exists in Australia a voluntary 'Green Power' trading programme which established prior to all the existing legislative schemes in 1997 by the NSW Sustainable Energy Development Authority (SEDA).[14] The programme is designed to facilitate the installation of new renewable energy generators, by increasing customer awareness of, and confidence in, renewable energy products. This programme commenced in NSW but has now been introduced nationally. It has no legislative foundation and is not compliance driven. Rather, the Green Power programme has developed as an environmental certification scheme, for energy retailers who wish to differentiate their energy products due to their environmental attributes.

Under the Green Power programme, renewable energy products, which meet the specified accreditation criteria, earn the right to use a Green Power product certification logo, which enables retailers to sell the product at a higher price than electricity derived from the combustion of fossil fuels. The scheme is consumer demand driven and imposes obligations on electricity retailers to ensure renewable energy, attracting the higher tariff structure, is produced from Green Power approved electricity generators.

A 'Green Power Right' is defined under SEDA's Green Power programme as 'the right to claim any eligible Green Power generation (or a portion of generation) from a Green Power generator that has occurred during a defined period, under the Green Power program'.[15] There is no legislative framework in which such rights must be purchased or traded. Rather, participants in the Green Power programme who wish to on-sell SEDA-accredited Green Power to their customers at a premium price, must acquire a number of Green Power Rights equivalent to the megawatt hours of SEDA-accredited Green Power they sell.

[14] <http://www.greenpower.com.au> accessed 28 May 2004.

[15] See, *About the Green Power Accreditation Program*, accessible at <http://www.greenpower.com.au/GPTherules.shtml> accessed 3 May 2004.

One Green Power Right represents the right to sell one MWh of SEDA-accredited Green Power (at a premium to customers who wish to pay for environmentally friendly electricity). Once a Green Power Right is sold, that power cannot be used by a third party to create SEDA-accredited Green Power or RECs. Trades are conducted on the basis of contractual arrangements between the buyer and seller of the Green Power Rights. However, purchasers of Green Power Rights bear the responsibility for ensuring that the rights are valid.

Green Power Rights may be bought or sold at any time by any party, including generators, retailers, and independent traders. Retailers may purchase and on-sell Green Power Rights separately from the associated electricity produced by a Green Power generator, though the rights will only be valid where it can be demonstrated that the electricity to which the Green Power Rights are associated has in fact been generated. Green Power Rights are only valid during the settlement period in which the associated generation occurred, unless the Green Power Programme Manager specifically approves a carryover.

5 State Based Trading Schemes

The Australian Federal Government alone holds the constitutional power to adopt international treaties. However, it is the Australian State Governments that hold the power to make laws for, and possess jurisdiction to enforce compliance over, many key sectors of the Australian economy responsible for a large proportion of the national GHG inventory, such as energy, land ownership and use, planning, infrastructure, and natural resources. Each of the six States and two Territories in Australia has their own democratically elected lawmaking Parliaments. Additionally, the Australian State Governments have a broad plenary power to legislate for the good governance of their jurisdiction.

Australian States do not have constitutional power to enter into international treaties or directly influence Australia's national foreign policy regarding the UNFCCC, the Protocol, or any other international treaty. Notwithstanding this lack of direct jurisdiction the Australian States have all taken an active and vocal role in the national climate change debate and have each introduced a broad range of policies and measures seeking to address climate change issues. Some of these policies, regulations and measures seek to create a form of tradable 'emissions right' consistent with the notion of emissions rights under the Protocol. One emissions trading scheme has been developed at a State level in NSW that is designed to comply with the definitions and structures within the UNFCCC and the Protocol.

State Government initiatives to address climate change include:

(i) A range of State based legislation giving legal recognition to the carbon sequestration properties of forestry plantation activities and creates a range of tradable state based Carbon Sequestration Rights, which can be traded separately from land and forestry right interests;[16]

(ii) A GHG emissions trading scheme established by NSW for electricity retailers and large electricity users;[17]

(iii) A gas trading scheme designed to promote the increased use of gas for electricity generations in the State of Queensland; and[18]

(iv) Other policies, plans and schemes that are consistent with the National Greenhouse Strategy.[19]

Given that the Kyoto Protocol has not yet come into effect, and the prevailing policy position of Australia's Federal Government not to ratify the Protocol, Australian companies cannot directly participate in JI or Clean Development Mechanism (CDM) project activities as defined under Articles 6 and 12 of the Protocol or International Emissions Trading established under Article 17 under the Protocol (see Chapter 13 above for a further discussion of the ability of non-Party companies to participate in the flexible mechanisms).

The State initiatives have created, and will continue to create, local Australian markets for the trade of carbon rights and emission reduction rights. However, notwithstanding the symmetry in design of the State-based policies and carbon sequestration laws to mirror the Protocol requirements, the rights created will not be eligible instruments for trade within the Protocol framework without a change of national policy regarding the ratification of the Kyoto Protocol.

5.1 New South Wales Emissions Trading Scheme

The New South Wales Government has taken a strong stance in support of Australia's accession to the Protocol. New South Wales is currently the only State

[16] Forestry Rights (Amendment) Act 2000 (Vic), Forestry and Land Title Amendment Act 2001 (Qld), Forestry Rights Registration Act 1990 (Tas), Carbon Rights Act 2002 (WA), Tree Plantation Agreements Act 2002 (WA), Acts Amendment (Carbon Rights and Tree Plantation Agreements) Act 2002 (WA), Forest Property Act 2000 (SA), Carbon Rights Legislation Amendment Act 1998 (NSW).

[17] Electricity Supply Amendment (Greenhouse Gas Emission Reduction) Act 2002 (NSW), Electricity Supply (General) Regulation 2001 (NSW), Greenhouse Gas Benchmark Rule (Compliance) No 1 of 2003, Greenhouse Gas Benchmark Rule (Generation) No 2 of 2003, Greenhouse Gas Benchmark Rule (Demand Side Abatement) No 3 of 2003, Greenhouse Gas Benchmark Rule (Large User Abatement Certificates) No 4 of 2003, Greenhouse Gas Benchmark Rule (Carbon Sequestration) No 5 of 2003.

[18] The Queensland 13 per cent Gas Scheme, introduced by the Queensland Government through *Queensland Energy Policy—A Cleaner Energy Strategy*, <http://www.energy.qld.gov.au/gas/13percentgasscheme.asp> accessed 3 May 2004.

[19] For instance, see the Greenhouse Policies of the Australian Capital Territory, the Northern Territory, South Australia and Western Australia implemented pursuant to the National Greenhouse Strategy National Greenhouse Strategy: Strategic Framework for Advancing Australia's Greenhouse Response (release date: 26 November 1998), <http://www.greenhouse.gov.au/pubs/ngs/ngs.html> accessed 3 May 2004.

or Territory to have implemented mandatory GHG emissions reduction legislation targeting the energy sector.

In December 2002, the NSW State Parliament passed legislation which amended the Electricity Supply Act 1995 (NSW) to place an obligation upon electricity retailers and certain entities including large electricity users to meet mandatory targets for the reduction of emission of GHGs from electricity production and use. The amendments were effected through the Electricity Supply Amendment (Greenhouse Gas Emission Reduction) Act 2002 (NSW).[20] The Act and related guidelines, rules and regulations, made mandatory a scheme that had previously been functioning on a voluntary basis. The emissions trading scheme is known as the NSW Greenhouse Gas Abatement Scheme. It commenced operation on 1 January 2003. It will remain in force until 2012.

The two main goals of the new scheme are:

(i) the reduction of GHG emissions associated with the production and use of electricity; and

(ii) the development and encouragement of activities that offset the production of GHG emissions.

These goals are to be met through:

(i) establishing State GHG benchmarks and individual GHG benchmarks for certain participants in the electricity industry and large users of electricity;

(ii) imposing a penalty on relevant entities which fail to meet GHG benchmarks in any year;

(iii) recognizing activities that reduce or promote the reduction of GHG emissions; and

(iv) enabling trading in, and use of, certificates created as a result of those activities for the purpose of meeting GHG benchmarks.

The Act provides a broad legal framework for the operation of the scheme. Most matters of detail, and in particular, specific provisions relating to the creation of tradable Abatement Certificates (ACs), are set out in the Regulations, the Guidelines, and the five Greenhouse Gas Benchmark Rules relating to compliance, generation, demand side abatement, large user abatement, and carbon sequestration.[21]

[20] Electricity Supply Amendment (Greenhouse Gas Emission Reduction) Act 2002 (NSW), <http://www.greenhousegas.nsw.gov.au>. See also, *Overview of the NSW Greenhouse Gas Abatement Scheme*, <http://www.greenhousegas.nsw.gov.au> accessed 3 May 2004.

[21] Electricity Supply Amendment (Greenhouse Gas Emission Reduction) Act 2002 (NSW), Electricity Supply (General) Regulation 2001 (NSW), Greenhouse Gas Benchmark Rule (Compliance) No 1 of 2003, Greenhouse Gas Benchmark Rule (Generation) No 2 of 2003, Greenhouse Gas Benchmark Rule (Demand Side Abatement) No 3 of 2003, Greenhouse Gas Benchmark Rule (Large User Abatement Certificates) No 4 of 2003, Greenhouse Gas Benchmark Rule (Carbon Sequestration) No 5 of 2003.

The Act establishes increasing State GHG benchmarks (in tonnes of carbon dioxide equivalent (CO_2e) per capita) for the years commencing 1 January 2003 and ending 31 December 2012. The following entities are identified as 'benchmark participants' with obligations under the Scheme:

(i) a NSW retail electricity supplier;

(ii) a NSW electricity generator prescribed by the regulations;

(iii) a 'market customer'[22] in respect of an electricity load over a certain size for use in NSW;

(iv) a large customer who has made an election to be subject to a GHG benchmark; and

(v) a person engaged in carrying out 'State significant development'[23] that has made an election to be subject to a GHG benchmark.

Each individual participant's benchmark is calculated by:

(i) multiplying the State population for a year by the State GHG benchmark per head of population to determine the electricity sector benchmark;

(ii) determining the proportion of the total electricity demand in the State that was used by the participant for that year; and

(iii) applying the proportion to the electricity sector benchmark to calculate the number of tonnes of CO_2e that is the participant's benchmark.

The number of tonnes of abatement for which a participant is actually responsible is determined by multiplying the total amount of electricity supplied, purchased or used by the participant in that year by the NSW pool coefficient for GHG emissions, and subtracting from that figure the number of tonnes of GHG emissions actually abated by the participant.

Failure to comply with a compliance target in a given year results in a GHG shortfall and the incurring of a GHG penalty for that year unless the liable entity can produce sufficient ACs to offset the shortfall. The legislation enables a participant to carry forward an amount of GHG shortfall (not exceeding 10 per cent of its GHG benchmark for a year) to the next year.

Retail suppliers of electricity are required to comply with their GHG emissions benchmark and the relevant GHG benchmark rules as a condition of their electricity retailer's licence to operate issued under the Electricity Supply Act 1995 (NSW).

A benchmark participant who fails to meet its GHG benchmark for a year will be liable to pay a penalty of AUD$10.50 per tonne of CO_2e of GHG shortfall for the year concerned.

[22] A market customer means a customer that has classified any of its electricity loads as a market load and that is registered with the Australian electricity regulatory authority as a market customer under the National Electricity Code: see Electricity Supply Amendment (Greenhouse Gas Emission Reduction) Act 2002 (NSW), s 97A.

[23] As determined under the Environmental Planning and Assessment Act 1979 (NSW).

5.1.1 *Creating tradable Abatement Certificates*

The legislation also makes provision for the creation of ACs by accredited Abatement Certificate providers. Each AC represents one tonne of CO_2e of GHG emissions abated.

The regulations and benchmark rules establish details concerning eligibility for accreditation and the activities in respect of which an entity may be accredited, including:

(i) low emission generation of electricity;
(ii) activities that result in reduced consumption of electricity;
(iii) the capture of carbon dioxide and storage in tree biomass in forests;
(iv) activities carried out by elective participants that reduce on-site emissions not directly related to electricity consumption.

Abatement Certificates must be created within six months after the end of the year in which the abatement activity takes place.

5.1.2 *Trading and use of Abatement Certificates*

The legislation allows the creation of an AC to be registered by the Scheme Administrator, thereby bringing the AC into effect. Once an AC has been registered, all subsequent transfers must also be registered with the Scheme Administrator. Registration with the Scheme Administrator is proof of ownership.

Depending on which activity created the ACs, they can be either transferable or non-transferable. Voluntary benchmark participants are only eligible to create non-transferable ACs on-site. These 'Large User Abatement Certificates' can be created through modifications to the manufacturing process, which reduce GHG emissions. They can only be used to meet the benchmark target of the large user who creates them. By creating their own Large User Abatement Certificates, large users can negotiate a reduction in energy prices with electricity retailers to avoid any 'pass through' of costs associated with the electricity retailer's liability under the scheme for the electricity sold to the large user.

Accredited entities eligible to create ACs from other activities (reducing emissions associated with electricity generation, demand side abatement and carbon sequestration) can create transferable ACs, which can be freely traded to third parties on the market.

The scheme includes provision for banking and borrowing of certificates. Banking occurs when a participant has over-complied early in the life of the scheme. That is, a participant is entitled to redeem excess certificates at a later stage. The legislation does not place a limit on how long surplus certificates can be banked. Borrowing, on the other hand, occurs when a retailer fails to fully meet its target in a particular year. Retailers are allowed a shortfall of up to 10 per cent of their target for any year. This shortfall must then be made up in the following year. There are no shortfalls allowed in 2007, so the final target must be met.

The legislation confers strong powers on the tribunal and Scheme Administrator to audit the legitimacy of the certificates created. If an AC is found to be invalid then the liability falls on the creator of the certificate, not on the purchaser. This provides some assurance for purchasers making financial outlay in purchasing ACs. If in such circumstances the creator is unable to meet its liabilities, the legislation provides an ability to gain access to financial assurances (lodged at the time of certificate creation).

5.1.3 State emissions trading schemes: potential future horizons

The NSW Greenhouse Benchmark Scheme is in its infancy. The legislation, regulation and rules, which establish and regulate the scheme, are complex and details set out in the scheme rules continue to be refined.

Notwithstanding some early implementation problems that are bound to arise in regulating such a complex scheme, the NSW Greenhouse Benchmark Scheme is a serious attempt by an Australian State Government to implement an emissions trading regime based closely on methodologies and principles of the Protocol as a means to regulate the State's rising greenhouse emissions trajectory in the energy sector. The scheme provides a high level of detail for the calculation of emissions, abatement, and the creation of a working trading regime and has undergone some fine-tuning to address practical obstacles experienced by participants during the early operation of the scheme. Other Australian States are considering various emissions trading options. A key concern is how emissions trading may assist the objective of encouraging abatement of GHG emissions from the supply and use of energy. A multi-state working group has been established to explore various emissions trading models and options.

5.2 Queensland 13 per cent Gas Scheme

In May 2000, the Queensland Government released the Queensland Energy Policy, which aims to diversify Queensland's energy mix from a large reliance on coal towards greater use of natural gas and renewable energy resources. The policy includes a plan to introduce a permit and licence scheme, which will require electricity retailers to source 15 per cent of the electricity, which they sell from gas-fired, or renewable generation. At least 13 per cent must be sourced from gas-fired plants. This has become known as the 13 per cent Gas Scheme.[24]

The scheme is scheduled to commence on 1 January 2005 and will be in place for 15 years until 31 December 2019. The 13 per cent Gas Scheme was developed with the expectation that a major natural gas pipeline would be constructed from major gas reserves located in Papua New Guinea (PNG Pipeline). At the date this chapter is prepared, the construction of the PNG Pipeline now looks in doubt,

[24] *Queensland Energy Policy—A Cleaner Energy Strategy*, <http://www.energy.qld.gov.au/gas/13percentgasscheme.asp> accessed 3 May 2004.

which will remove a major source of gas from the Queensland market.[25] However, no announcement has been made regarding varying the structure of the 13 per cent Gas Scheme in light of the possible absence of the gas which the PNG Gas Pipeline would have provided to the Queensland electricity generators.

The 13 per cent Gas Scheme is a tradable certificate based scheme. Accredited Parties (generators of eligible gas-fired electricity) may create Gas Electricity Certificates (GECs) which, like RECs and ACs under the Federal and NSW regulatory schemes, have value and can be traded separately to the electricity to which they relate. In order to be eligible to create GECs, a generator must:

(i) register with the Regulator of the Scheme;

(ii) generate electricity from an eligible fuel source (the eligible fuel sources are natural gas, coal seam gas, waste coal mine gas and waste gases associated with petroleum refining);

(iii) generate electricity above its baseline; and

(iv) generate electricity, which supports energy load in Queensland.

Each GEC will be equivalent to 1 MWh of electricity. If large electricity retailers and others that sell electricity to end users connected to a major grid (exceeding 100 MW) fail to source at least 13 per cent of the electricity, which they sell in Queensland, from gas, they must surrender GECs to the Regulator of the scheme.

Once a GEC has been created, it will be valid for only the year in which it is created and the following two calendar years. Following the expiry of this period, the GEC cannot be surrendered to discharge obligations under the scheme and it becomes worthless.

Another aspect of the scheme is the 'Queensland Development Exemption'. This exemption aims to take into account the requirements of very large, energy intensive consumers that provide substantial economic development benefits to the State. An existing or new consumer which has had a total electricity load greater than 750 GWh per year at a single site in Queensland will be eligible to obtain a Queensland Development Exemption for any portion of their electricity load that is not sourced from eligible gas-fired generation. This is subject to consultation with the scheme Coordinator-General regarding the strategic and economic benefits of the project to Queensland.

The scheme also provides a 'Renewable Energy Exemption', whereby parties can obtain an exemption from liability for electricity generation when they can demonstrate that the electricity:

(i) is physically and directly supplied to the end user; and

(ii) is generated from an eligible renewable energy source.

A review of the scheme will be undertaken if significant national or international measures aimed at greenhouse abatement, such as emissions trading, are introduced.

[25] See *PNG to Australia Gas Pipeline Faces New Deadline*, ABC Radio Australia, 31 March 2004, <http://www.abc.net.au/ra/newstories/RANewsStories_1077489.htm> accessed 3 May 2004.

6 Interaction Between Australian Trading Schemes

One of the major issues that has arisen due to the number of discrete trading schemes being developed on State and national levels to reduce Australia's greenhouse signature is the fact that certain activities (for example, renewable energy generation) have the potential to create a number of different assets such as RECs, ACs, and Green Power Rights. A critical issue in development of these schemes is whether the same MWh of renewable energy can 'double dip' to create more than one right (that is, whether the same MWh used to create a REC may also be used to create an Abatement Certificate under the NSW emissions trading scheme and/or a Green Power Right).

Companies with liability under the NSW Benchmark Scheme may count a proportion of any RECs surrendered by the same company to the ORER to comply with the Commonwealth renewable energy scheme, in their calculation of Attributable Emissions for their NSW compliance. This proportion (the Renewable Energy Percentage) is determined yearly by the ORER under the Regulations to the REEA. In contrast, however, the Green Power Programme specifically states that renewable energy purchased to make Green Power sales may not also be used by retailers to meet their Federal MRET obligations. As a result, RECs surrendered to meet Green Power compliance obligations cannot also be used towards NSW greenhouse benchmark obligations.

The Federal Renewable Energy law does not currently make any specific reference to its relationship with the NSW regulatory regime or other State based regimes, and nothing in the Commonwealth's supporting documentation suggests that the Federal laws intended to exclude renewable energy sourced under State based regimes. However, the AGO has indicated that their preferred view is that the state based schemes will be additional to the Federal Renewable Energy law (namely the renewable energy will not be interchangeable) and it should be up to the States to determine whether, and how, their own schemes fit within the national renewable energy regime. It almost goes without saying that the various regimes create an onerous mix of regulatory and compliance issues for businesses operating in Australia across State borders.

As a general contractual issue even where an emissions trading scheme does not deal with the issue of 'double dipping', buyers should be reluctant to acquire rights or entitlements under any scheme which have been used or sold for other purposes as this negates the 'additionality' of these rights which is required under general emissions trading principles.

7 State Based Tradable Carbon Sequestration Rights

Due to Australia's large landmass, it is a prime location for forestry projects and has a great potential for carbon sequestration. As discussed above, the regulation of the

Table 27.1: Australian State's Carbon Sequestration Legislation

State	Title of Act	Status of Act	Nature of Carbon Sequestration Right
New South Wales	Carbon Rights Legislation Amendment Act 1998 (which amended the Conveyancing Act 1919)	In force	Specified as a *profit à prendre* that can be registered at the Land Titles Office. Assignable independently of other rights over land or timber.
South Australia	Forest Property Act 2000 (which will amend the Real Property Act 1886)	In force	Specified as a *profit à prendre* that can be registered at the Land Titles Office. Assignable independently of other rights over land or timber.
Western Australia	Conservation and Land Management Act 1984	In force	The Act provides the right to harvest a crop of trees, including forest produce from the crop, as a *profit à prendre*. Assignable independently of other rights over land. Though not specified in the Act, the Government has indicated that a carbon sequestration right is a *profit à prendre* for the purposes of this Act.
	Carbon Rights Act 2002 Tree Plantation Agreements Act 2002; Acts Amending (Carbon Rights and Tree Plantation Agreements) Act 2002	In force	The legislation will allow carbon sequestration rights to be registered on title as a separate registrable interest in the land. A proprietor of a carbon right has the legal and commercial benefits (and risks) arising from changes to the atmosphere that are caused by carbon sequestration and carbon release occurring in or on land in respect of which the carbon right is registered.
Tasmania	Forestry Rights Registration Act 1990	In force	A subset of a 'forestry right'. A forestry right is deemed to be a *profit à prendre* and can be registered and assigned.
Queensland	Forestry and Land Title Amendment Act 2001 (which amends the Forestry Act 1959 and the Land Title Act 1994)	In force	Right to access, maintain or harvest a 'natural resource product' on land. Natural resource product includes carbon sequestered from trees. Right to natural resource specified as a *profit à prendre*.
Victoria	Forestry Rights (Amendment) Act 2000 (which amends the Forestry Rights Act 1996)	In force.	A subset of a 'forest property right'—'a right to commercially exploit carbon sequestered by trees.' May enter into an agreement to grant a person a carbon sequestration right regarding forest property on land.

forestry sector falls within the jurisdiction of the Australian State rather than Federal Government. Several Australian State Governments have recognized that carbon sequestered in trees may have value additional to, and separate from, the actual trees themselves or the land on which they are grown. Although a liquid trading market in carbon sequestration rights has not emerged, most Australian States now enable the registration of forestry rights (including carbon sequestration rights) on land titles, providing greater certainty for investors in Australian forests for their carbon value.

The Australian States have a long history of developing law and policy at a State level, which is similar to, but not always entirely consistent with, other Australian States. Unfortunately, this has resulted in some administrative burdensome issues by businesses confronted by different regulations in each State in which they operate. One area where this is beginning to be seen in the climate change area is with regard to carbon sequestration in plantation forestry.

In most Australian States, specific legal rights can be created and registered to carbon sequestered from forestry activities. However, each State based methodology deals with the identification, recording and transfer of carbon sequestration rights in slightly different ways. This may affect their enforceability for investors. For example, if the carbon right cannot be registered on the land, the owner of the carbon right may be adversely affected if the owner of the land sells the land (and the forest on it) to a third party.

A very brief summary of each State's carbon sequestration legislation is set out in Table 27.1.

Carbon sequestration rights which are registered on title pursuant to the NSW law above form the basis of the sequestration-related tradable Abatement Certificate under the NSW GHG Benchmark Scheme. Should other Australian States develop their own emissions trading schemes, or should a decision be made for a multi-State scheme to be developed on the recommendation of the working party, it is likely that the above laws will require a degree of harmonization in order to permit uniform rules regarding the property rights foundation of tradable carbon sequestration instruments.

8 Conclusion

Australia is a unique country, vulnerable to the predicted changes in climate over the next fifty years. As a signatory to the UNFCCC it has a commitment to work towards the ultimate objective of the UNFCCC being, 'the stabilization of [GHG] concentrations in the atmosphere that would prevent dangerous anthropogenic interference with the climate system'.[26]

[26] Art 2 UNFCCC.

At the Federal level, it is current Government policy to seek to contribute towards the achievement of this objective outside the Kyoto Protocol framework. The Federal Government has also resisted implementing a national emissions trading scheme under Federal law. The Federal Government has introduced a national renewable energy trading scheme which has been successful in stimulating growth in renewable energy generation capacity in Australia.

The current development of emissions trading regimes in Australia is taking place at a State level. New South Wales introduced its Greenhouse Gas Benchmark Scheme in 2003 targeting the energy sector in that State. Other States are also currently considering emissions trading as a policy tool. Considerable analysis is currently being commissioned to develop an option for an inter-jurisdictional GHG emissions trading scheme allowing trading to occur between Australian States and Territories in the absence of a Federal national emissions trading regime. This is a challenging task as Australian political and legal history is littered with failed attempts by States to agree on issues of national significance.

Notwithstanding the difficult policy and legislative path ahead, the State Governments of Australia are to be admired for their preparedness to seek to lead a national emissions trading response to GHG abatement.

28

Canada's Experience in Emissions Trading and Participating in the Kyoto Mechanism

*Markus Gehring and Bradnee Chambers**

1 Introduction

On 17 December 2002, Canada became the ninety-ninth country to ratify the Kyoto Protocol. Early in the Kyoto negotiations, Canadian policymakers realized that achieving its 6 per cent reduction of greenhouse gas (GHG) emissions from 1990 levels would require a serious commitment—one with potentially significant economic implications on its energy, transport, and manufacturing sectors. With this in mind, an advanced Canadian strategy was to develop experience in cost-effective, market-based systems. Particular emphasis was placed on emissions trading, which could lay the groundwork and prepare Canada for the eventuality of a future comprehensive implementation programme. This timely planning led Canada to create national and provincial voluntary trading programmes such as Greenhouse Gas Emission Reduction Trading (GERT)[1] and Pilot Emission Reduction Trading (PERT)[2] that eventually provided valuable experience and insight for prospective legislation. Given the Canadian planning, pilot phase initiatives, and its overall emphasis on emission trading in the national implementation of its Kyoto commitment, the Canadian case offers a valuable example for other countries seeking to participate in emission trading under the Protocol.

This chapter provides a legal analysis of the Canadian experience on emissions trading from its early days before the final Kyoto regime was complete, up until the ratification of the Protocol. Section 2 of the chapter details some of the pilot work Canada undertook as voluntary schemes. Section 3 examines a Canadian regula-

* Markus Gehring is Lead Counsel and Bradnee Chambers a senior legal research fellow at the Centre for International Sustainable Development Law. The views expressed in this chapter are the views of the authors and do not necessarily represent the views of their employer. The authors would like to thank Hilary Stedwill, Researcher, Centre for International Sustainable Development Law (CISDL), for his invaluable research and Jessica Green for her comments. Any errors are those of the authors alone.

[1] See <http://www.gert.org> accessed 22 March 2004.

[2] See <http://www.cleanaircanada.org> accessed 22 March 2004.

tory example of emission trading in air pollution in Ontario, and some of the early lessons gleaned from this market. Section 4 gives an overview of the proposed Canadian emissions trading legislation and discusses some of the more serious problems it faces for enacting this law. The final section of this case study offers some insights and highlights some of the challenges that Canada must face and eventually overcome if it plans to support the future of the Kyoto process.

2 Early Carbon Emission Reductions and Trading: GERT and PERT

Canadian companies and provincial entities have been among the earliest to engage in voluntary certification and trading of carbon emission reductions.[3] Since the mid-1990s they have participated in a carbon market with commercial deals concerning emission rights that did not fulfil the criteria of any unit defined under the Kyoto Protocol. Most of the voluntary trading took place in the context of two pilot programmes: GERT and PERT.[4]

The legal frameworks for early carbon emission trading in Canada prior to the Kyoto Protocol were public–private partnership programmes. PERT and GERT were both organized jointly by the Federal Government and large energy producers to test and understand the functioning of an emissions trading scheme. Both were based on a memorandum or letter of understanding (MOU) between the federal government, the Ontario Provincial Government, and private enterprises. The basic tenets of the programmes amounted to an agreement by the participating government to award carbon credits for verifiable emissions reductions by the participants and a registry where the transactions could be recorded. The emission reductions earned were intended to comply with future emission obligations that would eventually be imposed by the participating provincial governments. Credits could also be traded or sold by the organizations that created the reductions.[5]

GERT began 1 January 1997; projects were accepted from 3 June 1998, until 31 December 2002. An MOU setting out the objectives, structure, roles of government partners, and the context in which government partners would recognize emission reductions achieved through the project was signed on 13 October 1998 under the auspices of the Government of British Columbia.[6]

[3] R Rosenzweig and M Varilek, *Greenhouse Gas Trading in Canada, Greenhouse Gas Market 2003* (International Emission Trading Association (IETA), Geneva, 2003) 31.

[4] See the homepages of GERT and PERT, n 1 and n 2 above respectively.

[5] BC Government, *Greenhouse Gas Emissions Reduction Trading Program*, <http://www.gert.org> accessed 22 March 2004.

[6] GERT governmental members: Environment Canada, Natural Resources Canada, BC Ministry of Water, Lands and Air Protection, BC Ministry of Energy and Mines, Alberta Environment, Alberta Department of Resource Development, Saskatchewan Energy & Mines, Manitoba Department of Energy and Mines, Quebec Ministry of Natural Resources, Nova Scotia Department of Natural Resources, Greater Vancouver Regional District and non-governmental members: Canada's Climate

The GERT pilot was specifically intended to provide practical experience for the participants, to assess the environmental and economic benefits of emissions trading, and to test technical, administrative, and legal elements of an Emission Trading Scheme (ETS). It also tried to maximize involvement of the private sector and encourage investment in GHG reduction projects and to build foundations for a possible future obligatory ETS.

GERT projects had a steering committee comprised of a representative from each partner, a pilot manager and a technical committee, a multi-stakeholder body to review projects and trades in order to assess whether they result in actual emission reductions that are measurable, verifiable and over and above what is required by law, and follow the other rules established by the GERT Pilot. The MOU contained the obligations for government partners to credit, to the extent possible, verifiable emission reductions from trades registered under GERT against future emission obligations. However, since there was hardly any trading and the private sector was very reluctant to engage, this obligation has de facto become obsolete.[7] The final report evaluating GERT pointed out that the overall legal uncertainty constituted a barrier to trading, as did the lack of an incentive for the participants.[8] As such the main experience for the project members was of a technical administrative nature. Participants also discussed legal issues around emission reduction projects.[9] One of the main outcomes of these discussions was that ownership of emission rights under the system required a concrete definition so as to minimize ownership related risks.

On the question of ownership, which might be particularly relevant to the future ETS in Canada, the GERT approach was unique because there were no legal rules, case law, or government policy. In the term sheet, which forms the basis for any transaction, ownership of the emission rights was to be a mandatory criterion, to be determined to the best of the abilities of the reviewer/s and transacting parties, and to be warranted by the seller. Although GERT was acting as a reviewer of the transactions, it had no mandate or capacity to adjudicate or function as an arbitrator in case of doubtful ownership. This 'buyer beware' approach was regarded as successful.[10] In the project reviews, the Technical Committee made no

Change, Voluntary Challenge and Registry Inc, Canadian Association of Petroleum Producers, Canadian Electricity Association, Canadian Energy Pipeline Association, Canadian Gas Association, Canadian Pulp and Paper Association, Canadian Wind Energy Association, Greenhouse Gas Emissions Management Consortium, Pembina Institute for Appropriate Development, West Coast Environmental Law Association.

[7] Only five projects underwent GERT review, see <http://www.gert.org/projects/index.htm>.

[8] *Greenhouse Gas Emission Reduction Trading Pilot (GERT)—Final Report*, <http://www.gert.org/Final/020810%20GERT%20Final%20Report.pdf>, 18.

[9] *Outcomes from Climate Change Central's Greenhouse Gas Emissions Trading Workshop* (Calgary, Alberta, 21 February 2002), prepared by R Ezekial and J Harris, Fasken Martineau, Vancouver, BC. This report was co-sponsored by GERT and the federal-provincial Domestic Emission Trading Working Group of the National Air Issues Coordinating Committee—Climate Change and is available at <http://www.fasken.com> accessed 22 March 2004.

[10] *Greenhouse Gas Emission Reduction Trading Pilot (GERT)—Final Report*, 15.

determinations as to who owned the emission reductions, and GERT expected that the project proponent would provide some information supporting its claim of ownership. Also, the GERT Pilot Rules provided that if the Technical Committee became aware of what it considered to be an obvious or significant ownership issue, it could delay or refuse to proceed with the review of a project until the matter was resolved to the satisfaction of the disputing parties. While this approach served GERT adequately, the evaluation of the pilot programme recommended that these issues would need to be addressed in the future Canadian ETS.[11]

In addition, the evaluation recommended that contract negotiations in the future should include issues concerning surpluses, that is, the reduction that exceeds what was expected by the buyer and seller, *force majeure*, vintage[12] and international requirements.

The PERT Project was formed in 1996. The key element was to form an industry-led, multi-stakeholder group to set the standard for the development and implementation of a robust emission reduction trading system. PERT was originally designed to reduce smog and acid rain emissions of nitric oxide (NO_x) and sulphur dioxide (SO_2), but as climate change concerns grew, it was extended to include CO_2 emissions. PERT began to review and register CO_2 emission reductions in 1998, and later, the Ontario Ministry of Environment became a partner to the project. A Letter of Understanding[13] was negotiated in order to provide an industry incentive for early action and contains a similar regulation about recognition of credits to the one of the GERT project.[14] CleanAir Canada,[15] which PERT eventually morphed into, is currently seeking to renew the Letter of Understanding to safeguard early carbon reductions for the future Canadian ETS.

The emission reduction goal, or GHG emission cap, in the PERT system turned out to be key for the functioning of the trading. It was the cap that provided the incentive and stimulated the market forces.[16] Since the Federal Government announced that future GHG emissions reduction credit and emission right trading in Canada will be part of a legal regime, efforts to create more voluntary programmes have ceased, and there are doubts as to whether voluntary efforts can be integrated into the future trading system.[17]

Most GHG emission reduction credits (ERC) purchased by Canadian companies were generated by projects located on the North American continent. Non-compliance reductions units have traded for approximately US$0.60 to $1.50 per tonne of carbon dioxide equivalent (CO_2e). While most transactions involved outright purchases of reductions, some involved financial derivatives such as 'call

[11] *Greenhouse Gas Emission Reduction Trading Pilot (GERT)—Final Report*, 20.

[12] Vintage means the exact amount of reductions realized after the project has started.

[13] See <http://www.cleanaircanada.org/notes/letterofunderstanding.pdf>.

[14] See n 2 above.

[15] A new not-for-profit, federally incorporated organization, see <http://www.cleanaircanada.org> accessed 22 March 2004.

[16] GE Taylor, *Role of Emissions Trading in Canada's Kyoto Climate Change Plan, Greenhouse Gas Market 2003* (IETA, Geneva, 2003) 36.

[17] ibid 37.

options', whereby a company buys the right, but not the obligation to purchase a specific quantity of emission reduction credits at a fixed price in the future.[18] All these transactions were based on normal private purchase contracts mainly under common law.

Both GERT and PERT have been largely superseded by the emerging Kyoto Protocol regime. Yet at the time, their approaches were welcomed to respond to the pressure within the Kyoto negotiations to develop real-life projects, and build a body of expertise with respect to issues of verification, additionality and ownership of reductions.[19]

3 Canadian Regulatory Examples for Emission Trading[20]

Ontario's ETS is the first provincial air emissions trading programme in Canada and it is expected that the future Canadian carbon ETS will be informed by the implementation of this flexible mechanism. The scheme includes the Emissions Trading Code, which introduces the use of market-based instruments for reducing emissions of NO_x and SO_2 in the province.[21] The Code is intended to supplement Ontario Regulation 397/01, which governs emissions trading under the Ontario Environmental Protection Act.[22] Emissions trading officially began on 31 December 2001.

The Ontario system is designed as a 'cap, credit and trade' variety, which combines 'cap-and-trade' features with those of a 'baseline-and-credit' system.[23] As such, it offers some ideas of how programmes based on *allowances* can fit together with programmes based on *credits*, as well as the challenges that the linking of the different type of systems implies.

Currently, only the six fossil fuel-fired power plants of the provincially-owned Ontario Power Generation (OPG) are currently included in the scheme; however, beginning in 2004, the Government plans to expand the system to include other emitters, such as electricity generators as well as major industrial emitters (that is, pulp and paper sector, cement and concrete manufacturers, iron and steel

[18] R Rosenzweig and M Varilek, *Greenhouse Gas Trading in Canada, Greenhouse Gas Market 2003* (IETA, Geneva, 2003) 31.

[19] See Taylor, n 16 above, 36.

[20] This part shares thoughts with M Gehring and C Streck, 'Emissions Trading—Legal Nature, Title, Transfer and Taxation of Emission Allowances and Credits' forthcoming in the Tulane Journal of Environmental Law.

[21] Future considerations include adding GHGs such as CO_2 to the list, as well as toxic and cancer-causing compounds such as mercury, cadmium, and arsenic.

[22] The regulation can be found at <http://www.ene.gov.on.ca/envision/env_reg/er/documents/2001/RA01E0020-A.pdf>; technical descriptions of NO_x/SO_2 emission reductions: <http://www.ene.gov.on.ca/envision/env_reg/er/documents/2001/RA01E0020-C.pdf> and the Ontario Emissions Trading Code: <http://www.ene.gov.on.ca/envision/env_reg/er/documents/2003/XA03E0001.pdf>.

[23] The UK Carbon Trading Scheme also includes a baseline-and credit system which complements the cap-and-trade system.

manufacturers, petroleum refineries, chemical manufacturers, and non-iron smelters).

During the first phase of the programme, the Ontario Ministry of the Environment was responsible for allocating NO_x and SO_2 emission allowances to OPG, which then distributed these allowances among their stations, free of charge. Beginning in 2004 for SO_2, and in 2008 for NO_x, all electricity generators in the province will apply individually for allowance allocations based on their production estimates for the coming year. With a cap on the total number of emissions, allowances will be allocated by the Ontario Government to these additional generators at the end of the year based on each station's electricity production during the year.

Unlike the allowances, which are emissions rights assigned by the provincial government, ERCs are generated by the firms themselves. In order to obtain ERCs for use or trade, companies must implement an emission reduction project in accordance with approved standard methods (SM). Companies can either use one of the pre-approved SMs,[24] or they can propose a new method, which must be approved by the Ministry. This approach is intended to enable the Ministry to ensure that proponents' projects are consistently evaluated, and that the process is predictable and transparent. Both types of credits can be used for compliance purposes.

A company may only use a limited amount of ERCs. The Regulation sets the ratio between allowances and credits in the following way: the maximum allowable use of ERCs to an emitter will be limited to 33 per cent of allowances used for NO_x and 10 per cent of allowances used for SO_2. In addition, there will be a 10 per cent discount of all retired (used) ERCs for the benefit of the environment. In other words not more than 33 per cent of the NO_x obligation can be met by created ERCs and not more than 10 per cent of the SO_2. This limitation was set to ensure that the local air quality actually benefits from the system. It still leaves 90 per cent of the generated ERCs to be used by emitters to meet compliance obligations.[25] So a new SM would need to be approved by the Ministry, but would then generate a set of credits for the firm in question. Ten per cent of the emissions credits within the project would be retired, while the rest could be traded or sold on the SO_2 and NO_x markets.

[24] For example, the installation of low NO_x burners on electricity generators. These are designed to reduce NO production in boilers by controlling air/fuel mixing to affect combustion reaction.

[25] The provision reads rather complicated: s 26 of the Ontario Regulation on emissions trading: Limits on retirement of emission reduction credits—ratio of credits to allowances limits:

'(1) The owner of a facility may not retire an amount of nitric oxide emission reduction credits in a year for the purpose of subsection 20(4) [use for compliance] unless the following statement is true: $(A - B) \times 0.9 = C \times 0.33$ where, A = the total amount of nitric oxide emission reduction credits that are retired in the year for the purpose of subsection 20(4), B = the amount of nitric oxide emitted from the facility in the previous year because of reliability must run contracts as defined in the market rules or directions given by the IMO under the authority of the market rules, C = the total amount of nitric oxide emission allowances that are retired in the year for the purpose of subsection 20(4).'

3.1 Legal Nature of Allowances and Credits

A rather circular definition of allowances and credits for the Ontario scheme is found in section 1 of the Emissions Trading Regulation (Regulation 397/01):

Definitions—In this Regulation, [. . .]

'emission allowances' means nitric oxide emission allowances or sulphur dioxide emission allowances;

'emission reduction credits' means nitric oxide emission reduction credits or sulphur dioxide emission reduction credits [. . .][26]

The combination of an allowance cap-and-trade and a baseline-and-credit system is particularly interesting for our analysis. The use of credits to fulfil emission reduction obligations is innovative, as the two systems are normally kept separate because usually the cap-and-trade system does not involve elements of an offset system. The goal of air quality protection clearly remains at the heart of the Ontario scheme; the likely result of massive credit creation would mean a substantial increase in the number of energy saving projects, but also substantially lessen the desired improvement of air quality.

3.2 International Transfers

The Ontario legislature has limited the creation of credits in foreign countries by recognizing only those credits created through approved activities using standard methods and only those created in Ontario or the Ozone Transport Commission (OTC)[27] states.[28] Though limited, there seems to be a principle emerging that only emission credits from countries or sub-national entities with similar laws in place will be recognized. This is highly relevant for the shape of the future carbon emission trading scheme: If the Kyoto Protocol does not enter into force, reciprocity might provide the necessary framework for international emissions trading.

However, international transfers have raised some issues that are yet unresolved. Though not explicitly stated in the Ontario Regulation, there seems to be some understanding that after the relatively complicated approval process, the owner of the reduction project holds title to these credits. The restriction on the US states of origin, preventing them from being permitted to generate recognized credits, is currently being challenged in an ongoing court case. An applicant is trying to receive recognition of very significant NO_x reductions in the form of credits from a plant in a US state that is not part of the OTC.[29] Relying on section 18(3) of the

[26] Ontario Regulation made under the Environmental Protection Act—Emissions Trading, at <http://www.ene.gov.on.ca/envision/env_reg/er/documents/2001/RA01E0020-A.pdf>.

[27] The OTC is comprised of the states of Maine, New Hampshire, Vermont, Massachusetts, Connecticut, Rhode Island, New York, New Jersey, Pennsylvania, Maryland, Delaware, the northern counties of Virginia, and the District of Columbia

[28] See 42 USC §7651g(i).

[29] See n 27 above.

Regulation, the applicant submitted that the Ontario Government cannot refuse to recognize the credits simply because they were created outside the preferred areas:

> Despite subsection (2), emission reduction credits may be created as a result of an emission reduction project undertaken outside the areas referred to in subsection (2) [all OTC States] if the Director is satisfied that reductions of nitric oxide emissions or sulphur dioxide emissions achieved by the project have a measurable effect in Ontario.[30]

However, rather than questioning the alleged effects in Ontario, the Ministry disputed the ownership of the credits under section 18(1) and (2). Section 18 states that:

> (1) Subject to subsections (2) and (3), a person may create emission reduction credits in accordance with the Ontario Emissions Trading Code

and

> (2) Emission reduction credits may be created only as a result of an emission reduction project undertaken in Ontario the District of Columbia or one of the following states: 1. Delaware. 2. Illinois. 3. Indiana. 4. Kentucky. 5. Maryland. 6. Michigan. 7. New Jersey. 8. New York. 9. Ohio. 10. Pennsylvania. 11. West Virginia. 12. Wisconsin.[31]

The Ontario Ministry of the environment raised the question of who actually *created* the credits—the owner of the facility, the applicant or somebody else. The case has not yet been resolved but it appears that the applicant will likely turn out to be the 'creator' of the new ERCs. The mere interest of the applicant in 'creating' credits for the purpose of the Ontario regulation can probably not be sufficient. This is an interesting development because the ownership of carbon credits is also open to some definitional difficulties, and these would certainly be invoked by traders.

Both allowances and ERCs can be traded freely internationally, within the limits of certain specific restrictions applying mainly to the credits. First, the use of ERCs to meet emissions obligations is limited to those whose emissions originated in Ontario, the twelve key States named in the Ozone Annex[32] or the District of Columbia. Secondly, credits can only be claimed for a twelve-month accumulation period, and allowances and ERCs may only be used once. Thirdly, while unused allowances and ERCs may be banked for future use, they are subject to a discount factor.

The Ontario Emissions Trading Regulation also contains rules for the Emissions Trading Registry.[33]

Box 1. Ontario Emissions Trading Registry Functions

- Provides notice to the public of the distribution of NO and SO_2 Emission Allowances;
- Provides notice of the application for and the creation of NO and SO_2 ERCs;
- Provides public access to all documentation submitted in support of an application to create an ERC; and
- Records decisions about credit creation and credit and allowance retirement.

[30] See n 26 above. [31] ibid. [32] See above under NO_x Budget Emission Trading.

[33] See Ontario Ministry of the Environment and Energy website <http://www.ene.gov.on.ca>.

There are some discussions as to whether to expand this existing and functioning registry for GHG emissions, but no definite decision has been taken.[34]

4 Kyoto Ratification

After contentious debates in the House of Commons and an open rift on the issue with some of the oil rich provinces such as Alberta, the Canadian Prime Minister Jean Chrétien made what was perceived by many to be a surprise commitment on the eve of the World Summit on Sustainable Development to ratify the Kyoto Protocol. Roughly five months later the Protocol became national law in Canada.[35] But despite the ratification, the implementation of the Protocol presents a number of challenges that will still need to be resolved. The following section explores some of these issues in the context of emissions trading.

4.1 Constitutional Questions and Federalism

The first of these challenges was that Canada's ratification of the Kyoto Protocol raised questions under the Canadian Constitution. The level of government responsibility for the implementation of the Protocol needs to be determined, particularly with regard to the future carbon emission trading scheme. The Federal Government's authority to sign and bind Canada to international treaty obligations is well established and uncontested. Despite this authority, the Federal Government is not necessarily able to implement the treaties it signs where they contain obligations in areas of provincial jurisdiction.[36] Consequently, there is a need to determine whether the jurisdiction for any proposed legal regime to create an effective and efficient ETS is federal, provincial, or requires their cooperation.

Jurisdiction over environmental matters was never assigned to any particular level of government. For a domestic carbon emission trading scheme there are several aspects that the Federal Government is able to legislate and regulate: it can pass criminal laws, taxation laws, trade and commerce laws, and laws to protect the peace, order and good government of Canada.[37] Provincial governments have competence and authority to pass laws affecting property and civil rights, local matters, direct taxes, intra-provincial undertakings and forest resources.[38] The dominant theory seems to be that a domestic emissions credit trading system

[34] On the new Federal Canadian Law, see below.

[35] Ratification 17 December 2002, see <http://www.treaty-accord.gc.ca/Treaties_CLF/Details.asp?Treaty_ID=103129> accessed 18 April 2004.

[36] *AG Canada v AG Ontario (Re: Labour Conventions)* [1937] 1 DLR 673 (JC). While this decision is now more than sixty years old and has been criticized, it remains the leading decision on the matter of federal treaty implementation activities in provincial jurisdiction.

[37] Constitution Act, 1867 (UK) 30 & 31 Vict. c 3, s 91.

[38] ibid, s 92. Note also that Canadian municipal governments enjoy whatever powers are delegated to them from provincial governments.

would be considered a matter of 'national concern', for which the Federal Government would be justified in passing laws for the peace, order and good government of Canada. Recently, the Federal Government has announced that it will seek cooperation with provincial and territorial governments.[39]

The leading case on questions of jurisdiction to legislate environmental matters is the Supreme Court of Canada decision, *R v Crown Zellerbach Canada*,[40] in which the bench found (albeit by the smallest majority), that dumping pollution into *marine* waters was a 'national concern'; therefore the Federal Government should be allowed to exercise its powers for the peace, order, and good government of Canada. The majority explained that the matter to be regulated must either (1) be a matter that did not exist at confederation, or be a local matter that has grown to national proportions, or (2) have 'a singleness, distinctiveness and indivisibility that clearly distinguishes it from matters of provincial concern and a scale of impact on provincial jurisdiction that is reconcilable with the fundamental distribution of legislative power . . . '[41] The Court felt the difficulty of provinces to control impacts on other provinces effectively was a strong indicator of a matter being of 'national concern'.

Despite this strong precedent in Canadian constitutional law, the bench in *R v Crown Zellerbach Canada* nevertheless leaves room for debate. It is difficult to determine the boundaries to delimit the extent to which federal activities can encroach upon otherwise provincial matters in dealing with a national concern. Thus, some have argued that federal laws affecting energy efficiency standards and controlling actual GHG emissions would be constitutional, while regulating urban growth, forests on provincial lands, and domestic transit policies would not.[42] This concern probably explains the proposal by the Ministry of Natural Resources that the Federal Government will only keep account of emissions credits, but otherwise not regulate contracts between private parties.

Presently, there are provincial regulations that would implement the obligations under the Kyoto Protocol, although some provinces may be heading that way. The provinces of British Columbia, Quebec, and Alberta have expressed some dissatisfaction with federal plans, which could drive them to create provincial systems. Ontario experimented with credit systems to reduce emissions of other pollutants with some success. British Columbia and Quebec invested heavily in hydroelectric projects, which have much lower GHG emission intensities than coal, oil, and gas power plants in other provinces. Both provinces would therefore prefer a federal emission trading scheme and object to 'grandfathering' of existing carbon emitting facilities, since this could negate the competitive advantage these provinces have gained through the already successful reduction of their GHG emissions. They

[39] See Governor General, Speech from the Throne, 2 February 2004, at <http://www.pm.gc.ca/grfx/docs/sft_fe2004_e.pdf>.

[40] [1988] 1 SCR 401.

[41] ibid, 433.

[42] C Rolfe, *Turning Down the Heat: Emissions Trading and Canadian Implementation of the Kyoto Protocol* (West Coast Environmental Law Research Foundation, Vancouver, 1998) 353.

advocate an ETS which reflects their advantage.[43] Alberta, on the other hand, has relied heavily on the development of fossil fuels to drive its economy; its tar sands projects promise to increase the province's overall GHG emissions in the near future. Alberta sees the federal plan as a threat to its economy. Rather than lobby the Federal Government, Alberta drafted legislation (still pending) to reduce its GHG emissions by 50 per cent of 1990 levels by 2020 using a unique 'Alberta-centric' method of defining credits and eligible offsets for credits.[44] Ontario has not dissented about federal plans, because the federal plan draws lessons from the significant experience with emission trading schemes.[45]

4.2 Carbon Emission Trading in the Climate Change Plan for Canada

On 21 November 2002 the Government of Canada released the Climate Change Plan for Canada.[46]

It proposes a number of voluntary measures in areas such as transportation or renewable energies, mostly on an incentive and information basis; there is very little in terms of mandatory requirements. The Plan views the Federal Government as a promoter of best practices; it has outlined a so-called 'house in order' approach, whereby the eleven departments and agencies which account for 95 per cent of federal emissions have agreed to collectively meet a target of reducing GHG emissions within their operations by 31 per cent from 1990 levels to 2010.[47]

For large 'final emitters', covenants and emission trading are key elements of the Climate Change Plan for Canada. Covenants are voluntary agreements with industry whereby specific carbon reduction targets and measures are agreed. Another pillar of the plan calls for a future carbon ETS, which is seen as a market-based instrument with 'tremendous potential to minimize the cost of meeting Canada's climate change objectives by exploiting the efficiencies of markets while encouraging innovation and strategic investments.'[48] The plan proposes that companies would be able to meet their reduction commitments through emission reductions in their domestic installations, other domestic emission reductions or international permits. The Plan proposes therefore to link the future domestic trading scheme with the international carbon market that is being created under the Kyoto Protocol. To meet this end the Federal Government commits itself to cooperation with industry, provinces, territories, and stakeholders.[49]

Despite many concessions, the Climate Change Plan has received lukewarm reviews from industry. The Federal Government made concessions to the petroleum

[43] G Taylor, 'Emissions Trading in Canada's Kyoto Climate Change Plan' in IETA (eds), *Greenhouse Gas Market 2003*, <http://www.ieta.org/Documents/New_Documents/ggm_2003.pdf>.

[44] ibid.

[45] See section 3 above.

[46] Climate Change Plan for Canada, <http://www.climatechange.gc.ca/plan_for_canada/plan/pdf/full_version.pdf>.

[47] Federal House in Order Initiative, <http://www.fhio-ifppe.gc.ca>.

[48] Climate Change Plan for Canada, 17.

[49] ibid.

industry, promising to keep the reduction commitments in that sector to no more than 15 per cent below the 'business as usual' baseline during the first commitment period of the Kyoto Protocol.[50] Furthermore, in response to industry lobbying, the federal government also ensured that the costs of GHG credits at CAN$15 per tonne after industry lobbying.[51] Natural Resources Canada (NRCan) reports that industry favours an 'intensity approach' to emissions credits by which credits are measured as reductions in emissions relative to units of production.[52] Industry also lobbied for ' . . . quickly eliminating . . . the regulatory uncertainty it faces under the large final emitters program in order to protect its continued access to capital.'[53] In response, NRCan is proposing the use of 'covenants' where industry sectors can negotiate terms with the federal government to deviate from GHG credit legislation and regulation requirements.[54] The department highlights that, 'the reasons for negotiating a covenant should be unambiguous and the implications of the resulting derogations [from any legislation] need to be clearly demonstrated.'[55]

4.3 Principles for the Canadian Carbon Market

This cooperative approach led to an agreement between industry representatives and NRCan in October 2003 on Principles for the Canadian Carbon Market.[56] They agreed that twelve principles should guide the future domestic carbon emission trading system. These principles are regarded as important tools to reduce GHG emission reductions by promoting price discovery, rewarding innovation, supporting competitiveness and providing incentives for low emission technologies.[57] Canadian industry members of the International Emissions Trading Association formed a Canadian Working Group on the Carbon Market with which the Federal Government is closely cooperating.

These twelve principles comprise legal, economic, and political tenets. The first principle is sustainability, and requires that carbon emission reductions comply with environmental integrity, economic efficiency and social equity. The next principles commit the future system to be linked to systems in other countries that ratified the Kyoto Protocol and to be consistent with the international commitments under the Protocol. Certainty in the domestic system is regarded important to minimize disruptions in the marketplace, while still being flexible enough for international changes. It is expected that domestic reductions instrument will be fully fungible with Kyoto units.

50 See Taylor, n 43 above, 36.

51 ibid.

52 NRCan, *Overview of Legislation*, <http://www.nrcan-rncan.gc.ca/lfeg-ggef/English/ol_en.pdf>.

53 NRCan, *Variation by Covenant*, <http://www.nrcan-rncan.gc.ca/lfeg-ggef/English/var_covenant_en.pdf>.

54 ibid.

55 ibid.

56 See Principles for the Canadian Carbon Market, <http://www.nrcan.gc.ca/media/newsreleases/2003/200389a_e.htm>.

57 ibid.

Other principles provide for transparency, liquidity, and low transaction costs. For example, one principle calls for a single carbon market in Canada, which should be established without domestic barriers to trading such as between provinces. Banking of emission rights should also be permitted without discount within the first Kyoto commitment period.[58] The roles of the private sector and of Government are defined by the principles; the Government's role is confined to providing the framework for trading, including creating the instrument and the registry. The private sector is responsible for the trading itself, using the necessary infrastructure.

While the principles seem relatively basic, the fact that they received governmental and industry approval makes them particularly relevant to the future emission trading scheme. The principles draw upon earlier experience from the pilots and the Ontario emission trading system, and commit Canada to an international carbon emission trading scheme which will likely function on the basis of reciprocity or in full compliance with the Marrakesh commitments.

4.4 Proposed Legislation

Federal legislation on large emitters, either as part of the Canadian Environmental Protection Act or as part of a Climate Change Implementation Act, must be drafted to ensure that the covenants operate effectively, and to ensure that there are incentives and penalties for compliance. The legislation will authorize the Canadian Government to set permitted levels of emissions as well as measurement, reporting, and compliance obligations for the regulations and covenants. Thus large final emitters[59] will be able to opt for negotiations or accept the emission reduction targets set out in the regulations. There will be financial penalties and/or other measures to induce adjustment in a firm's or sector's GHG emissions. Both types of companies, having agreed to a covenant or not, will have equal rights to participate in the emission trading scheme. Both will also be eligible for measures preventing high permit prices. Penalties for non-compliance will also be comparable.

The proposed legislation would include the following key elements:[60] large final emitters will have to quantify their GHG emissions, report them, and will be prohibited from emitting without a permit. Some permits[61] (85 per cent of forecast of 2010 emissions) will be provided free of charge. The rest will most likely have to be created through reductions or bought. Each large final emitter will need to remit an amount of permits equal to their actual GHG emissions.

[58] That is, the 'storage' of credits to use them in future years for compliance.

[59] See NRCan, n 52 above, 2: 'In this context, an emitter is a company with activities that generate GHG emissions in large industrial sectors that are defined on the basis of having emissions that meet the following two criteria: average annual greenhouse gas emissions per facility of 8 CO_2e kilotonnes or more; and average annual emissions per $1,000 output of 20 CO_2e kilogrammes (kg) or more. Large final emitters consist of companies active in three sectors: thermal electricity, oil and gas, and mining and manufacturing.'

[60] See NRCan, n 52 above.

[61] Or allowances; the terminology has not yet been decided but the proposal uses permit while other systems use the term 'allowance'.

NRCan also proposed specific legal obligations: large final emitters are required to measure or quantify their emissions of GHG and associate output in accordance with quantification protocols established by regulations.[62]

The legislation will also be binding on governmental entities that are considered such as public electricity facilities. The compliance period coincides with the compliance period defined under the Kyoto Protocol: it will start on 1 January 2008 and then be assessed periodically.[63] Per industry's request, the Government agreed to take an emission intensity approach in defining emission reduction targets, thus counting emissions per unit of output. The targets will be progressively tightened over time. Not yet decided is whether to take an ex ante distribution approach to encourage reductions. The emitter would receive a portion of the first allocation at the beginning of the year, and only later would it receive the rest of its allocated amount.

It is foreseen that the terms of the regulations would be varied by negotiated covenants between the Government and industry. Thus they would take into account concerns about competitiveness and early action for each sector. The flexibility provided by the covenant option means that the regulations themselves will be kept 'short, simple and transparent.'[64]

The Natural Resources Minister will have the authority to sign covenants with industry in which the emission intensity target or compliance periods could be adjusted.[65] Emitters that take early action should not be disadvantaged; this provision could be reflected in covenants. As agreed in the principles, the Government does not see any role for itself in the trading besides operating the registry, setting targets, and distributing the permits. As such, it could be forced to prevent international transactions because it is required to hold 90 per cent of its initial Assigned Amount Units (AAUs) in their national registry.[66] The covenant approach will also be taken in the provinces and territories. The Federal Minister will be able to enter into agreements with those territorial entities and ask for suspension of federal regulation in areas where agreement has been reached.[67]

The respect of commercially confidential information needs to be balanced with the public interest in disclosure of carbon emissions. Like in most areas of emission reporting, the interest of private companies not to reveal their emissions in order to protect, for example, trade secrets needs to balanced with the public interest in disclosure. Non-compliance with the law will be prosecuted. Minor breaches of its covenant could be permitted, whereby the covenant remains in force. However, a company that breaches its covenant in a serious way will be subject to the provisions in the legislation, and its covenant will be terminated to ensure consistency of treatment between the two regimes.[68]

[62] NRCan, *Legal Obligations* <http://www.nrcan-rncan.gc.ca/lfeg-ggef/English/legal_obligations_en.pdf>.

[63] ibid.

[64] ibid.

[65] See NRCan, n 53 above.

[66] Annex 6 of Decision 18/CP.7 UNFCCC.

[67] Federal and Provincial Governments Sign Climate Change Co-Operation Agreement Winnipeg, 19 March 2004, <http://www.climatechange.gc.ca/english/newsroom/2004/20040319_manitoba.asp>.

[68] ibid.

The definition of emission permits proposed by the Federal Government is similar to Ontario's ETS definition: 'An emissions permit should be characterized as an authorization to emit one tonne of carbon dioxide equivalent greenhouse gas'.[69] As agreed in the principles, the legislation will state explicitly that an emissions permit is transferable. In addition, the legislation proposes a provision making clear that the emitter does not have a property right in the emissions permit or in its annual allocations vis-à-vis the federal government.[70]

The permits distributed through the *gratis* allocation method could be defined as Kyoto Units, including AAUs or as a 'national carbon currency' as in Europe. In the first case this would mean AAUs and permits would be identical. The decision to classify them as such has not yet been taken, and depends on the analysis of the implications of Kyoto Protocol requirements and whether there are concerns related to delays in Canada's access to its AAUs, since Canada will receive its AAUs only after the UNFCCC review. The approval of the calculation of Canada's assigned amount and national inventory system, which will be linked to the numerous issues related to the establishment of our first commitment period allocation of Kyoto units (such as the level of sinks) will be required.[71]

All permits will be registered in the Canadian registry. Banking of unused permits between large final emitters compliance years should be automatic within the first compliance period and, if there is an ex ante distribution of permits, borrowing will not be allowed.[72] The permits will be transferable from one party's account to another. The transfer of permits identified by their serial number requires governmental notification to be completed. This notification shall also serve as a confirmation of the validity of a permit.

There shall also be a de minimis threshold emissions of 20 or 50 tonnes across all industry sectors. De minimis status will be calculated on the total company emissions from all sectors.[73] Government programmes will address carbon emissions from these companies with reduction incentive programmes or voluntary participation in the emission trading scheme.

The Government shall also have the authority to issue price assurance mechanism units.[74] These units would sell for CDN$15 during the first commitment period. They could be used for compliance but not for trade or banking. That

[69] NRCan, *Permit Definition* <http://www.nrcan-rncan.gc.ca/lfeg-ggef/English/permit_definition_en. pdf>.

[70] This is very similar to § 403(f) of the 1990 amendments to the CAA , 42 USC 7651b (Title 42: The Public Health and Welfare, Chapter 85: Air Pollution Prevention and Control, Subchapter IV-A: Acid Deposition Control, sec. 7651b: Sulfur dioxide allowance program for existing and new units): 'An allowance allocated under this subchapter is a limited authorization to emit sulfur dioxide in accordance with the provisions of this subchapter. Such allowance does not constitute a property right. Nothing in this subchapter or in any other provision of law shall be construed to limit the authority of the United States to terminate or limit such authorization . . . '.

[71] See NRCan, n 69 above.

[72] NRCan, *Canada's Emission Trading System* <http://www.nrcan-rncan.gc.ca/lfegggef/English/emission_trading_en.pdf>.

[73] NRCan, *De Minimis* < http://www.nrcan-rncan.gc.ca/lfeg-ggef/English/deminimis_en.pdf>.

[74] NRCan, *The Price Assurance Mechanism* < http://www.nrcan-rncan.gc.ca/lfeg-ggef/English/price_assurance_en.pdf>.

means that companies could apply to be assigned these PAM units to meet their obligations. The challenge will be to reduce the Government's contingent liability, that is, PAM units would have to be backed up with AAUs or other Kyoto credits by the Government. There are some serious questions as to whether these additional units could distort the market especially if it is linked to other trading schemes. The current proposal tries to mitigate these by using annual range forward contracts. These are annual agreements that would be made in advance between the Government and large emitters. A portfolio of call and put options would be constructed along with strike prices. A put option would require the LFE to sell some Kyoto compliant units to the Government at CDN$15 if the international price dropped below CDN$15.

4.5 Enacted Legislation on Reporting

The first part of the implementation legislation has recently been adopted. It was announced in a public notice published on 13 March 2004 pursuant to the Environmental Act of Canada:[75]

> Notice is hereby given, pursuant to subsection 46(1) of the Canadian Environmental Protection Act, 1999, that, with respect to emissions of GHGs identified in Schedule 1 [List of the six Kyoto GHGs] and for the purpose of conducting research, creating an inventory of data, formulating objectives and codes of practice, issuing guidelines or assessing or reporting on the state of the environment, any person who operates a facility described in Schedule 2 [Defines a 100 kilotonnes CO_2 equivalent threshold] to this notice during the 2004 calendar year and who possesses or who may reasonably be expected to have access to information described in Schedule 3 [lists the reportable information such as company name, process and emissions according to source] shall provide the Minister of the Environment with this information no later than June 1, 2005.

In the first phase, only those facilities that emit more than 100 kilotonnes of CO_2e per year are required to report their emissions. This includes major industrial facilities that produce electricity, heat or steam using fossil fuels, certain power generating facilities, integrated steel mills, facilities involved in smelting and refining metals, petroleum refineries, and chemical producers. Many of these have already begun registering their GHGs.[76]

Statistics Canada will be the implementing agency for this registry legislation. It was allegedly chosen because of its highly reliable and cost effective reporting, and its positive working relations with industry, federal, provincial and territorial governments and NGOs.[77] The 2004 reports will be due 1 June 2005.

[75] Notice with respect to reporting of GHGs for 2004, *Canada Gazette*, Vol 138, No 11, 2004-03-13, <http://canadagazette.gc.ca/partI/2004/20040313/html/notice-e.html#i3>.

[76] First Phase of Mandatory Greenhouse Gas Emissions Reporting [Backgrounder, 12 March 2004], <http://www.ec.gc.ca/press/2004/040312_b_e.htm>.

[77] Selection of Statistics Canada as the Vehicle for a Mandatory GHG Emissions Reporting System [Backgrounder, 12 March 2004], <http://www.nrcan-rncan.gc.ca/lfeg-ggef/English/reportingvehicle/index.htm>.

5 Canada's Future in the International Carbon Market

There are some questions about how to organize the linkage between the Canadian and other carbon emission schemes. The Ontario experience seems to be valuable in this regard, though some unresolved issues remain. The developers of a GHG emissions credit legislative regime anticipate that regulated industries will have access to credits from 'international sources',[78] although the Federal Government may restrict international transactions that risk compromising Canada's obligation to hold 90 per cent of its allocation in its national credit reserves.[79] This last restriction might be important if the forecast of some legal theorists is correct. There are some discussions that many companies in Canada that are owned by foreign (and mostly American) entities could be expected to transfer credits out of Canada to meet the emissions requirements in other jurisdictions.[80]

Another important question for the future is how the Canadian Government under Prime Minister Paul Martin will engage post-Kyoto negotiations. The Kyoto Protocol is designed to build up strength gradually through successive commitment rounds where nations can review their progress and achieve deeper cuts on fossil fuel emissions. Officially, negotiations on these new targets are not due to start until 2005, but in the global climate talks that recently ended in Italy, many countries were already in preliminary discussions.

At 1997 negotiations in Kyoto, Canada's GHG emission target was not decided by any real vision, but rather a last minute phone call. The then Environment Minister Christine Stewart called the Prime Minister to get his permission to follow a higher than expected target offered by the US under Bill Clinton's presidency. In the four years of negotiations that followed, Canada was at times unrecognizable in some of its positions. It attempted to create loopholes on land-use and forestry, suggested it receive credit toward reductions for its natural gas exports to the US, and, to the dismay of environmentalists, even seriously tried to have nuclear energy as an option in an offshore technology transfer scheme under the Kyoto Protocol called the Clean Development Mechanism.

The next commitment period will be a true litmus test. The Prime Minister's views so far have focused on the 240 megatonnes carbon reduction that Canada has promised to achieve in this commitment period (2008–12). This amount pales in comparison to what is to come in the Kyoto process.

A recent study based on each country's self-reporting of its emissions showed that actual GHG emissions from industrialized countries including Canada have not subsided but are in fact increasing and could grow as much as 8 per cent in the next six years. This spike in emissions, together with past increases, could bring

[78] See NRCan, n 52 above. [79] ibid.

[80] S Rollings Magnusson *et al.*, *The Kyoto Protocol: Implications of a Flawed but Important Environmental Policy* (Can. Pub. Pol., 2000) 26, 347.

overall emissions to 17 per cent higher than the promise Canada and other countries undertook by signing the Climate Convention in 1992. This can only mean that deeper commitments are required in future reduction periods if a serious dent is going to be made in the GHG concentrations in the atmosphere.

The world's largest emitter, the US, will not ratify without meaningful commitments from developing countries. Since developing nations are suspicious that targets are a means of impeding their economic growth and their right to development, encouraging additional countries to ratify becomes even more politically complex.

So, to be an honest broker and leader in climate affairs in the next round of target talks, the Canadian negotiators will have to convince both poor nations and stragglers such as the US that Canada is not in it for the money but as a genuine leader in protecting the global environment. This will mean leading by example through measures such as targets with deeper cuts in emissions, and national policies and measures that wean its industries from reliance on fossil fuels.

For Canada the road ahead on global environmental issues like climate change will be rough, at least until it can stimulate a greener economy, but Canada should stick with it, be a leader and make deeper cuts in the second period. This will earn it credibility and show a vision of Canada.

29

Linking Emissions Trading Schemes: The EU ETS and the 'Linking Directive'

*Jürgen Lefevere**

1 Introduction

Linking domestic or regional greenhouse gas (GHG) emissions trading schemes can provide important benefits to participants in such trading schemes. By expanding the market and increasing the availability of reduction opportunities it can increase the cost-effective functioning of the schemes involved and enhance their liquidity. Linking such trading schemes with project-based mechanisms such as the Clean Development Mechanism (CDM) and Joint Implementation (JI) can furthermore provide an important stimulus to the transfer of clean technology to countries or sectors not covered under a trading scheme. Linking emissions trading schemes can however also provide a threat to the schemes involved. Ill-designed linking of schemes can undermine their environmental effectiveness through negatively affecting the total reductions to be reached and allowing the circumvention of environmental guarantees built into these schemes. Ill-designed linking can also negatively impact competition between companies covered under the schemes and can have important impacts on the market price of emission reductions. Overly complicated 'gateways' between the linked schemes can unnecessarily increase the transaction costs of trades between these schemes.

With the entry into force of the emission allowance trading Directive[1] (ET Directive) on 13 October 2003, the European Union (EU) laid the foundation for the largest ever GHG trading scheme, which will start on 1 January 2005. The EU Emissions Trading Scheme (EU ETS) provides excellent examples of the various 'linking' challenges that emissions trading schemes face. The EU ETS itself can already be seen as a major linking exercise. Rather than establishing an EU-wide

* Jürgen Lefevere is Administrator, Climate, Ozone and Energy Unit, European Commission, Directorate General Environment. The views expressed in this chapter are the views of the author and do not necessarily represent the views of the European Commission.

[1] Directive 2003/87/EC of the European Parliament and of the Council of 13 October 2003 establishing a scheme for greenhouse gas emission allowance trading within the Community and amending Directive 96/61/EC. See ch 23 above.

trading scheme, the ET Directive requires the establishment of twenty-five domestic emissions trading schemes and provides the harmonized components to allow those domestic schemes to be linked into an EU-wide trading scheme covering over 12,000 installations in the EU's twenty-five Member States. The detail to which the various components of the trading schemes in the Member States have been harmonized, or will be further harmonized in future revisions of the scheme, provides an important lesson for linking schemes outside the EU, either with each other or with the EU ETS. Linking the EU ETS with trading schemes outside the EU will be a second important linking challenge. Developments in the EU ETS are closely followed by countries and companies outside the EU who seek to participate in the newly created EU GHG emission reduction market. As a result there is a growing interest in the possibility of linking schemes in 'third countries' with the EU ETS, a possibility that is explicitly provided for in Article 25 of the ET Directive. The third 'linking challenge' has received the most attention over the past year and concerns the linking of the EU ETS with the Kyoto Protocol's flexible mechanisms.

This contribution starts with a brief overview of these various 'linking' challenges. It then focuses on the issue of linking the EU ETS with the Kyoto Protocol's project-based mechanisms through the 'Linking Directive' that was recently agreed between the EU Parliament, Council of the EU, and the European Commission and which is expected to enter into force in the second half of 2004.

2 The EU ETS: Linking Challenges

It is tempting to argue that linking GHG emissions trading schemes should not pose major difficulties. After all, the necessary elements for linking trading schemes have been provided under the Kyoto Protocol. The Kyoto Protocol and the Marrakesh Accords provide for a cap, a trading 'currency', monitoring, reporting, and verification procedures and set out a procedure to address non-compliance by States. Since the Kyoto Protocol and the Marrakesh Accords explicitly allow States to authorize private entities to participate in the Kyoto mechanisms, albeit under the States' responsibility, such participation could involve a simple mandate to private entities that allows them to participate in the Kyoto mechanisms.

While mandating private entities to participate in Kyoto trades would indeed be easy, such a mandate usually goes hand in hand with introducing an obligation to reduce or limit emissions for those private entities. Without such obligation, the responsibility to achieve the target would remain upon the State's authorities, who would, themselves, be responsible for investing in the necessary reductions using public money.[2] The role of private entities in such system would be limited to

[2] Some States have opted to use public money to buy CDM and JI credits to achieve their Kyoto target. The Netherlands for instance has indicated that it will meet half of its reduction effort domestically

selling emission reductions to the government, rather than using emission trading for compliance with their own targets. The translation of a State's obligation to limit or reduce its GHG emissions under the Kyoto Protocol into GHG reduction limitation or reduction obligations for domestic emitters can take many shapes and forms, as is demonstrated by the wide divergence in approaches between countries that have ratified the Kyoto Protocol. It is precisely the differences in these approaches, that is, the differences in the design of the domestic emission reduction and limitation obligations, that provide for potential obstacles in linking trading schemes. It is also in the design of emission reduction and limitation obligations for private entities that much of the framework set out in the Marrakesh Accords and the Kyoto Protocol, which is designed to be applied at the level of states rather than companies, requires further elaboration at the national level.

As already stated in the introduction, the EU ETS is itself an example of linking domestic trading schemes. The ET Directive requires Member States to establish their domestic trading schemes and prescribes key components of those trading schemes. These key components include the key design of the trading scheme, including its coverage, timing, and key allocation and enforcement rules,[3] as well as harmonized rules for monitoring and reporting of emissions at installation level,[4] and a harmonized registry system. Member States however retain considerable freedom in the design of key parts of their trading scheme, including in particular the allocation of allowances to installations participating in the scheme, key aspects of enforcement and sanctions as well as verification. Interestingly, there are already increasing calls to further harmonize elements of the EU ETS across the Member States. In various fora, sectors of industry have voiced concerns over the differences in allocation methods and differences in coverage of the trading activities between Member States. Concerns are also being raised over the absence of harmonized rules for the accreditation of verifiers, allowing each Member State to set up a separate accreditation process. Concerns are furthermore being raised about the low level of harmonization of verification criteria and the more general enforcement approaches to ensure the operation of the trading scheme. The first few years of operation of the EU ETS are likely to bring forward a range of issues where more coordination between Member States in the implementation of the

and the other half through the use of the Kyoto Protocol's flexible mechanisms. Since May 2000, the Dutch Government has started two major investment programmes: ERUPT and CERUPT. Using these programmes the Government, through public tendering procedures, is purchasing CDM and JI credits. The Government has set aside a total of US$450 million for buying credits representing between 20 and 25 million tonnes of CO_2 emissions annually between 2008 and 2012. More information on the ERUPT and CERUPT programmes can be found at: <http://www.carboncredits.nl>, accessed 2 May 2004 and see ch 17 above.

[3] See n 1 above.

[4] Commission Decision 2004/156/EC of 29 January 2004 establishing guidelines for the monitoring and reporting of greenhouse gas emissions pursuant to Directive 2003/87/EC of the European Parliament and of the Council.

Directive is asked for and that may lead to further harmonization at EU level in the next revision of the ET Directive, which will in turn provide important lessons for efforts to link other emissions trading regimes.

Article 25 of the ET Directive states that 'agreements *should* be concluded with third countries listed in Annex B to the Kyoto Protocol which have ratified the Protocol to provide for the mutual recognition of allowances' between the EU ETS and trading schemes in those third countries. The change of the word 'may' in the Commission's original proposal to the word 'should' during the negotiations on the text of the ET Directive shows the importance that the majority of Member States attaches to closing such agreements.[5] Any such links are to be established through the use of Article 300 of the European Community Treaty. This article requires the Commission to make recommendations to the Council of the European Union, which authorizes the Commission to open these negotiations. The negotiations are conducted by the Commission in consultation with Council committees, and the final agreement is approved by Council by qualified majority. Once a linking agreement has been concluded, the Commission is to draw up any necessary provisions relating to the mutual recognition of allowances following the 'comitology' procedure referred to in Article 23(2) of the ET Directive (see below). While informal interest is being shown by a growing number of countries to establish links with the EU trading scheme under this provision, no formal negotiations have started, and it is unclear what the EU's requirements for such links will be. In view of the fact that, as mentioned above, the EU trading scheme is itself a linking of domestic schemes through harmonized components, it is not unlikely that if full linking is to take place, the third country will need to mirror to a large extent the requirements of the EU ETS. In absence of such large overlaps it is unlikely that full trading will be allowed to take place, but that some type of 'gateway' may need to be designed that regulates the trading between the schemes. The design of any such gateway will depend on the differences between the schemes and can include quantitative or qualitative restrictions, or a combination of both. Interestingly, the issue of linking with third countries also came up during the negotiations on the Linking Directive, discussed below. Parliament wanted to support the developers of regional trading schemes in the US and Australia by providing the possibility for regional trading schemes in countries that have not ratified the Kyoto Protocol to link with the EU ETS. As a result of Parliament's insistence, a new preambular paragraph was introduced into the Linking Directive that requires the Commission, following the entry into force of the Kyoto Protocol, to examine whether it could be possible to conclude agreements with third countries that have not yet ratified the Protocol, to provide for the mutual recognition of allowances between the EU ETS and mandatory cap and trade schemes within such countries.

The third and thus far most discussed linking issue under the EU ETS concerns the links with the Kyoto Protocol. When the Commission presented its proposal for

[5] During the negotiations a new recital 18 was also added to the ET Directive. This recital stresses the increased cost-effectiveness of the EU ETS as a result of links with regimes in third countries.

the ET Directive in October 2001,[6] it was criticized for not including an explicit link to the Kyoto Protocol's flexible mechanisms (JI, CDM, and International Emissions Trading (IET)). The original proposal was however intentionally designed to be able to operate independently from the Kyoto Protocol's flexible mechanisms, for a number of reasons. A first, practical, reason for this was that when the Commission issued the proposal, the negotiations on the Marrakesh Accords, spelling out the rules for the functioning of the flexible mechanisms, had not been finalized. These rules were only adopted in Marrakesh in November 2001, a month after the Commission had issued its proposal.[7] During the development of the Proposal it was felt that these international rules should not be pre-judged. It was, however, important that the proposal was issued before the finalization of those rules. The Community used the ET proposal, together with the proposal for a ratification Decision and the ECCP communication,[8] as an important political signal before the start of the negotiations in Marrakesh to show the world that the European Union was serious about its intentions to ratify and implement the Kyoto Protocol. Waiting until after the conclusion of the Marrakesh Accords to issue the proposal for the ET Directive would have significantly weakened this important signal.

But perhaps the most important reason for not including a direct link in the ET Directive was that the ET Directive was in the first instance designed as a domestic implementation measure. It was felt that creating a direct link in the original proposal could complicate and delay the adoption of the ET Directive because of the differing views between stakeholders on the desirability of linking, but also because of the uncertainty on the future of the Kyoto Protocol at the time of the elaboration of the proposal. The environmental NGOs, from the start of the emissions trading debate, strongly objected to any links with the Kyoto Protocol's flexible mechanisms.[9] Industry on the other hand has traditionally been a strong proponent of a direct link with the Protocol's mechanisms.[10] The European

[6] Commission Proposal for a Directive of the European Parliament and of the Council establishing a scheme for greenhouse gas emission allowance trading within the Community and amending Council Directive 96/61/EC, COM(2001)581 of 23 October 2001, <http://europa.eu.int/eur-lex/en/com/pdf/2001/en_501PC0581.pdf> accessed 16 June 2004.

[7] These texts are part of the Marrakesh Accords, adopted in November 2002, and can be found in FCCC/CP/2001/13/Add.2.

[8] Council Decision 2002/358/EC of 25 April 2002 concerning the conclusion, on behalf of the European Community, of the Kyoto Protocol to the United Nations Framework Convention on Climate Change and the joint fulfilment of commitments thereunder and the Communication from the Commission on the Implementation of the First Phase of the European Climate Change Programme, COM(2001)580, 23 October 2001.

[9] Climate Action Network Europe's position on the original Commission proposal, *Emissions trading in the EU: let's see some targets!*, 20 December 2001. More recently: CAN Europe press release of 23 July 2003, *Commission shoots its own emission trading system full of holes*. Both press releases are available through: <http://www.climnet.org> accessed 2 May 2004.

[10] See the letter by UNICE, a key industry lobby group in Brussels, from 10 December 2001 to the Belgian Council Presidency and more recently the UNICE letter to Margot Wallström, Commissioner, DG Environment of 4 July 2003, 'UNICE preliminary comments on linking the Kyoto Protocol mechanisms (JI and CDM) with the EU emission trading scheme'. Both letters are available through <http://www.unice.org> accessed 2 May 2004.

Parliament has throughout the negotiations maintained that any link with the Kyoto mechanisms would only be acceptable after 2008, as long as the projects do not include carbon sinks or sources of energy that use nuclear power. Although the Commission did find a link with the Kyoto flexible mechanisms to be desirable, such link was to be 'subject to the satisfactory resolution of outstanding issues regarding their environmental integrity.'[11] The Commission therefore preferred to discuss the link with the Kyoto mechanisms in the context of the elaboration and adoption of a separate legislative instrument.

The original Commission proposal did not include the flexible mechanisms, but it did refer to linking the EU scheme to these mechanisms in a number of places. The first reference was included in the review clause in the proposal's Article 26 (Article 30 in the final text), which included 'the use of credits from project mechanisms' as one of the issues to be considered in the review. The second reference was included in the proposal's Article 24 (Article 25 in the final text) which allowed the Community to conclude agreements with third countries to link with their trading regimes.

With both industry and a number of Member States strongly advocating a direct link with the Kyoto mechanisms, the Commission, throughout the negotiations on the Directive, made a number of concessions to show a stronger commitment to use the mechanisms in the European trading regime. As a result of the discussions with the European Parliament and the Council a new paragraph 19 was added to the preamble, which stresses the importance of the project-based mechanisms CDM and JI in increasing the cost-effectiveness of the Community scheme. Importantly, during the negotiations a new third paragraph was added to the review provision in Article 30, which states that:

> Linking the project-based mechanisms, including Joint Implementation (JI) and the Clean Development Mechanisms (CDM), with the Community scheme is desirable and important to achieve the goal of both reducing global greenhouse gas emissions and increasing the cost-effective functioning of the Community scheme. Therefore, the emission credits from the project-based mechanisms will be recognised for their use in this scheme subject to provisions adopted by the European Parliament and the Council on a proposal from the Commission, which should apply in parallel wit the Community scheme in 2005.

The issue of linking the EU ETS with international emissions trading under Article 17 of the Kyoto Protocol was addressed through the inclusion of the above-mentioned Article 25 of the ET Directive, which allows the linking of the EU ETS to third country regimes through mutual recognition agreements.

The following sections of this contribution will focus on the linking of the EU ETS with the Kyoto Protocol's project-based mechanisms JI and CDM.

[11] Explanatory Memorandum to the Commission Proposal, n 6 above, 17.

3 The 'Linking' Directive: From Proposal to Adoption

3.1 Elaboration of the Proposal

Soon after the adoption of the proposal for the ET Directive and the international agreement on the Marrakesh Accords in November 2001, the Commission placed the elaboration of a proposal for a Directive to linking the project-based mechanisms to the ET Directive (the proposal for a Linking Directive) high on its agenda. In January 2002 the Commission revived ECCP Working Group (WG) I, a meeting of which took place on 27 February 2002, which was followed by three further meetings.[12] These meetings in WG I were used to exchange ideas among stakeholders on a number of issues related to the possibility of linking project-based mechanisms with the trading regime set up under its ET proposal. Important issues which were discussed included, the timing of the inclusion of the various project-based mechanisms, in particular whether any of these mechanisms can be linked before 2008, as well as the question on which types of projects can be linked to the trading regime. In relation to the latter, environmental NGOs argued that should any projects be linked at all, which they opposed, these should be limited to specific types of projects that conform to high environmental standards.[13] In particular projects using carbon sinks or nuclear energy should be excluded. Industry continued to take the view that the Kyoto Mechanisms should be incorporated 'as is' in the ET Directive. The conclusions of the group, adopted at its fourth meeting in September 2002, called for 'the early adoption of legislation regarding the recognition of project credits' 'as a matter of particular priority' and stated that 'the Commission should aim to make its proposal for a Directive linking JI/CDM credits with the EU emissions trading scheme early in 2003' and that 'the Council and the European Parliament should aim at adopting this legislation so as to allow its implementation as from the commencement date of the EU emissions trading scheme.'[14]

3.2 The Commission Proposal

The Commission had planned the adoption of its proposal early in the first half of 2003, but the proposal was delayed in the adoption process within the Commission. One of the reasons for the delay was the disagreement between DG Environment and other Commission services on the need to limit the number of

[12] The terms of reference, agenda, minutes, and conclusions of these meetings are published on the internet at: <http://europa.eu.int/comm/environment/climat/ji_cdm.htm> accessed 2 May 2004.

[13] See, inter alia, CAN Europe's reaction to the ET proposal, n 9 above.

[14] Environmental NGOs, in a letter to the Commission, disagreed with the representative nature of these conclusions, objecting in particular to this paragraph. ECCP Working Group on JI/CDM, Conclusions, 15 November 2002, available at <http://europa.eu.int/comm/environment/climat/jicdm/jicdm_final_conclusions.pdf> accessed 2 May 2004.

credits from project-based mechanisms to be used in the EU trading regime. This disagreement will be further discussed below. The Commission finally issued its proposal on 23 July 2003.[15] The text of the Directive itself is short, little over five pages, and takes the form of an amendment to the ET Directive, thus fully integrating the use of the CDM and JI into the ET Directive.

The proposal allowed for the 'conversion' of JI and CDM credits into allowances for use in the EU ETS from 1 January 2008 onwards. Such conversion could take place for all CDM and JI credits (CERs and ERUs), except for credits resulting from land use, land-use change, and forestry (LULUCF or 'sinks') projects and credits resulting from nuclear facilities. The proposal did not foresee a limit on the amount of credits to be converted, but required the Commission to monitor this amount. When the amount reached 6 per cent, the Commission was to undertake an immediate review to consider whether 'a maximum of for example 8 per cent of the total quantity of allowances allocated by the Member States'[16] was to be introduced following a comitology procedure (see above). The proposal furthermore required Member States to ensure that baselines for projects undertaken in the Community respect the 'Community acquis communautaire', which covers all EU legislation in force and related case law of the courts. The Proposal also prohibited ERUs to be issued for reductions or limitations of GHG emissions from installations covered under the ET Directive, with the exception of JI projects that were approved before 31 December 2004 or the date of a state's accession to the EU, to avoid double counting of emissions reductions undertaken in such installations. Member States were furthermore required to ensure that CDM and JI projects take into account environmental and social impacts of such projects and that they result in real GHG mitigation benefits that are additional and that these projects lead to the transfer of environmentally safe and sound technology and know-how.

3.3 The Compromise Package

The proposal was issued at the start of the Italian Presidency of the EU, which made its adoption one of its key priorities. Although the proposal was very short, the wide spectrum of views on whether and how CDM and JI credits can be used under the EU trading regime caused lively discussions in Council and Parliament. With very divergent views between Member States, negotiations on the proposal progressed little during the second half of 2003. This changed in the beginning of 2004. In December 2003, Council debated the key elements of the Proposal and asked for its swift adoption.[17] During that same month, the European Parliament's rapporteur

[15] European Commission Proposal of 23 July 2003 for a Directive of the European Parliament and of the Council amending the Directive establishing a scheme for greenhouse gas emission allowance trading within the Community, in respect of the Kyoto Protocol's project mechanisms, COM(2003)403 (Proposal for a Linking Directive).

[16] Proposals for a Linking Directive, Explanatory Memorandum, 8.

[17] Minutes of the 2,556th meeting of the Council of the European Union (Environment), held in Brussels on 22 December 2003.

on the Proposal, Mr Alexander de Roo, had indicated in various meetings that he wanted an agreement on the Proposal before Parliament's elections in June 2004, and that for such agreement Parliament could be flexible on quantitative restrictions if qualitative requirements for CDM and JI credits coming into the EU trading scheme were sufficiently robust. The Irish, who took over the Council presidency in January 2004, picked up this message and pushed for agreement on the Directive before Parliament's elections. As a result of these efforts of the Parliament's rapporteur, the Irish presidency, and the Commission reached informal agreement on a compromise package on 5 April 2004. On 7 April 2004 this agreement was supported in the Council's Committee of permanent representatives (Coreper). On 20 April 2004 Parliament voted in favour of a series of amendments reflecting the compromise package agreed with the Council and the Commission. The Council formally agreed to the compromise package on 13 September 2004, and the Directive is expected to be published and enter into force in October 2004.

The compromise package is significantly different from the Commission's original proposal as a result of intense negotiations in Council and between Council, Commission and, Parliament. The final Directive will now allow CDM credits to be used in the EU ETS from 1 January 2005 onwards, whereas JI credits will only be able to come into the scheme from 2008 onwards. The concept of 'conversion' of CDM and JI credits into EU allowances was removed from the Directive. Instead companies under the EU ETS can use CDM and JI credits directly for compliance with their obligations under the EU ETS. The issue of 'supplementarity' proved the most controversial during the negotiations. Agreement was reached on an installation-specific cap from 1 January 2008 onwards, the level of which is to be defined by Member States in their national allocation plans for the periods from 2008 onwards. In addition to this installation-specific cap, the Commission is to monitor the levels of use and may make proposals to harmonize the cap across Member States. Credits from nuclear and sinks projects remain excluded in the EU ETS, although the exception for nuclear credits is now limited to 2012 and the Commission is explicitly required to report on the technical aspects of bringing sinks credits into the EU ETS in its mid-2006 review. The 'double counting' provision was significantly amended to provide more flexibility for Member States to choose whether to bring JI and CDM projects that directly or indirectly affect emissions from installations under the EU ETS inside the EU trading scheme or to allow ERUs and CERs to be issued subject to cancellation of an equal number of allowances. The final compromise also removes all extra checks of Member States over CDM and JI project activities, except the requirement that baselines must respect the *acquis communautaire* and that Member States must ensure that the participation of private and public entities in CDM and JI projects under their authority is in accordance with the rules set out in the Kyoto Protocol and the Marrakesh Accords. One additional requirement was included at the insistence of the European Parliament. Member States must now ensure that, when they approve hydro-power projects with a capacity over 20 MW that relevant international criteria and guidelines for such projects, including those elaborated by the World Commission on Dams, are respected during the development of such projects. A range of other issues, including the use of 'domestic projects' (unilateral JI), was moved to the review provisions in Article 30 of the ET Directive, as a result of which the Commission will have to consider a long list of issues in its mid-2006 review. Among the list of issues to be considered in the mid-2006 review also features, in the form of a new preambular paragraph, the Commission's assessment of the possibility to link with trading schemes within countries that have not yet ratified the Kyoto Protocol, albeit this assessment should only be made after the Protocol has entered into force. The Directive will enter into force on the day of its publication and must be transposed by the Member States at the latest twelve months after its entry into force.

4 CDM and JI in the EU ETS: Opportunities and Threats

The Linking Directive will establish a *direct* link between project-based mechanisms and the ET Directive. The link will allow participating entities in domestic emissions trading programmes under the ET Directive to use credits generated by the project-based mechanisms to offset reduction obligations under the ET Directive.

Before discussing the threats and opportunities for allowing these units to be used to comply with the obligations under the EU trading regime, two issues are worth pointing out. First, an *indirect* link between the ET Directive and these project-based mechanisms already exists. The ET Directive does not determine the size of the cap for the sectors covered by the Directive. The size of the cap is to be determined by the Member States on the basis of the criteria set out in Annex III to the ET Directive and, *inter alia*, depends on their GHG reduction commitments and the allocation of the responsibility to achieve these commitments across the different sources and sinks in their territory, including the sources covered by the ET Directive. During 2008–12, Member States can use Certified Emission Reductions (CERs) and Emission Reduction Units (ERUs) (together: Kyoto Units) to meet their commitments under the Protocol and the EU 'Bubble' agreement. The Member State could buy such units[18] and provide larger allocations to entities in the trading programme, subject to the criteria in Annex III to the ET Directive and the Commission review of the national allocation plans. Such use of Kyoto Units by Member States will affect the stringency of the cap, in particular the impact of the cap on the amount of domestic reductions that will be achieved under the ET Directive. A preliminary assessment of the first draft national allocation plans shows that a number of Member States indeed plan to use the Kyoto Mechanisms to limit the reductions to be made by installations under the EU

[18] For instance through initiatives such as the Dutch ERUPT and CERUPT programmes, see n 18 above.

ETS.[19] A second issue that is worth pointing out concerns the widely heard criticism during the negotiations on the Linking Directive, in particular from industry and a number of Member States, that the Directive restricts access to the Kyoto Mechanisms. It can be argued that this criticism is unfounded; the Linking Directive aims to do exactly the opposite. As already discussed above, the proposal for the Linking Directive, once in force, will for the first time allow legal entities to use the project-based mechanisms to comply with their obligations. Rather than limiting the market, this will create a whole new market for CDM and JI credits and provide a true incentive to the use of those mechanisms, even if the EU decides, for whatever reasons, not to allow all Kyoto Units to be used under the ET Directive.

There are a number of reasons for allowing entities to directly use project-based mechanisms for compliance with their targets under the ET Directive. The most frequently used reason is that the inclusion of project-based mechanisms can reduce the compliance costs for sectors covered under the ET Directive by broadening the range of opportunities to reduce emissions in another Member State or outside the EU at lower costs. A second reason is that the use of project-based mechanisms can engage sources and sinks not covered by the Directive in implementing cost-effective reduction options. Including project-based mechanisms can bring a net benefit to project developers who are not under an obligation to limit their emissions pursuant to the ET Directive, by giving them a financial incentive to reduce their emissions if those emissions can be reduced at a lower cost than the market price for allowances under the ET Directive. The use of the project-based mechanisms within the EU could thus be seen as a piecemeal and ad hoc way to extend of the coverage of the trading regime to gases and sectors not covered under the ET Directive, starting where the coverage of the ET Directive ends. Including CDM and JI in the ET Directive will also provide an important boost to the use of these instruments. Before the Linking Directive, there was very little interest from companies in investing in CDM and JI projects. The combination of emission caps under the ET Directive with the possibility to use CDM and JI will create an important incentive for companies to invest in these instruments and thus kick-start the CDM and JI markets.

There are also a number of important arguments that caution against creating a direct link. A first argument is that the ET Directive is a domestic implementation measure and should in the first place be used to reduce GHG emissions *inside* the EU, rather than embarking on a path which will make the EU's ability to achieve its emission reduction target dependent on buying sufficient credits from elsewhere. Linked to this argument are fears that opening the EU trading regime for CDM and JI credits may flood the EU market with these credits and thus avoid the need for intra-EU reductions. While the Kyoto Protocol allows private entity

[19] A list of national allocation plans can be found on the Commission's emissions trading website: <http://europa.eu.int/comm/environment/climat/emission_plans.htm> accessed 2 May 2004.

trading, it only does so under the responsibility of Parties and to the extent that Parties wish to allow this. The EU trading regime would be the first and hitherto only regime that would allow the use of these credits at the domestic level, and indeed creates an incentive for doing so by imposing a cap on the sources included in the regime. Since the regime only covers part of the sources within the EU and will initially concentrate on CO_2 only, exposing this regime to the full global CDM and JI market could be disproportionate and lead to a significant lowering of the market price of allowances, leading to little reductions to be made within the EU and the bulk of those reductions purchased from the Kyoto mechanisms. A further argument is that using JI to expand the coverage of the EU trading regime within the EU does not actually lead to reductions, as the credits earned by those investments will be used to avoid reductions in the sectors covered by the regime. Instead this broadening of coverage can provide an impetus for sectors to avoid other policies and measures limiting their emissions or avoid being brought directly into the EU trading regime. An important argument used by environmental NGOs against 'linking' are the doubts that some of the CDM and JI projects that are currently in the pipeline do not provide sustainable development benefits or represent real reductions compared to what would have happened otherwise (that is, these projects are not 'additional'). Using such credits inside the EU trading regime would undermine its environmental credibility.

5 Linking the EU ETS with CDM and JI—Key Issues

The following paragraphs provide a background to the five key elements of the Linking Directive in the form of questions that played a key role in the ongoing negotiations:

- When should a link be established?
- How should this link be established?
- Which projects should a link be established with?
- Should there be any quantitative limits on a link?
- How can double counting be avoided?

5.1 When to link

The EU ETS is scheduled to start on 1 January 2005. The Kyoto Protocol's first commitment period starts on 1 January 2008 and ends on 31 December 2012. Under the Kyoto Protocol and the Marrakesh Accords, JI credits (ERUs) can only be issued for emission reductions and carbon sequestration projects from 2008 onwards. JI credits could therefore not be used pre-2008, as they will not yet exist. Although a JI-like structure could be created pre-2008, such structure would be cumbersome to design and implement for only a three-year period, and its links with actual JI crediting under the Kyoto Protocol would not necessarily be guaranteed.

Under the Kyoto Protocol and as a consequence of the Marrakesh Accords' 'prompt start' of the CDM, CDM credits (CERs) can be issued for emission reductions or carbon sequestration projects starting from 2000. Since the 'prompt start' of the CDM is based on decisions adopted by the Conference of the Parties (COP) to the UNFCCC, rather than decisions adopted by the COP serving as the Meeting of the Parties (COP/MOP) to the Kyoto Protocol, CERs can even be issued prior to the Protocol's entry into force. CERs could therefore be used pre-2008. Although under the Kyoto Protocol CERs can only be used to fulfil commitments from 2008 onwards, their recognition for compliance purposes pre-2008 in the EU ETS would help the early start of the CDM by generating the interest of the private sector in these credits for the 2005–8 period. From a Member State perspective it could have been argued that it would have been more desirable to wait with the use of CERs in the EU ETS until after the start of the first commitment period, as this will reduce intra-EU emissions pre-2008 and ensure an increased availability of CERs during the Protocol's first commitment period, since CERs for reductions pre-2008 would not come onto the EU market until 2008. Industry however indicated that it wished to see CDM credits included in the 2005–7 period of the EU trading regime as this lowers costs for industry in the short term.[20]

The discussions on allowing CDM credits to enter the EU trading regime from 2005 onwards also raised the questions what would happen to the CERs once they have been surrendered by an installation to compensate for emissions during the 2005–8 period. Logically, they should be cancelled by the Member State authority, as they have already been used to offset GHG emissions. But since Member States have no emission limitation objective themselves pre-2008, strictly there is no need for them doing so. Member States could decide to hold on to these CERs and use them to offset emissions during the first commitment period, although by doing so they would use one CER to offset two tonnes of GHG emissions.

The Commission's original proposal only allowed CERs and ERUs to come into the EU ETS from 1 January 2008 onwards. In reaction to pressure from Member States and Parliament, the final compromise allows CERs to come in from 1 January 2005 onwards. CERs that are used in the pre-2008 period must however be cancelled and cannot be used again by Member States for compliance with their Kyoto target.

5.2 How to link

One of the key differences between the EU ETS and the Kyoto Protocol is that units traded under the EU trading regime are fully 'fungible' (all units are fully interchangeable and thus have the same value) and that Member States are required

[20] See n 10 above.

to accept all EU allowances for compliance, irrespective of the company that the allowance was originally allocated to and the country it was originally allocated by. All EU Allowances are furthermore treated the same under the ET Directive's banking rules. Under the Kyoto Protocol trading regime there are differences between the various credits (CERs, ERUs, AAUs, and RMUs) with regard to their generation, banking and use. Parties under the Kyoto Protocol are furthermore free to choose whether they wish to accept credits from another Party or project.

In the Commission's proposal for the Linking Directive, the reconciliation of the two different regimes was built into the 'entry point' of the Kyoto Units into the EU ETS. The Commission proposed the conversion of CERs and ERUs into newly issued allowances, which could then be further traded in the EU ETS. This conversion was to be done by a Member State, which, under the text of the proposal '*may*' issue one new allowance in exchange for one CER or ERU. The use of the word 'may' signifies that Member States would continue to keep the freedom to impose other criteria for the conversion of Kyoto credits into allowances. By converting the Kyoto Units into EU allowances, the Commission wanted to maintain the full fungibility of its units. The advantage of conversion would have been that once a Kyoto unit was brought into the EU ETS it would have been fully equivalent to the EU 'currency', giving business more certainty about the possibilities to use, and therefore, the value of these units. There was also a tactical reason for the Commission to propose conversion rather than allow direct use of Kyoto credits. Because through conversion Member States would have been forced to give up an allowance, which is backed by a Kyoto Protocol Assigned Amount Unit (AAU), Member States would have had an incentive to limit this conversion to avoid a large-scale 'swapping' of CERs and ERUs for allowances backed by AAU, which are subject to fewer restrictions under the Kyoto Protocol.

At the insistence of a majority of Member States, led by the UK, the Linking Directive no longer uses the conversion concept. Instead it allows the direct use of CERs and ERUs in the Community scheme by providing that Member States may allow operators to surrender these CERs and ERUs for compliance with their obligations under the EU ETS. Interestingly, the text of the Linking Directive maintains a reference to the exchange of those CERs and ERUs for an allowance that is issued and immediately cancelled. The reason for this reference was mostly technical, ensuring that only allowances are actually used in the EU ETS and allowing Member States to more easily manage their Kyoto Units. The technical details of the use and exchange process will be set out in the forthcoming Commission Regulation on a standardized registries system, which is to be adopted in the course of 2004.[21]

5.3 Which Projects to Link With

One of the most difficult issues to solve during the negotiations on the Linking Directive was whether all project credits generated under the Kyoto Protocol

[21] See also ch 7 above.

should be allowed to be used in the EU ETS, or whether it should be limited to a subset of these projects.

During the international negotiations on the rules on the project-based mechanisms leading up to the Marrakesh Accords, the EU put forward a number of proposals to strengthen the environmental integrity of the mechanisms. These proposals included a definition of 'supplementarity', excluding sinks credits in the CDM, prohibiting nuclear energy projects, as well as opposing Article 3(4) forestry credits in the first commitment period.[22] As part of the agreement on the Marrakesh Accords, the EU, like other major players, had to compromise on a number of its positions. Although the EU had agreed the Marrakesh Accords internationally, it has not decided whether it wanted to make full use of the Kyoto mechanisms internally.

There are a number of 'environmental integrity' arguments on the basis of which a selective inclusion of project-based mechanisms has been advocated. These arguments include:

(1) whether specific types of projects constitute real emission reductions;
(2) whether those can be measured or verified;
(3) the need to pursue long-term reductions rather than temporary storage; and
(4) the environmental and social impact of specific project types.

There are a number of approaches to selectively allow the use of JI and CDM credits for compliance with the obligations under the EU trading. A first possibility could be to limit the types of instruments to link with. The EU could decide not to link with CDM or JI. Individual EU Member States could have decided to adopt a policy of complying with their targets under the EU burden sharing agreement on the basis of domestic action alone, which in principle amounts to a decision not to use the Kyoto mechanisms. A second possibility is to limit the types of projects. Such restrictions could take the form of a positive list or a negative list. A positive list specifies the projects from which credits can be used. Thus a positive list could specify project types (for example, renewable energy), project sizes (for example, less than 100 ktCO_2) and other characteristics of projects that generate acceptable credits.[23] A negative list accepts all project-based credits except those specifically excluded.

Concerns related to whether the reductions from a project are real, measurable, and verifiable could be addressed through a more specific definition of the 'additionality' concept in the Linking Directive. A more specific definition would, however, be difficult to apply as it would mean that the underlying project of every credit surrendered for compliance with emission limitations under the EU ETS would need an additional separate test of additionality to comply with the EU requirements. This would require the EU to establish an administrative and

[22] Various EU position papers for the international negotiations can be found through <http://europa.eu.int/comm/environment/climat/cop.htm> accessed 2 May 2004.

[23] During the international negotiations the EU in fact advocated such a positive list.

institutional machinery similar to the Executive Board of the CDM, at least in relation to its baseline and methodologies panel. The application of different additionality tests by the EU and internationally by the CDM Executive Board would create significant additional transaction costs and further fragment the nascent carbon markets. Concerns related to the environmental and social impact of projects could be addressed through the application of existing EU and regional rules on environmental impacts, access to information and access to justice.[24] A key question here is whether the application of those rules should be limited to projects within the EU or whether these rules, or the concepts underlying them, should also be applied to projects outside the EU.

Effective enforcement of any type of limited access can be complicated or even impossible. Any restrictions would need to be implemented on an EU-wide level. The reason for this is that the ET Directive does not allow any restrictions on the transfer of EU allowances. Once a project credit is used in the EU ETS it frees up an allowance which can then be freely traded inside the EU ETS. This means that investors could circumvent any national rules by choosing to bring the project credit into the trading scheme in the Member State with the least restrictions and transferring the allowances that are freed up to the installation in the Member State where this credit was needed. This same possibility to circumvent EU-wide restrictions would also pose itself in case the EU trading regime was to be linked to trading regimes outside the EU.

The Commission Proposal for the Linking Directive aimed to tackle these issues through a combination of methods. As already seen above, the 'gateway' in the proposal stated that Member States 'may' convert CDM and JI credits into EU allowances, thus allowing Member States to define further criteria for conversion. The proposal furthermore contained a 'negative list' of types of projects inadmissible for converting into EU allowances. This list contained nuclear facilities and sinks projects. The proposal however qualified the exclusion of nuclear projects to projects by stating that this exclusion was 'in accordance with the Kyoto Protocol and subsequent decisions adopted thereunder.' This language was included in the proposal as certain Member States and Commission services did not want to see a categorical and indefinite exclusion of nuclear projects in view of the fact that these projects were not clearly excluded in the Marrakesh Accords.[25] The proposal to exclude sinks projects was in line with the EU's strong opposition to the inclusion

[24] These include Council Directive 85/337/EEC of 27 June 1985 on the assessment of the effects of certain public and private projects on the environment, Directive 2003/4/EC of the European Parliament and of the Council of 28 January 2003 on public access to environmental information; and Council Directive 96/61/EC of 24 September 1996 concerning integrated pollution prevention and control.

[25] The preambular paras of the Marrakesh decisions on JI and CDM include a statement that 'Parties included in Annex I to the Convention are to refrain from using emission reduction units generated from nuclear facilities to meet their commitments . . .', therefore not excluding nuclear energy CDM projects, but excluding the use of the resulting credits to meet the Kyoto commitments, see FCCC/CP/2001/13/Add.2, 5 and 20.

of sinks projects in the CDM during the international negotiations. When the Commission released its proposal it was technically difficult to include sinks projects, as the international rules for the inclusion of sinks in the CDM were only to be agreed by COP9 in December 2003.

The final text of the Linking Directive will maintain the exclusion of CERs and ERUs from nuclear facilities and sinks projects. Due to pressure in particular from France, the language on nuclear facilities was however significantly amended and now only excludes these projects until 2012. There was also significant pressure from a number of Member States to include sinks credits in the EU ETS. One of the reasons for this pressure was that at COP9 in Milan in December 2003 international agreement was reached on the rules for the inclusion of sinks in the CDM.[26] Parliament however remained opposed to the exclusion of sinks and the European Commission pointed out the practical difficulties of bringing temporary credits into a scheme based on permanent reductions. As a compromise the Commission is now to consider 'technical provisions' related to the various aspects of including sinks credits into the EU scheme from 2008 in its mid-2006 review and may put forward proposals to include sinks in the EU ETS.

Earlier unofficial versions of the Commission's proposal also excluded hydro-electric power production incompatible with the criteria and guidelines of the World Commission on Dams in its year 2000 Final Report. This exclusion was removed in the final Commission proposal, but instead it proposed to include the obligation for the Commission to consider in its mid-2006 review 'the impact of project mechanisms on host countries, particularly on their development objectives, including whether JI and CDM large hydroelectric power production projects have been established which have negative environmental and social impacts'.[27] Due to pressure from the European Parliament the final text of the Linking Directive goes one step further and requires Member States to ensure that when approving hydro-electric power production projects with a generating capacity exceeding 20 MW 'that relevant international criteria and guidelines, including those contained in the World Commission on Dams year 2000 Final Report, will be respected during the development of such project activities.'[28]

The Commission's proposal furthermore contained a number of guarantees in a new Article 11(*ter*) to be included in the ET Directive. Paragraph 1 of this proposal requires that accession countries take into account the *acquis communautaire* in the establishment of baselines for JI and CDM projects. This provision was included to avoid accession countries using CDM and JI projects to bring their infrastructure in line with EU requirements. As these countries are obliged to do so under their Accession Treaty, providing ERUs or CERs in return for such projects would clearly not be 'additional' under the rules of the Kyoto Protocol. Paragraph 4 of the

[26] Decision 19/CP.9 on Modalities and procedures for afforestation and reforestation project activities under the clean development mechanism in the first commitment period of the Kyoto Protocol.

[27] Proposal for a Linking Directive, amendment of Art 30 (2)(l) new of the EU ETS Directive.

[28] Final version of the Linking Directive (formally to be adopted), Art 11b (6).

new Article 11(*ter*) of the Commission Proposal requires Member States participating in CDM and JI projects outside the Community to take into account the environmental and social impacts of these projects. It also requires Member States to 'ensure that these projects are developed and implemented in such a manner to contribute to sustainable development, and to the specific development needs and objectives of the host countries.'[29] Paragraph 5 furthermore requires Member States to ensure that project activities result in:

(a) real, measurable, and long-term benefits related to the mitigation of climate change;
(b) reductions in emissions that are additional to any that would occur in the absence of the proposed project activity; and
(c) the transfer of environmentally safe and sound technology and know-how.

While each of these requirements is already included in the Kyoto Protocol and the Marrakesh Accords,[30] the proposal clarifies the obligation to respect them. Perhaps the most interesting part of the Commission's proposal was the possibility for the Commission to adopt further guidance on all the issues listed in the new Article 11(*ter*) through the comitology procedure. In the final text of the Directive these extra guarantees were significantly watered down, to prevent a duplication of the checks required under the Marrakesh Accords and reduce the transaction costs of using the mechanisms. While the requirement to respect the *acquis communautaire* remains, the other proposed provisions were replaced with a new Article 11b(5), which states that 'a Member State that authorises private or public entities to participate in project activities shall remain responsible for the fulfilment of its obligations under the UNFCCC and the Kyoto Protocol and shall ensure that such participation is consistent with the relevant guidelines, modalities and procedures adopted pursuant to the UNFCCC or the Kyoto Protocol.'[31]

During the negotiations on the Linking Directive, a number of Member States also proposed that so-called 'domestic projects' should be allowed in the EU ETS. Reasons for this included that some Member States felt including JI projects and not domestic projects would discriminate against purely domestic investments and that the exclusion of domestic projects could be easily circumvented through a 'mail-box' participant in a second Member State. Proponents of the inclusion of domestic projects also felt that their inclusion could engage sectors outside the trading scheme in finding cost-effective emission reductions. The Commission and the European Parliament strongly opposed the inclusion of domestic projects. The main reason for this was that it was felt that allowing domestic projects would give an incentive for sectors outside the trading scheme to avoid coming under the scheme's 'cap' and would also negatively affect any future emission reduction

[29] Proposal for a Linking Directive, Recital 10.

[30] These requirements are included in Art 12(5) Kyoto Protocol and in para 43 and the preamble of the CDM text, FCCC/CP/2001/13/Add.2, 36 and 20.

[31] Final version of the Linking Directive (formally to be adopted), Article 11b (5).

policies to be applied to those sectors. The Commission argued that rather than bringing those sectors in through domestic projects, Member States should consider bringing them in through the opt-in clause in Article 24 of the ET Directive, or that these sectors should be considered for inclusion in the mid-2006 review of the Directive. The final text of the Linking Directive retains the exclusion of domestic projects, but requires the Commission to consider 'the modalities and procedures for Member States' approval of domestic project activities and for the issuing of allowances in respect of emissions reductions or limitations resulting from such activities from 2008'[32] as part of its report for the mid-2006 review of the Directive.

5.4 How much to link

As already stated above, the discussion on the amount of CDM and JI credits coming into the EU trading regime was one of the reasons for the delay in the adoption of the linking proposal by the Commission and played a central role in the negotiations on the compromise package on the Linking Directive.

Fears have been expressed that an unlimited linking of the EU trading regime with the Kyoto Protocol's project-based mechanisms could have a significant downwards impact on the price of the allowances traded in the EU trading regime and thus reduce the incentives for companies to reduce their emissions within the EU. This would also undermine the EU's domestic implementation of the 'supplementarity' obligations in the Kyoto Protocol and the Marrakesh Accords.

One way to address these legitimate policy concerns is through a quantitative limitation on the amount of credits that are introduced into the EU trading regime. In the negotiations on the Kyoto Protocol and the subsequent Marrakesh Accords, the EU tried to introduce a quantitative limit on the use of units traded under the Kyoto mechanisms as a means to implement the supplementarity principle. Although no specific quantitative limit was introduced in respect of supplementarity, the Marrakesh Accords did include a reference to the supplementarity principle in a number of places, affirming that 'the use of the mechanisms shall be supplemental to domestic action and that domestic action shall thus constitute a significant element of the effort made by each Party included in Annex I' to meet its Kyoto target.[33] The Marrakesh Accords also introduced a number of quantitative restrictions relating to specific units:

(a) a cap on the amount of sinks credits for other projects than afforestation and reforestation projects that each Annex I country can issue under Article 3(4) of the Kyoto Protocol;[34]

[32] Final version of the Linking Directive (formally to be adopted), Amendment of Art 30(2)(n) (new) of the EU ETS Directive.

[33] See for instance the preamble of Decision 15/CP.7, FCCC/CP/2001/13/Add.2.

[34] Decision 19/CP.7, FCCC/CP/2001/13/Add.2, 63, para 28.

(b) a prohibition on the banking of RMUs and limits the banking of CERs and ERUs into the following commitment period to a maximum 2.5 per cent of a Party's assigned amount for each of these units;[35]

(c) a cap on the total amount of CERs from land use, land-use change and forestry project activities that each Annex I Party can acquire limited to one per cent of its assigned amount.[36]

These specific quantitative restrictions all concern credits that some Parties felt could, because of the uncertainty of the real reductions represented by these credits, and in particular the potentially large quantities in which these credits could be generated, undermine the Kyoto Protocol targets by providing cheap and plentiful reduction opportunities, thus providing a disincentive for domestic action.

An early internal Commission draft of the linking proposal that included a paragraph stating that 'Member States may convert CERs and ERUs from project activities for use in the Community-scheme up to 6 per cent of the total quantity of allowances allocated by the Member State'[37] for each of the ET Directive's compliance periods. Industry, as well as a number of Member States and other stakeholders, strongly lobbied the Commission delete this paragraph.[38] As a result of this opposition, the proposed cap was replaced by a provision that triggers a review at 6 per cent, on the basis of which a cap may be proposed through a comitology process of 'for example' 8 per cent.[39]

During the negotiations on the linking proposal a number of Member States, led by the UK, argued for an installation-specific cap on the use of project credits by installations. The argumentation for this was that such a cap would give up-front certainty to industry on their use of the Kyoto mechanisms for compliance with their targets under the EU trading scheme. Such certainty was not given in the original Commission proposal, which allowed for a cap to come in at a later stage and did not define the shape that any such cap would take. The agreed text of the Linking Directive contains a combination of the various proposed approaches. From 2008 onwards, Member States are to a apply a percentage of the allocation to each installation up to which CERs and ERUs can be used to meet the targets under the EU ETS. This percentage is to be defined in the national allocation plan (NAP) of each Member State. Annex III of the ET Directive, containing criteria for drafting of the NAPs and the Commission's assessment of these NAPs, will include a new criterion 12 which states that:

> The plan shall specify the maximum amount of CERs and ERUs which may be used by operators in the Community scheme as a percentage of the allocation of the allowances to

[35] Decision 19/CP.7, FCCC/CP/2001/13/Add.2, 61, paras 15 and 16.

[36] Decision 11/CP.7, FCCC/CP/2001/13/Add.1, 61, para 14.

[37] Proposal for a Linking Directive, Article 11(*bis*)(2).

[38] See for instance UNICE's letter of 4 July 2003, n 10 above.

[39] The 6 and 8 per cent figures were based on the study by P Criqui and A Kitous, *KPI Technical Report: Impacts of linking JI and CDM credits to the European Emission Allowance Trading Scheme (KPI-ETS)*, available at <http://europa.eu.int/comm/environment/climat/emission.htm> accessed 2 May 2004.

each installation. The percentage shall be consistent with the Member State's supplementarity obligations under the Kyoto Protocol and decision adopted pursuant to the UNFCCC or the Kyoto Protocol.

In addition to this installation-specific cap, the Commission is to monitor the levels of use and may make proposals to harmonize the cap across Member States.

5.5 Double Counting and JI in Accession Countries

The text of the ET Directive as adopted in October 2003 omits to address the issue of 'double counting'. 'Double counting' refers to a situation where CERs or ERUs are issued as a result of reductions that also lead to a reduction in emissions from an installation covered by the ET Directive. Double counting could, for instance, occur if an installation decides to stop its on-site power generation, and instead buys its power from an external source. By doing so it would reduce its own GHG emissions and free up allowances that it can then sell into the EU ETS market. The operator could decide to buy electricity generated by renewable energy. The operator of the renewable electricity power plant could however also attempt to obtain JI or CDM credits for the installation of renewable electricity capacity. By freeing up allowances through moving power generation off-site and by giving credits for the generation of renewable electricity that replaces this, the reduction of one tonne of GHG emissions would be credited twice. The issue of double counting is not limited to a situation under which the ET Directive and the project based mechanisms are directly linked. It can also occur without a link between the two regimes, although in case of such a link the double counting issue could be brought directly into the EU trading regime if the CDM or JI credits are subsequently used in the trading scheme.

The Commission's linking proposal included a new Article 11(*ter*)(2) to be introduced into the ET Directive to prevent double counting from occurring. This requires Member States to ensure that no ERUs or CERs are *issued* for reductions of anthropogenic emissions of GHGs or removals by sinks resulting from project activities that reduce or claim to reduce GHG emissions from installations covered under the ET Directive. By prohibiting the *issuance* of CERs and ERUs rather than their use in the trading scheme, this provision also prevents any units generated through such projects from being traded and used for compliance outside the scope of the ET Directive. This provision prohibits JI and CDM to be used for installations covered under the ET Directive. Prohibiting the issuance of ERUs and CERs for emissions for installations covered under the ET Directive could however pose a problem for specific JI projects that are currently being implemented in accession countries. The moment those countries become an EU Member State and are subject to the ET Directive, these projects would no longer be able to issue ERUs and would thus cease to exist as JI projects. While JI projects could in principle be converted into emission trades under the ET Directive, such conversion could create legal and contractual difficulties for projects that are already in the pipeline. The linking proposal proposed to solve this issue through allowing Member States to issue ERUs for JI projects that are approved before 31 December 2004, or the date of the Member State accession, but no allowances may be allocated for emission reductions resulting from the JI project. It is likely that the number of projects wishing to use such an exemption would have been only very small. Project developers could instead have opted to move into the ET Directive and sell the reductions achieved through their JI project under the EU trading regime. A key question in such situation would however be the relation between the number of allowances allocated to an installation and the baseline under the JI project.

During the negotiations on the Linking Directive the Commission's proposed approach to avoid double counting came under criticism from a number of accession countries and a number of Member States, as well as from Japan. Japan felt that, by practically excluding JI projects in the accession countries, the EU would claim the JI reduction opportunities in those countries for itself and would restrict countries like Japan from accessing this market. Japan's reasoning is strengthened by its argument that early informally circulated drafts of the registries Regulation prohibit the export of EU allowances outside the EU, thus de facto keeping those reductions within the EU. It can however be argued that the ET Directive creates and important new market for Japanese energy-efficient technology, especially in accession countries. Any reductions gained through investments in such technology will free up allowances that can be sold into the EU market. Although Japanese investors will thus not have ERUs 'in hand', they can easily buy such ERUs from the revenue of the sale of allowances, or even directly swap allowances against ERUs. A number of Member States and accession countries were worried about the effect that the proposed approach would have on JI projects in general, and in particular on renewable energy and energy efficiency projects. By excluding these projects from JI, Member States were prevented from giving an additional financial incentive to the development of such projects.

The 'double counting' provision in the final text of the Directive was significantly amended. Although the general prohibition to issue credits for JI credits remains, this was expanded to also apply to CDM credits, in view of the fact that two of the new Member States that acceded to the EU on 1 May 2004 (Cyprus and Malta) are not listed in Annex B to the Kyoto Protocol. Most importantly, the exceptions to the general prohibition were expanded. Member States may now issue CDM and JI credits for emission reductions or limitations in installations falling within the scope of the Directive up to 31 December 2012 as long as an equal number of allowances are cancelled by the operator of that installation. Member States may also issue CDM and JI credits for emission reductions or limitations that indirectly affect the emissions of installations under the EU ETS if an equal number of allowances are cancelled from the registry of the Member State where the CDM or JI credits originate. The implementation of the new double counting provisions is subject to the adoption of additional requirements through the comitology procedure.

6 Conclusion

The discussion on linking emissions trading schemes is still very much at its beginning. Nonetheless, the EU, through the development and implementation of the EU ETS, has already gained important experience with the linking of trading schemes. Much of this experience relates to the level of harmonization that is necessary to allow newly developed domestic trading schemes to operate an open and transparent emission allowance market. While the EU has thus far only harmonized key elements of the EU ETS, there are already calls for further harmonization of specific elements of the EU ETS. The Commission will have to evaluate the need for further harmonization in its mid-2006 review.

Important experience has also been gained in the development of the Linking Directive, in particular in relation to linking an entity-level emissions trading scheme with the Kyoto mechanisms that were primarily created to facilitate states to comply with their Kyoto targets. These lessons have shown that a one-on-one translation of the Kyoto rules to the entity level may not always be desirable or even possible. The Linking Directive has however not solved all issues, in particular the question of the desirability of and technical approaches to including sinks credits in the EU ETS was postponed until the mid-2006 review.

By being a 'first mover' in designing a large-scale domestic trading scheme, linking decisions made in the EU ETS will provide useful experiences for linking decisions made in other trading schemes. The EU ETS is however also likely to be at the centre of future linking discussions. Not only does the ET Directive explicitly call for links with third countries to be established, but, because of its sheer size, the EU ETS is also attractive for third countries to link into. Linking the EU ETS with regimes in third countries will be one of the important challenges, but also an important opportunity that the EU ETS will have to face over the next few years.

PART VIII

CONCLUSION

Summary and Outlook

David Freestone and Charlotte Streck

A number of key themes shine through the contributions to this collection. The first and overarching theme is *originality and innovation*—which arises from the very novelty of the regime created by the Kyoto Protocol. Not only does this originality extend to the mechanisms of the CDM, JI, and International Emissions Trading but to the respective roles of international institutions, governments, nongovernmental organizations, and the private sector. Secondly, and related, is the issue of experimentation or *'learning by doing'*—a phrase which appears in many of the contributions. The Kyoto venture is seen as such a new and exciting path to travel that the emerging regulatory frameworks appear as work in progress. The concept applies not simply to the international regime created by the UNFCCC but to national implementation and trading schemes in both Annex I and non-Annex I countries. Thirdly, despite the fact that there is still considerable *regulatory uncertainty*, the private sector as well as the public sector is looking at ways in which it can manage the risks involved in making investments in climate friendly technology in developing countries and countries with economies in transition. Many have already made the choice that the risk of doing nothing is greater than the uncertainty they face in moving forward. So inevitably a major preoccupation of the technical legal contributions focuses on *risk management*. We will look at each of these issues in turn.

1 Originality and Innovation

The challenge of climate change itself has forced new thinking on the international community. Not only is the Kyoto Protocol one of the first international instruments to envisage market mechanisms, as Freestone points out in the first chapter of this book, but the scale of the challenge facing industrialized countries in moving to a lower and ultimately to a no carbon economy is awesome. To achieve such an objective requires a paradigm shift by the major industrialized economies. The legally binding emission reduction targets established for Annex I countries by the Kyoto Protocol constitute such a paradigm shift—however, these targets can only represent the beginning of a much more ambitious effort to curb global climate change. The Kyoto Protocol by itself will not achieve the sort of reductions that would even arrest the increases in greenhouse gas emissions let alone achieve stabilization.

A key aspect of the Kyoto approach is harnessing the investment power and energies of the private sector, thereby transposing the polluter-pays principle from the international to the domestic level: from Annex I countries to private legal entities operating in these countries. Faced with the possibility of regulatory caps upon their carbon intensive activities the private sector in Europe, Canada, and Japan—and in some States of the US—is beginning to take a proactive approach to investments which can either reduce their own emissions—or utilizing the differences in marginal abatement costs between their own economies and those of transition economies and developing countries—to finance greenhouse gas reduction activities off-shore. Market studies are already pointing out the likely price differentials between carbon purchases today in what is the early market and those at the end of the Kyoto Protocol's first commitment period in 2012 or even in a possible second commitment period. Will chief executives escape shareholder criticism in the future for not taking early action to abate or set-off greenhouse gas emissions from the companies that they run? However, before we reach that point the new commodity of the emission reduction needs to be processed by sellers, buyers, traders and brokers.

Newcomers will doubtless find it difficult to find their way through the jungle of terms and acronyms describing different emission rights. Assigned Amount Units, Certified Emission Reductions, Emission Reduction Units, EU Allowances are new creatures entering legal language, describing different type of international, domestic, private, or regulatory rights. Wemaere and Streck try to take the reader by the hand and shed light on the different classes of emission rights and their ambiguous legal nature. Of direct practical relevance, Casamento describes the equally novel discussion surrounding the classification of emission rights in the context of international accounting standards.

The mechanism described in Article 17 of the Kyoto Protocol establishes the rules for the transfer and acquisition of the emission rights defined by the Protocol. De Witt Wijnen describes in his chapter the mechanism for International Emissions Rights against the background of general principles governing other forms of emissions trading. In the shadow of the Kyoto Protocol, recent years have seen the establishment of a number of national greenhouse gas trading schemes. Although many of these schemes are still suffering childhood 'teething troubles', the trend alone evidences the willingness of policy makers to experiment with cost-efficient ways of reducing GHG emissions. We have not tried to be exhaustive in our coverage of these schemes but simply to give a flavour of the variety and approach of some national emissions trading schemes that have arisen. These are often based on very different principles within countries as diverse as Chile (discussed by Main S.), Australia (Carmody and Willis), Canada (Gehring and

Chambers), and the UK (Dodwell). Overarching the European domestic schemes there is the European Emissions Trading Scheme (Dornau) which could represent a climate change mitigation regime even in the absence of Kyoto. The European Emissions Trading Scheme is currently in the process of being transposed into the twenty-five domestic emission trading schemes. Although emission trading schemes are all based on the creation of tradable emission rights, the rules that govern the different systems vary between the different domestic as well as between the domestic and international system. Marr gives us a flavour of the challenges that the integration of the scheme into national law brings when he describes how the European system is being integrated into the German legal system. Eventually of course, the domestic schemes of all the EU countries will be linked and will merge into one marketplace. The challenges that come with linking the different regimes are described by Lefevere. However, the issues that need to be considered when linking domestic regimes within the EU take on even greater complexity where the EU system allows for links with the Kyoto Protocol mechanisms.

2 Learning by Doing

In 1999 the World Bank collected substantial commitment from six governments and seventeen private sector participants in order to establish the Prototype Carbon Fund (PCF). With the PCF, the World Bank took on the pioneering challenge of developing the modalities for financing projects which might generate emission reductions which might be eligible for credit under both Articles 6 and 12 of the Kyoto Protocol. The first PCF Emission Reductions Purchase Agreement governing such project was negotiated in 2000. Around the same time the Dutch Government opened its first Emission Reduction Procurement Tender (ERUPT). Ever since then, the PCF and the ERUPT programme have pioneered contractual arrangements governing CDM and JI projects providing for the generation and transfer of emission credits (described by van der Weerd and Streck respectively). The PCF and ERUPT have followed different approaches in selecting and structuring the 'carbon' intervention in such projects, and yet share many of the same experiences: challenges in interpreting the—often ambiguous—rules, in exploring baseline methodologies—initially even before the establishment of the CDM Executive Board, in drafting contracts to share risks—while the international negotiations to finalize the rules were still taking place. These challenges have made the financing of projects under the Kyoto mechanisms a continuous and exciting learning experience.

The UNFCCC negotiators decided that the CDM should have a 'prompt start' and not wait for the finalization of all the rules—nor even the coming into force of the Protocol. Early implementation efforts had nothing else to rely on therefore but the text of Article 12. For the early 'implementers', trying to give effect to the intent of the negotiators of a mechanism which had been conceived and born during some long nights of negotiation in Kyoto in 1997, was not an easy task. It took four years of intense negotiations from the date of the adoption of the Kyoto Protocol until the Marrakesh Accords gave more detailed guidance on the modalities and procedures of the CDM. By then the implementation of the first CDM projects was already under way, and much of the understanding and experience gained in these projects was fed back to the negotiating table at Marrakesh.

In the spirit of the concepts that emerged from the 2001 World Summit on Sustainable Development, the CDM shows a promising path for new forms of cooperation between public and private entities and creates a fertile platform for partnerships among different sectors of the society. The agenda of the CDM includes contributing to a low carbon environment and sustainable development in this sense and one of its early impacts has been to bring together a wide variety of public and private actors. As Schmidt and Netto point out, the unique features of the CDM have resulted in a governance structure and a CDM project cycle which provide for a number of innovations for a multilateral mechanism. At an institutional level the international negotiations have developed public participation and disclosure requirements, which Eddy suggests are 'near to the state of the art', and concerns about not allowing the trading agenda to undermine the environmental integrity have resulted in a range of new procedures to balance these—as Werksman and Meijer discuss. These requirements also place a not insubstantial role on the host countries which need to approve and agree to projects and also to satisfy themselves that they achieve nationally agreed sustainable development criteria.

Many of the concepts included in the Protocol text for political reasons have triggered an extensive learning exercise within national systems. Huq and Reid describe the way that countries have had to appreciate and internalize the concept of benefit sharing and develop workable domestic definitions of sustainable development. New systems of domestic regulation are important but as O'Sullivan and Cormier show us, they should not be so bureaucratic as to deter private sector investors. A number of outstanding international difficulties for the Kyoto Mechanisms, such as the entry into force of the Protocol, are beyond the control of individual host countries. However, other challenges, such as ensuring that Annex I and non-Annex I countries accurately identify and efficiently carry out their responsibilities while minimizing conflicts of interest, are within each developing country's control. The market approach requires some rethinking of traditional approaches to environmental policy. There are a number of challenges and potential pitfalls which require careful analysis of national energy and environmental policies. However, the CDM can provide important and possible unrivalled opportunities for developing countries to gain access to clean and efficient technology and to attract environmentally and socially responsible foreign investment.

As if this were not enough, before the world community had begun to fully appreciate all the implications of this newly created commodity—the 'emission reduction'—the international negotiation process took a further major step. At the ninth session of the Parties to the UNFCCC in Milan in December 2003 a new sort of credit was created: as Scholz and Noble describe, CDM projects in the area of land use, land use change and forestry (LULUCF) pose a number of novel methodological problems and create carbon sequestration credits which actually expire. Questions regarding ownership, taxation, and accounting for these types of credits becomes even more important. While the UNFCCC negotiators are still deliberating on how these sequestration credits will actually work in practice, the first afforestation and reforestation projects are already under implementation. Bosquet describes for us how this regulatory uncertainty has been addressed in innovative contractual arrangements and also the way that a catalytic fund such as the newly established World Bank BioCarbon Fund can assist the Parties to the Convention and the Protocol to understand the practical implications of their negotiated agreements.

3 Uncertainty

Running thorough the volume is a central concern about dealing with uncertainty. It is describing a regime which has been developed to a remarkable degree of sophistication despite the fact that its central legal instrument—the Kyoto Protocol—has at the time of writing still not achieved sufficient ratifications to enter into force.

Joint Implementation, the other project-based mechanism defined by the Kyoto Protocol in its Article 6, does not provide for the prompt start which is envisaged for the CDM. The operation of JI depends on the entry into force of the Protocol. When deciding about the CDM in Marrakesh, the international community postponed the formulation of more refined guidelines on JI until the years closer to the start of the first commitment period of the Protocol in 2008. The Marrakesh Accords contain a number of guidelines for the implementation of Article 6 and recommend them for adoption by the first session of the Conference of the Parties of the UNFCCC serving as the Meeting of the Parties to the Kyoto Protocol after the entry into force of the Kyoto Protocol. Yet, when compared with the detailed modalities developed to govern the CDM, the JI text included in the Marrakesh Accords is still vague and incomplete. However the sort of investment necessary to reduce GHG emissions has a long lead time, so in order to have any significant impact the implementation of JI cannot wait. The result is that the often complex financing of such projects has to be structured in a way that addresses the regulatory uncertainties.

While JI projects can be implemented among any Annex I countries, transition countries are expected to play a key role as hosts in the implementation of JI. The Russian Federation and almost all of the Eastern European countries have large amounts of surplus AAUs available. The existence of such 'hot air' places most Economies in Transition in a position where they could sell their surplus quotas on the market. Alternatively, they could engage in JI or simply 'bank' these credits to provide for future growth. Väyrynen and Lecqoc point out that the first step for all potential JI host Parties is to develop a comprehensive strategy on how to manage their AAU assets and to assess which mechanism they will utilize to yield the maximum benefit to the global climate system and their own domestic economies. An AAU management system ensures that a country is able to explore all investment opportunities; it can promote energy efficiency and an increasing energy mix in a country. It can also attract foreign direct investments while ensuring long-term reductions in GHG emissions. Lastly and by no means least, it supports the Kyoto Protocol and builds infrastructure for International Emissions Trading.

4 Managing Risk

The continuously evolving regulatory system discussed above has resulted in a situation where emission reduction transactions have had to be implemented in an environment characterized by significant risks. Internationally, the Kyoto Protocol is still not in force, domestically, governments have yet to integrate transactions involving the transfer of emission reduction credits into their national legislation. Lawyers that are asked to draft contracts for CDM or JI transactions have to familiarize themselves rapidly with the special type of risks that come with such transactions. Apart from the risks arising from the regulatory system or the absence thereof, there is the need to secure continued host country support, the risk of failing to obtain international registration, the need to ensure credibility of the monitoring and verification system, the risk of changing baselines and problems arising from the trans-national and cross-border nature of the cooperation required. All these risks as well as the immature nature of the carbon market, Pogány reminds us, require a careful evaluation of the inherent risks of a carbon transaction. Due to the novelty of the business and the lack of experience in most actors, many of these risks are hard to manage, so that Singhania asks whether the time and the costs involved in negotiating and entering into an Emissions Reductions Purchase Agreement are justified at this point. Because of continued uncertainty about the returns from such investments the high transaction costs do not seem commensurate with the returns a project entity may expect. This is especially true when the payments for emission reductions under a carbon contract will only accrue to a project entity if and when the Kyoto Protocol has entered into force. To avoid bias or unfairness in contracts, they need to be based on a clear, understood and fair sharing of risks and benefits. Emission rights are a new commodity, and many existing contractual structures and risk mitigation techniques can be drawn upon in order for parties to agree and articulate commercial terms for their sale and purchase. However, the origin of emission reductions does require particular attention when defining the terms of their creation by emission reduction projects and sale and transfer between parties (Wilder, Willis, Guli). Robust contracts are the key for successful CDM or JI projects, as such transactions involving a long-term contractual relationship between the two parties which, like any other long-term contract, may require continuous cooperation over several years or even decades.

Wilder describes in a separate chapter, the additional risk and insecurities when non-Party public or private entities wish to participate in the Kyoto mechanisms. It is clear that governments from non-Party Annex I governments will not be able to participate directly in the flexible mechanisms. However, the rules do not exclude the possibility that Parties to the Protocol may authorize project participants from non-Party jurisdictions, or may enable companies from such jurisdictions to hold registry accounts within their national registries. Registries are indeed a central part of any sophisticated system of emissions trading whether International Emissions Trading pursuant to Article 17 of the Kyoto Protocol or regional trading such as will take place under the EU-wide Emissions Trading Scheme. A reliable registry system in itself is a risk mitigation tool as it accounts for any transfer of registered emission rights. Hobley and Hawkes make the case that a fully functional registry system is a key component for the development of a liquid and transparent emissions trading market.

The contributors to this book generally agree that despite the risk that the Kyoto Protocol may not enter into force, emissions reduction investment and trading, all in the name of sustainable development, is taking place and will continue to take place. Private parties and States or State agencies will be involved in these transactions, and will seek to mitigate their risk exposure through carefully drafted contracts. The nature of the business will bring together private and public actors coming from different legal systems and cultures, and the convergence of these systems could lead to different interpretations of contractual obligations—precipitating a dispute, and perhaps leading on to litigation or arbitration. Ratliff analyses in his chapter the potential for dispute and the general advantages of arbitration, including party autonomy to structure the process.

The overarching objective of this book has been to demonstrate the key issues that need to be addressed to make the operations of the Kyoto mechanisms a success. What has emerged also is the vital role that legal issues and lawyers need to play in this venture. As an international environmental law instrument the Kyoto Protocol is itself highly innovative in that way it imposes such onerous commitments on Annex I countries and the way that it envisages market mechanisms operating to reduce the overall costs to these countries in meeting these commitments. In so doing it opens up new opportunities to harness the particular talents of the private sector to use innovative carbon finance instruments to contribute to sustainable development in a range of ways—from the financing of clean energy production to sustainable land use techniques. However as we have seen the operation of the mechanisms themselves requires a whole range of new skills, including many commercial legal skills such as an understanding of trading regimes, securities and commodities regulation, and contractual risk sharing and mitigation.

At the time of finalization of this manuscript uncertainty still remains over the entry into force of the Kyoto Protocol itself, nevertheless it is clear that carbon trading is now a reality and this new area of legal practice is becoming increasingly mainstreamed. For the editors it is very gratifying to have gathered such a wealth of talent and insights within the covers of this book in an attempt to contribute to the clarification and development of the legal aspects of the complex issues involved in this very important area. Most climate scientists are agreed that anthropogenic impacts on the global climate system have already begun; the stakes are very high.

APPENDICES

APPENDIX 1

Kyoto Protocol to the United Nations Framework Convention On Climate Change

The Parties to this Protocol,

Being Parties to the United Nations Framework Convention on Climate Change, hereinafter referred to as 'the Convention',

In pursuit of the ultimate objective of the Convention as stated in its Article 2,

Recalling the provisions of the Convention,

Being guided by Article 3 of the Convention,

Pursuant to the Berlin Mandate adopted by decision 1/CP.1 of the Conference of the Parties to the Convention at its first session,

Have agreed as follows:

ARTICLE 1

For the purposes of this Protocol, the definitions contained in Article 1 of the Convention shall apply. In addition:

1. 'Conference of the Parties' means the Conference of the Parties to the Convention.

2. 'Convention' means the United Nations Framework Convention on Climate Change, adopted in New York on 9 May 1992.

3. 'Intergovernmental Panel on Climate Change' means the Intergovernmental Panel on Climate Change established in 1988 jointly by the World Meteorological Organization and the United Nations Environment Programme.

4. 'Montreal Protocol' means the Montreal Protocol on Substances that Deplete the Ozone Layer, adopted in Montreal on 16 September 1987 and as subsequently adjusted and amended.

5. 'Parties present and voting' means Parties present and casting an affirmative or negative vote.

6. 'Party' means, unless the context otherwise indicates, a Party to this Protocol.

7. 'Party included in Annex I' means a Party included in Annex I to the Convention, as may be amended, or a Party which has made a notification under Article 4, paragraph 2(g), of the Convention.

ARTICLE 2

1. Each Party included in Annex I, in achieving its quantified emission limitation and reduction commitments under Article 3, in order to promote sustainable development, shall:

(a) Implement and/or further elaborate policies and measures in accordance with its national circumstances, such as:

(i) Enhancement of energy efficiency in relevant sectors of the national economy;

(ii) Protection and enhancement of sinks and reservoirs of greenhouse gases not controlled by the Montreal Protocol, taking into account its commitments under relevant international environmental agreements; promotion of sustainable forest management practices, afforestation and reforestation;

(iii) Promotion of sustainable forms of agriculture in light of climate change considerations;

(iv) Research on, and promotion, development and increased use of, new and renewable forms of energy, of carbon dioxide sequestration technologies and of advanced and innovative environmentally sound technologies;

(v) Progressive reduction or phasing out of market imperfections, fiscal incentives, tax and duty exemptions and subsidies in all greenhouse gas emitting sectors that run counter to the objective of the Convention and application of market instruments;

(vi) Encouragement of appropriate reforms in relevant sectors aimed at promoting policies and measures which limit or reduce emissions of greenhouse gases not controlled by the Montreal Protocol;

(vii) Measures to limit and/or reduce emissions of greenhouse gases not controlled by the Montreal Protocol in the transport sector;

(viii) Limitation and/or reduction of methane emissions through recovery and use in waste management, as well as in the production, transport and distribution of energy;

(b) Cooperate with other such Parties to enhance the individual and combined effectiveness of their policies and measures adopted under this Article, pursuant to Article 4, paragraph 2(e)(i), of the Convention. To this end, these Parties shall take steps to share their experience and exchange information on such policies and measures, including developing ways of improving their comparability, transparency and effectiveness. The Conference of the Parties serving as the meeting of the Parties to this Protocol shall, at its first session or as soon as practicable thereafter, consider ways to facilitate such cooperation, taking into account all relevant information.

2. The Parties included in Annex I shall pursue limitation or reduction of emissions of greenhouse gases not controlled by the Montreal Protocol from aviation and marine bunker fuels, working through the International Civil Aviation Organization and the International Maritime Organization, respectively.

3. The Parties included in Annex I shall strive to implement policies and measures under this Article in such a way as to minimize adverse effects, including the adverse effects of climate change, effects on international trade, and social, environmental and economic impacts on other Parties, especially developing country Parties and in particular those identified in Article 4, paragraphs 8 and 9, of the Convention, taking into account Article 3 of the Convention. The Conference of the Parties serving as the meeting of the Parties to this Protocol may take further action, as appropriate, to promote the implementation of the provisions of this paragraph.

4. The Conference of the Parties serving as the meeting of the Parties to this Protocol, if it decides that it would be beneficial to coordinate any of the policies and measures in paragraph 1(a) above, taking into account different national circumstances and potential effects, shall consider ways and means to elaborate the coordination of such policies and measures.

ARTICLE 3

1. The Parties included in Annex I shall, individually or jointly, ensure that their aggregate anthropogenic carbon dioxide equivalent emissions of the greenhouse gases listed in Annex A do not exceed their assigned amounts, calculated pursuant to their quantified emission limitation and reduction commitments inscribed in Annex B and in accordance with the provisions of this Article, with a view to reducing their overall emissions of such gases by at least 5 per cent below 1990 levels in the commitment period 2008 to 2012.

2. Each Party included in Annex I shall, by 2005, have made demonstrable progress in achieving its commitments under this Protocol.

3. The net changes in greenhouse gas emissions by sources and removals by sinks resulting from direct human-induced land-use change and forestry activities, limited to afforestation, reforestation and deforestation since 1990, measured as verifiable changes in carbon stocks in each commitment period, shall be used to meet the commitments under this Article of each Party included in Annex I. The greenhouse gas emissions by sources and removals by sinks associated with those activities shall be reported in a transparent and verifiable manner and reviewed in accordance with Articles 7 and 8.

4. Prior to the first session of the Conference of the Parties serving as the meeting of the Parties to this Protocol, each Party included in Annex I shall provide, for consideration by the Subsidiary Body for Scientific and Technological Advice, data to establish its level of carbon stocks in 1990 and to enable an estimate to be made of its changes in carbon stocks in subsequent years. The Conference of the Parties serving as the meeting of the Parties to this Protocol shall, at its first session or as soon as practicable thereafter, decide upon modalities, rules and guidelines as to how, and which, additional human-induced activities related to changes in greenhouse gas emissions by sources and removals by sinks in the agricultural soils and the land-use change and forestry categories shall be added to, or subtracted from, the assigned amounts for Parties included in Annex I, taking into account uncertainties, transparency in reporting, verifiability, the methodological work of the Intergovernmental Panel on Climate Change, the advice provided by the Subsidiary Body for Scientific and Technological Advice in accordance with Article 5 and the decisions of the Conference of the Parties. Such a decision shall apply in the second and subsequent commitment periods. A Party may choose to apply such a decision on these additional human-induced activities for its first commitment period, provided that these activities have taken place since 1990.

5. The Parties included in Annex I undergoing the process of transition to a market economy whose base year or period was established pursuant to decision 9/CP.2 of the Conference of the Parties at its second session shall use that base year or period for the implementation of their commitments under this Article. Any other Party included in Annex I undergoing the process of transition to a market economy which has not yet submitted its first national communication under Article 12 of the Convention may also notify the Conference of the Parties serving as the meeting of the Parties to this Protocol that it intends to use an historical base year or period other than 1990 for the implementation of its commitments under this Article. The Conference of the Parties serving as the meeting of the Parties to this Protocol shall decide on the acceptance of such notification.

6. Taking into account Article 4, paragraph 6, of the Convention, in the implementation of their commitments under this Protocol other than those under this Article, a certain

degree of flexibility shall be allowed by the Conference of the Parties serving as the meeting of the Parties to this Protocol to the Parties included in Annex I undergoing the process of transition to a market economy.

7. In the first quantified emission limitation and reduction commitment period, from 2008 to 2012, the assigned amount for each Party included in Annex I shall be equal to the percentage inscribed for it in Annex B of its aggregate anthropogenic carbon dioxide equivalent emissions of the greenhouse gases listed in Annex A in 1990, or the base year or period determined in accordance with paragraph 5 above, multiplied by five. Those Parties included in Annex I for whom land-use change and forestry constituted a net source of greenhouse gas emissions in 1990 shall include in their 1990 emissions base year or period the aggregate anthropogenic carbon dioxide equivalent emissions by sources minus removals by sinks in 1990 from land-use change for the purposes of calculating their assigned amount.

8. Any Party included in Annex I may use 1995 as its base year for hydrofluorocarbons, perfluorocarbons and sulphur hexafluoride, for the purposes of the calculation referred to in paragraph 7 above.

9. Commitments for subsequent periods for Parties included in Annex I shall be established in amendments to Annex B to this Protocol, which shall be adopted in accordance with the provisions of Article 21, paragraph 7. The Conference of the Parties serving as the meeting of the Parties to this Protocol shall initiate the consideration of such commitments at least seven years before the end of the first commitment period referred to in paragraph 1 above.

10. Any emission reduction units, or any part of an assigned amount, which a Party acquires from another Party in accordance with the provisions of Article 6 or of Article 17 shall be added to the assigned amount for the acquiring Party.

11. Any emission reduction units, or any part of an assigned amount, which a Party transfers to another Party in accordance with the provisions of Article 6 or of Article 17 shall be subtracted from the assigned amount for the transferring Party.

12. Any certified emission reductions which a Party acquires from another Party in accordance with the provisions of Article 12 shall be added to the assigned amount for the acquiring Party.

13. If the emissions of a Party included in Annex I in a commitment period are less than its assigned amount under this Article, this difference shall, on request of that Party, be added to the assigned amount for that Party for subsequent commitment periods.

14. Each Party included in Annex I shall strive to implement the commitments mentioned in paragraph 1 above in such a way as to minimize adverse social, environmental and economic impacts on developing country Parties, particularly those identified in Article 4, paragraphs 8 and 9, of the Convention. In line with relevant decisions of the Conference of the Parties on the implementation of those paragraphs, the Conference of the Parties serving as the meeting of the Parties to this Protocol shall, at its first session, consider what actions are necessary to minimize the adverse effects of climate change and/or the impacts of response measures on Parties referred to in those paragraphs. Among the issues to be considered shall be the establishment of funding, insurance and transfer of technology.

ARTICLE 4

1. Any Parties included in Annex I that have reached an agreement to fulfil their commitments under Article 3 jointly, shall be deemed to have met those commitments provided that

their total combined aggregate anthropogenic carbon dioxide equivalent emissions of the greenhouse gases listed in Annex A do not exceed their assigned amounts calculated pursuant to their quantified emission limitation and reduction commitments inscribed in Annex B and in accordance with the provisions of Article 3. The respective emission level allocated to each of the Parties to the agreement shall be set out in that agreement.

2. The Parties to any such agreement shall notify the secretariat of the terms of the agreement on the date of deposit of their instruments of ratification, acceptance or approval of this Protocol, or accession thereto. The secretariat shall in turn inform the Parties and signatories to the Convention of the terms of the agreement.

3. Any such agreement shall remain in operation for the duration of the commitment period specified in Article 3, paragraph 7.

4. If Parties acting jointly do so in the framework of, and together with, a regional economic integration organization, any alteration in the composition of the organization after adoption of this Protocol shall not affect existing commitments under this Protocol. Any alteration in the composition of the organization shall only apply for the purposes of those commitments under Article 3 that are adopted subsequent to that alteration.

5. In the event of failure by the Parties to such an agreement to achieve their total combined level of emission reductions, each Party to that agreement shall be responsible for its own level of emissions set out in the agreement.

6. If Parties acting jointly do so in the framework of, and together with, a regional economic integration organization which is itself a Party to this Protocol, each member State of that regional economic integration organization individually, and together with the regional economic integration organization acting in accordance with Article 24, shall, in the event of failure to achieve the total combined level of emission reductions, be responsible for its level of emissions as notified in accordance with this Article.

ARTICLE 5

1. Each Party included in Annex I shall have in place, no later than one year prior to the start of the first commitment period, a national system for the estimation of anthropogenic emissions by sources and removals by sinks of all greenhouse gases not controlled by the Montreal Protocol. Guidelines for such national systems, which shall incorporate the methodologies specified in paragraph 2 below, shall be decided upon by the Conference of the Parties serving as the meeting of the Parties to this Protocol at its first session.

2. Methodologies for estimating anthropogenic emissions by sources and removals by sinks of all greenhouse gases not controlled by the Montreal Protocol shall be those accepted by the Intergovernmental Panel on Climate Change and agreed upon by the Conference of the Parties at its third session. Where such methodologies are not used, appropriate adjustments shall be applied according to methodologies agreed upon by the Conference of the Parties serving as the meeting of the Parties to this Protocol at its first session. Based on the work of, *inter alia*, the Intergovernmental Panel on Climate Change and advice provided by the Subsidiary Body for Scientific and Technological Advice, the Conference of the Parties serving as the meeting of the Parties to this Protocol shall regularly review and, as appropriate, revise such methodologies and adjustments, taking fully into account any relevant decisions by the Conference of the Parties. Any revision to methodologies or adjustments shall be used only for the purposes of ascertaining

compliance with commitments under Article 3 in respect of any commitment period adopted subsequent to that revision.

3. The global warming potentials used to calculate the carbon dioxide equivalence of anthropogenic emissions by sources and removals by sinks of greenhouse gases listed in Annex A shall be those accepted by the Intergovernmental Panel on Climate Change and agreed upon by the Conference of the Parties at its third session. Based on the work of, *inter alia*, the Intergovernmental Panel on Climate Change and advice provided by the Subsidiary Body for Scientific and Technological Advice, the Conference of the Parties serving as the meeting of the Parties to this Protocol shall regularly review and, as appropriate, revise the global warming potential of each such greenhouse gas, taking fully into account any relevant decisions by the Conference of the Parties. Any revision to a global warming potential shall apply only to commitments under Article 3 in respect of any commitment period adopted subsequent to that revision.

ARTICLE 6

1. For the purpose of meeting its commitments under Article 3, any Party included in Annex I may transfer to, or acquire from, any other such Party emission reduction units resulting from projects aimed at reducing anthropogenic emissions by sources or enhancing anthropogenic removals by sinks of greenhouse gases in any sector of the economy, provided that:

(a) Any such project has the approval of the Parties involved;

(b) Any such project provides a reduction in emissions by sources, or an enhancement of removals by sinks, that is additional to any that would otherwise occur;

(c) It does not acquire any emission reduction units if it is not in compliance with its obligations under Articles 5 and 7; and

(d) The acquisition of emission reduction units shall be supplemental to domestic actions for the purposes of meeting commitments under Article 3.

2. The Conference of the Parties serving as the meeting of the Parties to this Protocol may, at its first session or as soon as practicable thereafter, further elaborate guidelines for the implementation of this Article, including for verification and reporting.

3. A Party included in Annex I may authorize legal entities to participate, under its responsibility, in actions leading to the generation, transfer or acquisition under this Article of emission reduction units.

4. If a question of implementation by a Party included in Annex I of the requirements referred to in this Article is identified in accordance with the relevant provisions of Article 8, transfers and acquisitions of emission reduction units may continue to be made after the question has been identified, provided that any such units may not be used by a Party to meet its commitments under Article 3 until any issue of compliance is resolved.

ARTICLE 7

1. Each Party included in Annex I shall incorporate in its annual inventory of anthropogenic emissions by sources and removals by sinks of greenhouse gases not controlled by the Montreal Protocol, submitted in accordance with the relevant decisions of the Conference of

the Parties, the necessary supplementary information for the purposes of ensuring compliance with Article 3, to be determined in accordance with paragraph 4 below.

2. Each Party included in Annex I shall incorporate in its national communication, submitted under Article 12 of the Convention, the supplementary information necessary to demonstrate compliance with its commitments under this Protocol, to be determined in accordance with paragraph 4 below.

3. Each Party included in Annex I shall submit the information required under paragraph 1 above annually, beginning with the first inventory due under the Convention for the first year of the commitment period after this Protocol has entered into force for that Party. Each such Party shall submit the information required under paragraph 2 above as part of the first national communication due under the Convention after this Protocol has entered into force for it and after the adoption of guidelines as provided for in paragraph 4 below. The frequency of subsequent submission of information required under this Article shall be determined by the Conference of the Parties serving as the meeting of the Parties to this Protocol, taking into account any timetable for the submission of national communications decided upon by the Conference of the Parties.

4. The Conference of the Parties serving as the meeting of the Parties to this Protocol shall adopt at its first session, and review periodically thereafter, guidelines for the preparation of the information required under this Article, taking into account guidelines for the preparation of national communications by Parties included in Annex I adopted by the Conference of the Parties. The Conference of the Parties serving as the meeting of the Parties to this Protocol shall also, prior to the first commitment period, decide upon modalities for the accounting of assigned amounts.

ARTICLE 8

1. The information submitted under Article 7 by each Party included in Annex I shall be reviewed by expert review teams pursuant to the relevant decisions of the Conference of the Parties and in accordance with guidelines adopted for this purpose by the Conference of the Parties serving as the meeting of the Parties to this Protocol under paragraph 4 below. The information submitted under Article 7, paragraph 1, by each Party included in Annex I shall be reviewed as part of the annual compilation and accounting of emissions inventories and assigned amounts. Additionally, the information submitted under Article 7, paragraph 2, by each Party included in Annex I shall be reviewed as part of the review of communications.

2. Expert review teams shall be coordinated by the secretariat and shall be composed of experts selected from those nominated by Parties to the Convention and, as appropriate, by intergovernmental organizations, in accordance with guidance provided for this purpose by the Conference of the Parties.

3. The review process shall provide a thorough and comprehensive technical assessment of all aspects of the implementation by a Party of this Protocol. The expert review teams shall prepare a report to the Conference of the Parties serving as the meeting of the Parties to this Protocol, assessing the implementation of the commitments of the Party and identifying any potential problems in, and factors influencing, the fulfilment of commitments. Such reports shall be circulated by the secretariat to all Parties to the Convention. The secretariat shall list those questions of implementation indicated in such reports for further

consideration by the Conference of the Parties serving as the meeting of the Parties to this Protocol.

4. The Conference of the Parties serving as the meeting of the Parties to this Protocol shall adopt at its first session, and review periodically thereafter, guidelines for the review of implementation of this Protocol by expert review teams taking into account the relevant decisions of the Conference of the Parties.

5. The Conference of the Parties serving as the meeting of the Parties to this Protocol shall, with the assistance of the Subsidiary Body for Implementation and, as appropriate, the Subsidiary Body for Scientific and Technological Advice, consider:

(a) The information submitted by Parties under Article 7 and the reports of the expert reviews thereon conducted under this Article; and

(b) Those questions of implementation listed by the secretariat under paragraph 3 above, as well as any questions raised by Parties.

6. Pursuant to its consideration of the information referred to in paragraph 5 above, the Conference of the Parties serving as the meeting of the Parties to this Protocol shall take decisions on any matter required for the implementation of this Protocol.

ARTICLE 9

1. The Conference of the Parties serving as the meeting of the Parties to this Protocol shall periodically review this Protocol in the light of the best available scientific information and assessments on climate change and its impacts, as well as relevant technical, social and economic information. Such reviews shall be coordinated with pertinent reviews under the Convention, in particular those required by Article 4, paragraph 2(d), and Article 7, paragraph 2(a), of the Convention. Based on these reviews, the Conference of the Parties serving as the meeting of the Parties to this Protocol shall take appropriate action.

2. The first review shall take place at the second session of the Conference of the Parties serving as the meeting of the Parties to this Protocol. Further reviews shall take place at regular intervals and in a timely manner.

ARTICLE 10

All Parties, taking into account their common but differentiated responsibilities and their specific national and regional development priorities, objectives and circumstances, without introducing any new commitments for Parties not included in Annex I, but reaffirming existing commitments under Article 4, paragraph 1, of the Convention, and continuing to advance the implementation of these commitments in order to achieve sustainable development, taking into account Article 4, paragraphs 3, 5 and 7, of the Convention, shall:

(a) Formulate, where relevant and to the extent possible, cost-effective national and, where appropriate, regional programmes to improve the quality of local emission factors, activity data and/or models which reflect the socio-economic conditions of each Party for the preparation and periodic updating of national inventories of anthropogenic emissions by sources and removals by sinks of all greenhouse gases not controlled by the Montreal Protocol, using comparable methodologies to be agreed upon by the Conference of the Parties, and consistent with the guidelines for the preparation of national communications adopted by the Conference of the Parties;

(b) Formulate, implement, publish and regularly update national and, where appropriate, regional programmes containing measures to mitigate climate change and measures to facilitate adequate adaptation to climate change:

- (i) Such programmes would, *inter alia*, concern the energy, transport and industry sectors as well as agriculture, forestry and waste management. Furthermore, adaptation technologies and methods for improving spatial planning would improve adaptation to climate change; and
- (ii) Parties included in Annex I shall submit information on action under this Protocol, including national programmes, in accordance with Article 7; and other Parties shall seek to include in their national communications, as appropriate, information on programmes which contain measures that the Party believes contribute to addressing climate change and its adverse impacts, including the abatement of increases in greenhouse gas emissions, and enhancement of and removals by sinks, capacity building and adaptation measures;

(c) Cooperate in the promotion of effective modalities for the development, application and diffusion of, and take all practicable steps to promote, facilitate and finance, as appropriate, the transfer of, or access to, environmentally sound technologies, know-how, practices and processes pertinent to climate change, in particular to developing countries, including the formulation of policies and programmes for the effective transfer of environmentally sound technologies that are publicly owned or in the public domain and the creation of an enabling environment for the private sector, to promote and enhance the transfer of, and access to, environmentally sound technologies;

(d) Cooperate in scientific and technical research and promote the maintenance and the development of systematic observation systems and development of data archives to reduce uncertainties related to the climate system, the adverse impacts of climate change and the economic and social consequences of various response strategies, and promote the development and strengthening of endogenous capacities and capabilities to participate in international and intergovernmental efforts, programmes and networks on research and systematic observation, taking into account Article 5 of the Convention;

(e) Cooperate in and promote at the international level, and, where appropriate, using existing bodies, the development and implementation of education and training programmes, including the strengthening of national capacity building, in particular human and institutional capacities and the exchange or secondment of personnel to train experts in this field, in particular for developing countries, and facilitate at the national level public awareness of, and public access to information on, climate change. Suitable modalities should be developed to implement these activities through the relevant bodies of the Convention, taking into account Article 6 of the Convention;

(f) Include in their national communications information on programmes and activities undertaken pursuant to this Article in accordance with relevant decisions of the Conference of the Parties; and

(g) Give full consideration, in implementing the commitments under this Article, to Article 4, paragraph 8, of the Convention.

ARTICLE 11

1. In the implementation of Article 10, Parties shall take into account the provisions of Article 4, paragraphs 4, 5, 7, 8 and 9, of the Convention.

2. In the context of the implementation of Article 4, paragraph 1, of the Convention, in accordance with the provisions of Article 4, paragraph 3, and Article 11 of the Convention, and through the entity or entities entrusted with the operation of the financial mechanism of the Convention, the developed country Parties and other developed Parties included in Annex II to the Convention shall:

(a) Provide new and additional financial resources to meet the agreed full costs incurred by developing country Parties in advancing the implementation of existing commitments under Article 4, paragraph 1(a), of the Convention that are covered in Article 10, subparagraph (a); and

(b) Also provide such financial resources, including for the transfer of technology, needed by the developing country Parties to meet the agreed full incremental costs of advancing the implementation of existing commitments under Article 4, paragraph 1, of the Convention that are covered by Article 10 and that are agreed between a developing country Party and the international entity or entities referred to in Article 11 of the Convention, in accordance with that Article.

The implementation of these existing commitments shall take into account the need for adequacy and predictability in the flow of funds and the importance of appropriate burden sharing among developed country Parties. The guidance to the entity or entities entrusted with the operation of the financial mechanism of the Convention in relevant decisions of the Conference of the Parties, including those agreed before the adoption of this Protocol, shall apply *mutatis mutandis* to the provisions of this paragraph.

3. The developed country Parties and other developed Parties in Annex II to the Convention may also provide, and developing country Parties avail themselves of, financial resources for the implementation of Article 10, through bilateral, regional and other multilateral channels.

ARTICLE 12

1. A clean development mechanism is hereby defined.

2. The purpose of the clean development mechanism shall be to assist Parties not included in Annex I in achieving sustainable development and in contributing to the ultimate objective of the Convention, and to assist Parties included in Annex I in achieving compliance with their quantified emission limitation and reduction commitments under Article 3.

3. Under the clean development mechanism:

(a) Parties not included in Annex I will benefit from project activities resulting in certified emission reductions; and

(b) Parties included in Annex I may use the certified emission reductions accruing from such project activities to contribute to compliance with part of their quantified emission limitation and reduction commitments under Article 3, as determined by the Conference of the Parties serving as the meeting of the Parties to this Protocol.

4. The clean development mechanism shall be subject to the authority and guidance of the Conference of the Parties serving as the meeting of the Parties to this Protocol and be supervised by an executive board of the clean development mechanism.

5. Emission reductions resulting from each project activity shall be certified by operational entities to be designated by the Conference of the Parties serving as the meeting of the Parties to this Protocol, on the basis of:

(a) Voluntary participation approved by each Party involved;

(b) Real, measurable, and long-term benefits related to the mitigation of climate change; and

(c) Reductions in emissions that are additional to any that would occur in the absence of the certified project activity.

6. The clean development mechanism shall assist in arranging funding of certified project activities as necessary.

7. The Conference of the Parties serving as the meeting of the Parties to this Protocol shall, at its first session, elaborate modalities and procedures with the objective of ensuring transparency, efficiency and accountability through independent auditing and verification of project activities.

8. The Conference of the Parties serving as the meeting of the Parties to this Protocol shall ensure that a share of the proceeds from certified project activities is used to cover administrative expenses as well as to assist developing country Parties that are particularly vulnerable to the adverse effects of climate change to meet the costs of adaptation.

9. Participation under the clean development mechanism, including in activities mentioned in paragraph 3(a) above and in the acquisition of certified emission reductions, may involve private and/or public entities, and is to be subject to whatever guidance may be provided by the executive board of the clean development mechanism.

10. Certified emission reductions obtained during the period from the year 2000 up to the beginning of the first commitment period can be used to assist in achieving compliance in the first commitment period.

ARTICLE 13

1. The Conference of the Parties, the supreme body of the Convention, shall serve as the meeting of the Parties to this Protocol.

2. Parties to the Convention that are not Parties to this Protocol may participate as observers in the proceedings of any session of the Conference of the Parties serving as the meeting of the Parties to this Protocol. When the Conference of the Parties serves as the meeting of the Parties to this Protocol, decisions under this Protocol shall be taken only by those that are Parties to this Protocol.

3. When the Conference of the Parties serves as the meeting of the Parties to this Protocol, any member of the Bureau of the Conference of the Parties representing a Party to the Convention but, at that time, not a Party to this Protocol, shall be replaced by an additional member to be elected by and from amongst the Parties to this Protocol.

4. The Conference of the Parties serving as the meeting of the Parties to this Protocol shall keep under regular review the implementation of this Protocol and shall make, within

its mandate, the decisions necessary to promote its effective implementation. It shall perform the functions assigned to it by this Protocol and shall:

(a) Assess, on the basis of all information made available to it in accordance with the provisions of this Protocol, the implementation of this Protocol by the Parties, the overall effects of the measures taken pursuant to this Protocol, in particular environmental, economic and social effects as well as their cumulative impacts and the extent to which progress towards the objective of the Convention is being achieved;

(b) Periodically examine the obligations of the Parties under this Protocol, giving due consideration to any reviews required by Article 4, paragraph 2(d), and Article 7, paragraph 2, of the Convention, in the light of the objective of the Convention, the experience gained in its implementation and the evolution of scientific and technological knowledge, and in this respect consider and adopt regular reports on the implementation of this Protocol;

(c) Promote and facilitate the exchange of information on measures adopted by the Parties to address climate change and its effects, taking into account the differing circumstances, responsibilities and capabilities of the Parties and their respective commitments under this Protocol;

(d) Facilitate, at the request of two or more Parties, the coordination of measures adopted by them to address climate change and its effects, taking into account the differing circumstances, responsibilities and capabilities of the Parties and their respective commitments under this Protocol;

(e) Promote and guide, in accordance with the objective of the Convention and the provisions of this Protocol, and taking fully into account the relevant decisions by the Conference of the Parties, the development and periodic refinement of comparable methodologies for the effective implementation of this Protocol, to be agreed on by the Conference of the Parties serving as the meeting of the Parties to this Protocol;

(f) Make recommendations on any matters necessary for the implementation of this Protocol;

(g) Seek to mobilize additional financial resources in accordance with Article 11, paragraph 2;

(h) Establish such subsidiary bodies as are deemed necessary for the implementation of this Protocol;

(i) Seek and utilize, where appropriate, the services and cooperation of, and information provided by, competent international organizations and intergovernmental and non-governmental bodies; and

(j) Exercise such other functions as may be required for the implementation of this Protocol, and consider any assignment resulting from a decision by the Conference of the Parties.

5. The rules of procedure of the Conference of the Parties and financial procedures applied under the Convention shall be applied *mutatis mutandis* under this Protocol, except as may be otherwise decided by consensus by the Conference of the Parties serving as the meeting of the Parties to this Protocol.

6. The first session of the Conference of the Parties serving as the meeting of the Parties to this Protocol shall be convened by the secretariat in conjunction with the first session of the Conference of the Parties that is scheduled after the date of the entry into force of this Protocol. Subsequent ordinary sessions of the Conference of the Parties serving as the meeting of the Parties to this Protocol shall be held every year and in conjunction with ordinary sessions of the Conference of the Parties, unless otherwise decided by the Conference of the Parties serving as the meeting of the Parties to this Protocol.

7. Extraordinary sessions of the Conference of the Parties serving as the meeting of the Parties to this Protocol shall be held at such other times as may be deemed necessary by the Conference of the Parties serving as the meeting of the Parties to this Protocol, or at the written request of any Party, provided that, within six months of the request being communicated to the Parties by the secretariat, it is supported by at least one third of the Parties.

8. The United Nations, its specialized agencies and the International Atomic Energy Agency, as well as any State member thereof or observers thereto not party to the Convention, may be represented at sessions of the Conference of the Parties serving as the meeting of the Parties to this Protocol as observers. Any body or agency, whether national or international, governmental or non-governmental, which is qualified in matters covered by this Protocol and which has informed the secretariat of its wish to be represented at a session of the Conference of the Parties serving as the meeting of the Parties to this Protocol as an observer, may be so admitted unless at least one third of the Parties present object. The admission and participation of observers shall be subject to the rules of procedure, as referred to in paragraph 5 above.

ARTICLE 14

1. The secretariat established by Article 8 of the Convention shall serve as the secretariat of this Protocol.

2. Article 8, paragraph 2, of the Convention on the functions of the secretariat, and Article 8, paragraph 3, of the Convention on arrangements made for the functioning of the secretariat, shall apply *mutatis mutandis* to this Protocol. The secretariat shall, in addition, exercise the functions assigned to it under this Protocol.

ARTICLE 15

1. The Subsidiary Body for Scientific and Technological Advice and the Subsidiary Body for Implementation established by Articles 9 and 10 of the Convention shall serve as, respectively, the Subsidiary Body for Scientific and Technological Advice and the Subsidiary Body for Implementation of this Protocol. The provisions relating to the functioning of these two bodies under the Convention shall apply *mutatis mutandis* to this Protocol. Sessions of the meetings of the Subsidiary Body for Scientific and Technological Advice and the Subsidiary Body for Implementation of this Protocol shall be held in conjunction with the meetings of, respectively, the Subsidiary Body for Scientific and Technological Advice and the Subsidiary Body for Implementation of the Convention.

2. Parties to the Convention that are not Parties to this Protocol may participate as observers in the proceedings of any session of the subsidiary bodies. When the subsidiary bodies serve as the subsidiary bodies of this Protocol, decisions under this Protocol shall be taken only by those that are Parties to this Protocol.

3. When the subsidiary bodies established by Articles 9 and 10 of the Convention exercise their functions with regard to matters concerning this Protocol, any member of the Bureaux of those subsidiary bodies representing a Party to the Convention but, at that time, not a party to this Protocol, shall be replaced by an additional member to be elected by and from amongst the Parties to this Protocol.

ARTICLE 16

The Conference of the Parties serving as the meeting of the Parties to this Protocol shall, as soon as practicable, consider the application to this Protocol of, and modify as appropriate, the multilateral consultative process referred to in Article 13 of the Convention, in the light of any relevant decisions that may be taken by the Conference of the Parties. Any multilateral consultative process that may be applied to this Protocol shall operate without prejudice to the procedures and mechanisms established in accordance with Article 18.

ARTICLE 17

The Conference of the Parties shall define the relevant principles, modalities, rules and guidelines, in particular for verification, reporting and accountability for emissions trading. The Parties included in Annex B may participate in emissions trading for the purposes of fulfilling their commitments under Article 3. Any such trading shall be supplemental to domestic actions for the purpose of meeting quantified emission limitation and reduction commitments under that Article.

ARTICLE 18

The Conference of the Parties serving as the meeting of the Parties to this Protocol shall, at its first session, approve appropriate and effective procedures and mechanisms to determine and to address cases of non-compliance with the provisions of this Protocol, including through the development of an indicative list of consequences, taking into account the cause, type, degree and frequency of non-compliance. Any procedures and mechanisms under this Article entailing binding consequences shall be adopted by means of an amendment to this Protocol.

ARTICLE 19

The provisions of Article 14 of the Convention on settlement of disputes shall apply *mutatis mutandis* to this Protocol.

ARTICLE 20

1. Any Party may propose amendments to this Protocol.

2. Amendments to this Protocol shall be adopted at an ordinary session of the Conference of the Parties serving as the meeting of the Parties to this Protocol. The text of any proposed amendment to this Protocol shall be communicated to the Parties by the

secretariat at least six months before the meeting at which it is proposed for adoption. The secretariat shall also communicate the text of any proposed amendments to the Parties and signatories to the Convention and, for information, to the Depositary.

3. The Parties shall make every effort to reach agreement on any proposed amendment to this Protocol by consensus. If all efforts at consensus have been exhausted, and no agreement reached, the amendment shall as a last resort be adopted by a three-fourths majority vote of the Parties present and voting at the meeting. The adopted amendment shall be communicated by the secretariat to the Depositary, who shall circulate it to all Parties for their acceptance.

4. Instruments of acceptance in respect of an amendment shall be deposited with the Depositary. An amendment adopted in accordance with paragraph 3 above shall enter into force for those Parties having accepted it on the ninetieth day after the date of receipt by the Depositary of an instrument of acceptance by at least three fourths of the Parties to this Protocol.

5. The amendment shall enter into force for any other Party on the ninetieth day after the date on which that Party deposits with the Depositary its instrument of acceptance of the said amendment.

ARTICLE 21

1. Annexes to this Protocol shall form an integral part thereof and, unless otherwise expressly provided, a reference to this Protocol constitutes at the same time a reference to any annexes thereto. Any annexes adopted after the entry into force of this Protocol shall be restricted to lists, forms and any other material of a descriptive nature that is of a scientific, technical, procedural or administrative character.

2. Any Party may make proposals for an annex to this Protocol and may propose amendments to annexes to this Protocol.

3. Annexes to this Protocol and amendments to annexes to this Protocol shall be adopted at an ordinary session of the Conference of the Parties serving as the meeting of the Parties to this Protocol. The text of any proposed annex or amendment to an annex shall be communicated to the Parties by the secretariat at least six months before the meeting at which it is proposed for adoption. The secretariat shall also communicate the text of any proposed annex or amendment to an annex to the Parties and signatories to the Convention and, for information, to the Depositary.

4. The Parties shall make every effort to reach agreement on any proposed annex or amendment to an annex by consensus. If all efforts at consensus have been exhausted, and no agreement reached, the annex or amendment to an annex shall as a last resort be adopted by a three-fourths majority vote of the Parties present and voting at the meeting. The adopted annex or amendment to an annex shall be communicated by the secretariat to the Depositary, who shall circulate it to all Parties for their acceptance.

5. An annex, or amendment to an annex other than Annex A or B, that has been adopted in accordance with paragraphs 3 and 4 above shall enter into force for all Parties to this Protocol six months after the date of the communication by the Depositary to such Parties of the adoption of the annex or adoption of the amendment to the annex, except for those Parties that have notified the Depositary, in writing, within that period of their non-acceptance of the annex or

amendment to the annex. The annex or amendment to an annex shall enter into force for Parties which withdraw their notification of non-acceptance on the ninetieth day after the date on which withdrawal of such notification has been received by the Depositary.

6. If the adoption of an annex or an amendment to an annex involves an amendment to this Protocol, that annex or amendment to an annex shall not enter into force until such time as the amendment to this Protocol enters into force.

7. Amendments to Annexes A and B to this Protocol shall be adopted and enter into force in accordance with the procedure set out in Article 20, provided that any amendment to Annex B shall be adopted only with the written consent of the Party concerned.

ARTICLE 22

1. Each Party shall have one vote, except as provided for in paragraph 2 below.

2. Regional economic integration organizations, in matters within their competence, shall exercise their right to vote with a number of votes equal to the number of their member States that are Parties to this Protocol. Such an organization shall not exercise its right to vote if any of its member States exercises its right, and vice versa.

ARTICLE 23

The Secretary-General of the United Nations shall be the Depositary of this Protocol.

ARTICLE 24

1. This Protocol shall be open for signature and subject to ratification, acceptance or approval by States and regional economic integration organizations which are Parties to the Convention. It shall be open for signature at United Nations Headquarters in New York from 16 March 1998 to 15 March 1999. This Protocol shall be open for accession from the day after the date on which it is closed for signature. Instruments of ratification, acceptance, approval or accession shall be deposited with the Depositary.

2. Any regional economic integration organization which becomes a Party to this Protocol without any of its member States being a Party shall be bound by all the obligations under this Protocol. In the case of such organizations, one or more of whose member States is a Party to this Protocol, the organization and its member States shall decide on their respective responsibilities for the performance of their obligations under this Protocol. In such cases, the organization and the member States shall not be entitled to exercise rights under this Protocol concurrently.

3. In their instruments of ratification, acceptance, approval or accession, regional economic integration organizations shall declare the extent of their competence with respect to the matters governed by this Protocol. These organizations shall also inform the Depositary, who shall in turn inform the Parties, of any substantial modification in the extent of their competence.

ARTICLE 25

1. This Protocol shall enter into force on the ninetieth day after the date on which not less than 55 Parties to the Convention, incorporating Parties included in Annex I which accounted in total for at least 55 per cent of the total carbon dioxide emissions for 1990 of the Parties included in Annex I, have deposited their instruments of ratification, acceptance, approval or accession.

2. For the purposes of this Article, 'the total carbon dioxide emissions for 1990 of the Parties included in Annex I' means the amount communicated on or before the date of adoption of this Protocol by the Parties included in Annex I in their first national communications submitted in accordance with Article 12 of the Convention.

3. For each State or regional economic integration organization that ratifies, accepts or approves this Protocol or accedes thereto after the conditions set out in paragraph 1 above for entry into force have been fulfilled, this Protocol shall enter into force on the ninetieth day following the date of deposit of its instrument of ratification, acceptance, approval or accession.

4. For the purposes of this Article, any instrument deposited by a regional economic integration organization shall not be counted as additional to those deposited by States members of the organization.

ARTICLE 26

No reservations may be made to this Protocol.

ARTICLE 27

1. At any time after three years from the date on which this Protocol has entered into force for a Party, that Party may withdraw from this Protocol by giving written notification to the Depositary.

2. Any such withdrawal shall take effect upon expiry of one year from the date of receipt by the Depositary of the notification of withdrawal, or on such later date as may be specified in the notification of withdrawal.

3. Any Party that withdraws from the Convention shall be considered as also having withdrawn from this Protocol.

ARTICLE 28

The original of this Protocol, of which the Arabic, Chinese, English, French, Russian and Spanish texts are equally authentic, shall be deposited with the Secretary-General of the United Nations.

DONE at Kyoto this eleventh day of December one thousand nine hundred and ninety-seven.

IN WITNESS WHEREOF the undersigned, being duly authorized to that effect, have affixed their signatures to this Protocol on the dates indicated.

ANNEX A

Greenhouse gases

Carbon dioxide (CO_2)
Methane (CH_4)
Nitrous oxide (N_2O)
Hydrofluorocarbons (HFCs)
Perfluorocarbons (PFCs)
Sulphur hexafluoride (SF_6)

Sectors/source categories

Energy
Fuel combustion
Energy industries
Manufacturing industries and construction
Transport
Other sectors
Other
Fugitive emissions from fuels
Solid fuels
Oil and natural gas
Other
Industrial processes
Mineral products
Chemical industry
Metal production
Other production
Production of halocarbons and sulphur hexafluoride
Consumption of halocarbons and sulphur hexafluoride
Other
Solvent and other product use
Agriculture
Enteric fermentation
Manure management
Rice cultivation
Agricultural soils
Prescribed burning of savannas
Field burning of agricultural residues
Other
Waste

Solid waste disposal on land
Wastewater handling
Waste incineration
Other

ANNEX B

Party Quantified emission limitation or reduction commitment

(percentage of base year or period)
Australia 108
Austria 92
Belgium 92
Bulgaria* 92
Canada 94
Croatia* 95
Czech Republic* 92
Denmark 92
Estonia* 92
European Community 92
Finland 92
France 92
Germany 92
Greece 92
Hungary* 94
Iceland 110
Ireland 92
Italy 92
Japan 94
Latvia* 92
Liechtenstein 92
Lithuania* 92
Luxembourg 92
Monaco 92
Netherlands 92
New Zealand 100
Norway 101
Poland* 94
Portugal 92
Romania* 92
Russian Federation* 100
Slovakia* 92
Slovenia* 92
Spain 92

Sweden 92
Switzerland 92
Ukraine* 100
United Kingdom of Great Britain and Northern Ireland 92
United States of America 93

* Countries that are undergoing the process of transition to a market economy.

APPENDIX 2

CONFERENCE OF THE PARTIES

REPORT OF THE CONFERENCE OF THE PARTIES ON ITS SEVENTH SESSION, HELD AT MARRAKESH FROM 29 OCTOBER TO 10 NOVEMBER 2001

Addendum

PART TWO: ACTION TAKEN BY THE CONFERENCE OF THE PARTIES

Volume II

CONTENTS

Page

GE.02-60229

DECISION 15/CP.7

PRINCIPLES, NATURE AND SCOPE OF THE MECHANISMS PURSUANT TO ARTICLES 6, 12 AND 17 OF THE KYOTO PROTOCOL

The Conference of the Parties,

Recalling its decision 1/CP.3, in particular paragraphs 5(b), (c) and (e),

Further recalling its decisions 7/CP.4, 8/CP.4, 9/CP.4, 14/CP.5, and 5/CP.6 containing the Bonn Agreements on the implementation of the Buenos Aires Plan of Action, as appropriate,

Recalling also the preamble to the Convention,

Recognizing that, in using the mechanisms, Parties shall be guided by the objective and principles contained in Articles 2 and 3 and by Article 4, paragraph 7, of the Convention,

Further recognizing that the Kyoto Protocol has not created or bestowed any right, title or entitlement to emissions of any kind on Parties included in Annex I,

Emphasizing that the Parties included in Annex I shall implement domestic action in accordance with national circumstances and with a view to reducing emissions in a manner conducive to narrowing per capita differences between developed and developing country Parties while working towards achievement of the ultimate objective of the Convention,

Affirming that the use of the mechanisms shall be supplemental to domestic action and that domestic action shall thus constitute a significant element of the effort made by each Party included in Annex I to meet its quantified emission limitation and reduction commitments under Article 3, paragraph 1,

Further emphasizing that environmental integrity is to be achieved through sound modalities, rules and guidelines for the mechanisms, sound and strong principles and rules governing land use, land-use change and forestry activities and a strong compliance regime,

Aware of its decisions 11/CP.7, 16/CP.7, 17/CP.7, 18/CP.7, 19/CP.7, 20/CP.7, 21/CP.7, 22/CP.7, 23/CP.7 and 24/CP.7,

Recommends that the Conference of the Parties serving as the meeting of the Parties to the Kyoto Protocol, at its first session, adopt the draft decision below.

DRAFT DECISION -/CMP.1 (*MECHANISMS*)

PRINCIPLES, NATURE AND SCOPE OF THE MECHANISMS PURSUANT TO ARTICLES 6, 12 AND 17 OF THE KYOTO PROTOCOL

The Conference of the Parties serving as the meeting of the Parties to the Kyoto Protocol,

Recalling decision 1/CP.3, in particular paragraphs 5(b), (c) and (e),

Further recalling decisions 7/CP.4, 8/CP.4, 9/CP.4, 14/CP.5, 5/CP.6 containing the Bonn Agreements on the implementation of the Buenos Aires Plan of Action, 11/CP.7, 16/CP.7, 17/CP.7, 18/CP.7, 19/CP.7, 20/CP.7, 21/CP.7, 22/CP.7, 23/CP.7 and 24/CP.7, as appropriate,

Recalling also the preamble to the Convention,

Recognizing that, in using the mechanisms, Parties shall be guided by the objective and principles contained in Articles 2 and 3 and by Article 4, paragraph 7, of the Convention,

Further recognizing that the Kyoto Protocol has not created or bestowed any right, title or entitlement to emissions of any kind on Parties included in Annex I,

Emphasizing that the Parties included in Annex I shall implement domestic action in accordance with national circumstances and with a view to reducing emissions in a manner conducive to narrowing per capita differences between developed and developing country Parties while working towards achievement of the ultimate objective of the Convention,

Further emphasizing that environmental integrity is to be achieved through sound modalities, rules and guidelines for the mechanisms, sound and strong principles and rules governing land use, land-use change and forestry activities, and a strong compliance regime,

Aware of its decisions -/CMP.1 (*Article 6*), -/CMP.1 (*Article 12*), -/CMP.1 (*Article 17*), -/CMP.1 (*Land use, land-use change and forestry*), -/CMP.1 (*Modalities for the accounting of assigned amounts*), -/CMP.1 (*Article 5.1*), -/CMP.1 (*Article 5.2*), -/CMP.1 (*Article 7*) and -/CMP.1 (*Article 8*) and decision 24/CP.7,

1. *Decides* that the use of the mechanisms shall be supplemental to domestic action and that domestic action shall thus constitute a significant element of the effort made by each Party included in Annex I to meet its quantified emission limitation and reduction commitments under Article 3, paragraph 1;

2. *Requests* the Parties included in Annex I to provide relevant information in relation to paragraph 1 above, in accordance with Article 7 of the Kyoto Protocol, for review under its Article 8;

3. *Decides* that the provision of such information shall take into account reporting on demonstrable progress as contained in decision -/CMP.1 (*Article 7*);

4. *Requests* the facilitative branch of the compliance committee to address questions of implementation with respect to paragraphs 2 and 3 above;

5. *Decides* that the eligibility to participate in the mechanisms by a Party included in Annex I shall be dependent on its compliance with methodological and reporting requirements under Article 5, paragraphs 1 and 2, and Article 7, paragraphs 1 and 4, of the Kyoto Protocol. Oversight of this provision will be provided by the enforcement branch of the compliance committee, in accordance with the procedures and mechanisms relating to compliance as contained in decision 24/CP.7, assuming approval of such procedures and mechanisms by the Conference of the Parties serving as the meeting of the Parties to the Kyoto Protocol in decision form in addition to any amendment entailing legally binding consequences, noting that it is the prerogative of the Conference of the Parties serving as the meeting of the Parties to the Kyoto Protocol to decide on the legal form of the procedures and mechanisms relating to compliance;

6. *Decides* that certified emission reductions, emission reduction units and assigned amount units under Articles 6, 12 and 17, as well as removal units resulting from activities under Article 3, paragraphs 3 and 4, may be used to meet commitments under Article 3, paragraph 1, of the Parties included in Annex I, and can be added as provided for in Article 3, paragraphs 10, 11 and 12 of the Kyoto Protocol and in conformity with the provisions contained in decision -/CMP.1 (*Modalities for the accounting of assigned amounts*), and that emission reduction units, assigned amount units and removal units can be subtracted as provided for in Article 3, paragraphs 10 and 11, and in conformity with the provisions

contained in decision -/CMP.1 (*Modalities for the accounting of assigned amounts*), without altering the quantified emission limitation and reduction commitments inscribed in Annex B to the Kyoto Protocol.

DECISION 16/CP.7

GUIDELINES FOR THE IMPLEMENTATION OF ARTICLE 6 OF THE KYOTO PROTOCOL

The Conference of the Parties,

Recalling its decision 5/CP.6 containing the Bonn Agreements on the implementation of the Buenos Aires Plan of Action,

Aware of its decisions 3/CP.7, 11/CP.7, 15/CP.7, 17/CP.7, 18/CP.7, 19/CP.7, 20/CP.7, 21/CP.7, 22/CP.7, 23/CP.7 and 24/CP.7,

Affirming that it is the host Party's prerogative to confirm whether an Article 6 project activity assists it in achieving sustainable development,

Recognizing that Parties included in Annex I to the Convention are to refrain from using emission reduction units generated from nuclear facilities to meet their commitments under Article 3, paragraph 1,

1. *Urges* the Parties included in Annex II to the Convention to facilitate the participation in projects under Article 6 of Parties included in Annex I with commitments inscribed in Annex B that are undergoing the process of transition to a market economy;

2. *Invites* Parties included in Annex I to finance the administrative expenses for operating joint implementation under Article 6 by making contributions to the UNFCCC Trust Fund for Supplementary Activities in order to facilitate preparatory work by the secretariat, if necessary;

3. *Recommends* that the Conference of the Parties serving as the meeting of the Parties to the Kyoto Protocol, at its first session, adopt the draft decision below.

DRAFT DECISION -/CMP.1 (*ARTICLE 6*)

GUIDELINES FOR THE IMPLEMENTATION OF ARTICLE 6 OF THE KYOTO PROTOCOL

The Conference of the Parties serving as the meeting of the Parties to the Kyoto Protocol,

Aware of its decisions -/CMP.1 (*Mechanisms*), -/CMP.1 (*Article 12*), -/CMP.1 (*Article 17*), -/CMP.1 (*Land use, land-use change and forestry*), -/CMP.1 (*Modalities for the accounting of assigned amounts*), -/CMP.1 (*Article 5.1*), -/CMP.1 (*Article 5.2*), -/CMP.1 (*Article 7*) and -/CMP.1 (*Article 8*), and decisions 3/CP.7 and 24/CP.7,

1. *Decides* to confirm and give full effect to any actions taken pursuant to decision 16/CP.7 and to any other relevant decisions by the Conference of the Parties, as appropriate;

2. *Decides* to adopt the guidelines for the implementation of Article 6 of the Kyoto Protocol contained in the annex below;

3. *Decides* to establish the Article 6 supervisory committee, at its first session, to supervise, *inter alia*, the verification of ERUs generated by Article 6 projects;

4. *Decides* that projects under Article 6 aimed at enhancing anthropogenic removals by sinks shall conform to definitions, accounting rules, modalities and guidelines under Article 3, paragraphs 3 and 4, of the Kyoto Protocol;

5. *Decides* that projects starting as of the year 2000 may be eligible as Article 6 projects if they meet the requirements of the guidelines for the implementation of Article 6 of the Kyoto Protocol as set out in the annex below and that ERUs shall only be issued for a crediting period starting after the beginning of the year 2008;

6. *Urges* the Parties included in Annex II to facilitate the participation in Article 6 projects of Parties included in Annex I with commitments inscribed in Annex B that are undergoing the process of transition to a market economy;

7. *Decides* that any administrative costs arising from procedures contained in the annex below relating to the functions of the Article 6 supervisory committee shall be borne by both the Parties included in Annex I and the project participants according to specifications set out in a decision by the Conference of the Parties serving as the meeting of the Parties to the Kyoto Protocol at its first session;

8. *Decides further* that any future revision of the guidelines for the implementation of Article 6 shall be decided in accordance with the rules of procedure of the Conference of the Parties serving as the meeting of the Parties to the Kyoto Protocol, as applied. The first review shall be carried out no later than one year after the end of the first commitment period, based on recommendations by the Article 6 supervisory committee and by the Subsidiary Body for Implementation drawing on technical advice of the Subsidiary Body for Scientific and Technological Advice, as needed. Further reviews shall be carried out periodically thereafter. Any revision of the decision shall not affect ongoing Article 6 projects.

ANNEX

GUIDELINES FOR THE IMPLEMENTATION OF ARTICLE 6 OF THE KYOTO PROTOCOL

A. Definitions

1. For the purpose of the present annex the definitions contained in Article 1[1] and the provisions of Article 14 shall apply. Furthermore:

(a) An 'emission reduction unit' or 'ERU' is a unit issued pursuant to the relevant provisions in the annex to decision -/CMP.1 (*Modalities for the accounting of assigned amounts*) and is equal to one metric tonne of carbon dioxide equivalent, calculated using global warming potentials defined by decision 2/CP.3 or as subsequently revised in accordance with Article 5;

(b) A 'certified emission reduction' or 'CER' is a unit issued pursuant to Article 12 and requirements thereunder, as well as the relevant provisions in the annex to decision -/CMP.1 (*Article 12*), and is equal to one metric tonne of carbon dioxide equivalent, calculated using global warming potentials defined by decision 2/CP.3 or as subsequently revised in accordance with Article 5;

(c) An 'assigned amount unit' or 'AAU' is a unit issued pursuant to the relevant provisions in the annex to decision -/CMP.1 (*Modalities for the accounting of assigned amounts*) and is equal to one metric tonne of carbon dioxide equivalent, calculated using global warming potentials defined by decision 2/CP.3 or as subsequently revised in accordance with Article 5;

(d) A 'removal unit' or 'RMU' is a unit issued pursuant to the relevant provisions in the annex to decision -/CMP.1 (*Modalities for the accounting of assigned amounts*) and is equal to one metric tonne of carbon dioxide equivalent, calculated using global warming potentials defined by decision 2/CP.3 or as subsequently revised in accordance with Article 5;

(e) 'Stakeholders' means the public, including individuals, groups or communities affected, or likely to be affected, by the project.

B. Role of the Conference of the Parties serving as the meeting of the Parties to the Kyoto Protocol

2. The Conference of the Parties serving as the meeting of the Parties to the Kyoto Protocol (COP/MOP) shall provide guidance regarding the implementation of Article 6 and exercise authority over the Article 6 supervisory committee.

C. Article 6 supervisory committee

3. The Article 6 supervisory committee shall supervise, *inter alia*, the verification of ERUs generated by Article 6 project activities, referred to in section E below, and be responsible for:

(a) Reporting on its activities to each session of the COP/MOP;

[1] In the context of this annex, 'Article' refers to an Article of the Kyoto Protocol, unless otherwise specified.

(b) The accreditation of independent entities in accordance with standards and procedures contained in Appendix A below;

(c) The review of standards and procedures for the accreditation of independent entities in Appendix A below, giving consideration to relevant work of the executive board of the clean development mechanism (CDM) and, as appropriate, making recommendations to the COP/MOP on revisions to these standards and procedures;

(d) The review and revision of reporting guidelines and criteria for baselines and monitoring in Appendix B below, for consideration by the COP/MOP, giving consideration to relevant work of the executive board of the CDM, as appropriate;

(e) The elaboration of the Article 6 project design document, for consideration by the COP/MOP, taking into consideration Appendix B of the Annex on modalities and procedures for a clean development mechanism and giving consideration to relevant work of the executive board of the CDM, as appropriate;

(f) The review procedures set out in paragraphs 35 and 39 below;

(g) The elaboration of any rules of procedure additional to those contained in the present annex, for consideration by the COP/MOP.

4. The Article 6 supervisory committee shall comprise ten members from Parties to the Kyoto Protocol, as follows:

(a) Three members from Parties[2] included in Annex I that are undergoing the process of transition to a market economy;

(b) Three members from Parties included in Annex I not referred to in subparagraph (a) above;

(c) Three members from Parties not included in Annex I;

(d) One member from the small island developing States.

5. Members, including alternate members, of the Article 6 supervisory committee shall be nominated by the relevant constituencies referred to in paragraph 4 above and be elected by the COP/MOP. The COP/MOP shall elect to the Article 6 supervisory committee five members and five alternate members for a term of two years and five members and five alternate members for a term of three years. Thereafter, the COP/MOP shall elect, every year, five new members and five alternate members for a term of two years. Appointment pursuant to paragraph 12 below shall count as one term. The members and alternate members shall remain in office until their successors are elected.

6. Members of the Article 6 supervisory committee may be eligible to serve a maximum of two consecutive terms. Terms as alternate members do not count.

7. The Article 6 supervisory committee shall elect annually a chairperson and vicechairperson from among its members, with one being from a Party included in Annex I and the other being from a Party not included in Annex I. The positions of chairperson and vice-chairperson shall alternate annually between a member from a Party included in Annex I and a member from a Party not included in Annex I.

8. The COP/MOP shall elect an alternate member for each member of the Article 6 supervisory committee based on the criteria in paragraphs 4, 5 and 6 above. The nomination by a constituency of a candidate member shall be accompanied by a nomination of a candidate alternate member from the same constituency.

9. The Article 6 supervisory committee shall meet at least two times each year, whenever possible in conjunction with the meetings of the subsidiary bodies, unless decided

[2] In the context of this annex, 'Party' refers to a Party to the Kyoto Protocol, unless otherwise specified.

otherwise. All documentation for the Article 6 supervisory committee meetings shall be made available to alternate members.

10. Members, including alternate members, of the Article 6 supervisory committee shall:

(a) Serve in their personal capacities and shall have recognized competence relating to climate change issues and in relevant technical and policy fields. The cost of participation of members and of alternate members from developing country Parties and other Parties eligible under UNFCCC practice shall be covered by the budget for the Article 6 supervisory committee;

(b) Have no pecuniary or financial interest in any aspect of an Article 6 project;

(c) Subject to their responsibility to the Article 6 supervisory committee, not disclose any confidential or proprietary information coming to their knowledge by reason of their duties for the Article 6 supervisory committee. The duty of a member, including an alternate member, not to disclose confidential information constitutes an obligation in respect to that member, including an alternate member, and shall remain an obligation after the expiration or termination of that member's, including an alternate member's, function for the Article 6 supervisory committee;

(d) Be bound by the rules of procedure of the Article 6 supervisory committee;

(e) Take a written oath of service witnessed by the Executive Secretary of the UNFCCC or his/her authorized representative before assuming his or her duties.

11. The Article 6 supervisory committee may suspend and recommend to the COP/MOP the termination of the membership of a particular member, including an alternate member, for cause including, *inter alia*, breach of the conflict of interest provisions, breach of the confidentiality provisions, or failure to attend two consecutive meetings of the Article 6 supervisory committee without proper justification.

12. If a member, or an alternate member, of the Article 6 supervisory committee resigns or is otherwise unable to complete the assigned term of office or to perform the functions of that office, the Article 6 supervisory committee may decide, bearing in mind the proximity of the next session of the COP/MOP, to appoint another member, or an alternate member, from the same constituency to replace the said member for the remainder of that member's mandate. In such a case, the Article 6 supervisory committee shall take into account any views expressed by the group that had nominated the member.

13. The Article 6 supervisory committee shall draw on the expertise necessary to perform its functions, in particular taking into account national accreditation procedures.

14. At least two thirds of the members of the Article 6 supervisory committee, representing a majority of members from Parties included in Annex I and a majority of members from Parties not included in Annex I, must be present to constitute a quorum.

15. Decisions by the Article 6 supervisory committee shall be taken by consensus, whenever possible. If all efforts at reaching a consensus have been exhausted and no agreement has been reached, decisions shall as a last resort be adopted by a three-fourths majority vote of the members present and voting at the meeting. Members abstaining from voting shall be considered as not voting.

16. The full text of all decisions of the Article 6 supervisory committee shall be made publicly available. Decisions shall be made available in all six official languages of the United Nations.

17. The working language of the Article 6 supervisory committee shall be English.

18. Meetings of the Article 6 supervisory committee shall be open to attendance, as observers, by all Parties and by all UNFCCC accredited observers and stakeholders, except where otherwise decided by the Article 6 supervisory committee.

19. The secretariat shall service the Article 6 supervisory committee.

D. Participation requirements

20. A Party involved in an Article 6 project shall inform the secretariat of:

(a) Its designated focal point for approving projects pursuant to Article 6, paragraph 1(a);

(b) Its national guidelines and procedures for approving Article 6 projects, including the consideration of stakeholders' comments, as well as monitoring and verification.

21. Subject to the provisions of paragraph 22 below, a Party included in Annex I with a commitment inscribed in Annex B is eligible to transfer and/or acquire ERUs issued in accordance with the relevant provisions, if it is in compliance with the following eligibility requirements:

(a) It is a Party to the Kyoto Protocol;

(b) Its assigned amount pursuant to Article 3, paragraphs 7 and 8, has been calculated and recorded in accordance with decision -/CMP.1 (*Modalities for the accounting of assigned amounts*);

(c) It has in place a national system for the estimation of anthropogenic emissions by sources and anthropogenic removals by sinks of all greenhouse gases not controlled by the Montreal Protocol, in accordance with Article 5, paragraph 1, and the requirements in the guidelines decided thereunder;

(d) It has in place a national registry in accordance with Article 7, paragraph 4, and the requirements in the guidelines decided thereunder;

(e) It has submitted annually the most recent required inventory, in accordance with Article 5, paragraph 2, and Article 7, paragraph 1, and the requirements in the guidelines decided thereunder, including the national inventory report and the common reporting format. For the first commitment period, the quality assessment needed for the purpose of determining eligibility to use the mechanisms shall be limited to the parts of the inventory pertaining to emissions of greenhouse gases from sources/sector categories from Annex A to the Kyoto Protocol and the submission of the annual inventory on sinks;

(f) It submits the supplementary information on assigned amount in accordance with Article 7, paragraph 1, and the requirements in the guidelines decided thereunder and makes any additions to, and subtractions from, assigned amount pursuant to Article 3, paragraphs 7 and 8, including for the activities under Article 3, paragraphs 3 and 4, in accordance with Article 7, paragraph 4, and the requirements in the guidelines decided thereunder.

22. A Party included in Annex I with a commitment inscribed in Annex B shall be considered:

(a) To meet the eligibility requirements referred to in paragraph 21 above after 16 months have elapsed since the submission of its report to facilitate the calculation of its assigned amount pursuant to Article 3, paragraphs 7 and 8, and to demonstrate its capacity to account for its emissions and assigned amount, in accordance with the modalities adopted for the accounting of assigned amount under Article 7, paragraph 4, unless the enforcement branch of the compliance committee finds in accordance with decision 24/CP.7 that the Party does not meet these requirements, or, at an earlier date, if the enforcement branch of the compliance committee has decided that it is not proceeding with any questions of implementation relating to these requirements indicated in reports of the expert review teams under Article 8 of the Kyoto Protocol, and has transmitted this information to the secretariat;

(b) To continue to meet the eligibility requirements referred to in paragraph 21 above unless and until the enforcement branch of the compliance committee decides that the

Party does not meet one or more of the eligibility requirements, has suspended the Party's eligibility, and has transmitted this information to the secretariat.

23. Where it is considered to meet the eligibility requirements set out in paragraph 21 above, a host Party may verify reductions in anthropogenic emissions by sources or enhancements of anthropogenic removals by sinks from an Article 6 project as being additional to any that would otherwise occur, in accordance with Article 6, paragraph 1 (b). Upon such verification, the host Party may issue the appropriate quantity of ERUs in accordance with the relevant provisions of decision -/CMP.1 (*Modalities for the accounting of assigned amounts*).

24. Where a host Party does not meet the eligibility requirements set out in paragraph 21 above, the verification of reductions in anthropogenic emissions by sources or enhancements of anthropogenic removals by sinks from an Article 6 project as being additional to any that would otherwise occur, in accordance with Article 6, paragraph 1(b), shall occur through the verification procedure under the Article 6 supervisory committee, as set out in section E below. The host Party may however only issue and transfer ERUs upon meeting the requirements in paragraphs 21 (a), (b) and (d) above.

25. A host Party which meets the requirements in paragraph 21 above may at any time elect to use the verification procedure under the Article 6 supervisory committee.

26. The provisions in Article 6, paragraph 4, shall pertain, *inter alia*, to the requirements of paragraph 21 above.

27. The secretariat shall maintain a publicly accessible list of Parties that meet the eligibility requirements and that have been suspended in accordance with relevant provisions contained in decision 24/CP.7.

28. A Party hosting an Article 6 project shall make publicly available, directly or through the secretariat, information on the project in accordance with the reporting guidelines set out in Appendix B below and the requirements contained in decision -/CMP.1 (*Modalities for the accounting of assigned amounts*).

29. A Party that authorizes legal entities to participate in Article 6 projects shall remain responsible for the fulfilment of its obligations under the Kyoto Protocol and shall ensure that such participation is consistent with the present annex. Legal entities may only transfer or acquire ERUs if the authorizing Party is eligible to do so at that time.

E. Verification procedure under the Article 6 supervisory committee

30. The verification procedure under the Article 6 supervisory committee is the determination by an independent entity, accredited pursuant to Appendix A below, of whether a project and the ensuing reductions of anthropogenic emissions by sources or enhancements of anthropogenic removals by sinks meet the relevant requirements of Article 6 and these guidelines.

31. Project participants shall submit to an accredited independent entity a project design document that contains all information needed for the determination of whether the project:

(a) Has been approved by the Parties involved;

(b) Would result in a reduction of anthropogenic emissions by sources or an enhancement of anthropogenic removals by sinks that is additional to any that would otherwise occur;

(c) Has an appropriate baseline and monitoring plan in accordance with the criteria set out in Appendix B below.

32. The accredited independent entity shall make the project design document publicly available through the secretariat, subject to confidentiality provisions set out in paragraph 40 below, and receive comments from Parties, stakeholders and UNFCCC accredited observers on the project design document and any supporting information for 30 days from the date the project design document is made publicly available.

33. The accredited independent entity shall determine whether:

(a) The project has been approved by the Parties involved;

(b) The project would result in a reduction of anthropogenic emissions by sources or an enhancement of anthropogenic removals by sinks that is additional to any that would otherwise occur;

(c) The project has an appropriate baseline and monitoring plan in accordance with the criteria set out in Appendix B below;

(d) Project participants have submitted to the accredited independent entity documentation on the analysis of the environmental impacts of the project activity, including transboundary impacts, in accordance with procedures as determined by the host Party, and, if those impacts are considered significant by the project participants or the host Party, have undertaken an environmental impact assessment in accordance with procedures as required by the host Party.

34. The accredited independent entity shall make its determination publicly available through the secretariat, together with an explanation of its reasons, including a summary of comments received and a report of how due account was taken of these.

35. The determination regarding a project design document shall be deemed final 45 days after the date on which the determination is made public, unless a Party involved in the project or three of the members of the Article 6 supervisory committee request a review by the Article 6 supervisory committee. If such a review is requested, the Article 6 supervisory committee shall finalize the review as soon as possible, but no later than six months or at the second meeting following the request for review. The Article 6 supervisory committee shall communicate its decision on the determination and the reasons for it to the project participants and the public. Its decision shall be final.

36. Project participants shall submit to an accredited independent entity a report in accordance with the monitoring plan on reductions in anthropogenic emissions by sources or enhancements of anthropogenic removals by sinks that have already occurred. The report shall be made publicly available.

37. The accredited independent entity shall, upon receipt of a report referred to under paragraph 36 above, make a determination of the reductions in anthropogenic emissions by sources or enhancements of anthropogenic removals by sinks reported by project participants in accordance with Appendix B below, provided that they were monitored and calculated in accordance with paragraph 33 above.

38. The accredited independent entity shall make its determination under paragraph 37 above publicly available through the secretariat, together with an explanation of its reasons.

39. The determination regarding reported reductions in anthropogenic emissions by sources or enhancements of anthropogenic removals by sinks shall be deemed final 15 days after the date on which it is made public, unless a Party involved in the project or three of the members of the Article 6 supervisory committee request a review by the Article 6 supervisory committee. If such a review is requested, the Article 6 supervisory committee shall:

(a) At its next meeting or no later than 30 days after the formal request for the review decide on its course of action. If it decides that the request has merit, it shall perform a review;

(b) Complete its review within 30 days following its decision to perform the review;

(c) Inform the project participants of the outcome of the review, and make public its decision and the reasons for it.

40. Information obtained from project participants marked as proprietary or confidential shall not be disclosed without the written consent of the provider of the information, except as required by applicable national law of the host Party. Information used to determine whether reductions in anthropogenic emissions by sources or enhancements of anthropogenic removals by sinks are additional, to describe the baseline methodology and its application, and to support an environmental impact assessment referred to in paragraph 33(d) above, shall not be considered as proprietary or confidential.

41. Any provisions relating to the commitment period reserve or other limitations to transfers under Article 17 shall not apply to transfers by a Party of ERUs issued into its national registry that were verified in accordance with the verification procedure under the Article 6 supervisory committee.

42. The Article 6 supervisory committee shall suspend or withdraw the accreditation of an independent entity if it has carried out a review and found that the entity no longer meets the accreditation standards laid down in Appendix A. The Article 6 supervisory committee may suspend or withdraw accreditation only after the accredited independent entity has had the opportunity of a hearing and depending on the outcome of the hearing. The suspension or withdrawal is with immediate effect. The affected entity shall be notified, immediately and in writing, once the Article 6 supervisory committee has decided upon its suspension or withdrawal. The decision by the Article 6 supervisory committee on such a case shall be made public.

43. Verified projects shall not be affected by the suspension or withdrawal of the accreditation of an accredited independent entity unless significant deficiencies are identified in the determination referred to in paragraphs 33 or 37 above for which the entity was responsible. In this case, the Article 6 supervisory committee shall decide whether a different accredited independent entity shall be appointed to assess and, where appropriate, correct such deficiencies. If such an assessment reveals that excess ERUs have been transferred as a result of the deficiencies identified in the determination referred to in paragraphs 33 or 37 above, the independent entity whose accreditation has been withdrawn or suspended shall acquire an equivalent amount of AAUs and ERUs and place them in the holding account of the Party hosting the project within 30 days from the assessment mentioned above.

44. Any suspension or withdrawal of an accredited independent entity that adversely affects verified projects shall be decided on by the Article 6 supervisory committee only after the affected project participants have had the opportunity of a hearing.

45. Any costs related to the assessment referred to in paragraph 44 above shall be borne by the accredited independent entity whose accreditation has been withdrawn or suspended.

APPENDIX A

Standards and procedures for the accreditation of independent entities

1. An independent entity shall:

(a) Be a legal entity (either a domestic legal entity or an international organization) and provide documentation of this status;

(b) Employ a sufficient number of persons having the necessary competence to perform all necessary functions relevant to the verification of ERUs generated by Article 6 projects relating to the type, range and volume of work performed, under a responsible senior executive;

(c) Have the financial stability, insurance coverage and resources required for its activities;

(d) Have sufficient arrangements to cover legal and financial liabilities arising from its activities;

(e) Have documented internal procedures for carrying out its functions including, *inter alia*, procedures for the allocation of responsibilities within the organization and for handling complaints. These procedures shall be made publicly available;

(f) Have the necessary expertise to carry out the functions specified in this and relevant decisions by the COP/MOP, and, in particular, have sufficient knowledge and understanding of:

- (i) The guidelines for the implementation of Article 6 of the Kyoto Protocol, relevant decisions of the COP/MOP and of the Article 6 supervisory committee;
- (ii) Environmental issues relevant to the verification of Article 6 projects;
- (iii) The technical aspects of Article 6 activities relevant to environmental issues, including expertise in the setting of baselines and monitoring of emissions and other environmental impacts;
- (iv) Relevant environmental auditing requirements and methodologies;
- (v) Methodologies for the accounting of anthropogenic emissions by sources and/or anthropogenic removals by sinks;

(g) Have a management structure that has overall responsibility for performance and implementation of the entity's functions, including quality assurance procedures, and all relevant decisions relating to verification. The applicant independent entity shall make available:

- (i) The names, qualifications, experience and terms of reference of the senior executive, board members, senior officers and other relevant personnel;
- (ii) An organizational chart showing lines of authority, responsibility and allocation of functions stemming from the senior executive;
- (iii) Its quality assurance policy and procedures;
- (iv) Administrative procedures, including document control;
- (v) Its policy and procedures for the recruitment and training of independent entity personnel, for ensuring their competence for all necessary functions and for monitoring their performance;
- (vi) Its procedures for handling complaints, appeals and disputes;

(h) Not have pending any judicial process for malpractice, fraud and/or other activity incompatible with its functions as an accredited independent entity.

2. An applicant independent entity shall meet the following operational requirements:

(a) Work in a credible, independent, non-discriminatory and transparent manner, complying with applicable national law and meeting, in particular, the following requirements:

- (i) An applicant independent entity shall have a documented structure, which safeguards impartiality, including provisions to ensure the impartiality of its operations;
- (ii) If it is part of a larger organization, and where parts of that organization are, or may become, involved in the identification, development or financing of any Article 6 project, the applicant independent entity shall:
  - Make a declaration of all the organization's actual and potential Article 6 activities;

- Clearly define the links with other parts of the organization, demonstrating that no conflicts of interest exist;
- Demonstrate that no actual or potential conflict of interest exists between its functions as an accredited independent entity and any other functions that it may have, and demonstrate how business is managed to minimize any identified risk to impartiality. The demonstration shall cover all potential sources of conflict of interest, whether they arise from within the applicant independent entity or from the activities of related bodies;
- Demonstrate that it, together with its senior executive and staff, is not involved in any commercial, financial or other processes which might influence its judgement or endanger trust in its independence of judgement and integrity in relation to its activities, and that it complies with any rules applicable in this respect;

(b) Have adequate arrangements to safeguard confidentiality of the information obtained from Article 6 project participants in accordance with provisions contained in the annex on guidelines for the implementation of Article 6.

APPENDIX B

Criteria for baseline setting and monitoring

Criteria for baseline setting

1. The baseline for an Article 6 project is the scenario that reasonably represents the anthropogenic emissions by sources or anthropogenic removals by sinks of greenhouse gases that would occur in the absence of the proposed project. A baseline shall cover emissions from all gases, sectors and source categories listed in Annex A, and anthropogenic removals by sinks, within the project boundary.

2. A baseline shall be established:

(a) On a project-specific basis and/or using a multi-project emission factor;

(b) In a transparent manner with regard to the choice of approaches, assumptions, methodologies, parameters, data sources and key factors;

(c) Taking into account relevant national and/or sectoral policies and circumstances, such as sectoral reform initiatives, local fuel availability, power sector expansion plans, and the economic situation in the project sector;

(d) In such a way that ERUs cannot be earned for decreases in activity levels outside the project activity or due to *force majeure;*

(e) Taking account of uncertainties and using conservative assumptions.

3. Project participants shall justify their choice of baseline.

Monitoring

4. Project participants shall include, as part of the project design document, a monitoring plan that provides for:

(a) The collection and archiving of all relevant data necessary for estimating or measuring anthropogenic emissions by sources and/or anthropogenic removals by sinks of greenhouse gases occurring within the project boundary during the crediting period;

(b) The collection and archiving of all relevant data necessary for determining the baseline of anthropogenic emissions by sources and/or anthropogenic removals by sinks of greenhouse gases within the project boundary during the crediting period;

(c) The identification of all potential sources of, and the collection and archiving of data on increased anthropogenic emissions by sources and/or reduced anthropogenic removals by sinks of greenhouse gases outside the project boundary that are significant and reasonably attributable to the project during the crediting period. The project boundary shall encompass all anthropogenic emissions by sources and/or removals by sinks of greenhouse gases under the control of the project participants that are significant and reasonably attributable to the Article 6 project;

(d) The collection and archiving of information on environmental impacts, in accordance with procedures as required by the host Party, where applicable;

(e) Quality assurance and control procedures for the monitoring process;

(f) Procedures for the periodic calculation of the reductions of anthropogenic emissions by sources and/or enhancements of anthropogenic removals by sinks by the proposed Article 6 project, and for leakage effects, if any. Leakage is defined as the net change of anthropogenic emissions by sources and/or removals by sinks of greenhouse gases which occurs outside the project boundary, and that is measurable and attributable to the Article 6 project;

(g) Documentation of all steps involved in the calculations referred to in subparagraphs (b) and (f) above.

5. Revisions, if any, to the monitoring plan to improve its accuracy and/or completeness of information shall be justified by project participants and shall be submitted for the determination referred to in paragraph 37 of the annex on guidelines for the implementation of Article 6 of the Kyoto Protocol by the accredited independent entity.

6. The implementation of the monitoring plan and its revisions, as applicable, shall be a condition for verification.

DECISION 17/CP.7

MODALITIES AND PROCEDURES FOR A CLEAN DEVELOPMENT MECHANISM AS DEFINED IN ARTICLE 12 OF THE KYOTO PROTOCOL

The Conference of the Parties,

Recalling Article 12 of the Kyoto Protocol which provides that the purpose of the clean development mechanism shall be to assist Parties not included in Annex I to the Convention in achieving sustainable development and in contributing to the ultimate objective of the Convention, and to assist Parties included in Annex I in achieving compliance with their quantified emission limitation and reduction commitments under Article 3 of the Kyoto Protocol,

Recalling also its decision 5/CP.6 containing the Bonn Agreements on the implementation of the Buenos Aires Plan of Action,

Aware of its decisions 2/CP.7, 11/CP.7, 15/CP.7, 16/CP.7, 18/CP.7, 19/CP.7, 20/CP.7, 21/CP.7, 22/CP.7, 23/CP.7, 24/CP.7 and 38/CP.7,

Affirming that it is the host Party's prerogative to confirm whether a clean development mechanism project activity assists it in achieving sustainable development,

Recognizing that Parties included in Annex I are to refrain from using certified emission reductions generated from nuclear facilities to meet their commitments under Article 3, paragraph 1,

Bearing in mind the need to promote equitable geographic distribution of clean development mechanism project activities at regional and subregional levels,

Emphasizing that public funding for clean development mechanism projects from Parties in Annex I is not to result in the diversion of official development assistance and is to be separate from and not counted towards the financial obligations of Parties included in Annex I,

Further emphasizing that clean development mechanism project activities should lead to the transfer of environmentally safe and sound technology and know-how in addition to that required under Article 4, paragraph 5, of the Convention and Article 10 of the Kyoto Protocol,

Recognizing the need for guidance for project participants and designated operational entities, in particular for establishing reliable, transparent and conservative baselines, to assess whether clean development mechanism project activities are in accordance with the additionality criterion in Article 12, paragraph 5(c), of the Kyoto Protocol,

1. *Decides* to facilitate a prompt start for a clean development mechanism by adopting the modalities and procedures contained in the annex below;

2. *Decides* that, for the purposes of the present decision, the Conference of the Parties shall assume the responsibilities of the Conference of the Parties serving as the meeting of the Parties to the Kyoto Protocol as set out in the annex below on modalities and procedures;

3. *Invites* nominations for membership in the executive board:

(a) For facilitating the prompt start of the clean development mechanism, from Parties to the Convention to be submitted to the President of the Conference of the Parties at its present session, with a view to the Conference of the Parties electing the members of the executive board at that session;

(b) Upon the entry into force of the Kyoto Protocol, to replace any member of the executive board of the clean development mechanism whose country has not ratified or acceded to the Kyoto Protocol. Such new members shall be nominated by the same constituencies and elected at the first session of the Conference of the Parties serving as the meeting of the Parties to the Kyoto Protocol;

4. *Decides* that, prior to the first session of the Conference of the Parties serving as the meeting of the Parties to the Kyoto Protocol, the executive board and any designated operational entities shall operate in the same manner as the executive board and designated operational entities of the clean development mechanism as set out in the annex below;

5. *Decides* that the executive board shall convene its first meeting immediately upon the election of its members;

6. *Decides* that the executive board shall include in its work plan until the eighth session of the Conference of the Parties, *inter alia*, the following tasks:

(a) To develop and agree on its rules of procedure and recommend them to the Conference of the Parties for adoption, applying draft rules until then;

(b) To accredit operational entities and designate them, on a provisional basis, pending the designation by the Conference of the Parties at its eighth session;

(c) To develop and recommend to the Conference of the Parties, at its eighth session, simplified modalities and procedures for the following small-scale clean development mechanism project activities:

(i) Renewable energy project activities with a maximum output capacity equivalent of up to 15 megawatts (or an appropriate equivalent);

(ii) Energy efficiency improvement project activities which reduce energy consumption, on the supply and/or demand side, by up to the equivalent of 15 gigawatt/hours per year;

(iii) Other project activities that both reduce anthropogenic emissions by sources and directly emit less than 15 kilotonnes of carbon dioxide equivalent annually;

(d) To prepare recommendations on any relevant matter, including on Appendix C to the annex below, for consideration by the Conference of the Parties at its eighth session;

(e) To identify modalities for seeking collaboration with the Subsidiary Body for Scientific and Technological Advice on methodological and scientific issues;

7. *Decides*:

(a) That the eligibility of land use, land-use change and forestry project activities under the clean development mechanism is limited to afforestation and reforestation;

(b) That for the first commitment period, the total of additions to a Party's assigned amount resulting from eligible land use, land-use change and forestry project activities under the clean development mechanism shall not exceed one per cent of base year emissions of that Party, times five;

(c) That the treatment of land use, land-use change and forestry project activities under the clean development mechanism in future commitment periods shall be decided as part of the negotiations on the second commitment period;

8. *Requests* the secretariat to organize a workshop before the sixteenth session of the Subsidiary Body for Scientific and Technological Advice with the aim of recommending terms of reference and an agenda for the work to be conducted under paragraph 10(b) below on the basis of, *inter alia*, submissions by Parties referred to in paragraph 9 below;

9. *Invites* Parties to provide submissions to the secretariat by 1 February 2002 on the organization of the workshop referred to in paragraph 8 above, and to express their views on the terms of reference and the agenda for the work to be conducted under paragraph 10(b) below;

10. *Requests* the Subsidiary Body for Scientific and Technological Advice:

(a) To develop at its sixteenth session terms of reference and an agenda for the work to be conducted under subparagraph (b) below, taking into consideration, *inter alia*, the outcome of the workshop mentioned in paragraph 8 above;

(b) To develop definitions and modalities for including afforestation and reforestation project activities under the clean development mechanism in the first commitment period, taking into account the issues of non-permanence, additionality, leakage, uncertainties and socio-economic and environmental impacts, including impacts on biodiversity and natural ecosystems, and being guided by the principles in the preamble to decision -/CMP.1 (*Land use, land-use change and forestry*) and the terms of reference referred to in subparagraph (a) above, with the aim of adopting a decision on these definitions and modalities at the ninth session of the Conference of the Parties, to be forwarded to the Conference of the Parties serving as the meeting of the Parties to the Kyoto Protocol at its first session;

11. *Decides* that the decision by the Conference of the Parties at its ninth session, on definitions and modalities for inclusion of afforestation and reforestation project activities under the clean development mechanism, for the first commitment period, referred to in paragraph 10 (b) above, shall be in the form of an annex on modalities and procedures for afforestation and reforestation project activities for a clean development mechanism reflecting, *mutatis mutandis*, the annex to the present decision on modalities and procedures for a clean development mechanism;

12. *Decides* that certified emission reductions shall only be issued for a crediting period starting after the date of registration of a clean development mechanism project activity;

13. *Further decides* that a project activity starting as of the year 2000, and prior to the adoption of this decision, shall be eligible for validation and registration as a clean development mechanism project activity if submitted for registration before 31 December 2005. If registered, the crediting period for such project activities may start prior to the date of its registration but not earlier than 1 January 2000;

14. *Requests* Parties included in Annex I to start implementing measures to assist Parties not included in Annex I, in particular the least developed and small island developing States among them, with building capacity in order to facilitate their participation in the clean development mechanism, taking into account relevant decisions by the Conference of the Parties on capacity-building and on the financial mechanism of the Convention;

15. *Decides*:

(a) That the share of proceeds to assist developing country Parties that are particularly vulnerable to the adverse effects of climate change to meet the costs of adaptation, as referred to in Article 12, paragraph 8, of the Kyoto Protocol, shall be two per cent of the certified emission reductions issued for a clean development mechanism project activity;

(b) That clean development mechanism project activities in least developed country Parties shall be exempt from the share of proceeds to assist with the costs of adaptation;

16. *Decides* that the level of the share of proceeds to cover administrative expenses of the clean development mechanism shall be determined by the Conference of the Parties upon the recommendation of the executive board;

17. *Invites* Parties to finance the administrative expenses for operating the clean development mechanism by making contributions to the UNFCCC Trust Fund for Supplementary Activities. Such contributions shall be reimbursed, if requested, in accordance with procedures and a timetable to be determined by the Conference of the Parties upon the recommendation of the executive board. Until the Conference of the Parties determines a percentage for the share of proceeds for the administrative expenses, the executive board shall charge a fee to recover any project related expenses;

18. *Requests* the secretariat to perform any functions assigned to it in the present decision and in the annex below;

19. *Decides* to assess progress made regarding the clean development mechanism and to take appropriate action, as necessary. Any revision of the decision shall not affect clean development mechanism project activities already registered;

20. *Recommends* that the Conference of the Parties serving as the meeting of the Parties to the Kyoto Protocol, at its first session, adopt the draft decision below.

DRAFT DECISION -/CMP.1 (*ARTICLE 12*)

MODALITIES AND PROCEDURES FOR A CLEAN DEVELOPMENT MECHANISM AS DEFINED IN ARTICLE 12 OF THE KYOTO PROTOCOL

The Conference of the Parties serving as the meeting of the Parties to the Kyoto Protocol,

Recalling the provisions of Articles 3 and 12 of the Kyoto Protocol,

Bearing in mind that, in accordance with Article 12, the purpose of the clean development mechanism is to assist Parties not included in Annex I to the Convention in achieving sustainable development and in contributing to the ultimate objective of the Convention, and to assist Parties included in Annex I in achieving compliance with their quantified emission limitation and reduction commitments under Article 3 of the Kyoto Protocol,

Aware of its decisions -/CMP.1 (*Mechanisms*), -/CMP.1 (*Article 6*), -/CMP.1 (*Article 17*), -/CMP.1 (*Land use, land-use change and forestry*), -/CMP.1 (*Modalities for the accounting of assigned amounts*), -/CMP.1 (*Article 5.1*), -/CMP.1 (*Article 5.2*), -/CMP.1 (*Article 7*) and -/CMP.1 (*Article 8*), and decisions 2/CP.7 and 24/CP.7, *Cognizant* of decision 17/CP.7 on modalities and procedures for a clean development mechanism as defined in Article 12 of the Kyoto Protocol,

1. *Decides* to confirm, and give full effect to any actions taken pursuant to, decision 17/CP.7 and to any other relevant decisions by the Conference of the Parties, as appropriate;
2. *Adopts* the modalities and procedures for a clean development mechanism contained in the annex below;
3. *Invites* the executive board to review the simplified modalities, procedures and the definitions of small-scale project activities referred to in paragraph 6(c) of decision 17/CP.7 and, if necessary, make appropriate recommendations to the Conference of the Parties serving as the meeting of the Parties to the Kyoto Protocol;
4. *Decides* further that any future revision of the modalities and procedures for a clean development mechanism shall be decided in accordance with the rules of procedure of the Conference of the Parties serving as the meeting of the Parties to the Kyoto Protocol, as applied. The first review shall be carried out no later than one year after the end of the first commitment period, based on recommendations by the executive board and by the Subsidiary Body for Implementation drawing on technical advice from the Subsidiary Body for Scientific and Technological Advice, as needed. Further reviews shall be carried out periodically thereafter. Any revision of the decision shall not affect clean development mechanism project activities already registered.

ANNEX

MODALITIES AND PROCEDURES FOR A CLEAN DEVELOPMENT MECHANISM

A. Definitions

1. For the purposes of the present annex the definitions contained in Article 1[1] and the provisions of Article 14 shall apply. Furthermore:

(a) An 'emission reduction unit' or 'ERU' is a unit issued pursuant to the relevant provisions in the annex to decision -/CMP.1 (*Modalities for the accounting of assigned amounts*) and is equal to one metric tonne of carbon dioxide equivalent, calculated using global warming potentials defined by decision 2/CP.3 or as subsequently revised in accordance with Article 5;

(b) A 'certified emission reduction' or 'CER' is a unit issued pursuant to Article 12 and requirements thereunder, as well as the relevant provisions in these modalities and procedures, and is equal to one metric tonne of carbon dioxide equivalent, calculated using global warming potentials defined by decision 2/CP.3 or as subsequently revised in accordance with Article 5;

(c) An 'assigned amount unit' or 'AAU' is a unit issued pursuant to the relevant provisions in the annex to decision -/CMP.1 (*Modalities for the accounting of assigned amounts*) and is equal to one metric tonne of carbon dioxide equivalent, calculated using global warming potentials defined by decision 2/CP.3 or as subsequently revised in accordance with Article 5;

(d) A 'removal unit' or 'RMU' is a unit issued pursuant to the relevant provisions in the annex to decision -/CMP.1 (*Modalities for the accounting of assigned amounts*) and is equal to one metric tonne of carbon dioxide equivalent, calculated using global warming potentials defined by decision 2/CP.3 or as subsequently revised in accordance with Article 5;

(e) 'Stakeholders' means the public, including individuals, groups or communities affected, or likely to be affected, by the proposed clean development mechanism project activity.

B. Role of the Conference of the Parties serving as the meeting of the Parties to the Kyoto Protocol

2. The Conference of the Parties serving as the meeting of the Parties to the Kyoto Protocol (COP/MOP) shall have authority over and provide guidance to the clean development mechanism (CDM).

3. The COP/MOP shall provide guidance to the executive board by taking decisions on:

(a) The recommendations made by the executive board on its rules of procedure;

(b) The recommendations made by the executive board, in accordance with provisions of decision 17/CP.7, the present annex and relevant decisions of the COP/MOP;

(c) The designation of operational entities accredited by the executive board in accordance with Article 12, paragraph 5, and accreditation standards contained in Appendix A below.

[1] In the context of this annex, 'Article' refers to an Article of the Kyoto Protocol, unless otherwise specified.

4. The COP/MOP shall further:

(a) Review annual reports of the executive board;

(b) Review the regional and subregional distribution of designated operational entities and take appropriate decisions to promote accreditation of such entities from developing country Parties.[2]

(c) Review the regional and subregional distribution of CDM project activities with a view to identifying systematic or systemic barriers to their equitable distribution and take appropriate decisions, based, *inter alia*, on a report by the executive board;

(d) Assist in arranging funding of CDM project activities, as necessary.

C. Executive board

5. The executive board shall supervise the CDM, under the authority and guidance of the COP/MOP, and be fully accountable to the COP/MOP. In this context, the executive board shall:

(a) Make recommendations to the COP/MOP on further modalities and procedures for the CDM, as appropriate;

(b) Make recommendations to the COP/MOP on any amendments or additions to rules of procedure for the executive board contained in the present annex, as appropriate;

(c) Report on its activities to each session of the COP/MOP;

(d) Approve new methodologies related to, *inter alia*, baselines, monitoring plans and project boundaries in accordance with the provisions of Appendix C below;

(e) Review provisions with regard to simplified modalities, procedures and the definitions of small scale project activities and make recommendations to the COP/MOP;

(f) Be responsible for the accreditation of operational entities, in accordance with accreditation standards contained in Appendix A below, and make recommendations to the COP/MOP for the designation of operational entities, in accordance with Article 12, paragraph 5. This responsibility includes:

(i) Decisions on re-accreditation, suspension and withdrawal of accreditation;

(ii) Operationalization of accreditation procedures and standards;

(g) Review the accreditation standards in Appendix A below and make recommendations to the COP/MOP for consideration, as appropriate;

(h) Report to the COP/MOP on the regional and subregional distribution of CDM project activities with a view to identifying systematic or systemic barriers to their equitable distribution;

(i) Make publicly available relevant information, submitted to it for this purpose, on proposed CDM project activities in need of funding and on investors seeking opportunities, in order to assist in arranging funding of CDM project activities, as necessary;

(j) Make any technical reports commissioned available to the public and provide a period of at least eight weeks for public comments on draft methodologies and guidance before documents are finalized and any recommendations are submitted to the COP/MOP for their consideration;

(k) Develop, maintain and make publicly available a repository of approved rules, procedures, methodologies and standards;

(l) Develop and maintain the CDM registry as defined in Appendix D below;

[2] In the context of this annex, 'Party' refers to a Party to the Kyoto Protocol, unless otherwise specified.

(m) Develop and maintain a publicly available database of CDM project activities containing information on registered project design documents, comments received, verification reports, its decisions as well as information on all CERs issued;

(n) Address issues relating to observance of modalities and procedures for the CDM by project participants and/or operational entities, and report on them to the COP/MOP;

(o) Elaborate and recommend to the COP/MOP for adoption at its next session procedures for conducting the reviews referred to in paragraphs 41 and 65 below including, *inter alia*, procedures to facilitate consideration of information from Parties, stakeholders and UNFCCC accredited observers. Until their adoption by the COP/MOP, the procedures shall be applied provisionally;

(p) Carry out any other functions ascribed to it in decision 17/CP.7, the present annex and relevant decisions of the COP/MOP.

6. Information obtained from CDM project participants marked as proprietary or confidential shall not be disclosed without the written consent of the provider of the information, except as required by national law. Information used to determine additionality as defined in paragraph 43 below, to describe the baseline methodology and its application, and to support an environmental impact assessment referred to in paragraph 37(c) below, shall not be considered as proprietary or confidential.

7. The executive board shall comprise ten members from Parties to the Kyoto Protocol, as follows: one member from each of the five United Nations regional groups, two other members from the Parties included in Annex I, two other members from the Parties not included in Annex I, and one representative of the small island developing States, taking into account the current practice in the Bureau of the Conference of the Parties.

8. Members, including alternate members, of the executive board shall:

(a) Be nominated by the relevant constituencies referred to in paragraph 7 above and be elected by the COP/MOP. Vacancies shall be filled in the same way;

(b) Be elected for a period of two years and be eligible to serve a maximum of two consecutive terms. Terms as alternate members do not count. Five members and five alternate members shall be elected initially for a term of three years and five members and five alternate members for a term of two years. Thereafter, the COP/MOP shall elect, every year, five new members, and five new alternate members, for a term of two years. Appointment pursuant to paragraph 11 below shall count as one term. The members, and alternate members, shall remain in office until their successors are elected;

(c) Possess appropriate technical and/or policy expertise and shall act in their personal capacity. The cost of participation of members, and of alternate members, from developing country Parties and other Parties eligible under UNFCCC practice shall be covered by the budget for the executive board;

(d) Be bound by the rules of procedure of the executive board;

(e) Take a written oath of service witnessed by the Executive Secretary of the UNFCCC or his/her authorized representative before assuming his or her duties;

(f) Have no pecuniary or financial interest in any aspect of a CDM project activity or any designated operational entity;

(g) Subject to their responsibilities to the executive board, not disclose any confidential or proprietary information coming to their knowledge by reason of their duties for the executive board. The duty of the member, including alternate member, not to disclose confidential information constitutes an obligation in respect of that member, and alternate member, and shall remain an obligation after the expiration or termination of that member's function for the executive board.

9. The COP/MOP shall elect an alternate for each member of the executive board based on the criteria in paragraphs 7 and 8 above. The nomination by a constituency of a candidate member shall be accompanied by a nomination for a candidate alternate member from the same constituency.

10. The executive board may suspend and recommend to the COP/MOP the termination of the membership of a particular member, including an alternate member, for cause including, *inter alia*, breach of the conflict of interest provisions, breach of the confidentiality provisions, or failure to attend two consecutive meetings of the executive board without proper justification.

11. If a member, or an alternate member, of the executive board resigns or is otherwise unable to complete the assigned term of office or to perform the functions of that office, the executive board may decide, bearing in mind the proximity of the next session of the COP/MOP, to appoint another member, or an alternate member, from the same constituency to replace thesaid member for the remainder of that member's mandate.

12. The executive board shall elect its own chairperson and vice-chairperson, with one being a member from a Party included in Annex I and the other being from a Party not included in Annex I. The positions of chairperson and vice-chairperson shall alternate annually between a member from a Party included in Annex I and a member from a Party not included in Annex I.

13. The executive board shall meet as necessary but no less than three times a year, bearing in mind the provisions of paragraph 41 below. All documentation for executive board meetings shall be made available to alternate members.

14. At least two thirds of the members of the executive board, representing a majority of members from Parties included in Annex I and a majority of members from Parties not included in Annex I, must be present to constitute a quorum.

15. Decisions by the executive board shall be taken by consensus, whenever possible. If all efforts at reaching a consensus have been exhausted and no agreement has been reached, decisions shall be taken by a three-fourths majority of the members present and voting at the meeting. Members abstaining from voting shall be considered as not voting.

16. Meetings of the executive board shall be open to attendance, as observers, by all Parties and by all UNFCCC accredited observers and stakeholders, except where otherwise decided by the executive board.

17. The full text of all decisions of the executive board shall be made publicly available. The working language of the executive board shall be English. Decisions shall be made available in all six official languages of the United Nations.

18. The executive board may establish committees, panels or working groups to assist it in the performance of its functions. The executive board shall draw on the expertise necessary to perform its functions, including from the UNFCCC roster of experts. In this context, it shall take fully into account the consideration of regional balance.

19. The secretariat shall service the executive board.

D. Accreditation and designation of operational entities

20. The executive board shall:

(a) Accredit operational entities which meet the accreditation standards contained in Appendix A below;

(b) Recommend the designation of operational entities to the COP/MOP;

(c) Maintain a publicly available list of all designated operational entities;

(d) Review whether each designated operational entity continues to comply with the accreditation standards contained in Appendix A below and on this basis confirm whether to reaccredit each operational entity every three years;

(e) Conduct spot-checking at any time and, on the basis of the results, decide to conduct the above-mentioned review, if warranted.

21. The executive board may recommend to the COP/MOP to suspend or withdraw the designation of a designated operational entity if it has carried out a review and found that the entity no longer meets the accreditation standards or applicable provisions in decisions of the COP/MOP. The executive board may recommend the suspension or withdrawal of designation only after the designated operational entity has had the possibility of a hearing. The suspension or withdrawal is with immediate effect, on a provisional basis, once the executive board has made a recommendation, and remains in effect pending a final decision by the COP/MOP. The affected entity shall be notified, immediately and in writing, once the executive board has recommended its suspension or withdrawal. The recommendation by the executive board and the decision by the COP/MOP on such a case shall be made public.

22. Registered project activities shall not be affected by the suspension or withdrawal of designation of a designated operational entity unless significant deficiencies are identified in the relevant validation, verification or certification report for which the entity was responsible. In this case, the executive board shall decide whether a different designated operational entity shall be appointed to review, and where appropriate correct, such deficiencies. If such a review reveals that excess CERs were issued, the designated operational entity whose accreditation has been withdrawn or suspended shall acquire and transfer, within 30 days of the end of review, an amount of reduced tonnes of carbon dioxide equivalent equal to the excess CERs issued, as determined by the executive board, to a cancellation account maintained in the CDM registry by the executive board.

23. Any suspension or withdrawal of a designated operational entity that adversely affects registered project activities shall be recommended by the executive board only after the affected project participants have had the possibility of a hearing.

24. Any costs related to the review referred to in paragraph 22 above shall be borne by the designated operational entity whose designation has been withdrawn or suspended.

25. The executive board may seek assistance in performing the functions in paragraph 20 above, in accordance with the provisions of paragraph 18 above.

E. Designated operational entities

26. Designated operational entities shall be accountable to the COP/MOP through the executive board and shall comply with the modalities and procedures in decision 17/CP.7, the present annex and relevant decisions of the COP/MOP and the executive board.

27. A designated operational entity shall:

(a) Validate proposed CDM project activities;

(b) Verify and certify reductions in anthropogenic emissions by sources of greenhouse gases;

(c) Comply with applicable laws of the Parties hosting CDM project activities when carrying out its functions referred to in subparagraph (e) below;

(d) Demonstrate that it, and its subcontractors, have no real or potential conflict of interest with the participants in the CDM project activities for which it has been selected to carry out validation or verification and certification functions;

(e) Perform one of the following functions related to a given CDM project activity: validation or verification and certification. Upon request, the executive board may, however, allow a single designated operational entity to perform all these functions within a single CDM project activity;

(f) Maintain a publicly available list of all CDM project activities for which it has carried out validation, verification and certification;

(g) Submit an annual activity report to the executive board;

(h) Make information obtained from CDM project participants publicly available, as required by the executive board. Information marked as proprietary or confidential shall not be disclosed without the written consent of the provider of the information, except as required by national law. Information used to determine additionality as defined in paragraph 43 below, to describe the baseline methodology and its application, and to support an environmental impact assessment referred to in paragraph 37(c) below, shall not be considered as proprietary or confidential.

F. Participation requirements

28. Participation in a CDM project activity is voluntary.

29. Parties participating in the CDM shall designate a national authority for the CDM.

30. A Party not included in Annex I may participate in a CDM project activity if it is a Party to the Kyoto Protocol.

31. Subject to the provisions of paragraph 32 below, a Party included in Annex I with a commitment inscribed in Annex B is eligible to use CERs, issued in accordance with the relevant provisions, to contribute to compliance with part of its commitment under Article 3, paragraph 1, if it is in compliance with the following eligibility requirements:

(a) It is a Party to the Kyoto Protocol;

(b) Its assigned amount pursuant to Article 3, paragraphs 7 and 8, has been calculated and recorded in accordance with decision -/CMP.1 (*Modalities for the accounting of assigned amounts*);

(c) It has in place a national system for the estimation of anthropogenic emissions by sources and anthropogenic removals by sinks of all greenhouse gases not controlled by the Montreal Protocol, in accordance with Article 5, paragraph 1, and the requirements in the guidelines decided thereunder;

(d) It has in place a national registry in accordance with Article 7, paragraph 4, and the requirements in the guidelines decided thereunder;

(e) It has submitted annually the most recent required inventory, in accordance with Article 5, paragraph 2, and Article 7, paragraph 1, and the requirements in the guidelines decided thereunder, including the national inventory report and the common reporting format. For the first commitment period, the quality assessment needed for the purpose of determining eligibility to use the mechanisms shall be limited to the parts of the inventory pertaining to emissions of greenhouse gases from sources/sector categories from Annex A to the Kyoto Protocol and the submission of the annual inventory on sinks;

(f) It submits the supplementary information on assigned amount in accordance with Article 7, paragraph 1, and the requirements in the guidelines decided thereunder and makes any additions to, and subtractions from, assigned amount pursuant to Article 3, paragraphs 7 and 8, including for the activities under Article 3, paragraphs 3 and 4, in accordance with Article 7, paragraph 4, and the requirements in the guidelines decided thereunder.

32. A Party included in Annex I with a commitment inscribed in Annex B shall be considered:

(a) To meet the eligibility requirements referred to in paragraph 31 above after 16 months have elapsed since the submission of its report to facilitate the calculation of its assigned amount pursuant to Article 3, paragraphs 7 and 8, and to demonstrate its capacity to account for its emissions and assigned amount, in accordance with the modalities adopted for the accounting of assigned amount under Article 7, paragraph 4, unless the enforcement branch of the compliance committee finds in accordance with decision 24/CP.7 that the Party does not meet these requirements, or, at an earlier date, if the enforcement branch of the compliance committee has decided that it is not proceeding with any questions of implementation relating to these requirements indicated in reports of the expert review teams under Article 8 of the Kyoto Protocol, and has transmitted this information to the secretariat;

(b) To continue to meet the eligibility requirements referred to in paragraph 31 above unless and until the enforcement branch of the compliance committee decides that the Party does not meet one or more of the eligibility requirements, has suspended the Party's eligibility, and has transmitted this information to the secretariat.

33. A Party that authorizes private and/or public entities to participate in Article 12 project activities shall remain responsible for the fulfilment of its obligations under the Kyoto Protocol and shall ensure that such participation is consistent with the present annex. Private and/or public entities may only transfer and acquire CERs if the authorizing Party is eligible to do so at that time.

34. The secretariat shall maintain publicly accessible lists of:

(a) Parties not included in Annex I which are Parties to the Kyoto Protocol;

(b) Parties included in Annex I that do not meet the requirements in paragraph 31 above or have been suspended.

G. *Validation and registration*

35. Validation is the process of independent evaluation of a project activity by a designated operational entity against the requirements of the CDM as set out in decision 17/CP.7, the present annex and relevant decisions of the COP/MOP, on the basis of the project design document, as outlined in Appendix B below.

36. Registration is the formal acceptance by the executive board of a validated project as a CDM project activity. Registration is the prerequisite for the verification, certification and issuance of CERs related to that project activity.

37. The designated operational entity selected by project participants to validate a project activity, being under a contractual arrangement with them, shall review the project design document and any supporting documentation to confirm that the following requirements have been met:

(a) The participation requirements as set out in paragraphs 28 to 30 above are satisfied;

(b) Comments by local stakeholders have been invited, a summary of the comments received has been provided, and a report to the designated operational entity on how due account was taken of any comments has been received;

(c) Project participants have submitted to the designated operational entity documentation on the analysis of the environmental impacts of the project activity, including transboundary impacts and, if those impacts are considered significant by the project participants or the host Party, have undertaken an environmental impact assessment in accordance with procedures as required by the host Party;

(d) The project activity is expected to result in a reduction in anthropogenic emissions by sources of greenhouse gases that are additional to any that would occur in the absence of the proposed project activity, in accordance with paragraphs 43 to 52 below;

(e) The baseline and monitoring methodologies comply with requirements pertaining to:

(i) Methodologies previously approved by the executive board; or

(ii) Modalities and procedures for establishing a new methodology, as set out in paragraph 38 below;

(f) Provisions for monitoring, verification and reporting are in accordance with decision 17/CP.7, the present annex and relevant decisions of the COP/MOP;

(g) The project activity conforms to all other requirements for CDM project activities in decision 17/CP.7, the present annex and relevant decisions by the COP/MOP and the executive board.

38. If the designated operational entity determines that the project activity intends to use a new baseline or monitoring methodology, as referred to in paragraph 37(e) (ii) above, it shall, prior to a submission for registration of this project activity, forward the proposed methodology, together with the draft project design document, including a description of the project and identification of the project participants, to the executive board for review. The executive board shall expeditiously, if possible at its next meeting but not later than four months, review the proposed new methodology in accordance with the modalities and procedures of the present annex. Once approved by the executive board it shall make the approved methodology publicly available along with any relevant guidance and the designated operational entity may proceed with the validation of the project activity and submit the project design document for registration. In the event that the COP/MOP requests the revision of an approved methodology, no CDM project activity may use this methodology. The project participants shall revise the methodology, as appropriate, taking into consideration any guidance received.

39. A revision of a methodology shall be carried out in accordance with the modalities and procedures for establishing new methodologies as set out in paragraph 38 above. Any revision to an approved methodology shall only be applicable to project activities registered subsequent to the date of revision and shall not affect existing registered project activities during their crediting periods.

40. The designated operational entity shall:

(a) Prior to the submission of the validation report to the executive board, have received from the project participants written approval of voluntary participation from the designated national authority of each Party involved, including confirmation by the host Party that the project activity assists it in achieving sustainable development;

(b) In accordance with provisions on confidentiality contained in paragraph 27(h) above, make publicly available the project design document;

(c) Receive, within 30 days, comments on the validation requirements from Parties, stakeholders and UNFCCC accredited non-governmental organizations and make them publicly available;

(d) After the deadline for receipt of comments, make a determination as to whether, on the basis of the information provided and taking into account the comments received, the project activity should be validated;

(e) Inform project participants of its determination on the validation of the project activity. Notification to the project participants will include:

(i) Confirmation of validation and date of submission of the validation report to the executive board; or

(ii) An explanation of reasons for non-acceptance if the project activity, as documented, is judged not to fulfil the requirements for validation;

(f) Submit to the executive board, if it determines the proposed project activity to be valid, a request for registration in the form of a validation report including the project design document, the written approval of the host Party as referred to in subparagraph (a) above, and an explanation of how it has taken due account of comments received;

(g) Make this validation report publicly available upon transmission to the executive board.

41. The registration by the executive board shall be deemed final eight weeks after the date of receipt by the executive board of the request for registration, unless a Party involved in the project activity or at least three members of the executive board request a review of the proposed CDM project activity. The review by the executive board shall be made in accordance with the following provisions:

(a) It shall be related to issues associated with the validation requirements;

(b) It shall be finalized no later than at the second meeting following the request for review, with the decision and the reasons for it being communicated to the project participants and the public.

42. A proposed project activity that is not accepted may be reconsidered for validation and subsequent registration, after appropriate revisions, provided that it follows the procedures and meets the requirements for validation and registration, including those related to public comments.

43. A CDM project activity is additional if anthropogenic emissions of greenhouse gases by sources are reduced below those that would have occurred in the absence of the registered CDM project activity.

44. The baseline for a CDM project activity is the scenario that reasonably represents the anthropogenic emissions by sources of greenhouse gases that would occur in the absence of the proposed project activity. A baseline shall cover emissions from all gases, sectors and source categories listed in Annex A within the project boundary. A baseline shall be deemed to reasonably represent the anthropogenic emissions by sources that would occur in the absence of the proposed project activity if it is derived using a baseline methodology referred to in paragraphs 37 and 38 above.

45. A baseline shall be established:

(a) By project participants in accordance with provisions for the use of approved and new methodologies, contained in decision 17/CP.7, the present annex and relevant decisions of the COP/MOP;

(b) In a transparent and conservative manner regarding the choice of approaches, assumptions, methodologies, parameters, data sources, key factors and additionality, and taking into account uncertainty;

(c) On a project-specific basis;

(d) In the case of small-scale CDM project activities which meet the criteria specified in decision 17/CP.7 and relevant decisions by the COP/MOP, in accordance with simplified procedures developed for such activities;

(e) Taking into account relevant national and/or sectoral policies and circumstances, such as sectoral reform initiatives, local fuel availability, power sector expansion plans, and the economic situation in the project sector.

46. The baseline may include a scenario where future anthropogenic emissions by sources are projected to rise above current levels, due to the specific circumstances of the host Party.

47. The baseline shall be defined in a way that CERs cannot be earned for decreases in activity levels outside the project activity or due to *force majeure*.

48. In choosing a baseline methodology for a project activity, project participants shall select from among the following approaches the one deemed most appropriate for the project activity, taking into account any guidance by the executive board, and justify the appropriateness of their choice:

(a) Existing actual or historical emissions, as applicable; or

(b) Emissions from a technology that represents an economically attractive course of action, taking into account barriers to investment; or

(c) The average emissions of similar project activities undertaken in the previous five years, in similar social, economic, environmental and technological circumstances, and whose performance is among the top 20 per cent of their category.

49. Project participants shall select a crediting period for a proposed project activity from one of the following alternative approaches:

(a) A maximum of seven years which may be renewed at most two times, provided that, for each renewal, a designated operational entity determines and informs the executive board that the original project baseline is still valid or has been updated taking account of new data where applicable; or

(b) A maximum of ten years with no option of renewal.

50. Reductions in anthropogenic emissions by sources shall be adjusted for leakage in accordance with the monitoring and verification provisions in paragraphs 59 and 62(f) below, respectively.

51. Leakage is defined as the net change of anthropogenic emissions by sources of greenhouse gases which occurs outside the project boundary, and which is measurable and attributable to the CDM project activity.

52. The project boundary shall encompass all anthropogenic emissions by sources of greenhouse gases under the control of the project participants that are significant and reasonably attributable to the CDM project activity.

H. Monitoring

53. Project participants shall include, as part of the project design document, a monitoring plan that provides for:

(a) The collection and archiving of all relevant data necessary for estimating or measuring anthropogenic emissions by sources of greenhouse gases occurring within the project boundary during the crediting period;

(b) The collection and archiving of all relevant data necessary for determining the baseline of anthropogenic emissions by sources of greenhouse gases within the project boundary during the crediting period;

(c) The identification of all potential sources of, and the collection and archiving of data on, increased anthropogenic emissions by sources of greenhouse gases outside the project boundary that are significant and reasonably attributable to the project activity during the crediting period;

(d) The collection and archiving of information relevant to the provisions in paragraph 37(c) above;

(e) Quality assurance and control procedures for the monitoring process;

(f) Procedures for the periodic calculation of the reductions of anthropogenic emissions by sources by the proposed CDM project activity, and for leakage effects;

(g) Documentation of all steps involved in the calculations referred to in paragraph 53(c) and (f) above.

54. A monitoring plan for a proposed project activity shall be based on a previously approved monitoring methodology or a new methodology, in accordance with paragraphs 37 and 38 above, that:

(a) Is determined by the designated operational entity as appropriate to the circumstances of the proposed project activity and has been successfully applied elsewhere;

(b) Reflects good monitoring practice appropriate to the type of project activity.

55. For small-scale CDM project activities meeting the criteria specified in decision 17/CP.7 and relevant decisions by the COP/MOP, project participants may use simplified modalities and procedures for small-scale projects.

56. Project participants shall implement the monitoring plan contained in the registered project design document.

57. Revisions, if any, to the monitoring plan to improve its accuracy and/or completeness of information shall be justified by project participants and shall be submitted for validation to a designated operational entity.

58. The implementation of the registered monitoring plan and its revisions, as applicable, shall be a condition for verification, certification and the issuance of CERs.

59. Subsequent to the monitoring and reporting of reductions in anthropogenic emissions, CERs resulting from a CDM project activity during a specified time period shall be calculated, applying the registered methodology, by subtracting the actual anthropogenic emissions by sources from baseline emissions and adjusting for leakage.

60. The project participants shall provide to the designated operational entity, contracted by the project participants to perform the verification, a monitoring report in accordance with the registered monitoring plan set out in paragraph 53 above for the purpose of verification and certification.

I. Verification and certification

61. Verification is the periodic independent review and *ex post* determination by the designated operational entity of the monitored reductions in anthropogenic emissions by sources of greenhouse gases that have occurred as a result of a registered CDM project activity during the verification period. Certification is the written assurance by the designated operational entity that, during a specified time period, a project activity achieved the reductions in anthropogenic emissions by sources of greenhouse gases as verified.

62. In accordance with the provisions on confidentiality in paragraph 27(h) above, the designated operational entity contracted by the project participants to perform the verification shall make the monitoring report publicly available, and shall:

(a) Determine whether the project documentation provided is in accordance with the requirements of the registered project design document and relevant provisions of decision 17/CP.7, the present annex and relevant decisions of the COP/MOP;

(b) Conduct on-site inspections, as appropriate, that may comprise, *inter alia*, a review of performance records, interviews with project participants and local stakeholders, collection of measurements, observation of established practices and testing of the accuracy of monitoring equipment;

(c) If appropriate, use additional data from other sources;

(d) Review monitoring results and verify that the monitoring methodologies for the estimation of reductions in anthropogenic emissions by sources have been applied correctly and their documentation is complete and transparent;

(e) Recommend to the project participants appropriate changes to the monitoring methodology for any future crediting period, if necessary;

(f) Determine the reductions in anthropogenic emissions by sources of greenhouse gases that would not have occurred in the absence of the CDM project activity, based on the data and information derived under subparagraph (a) above and obtained under subparagraph (b) and/or (c) above, as appropriate, using calculation procedures consistent with those contained in the registered project design document and in the monitoring plan;

(g) Identify and inform the project participants of any concerns related to the conformity of the actual project activity and its operation with the registered project design document. Project participants shall address the concerns and supply relevant additional information;

(h) Provide a verification report to the project participants, the Parties involved and the executive board. The report shall be made publicly available.

63. The designated operational entity shall, based on its verification report, certify in writing that, during the specified time period, the project activity achieved the verified amount of reductions in anthropogenic emissions by sources of greenhouse gases that would not have occurred in the absence of the CDM project activity. It shall inform the project participants, Parties involved and the executive board of its certification decision in writing immediately upon completion of the certification process and make the certification report publicly available.

J. Issuance of certified emission reductions

64. The certification report shall constitute a request for issuance to the executive board of CERs equal to the verified amount of reductions of anthropogenic emissions by sources of greenhouse gases.

65. The issuance shall be considered final 15 days after the date of receipt of the request for issuance, unless a Party involved in the project activity or at least three members of the executive board request a review of the proposed issuance of CERs. Such a review shall be limited to issues of fraud, malfeasance or incompetence of the designated operational entities and be conducted as follows:

(a) Upon receipt of a request for such a review, the executive board, at its next meeting, shall decide on its course of action. If it decides that the request has merit it shall perform a review and decide whether the proposed issuance of CERs should be approved;

(b) The executive board shall complete its review within 30 days following its decision to perform the review;

(c) The executive board shall inform the project participants of the outcome of the review, and make public its decision regarding the approval of the proposed issuance of CERs and the reasons for it.

66. Upon being instructed by the executive board to issue CERs for a CDM project activity, the CDM registry administrator, working under the authority of the executive board, shall, promptly, issue the specified quantity of CERs into the pending account of the executive board in the CDM registry, in accordance with Appendix D below. Upon such issuance, the CDM registry administrator shall promptly:

(a) Forward the quantity of CERs corresponding to the share of proceeds to cover administrative expenses and to assist in meeting costs of adaptation, respectively, in accordance with Article 12, paragraph 8, to the appropriate accounts in the CDM registry for the management of the share of proceeds;

(b) Forward the remaining CERs to the registry accounts of Parties and project participants involved, in accordance with their request.

APPENDIX A

Standards for the accreditation of operational entities

1. An operational entity shall:

(a) Be a legal entity (either a domestic legal entity or an international organization) and provide documentation of this status;

(b) Employ a sufficient number of persons having the necessary competence to perform validation, verification and certification functions relating to the type, range and volume of work performed, under a responsible senior executive;

(c) Have the financial stability, insurance coverage and resources required for its activities;

(d) Have sufficient arrangements to cover legal and financial liabilities arising from its activities;

(e) Have documented internal procedures for carrying out its functions including, among others, procedures for the allocation of responsibility within the organization and for handling complaints. These procedures shall be made publicly available;

(f) Have, or have access to, the necessary expertise to carry out the functions specified in modalities and procedures of the CDM and relevant decisions by the COP/MOP, in particular knowledge and understanding of:

- (i) The modalities and procedures and guidelines for the operation of the CDM, relevant decisions of the COP/MOP and of the executive board;
- (ii) Issues, in particular environmental, relevant to validation, verification and certification of CDM project activities, as appropriate;
- (iii) The technical aspects of CDM project activities relevant to environmental issues, including expertise in the setting of baselines and monitoring of emissions;
- (iv) Relevant environmental auditing requirements and methodologies;
- (v) Methodologies for accounting of anthropogenic emissions by sources;
- (vi) Regional and sectoral aspects;

(g) Have a management structure that has overall responsibility for performance and implementation of the entity's functions, including quality assurance procedures, and all relevant decisions relating to validation, verification and certification. The applicant operational entity shall make available:

- (i) The names, qualifications, experience and terms of reference of senior management personnel such as the senior executive, board members, senior officers and other relevant personnel;
- (ii) An organizational chart showing lines of authority, responsibility and allocation of functions stemming from senior management;

(iii) Its quality assurance policy and procedures;
(iv) Administrative procedures, including document control;
(v) Its policy and procedures for the recruitment and training of operational entity personnel, for ensuring their competence for all necessary functions for validation, verification and certification functions, and for monitoring their performance;
(vi) Its procedures for handling complaints, appeals and disputes;

(h) Not have pending any judicial process for malpractice, fraud and/or other activity incompatible with its functions as a designated operational entity.

2. An applicant operational entity shall meet the following operational requirements:

(a) Work in a credible, independent, non-discriminatory and transparent manner, complying with applicable national law and meeting, in particular, the following requirements:

(i) An applicant operational entity shall have a documented structure, which safeguards impartiality, including provisions to ensure impartiality of its operations;
(ii) If it is part of a larger organization, and where parts of that organization are, or may become, involved in the identification, development or financing of any CDM project activity, the applicant operational entity shall:
- Make a declaration of all the organization's actual and planned involvement in CDM project activities, if any, indicating which part of the organization is involved and in which particular CDM project activities;
- Clearly define the links with other parts of the organization, demonstrating that no conflicts of interest exist;
- Demonstrate that no conflict of interest exists between its functions as an operational entity and any other functions that it may have, and demonstrate how business is managed to minimize any identified risk to impartiality. The demonstration shall cover all sources of conflict of interest, whether they arise from within the applicant operational entity or from the activities of related bodies;
- Demonstrate that it, together with its senior management and staff, is not involved in any commercial, financial or other processes which might influence its judgement or endanger trust in its independence of judgement and integrity in relation to its activities, and that it complies with any rules applicable in this respect;

(b) Have adequate arrangements to safeguard confidentiality of the information obtained from CDM project participants in accordance with provisions contained in the present annex.

APPENDIX B

Project design document

1. The provisions of this appendix shall be interpreted in accordance with the annex above on modalities and procedures for a CDM.

2. The purpose of this appendix is to outline the information required in the project design document. A project activity shall be described in detail taking into account the provisions of the annex on modalities and procedures for a CDM, in particular, section G on

validation and registration and section H on monitoring, in a project design document which shall include the following:

(a) A description of the project comprising the project purpose, a technical description of the project, including how technology will be transferred, if any, and a description and justification of the project boundary;

(b) A proposed baseline methodology in accordance with the annex on modalities and procedures for a CDM including, in the case of the:

(i) Application of an approved methodology:
- Statement of which approved methodology has been selected;
- Description of how the approved methodology will be applied in the context of the project;

(ii) Application of a new methodology:
- Description of the baseline methodology and justification of choice, including an assessment of strengths and weaknesses of the methodology;
- Description of key parameters, data sources and assumptions used in the baseline estimate, and assessment of uncertainties;
- Projections of baseline emissions;
- Description of how the baseline methodology addresses potential leakage;

(iii) Other considerations, such as a description of how national and/or sectoral policies and circumstances have been taken into account and an explanation of how the baseline was established in a transparent and conservative manner;

(c) Statement of the estimated operational lifetime of the project and which crediting period was selected;

(d) Description of how the anthropogenic emissions of GHG by sources are reduced below those that would have occurred in the absence of the registered CDM project activity;

(e) Environmental impacts:

(i) Documentation on the analysis of the environmental impacts, including transboundary impacts;

(ii) If impacts are considered significant by the project participants or the host Party: conclusions and all references to support documentation of an environmental impact assessment that has been undertaken in accordance with the procedures as required by the host Party;

(f) Information on sources of public funding for the project activity from Parties included in Annex I which shall provide an affirmation that such funding does not result in a diversion of official development assistance and is separate from and is not counted towards the financial obligations of those Parties;

(g) Stakeholder comments, including a brief description of the process, a summary of the comments received, and a report on how due account was taken of any comments received;

(h) Monitoring plan:

(i) Identification of data needs and data quality with regard to accuracy, comparability, completeness and validity;

(ii) Methodologies to be used for data collection and monitoring including quality assurance and quality control provisions for monitoring, collecting and reporting;

(iii) In the case of a new monitoring methodology, provide a description of the methodology, including an assessment of strengths and weaknesses of the methodology and whether or not it has been applied successfully elsewhere;

(i) Calculations:

(i) Description of formulae used to calculate and estimate anthropogenic emissions by sources of greenhouse gases of the CDM project activity within the project boundary;

(ii) Description of formulae used to calculate and to project leakage, defined as: the net change of anthropogenic emissions by sources of greenhouse gases which occurs outside the CDM project activity boundary, and that is measurable and attributable to the CDM project activity;

(iii) The sum of (i) and (ii) above representing the CDM project activity emissions;

(iv) Description of formulae used to calculate and to project the anthropogenic emissions by sources of greenhouse gases of the baseline;

(v) Description of formulae used to calculate and to project leakage;

(vi) The sum of (iv) and (v) above representing the baseline emissions;

(vii) Difference between (vi) and (iii) above representing the emission reductions of the CDM project activity;

(j) References to support the above, if any.

APPENDIX C

Terms of reference for establishing guidelines on baselines and monitoring methodologies

The executive board, drawing on experts in accordance with the modalities and procedures for a CDM, shall develop and recommend to the COP/MOP, *inter alia*:

(a) General guidance on methodologies relating to baselines and monitoring consistent with the principles set out in those modalities and procedures in order to:

(i) Elaborate the provisions relating to baseline and monitoring methodologies contained in decision 17/CP.7, the annex above and relevant decisions of the COP/MOP;

(ii) Promote consistency, transparency and predictability;

(iii) Provide rigour to ensure that net reductions in anthropogenic emissions are real and measurable, and an accurate reflection of what has occurred within the project boundary;

(iv) Ensure applicability in different geographical regions and to those project categories which are eligible in accordance with decision 17/CP.7 and relevant decisions of the COP/MOP;

(v) Address the additionality requirement of Article 12, paragraph 5(c), and paragraph 43 of the above annex;

(b) Specific guidance in the following areas:

(i) Definition of project categories (e.g. based on sector, subsector, project type, technology, geographic area) that show common methodological characteristics for baseline setting, and/or monitoring, including guidance on the level of geographic aggregation, taking into account data availability;

(ii) Baseline methodologies deemed to reasonably represent what would have occurred in the absence of a project activity;

(iii) Monitoring methodologies that provide an accurate measure of actual reductions in anthropogenic emissions as a result of the project activity, taking into account the need for consistency and cost-effectiveness;

(iv) Decision trees and other methodological tools, where appropriate, to guide choices in order to ensure that the most appropriate methodologies are selected, taking into account relevant circumstances;

(v) The appropriate level of standardization of methodologies to allow a reasonable estimation of what would have occurred in the absence of a project activity wherever possible and appropriate. Standardization should be conservative in order to prevent any overestimation of reductions in anthropogenic emissions;

(vi) Determination of project boundaries including accounting for all greenhouse gases that should be included as a part of the baseline, and monitoring. Relevance of leakage and recommendations for establishing appropriate project boundaries and methods for the *ex post* evaluation of the level of leakage;

(vii) Accounting for applicable national policies and specific national or regional circumstances, such as sectoral reform initiatives, local fuel availability, power sector expansion plans, and the economic situation in the sector relevant to the project activity;

(viii) The breadth of the baseline, e.g. how the baseline makes comparisons between the technology/fuel used and other technologies/fuels in the sector;

(c) In developing the guidance in (a) and (b) above, the executive board shall take into account:

(i) Current practices in the host country or an appropriate region, and observed trends;

(ii) Least cost technology for the activity or project category.

APPENDIX D

Clean development mechanism registry requirements

1. The executive board shall establish and maintain a CDM registry to ensure the accurate accounting of the issuance, holding, transfer and acquisition of CERs by Parties not included in Annex I. The executive board shall identify a registry administrator to maintain the registry under its authority.

2. The CDM registry shall be in the form of a standardized electronic database which contains, *inter alia*, common data elements relevant to the issuance, holding, transfer and acquisition of CERs. The structure and data formats of the CDM registry shall conform to technical standards to be adopted by the COP/MOP for the purpose of ensuring the accurate, transparent and efficient exchange of data between national registries, the CDM registry and the independent transaction log.

3. The CDM registry shall have the following accounts:

(a) One pending account for the executive board, into which CERs are issued before being transferred to other accounts;

(b) At least one holding account for each Party not included in Annex I hosting a CDM project activity or requesting an account;

(c) At least one account for the purpose of cancelling ERUs, CERs, AAUs and RMUs equal to excess CERs issued, as determined by the executive board, where the accreditation of a designated operational entity has been withdrawn or suspended;

(d) At least one account for the purpose of holding and transferring CERs corresponding to the share of proceeds to cover administrative expenses and to assist in meeting costs of adaptation in accordance with Article 12, paragraph 8. Such an account may not otherwise acquire CERs.

4. Each CER shall be held in only one account in one registry at a given time.

5. Each account within the CDM registry shall have a unique account number comprising the following elements:

(a) Party/organization identifier: the Party for which the account is maintained, using the two-letter country code defined by the International Organization for Standardization (ISO 3166), or, in the cases of the pending account and an account for managing the CERs corresponding to the share of proceeds, the executive board or another appropriate organization;

(b) A unique number: a number unique to that account for the Party or organization for which the account is maintained.

6. Upon being instructed by the executive board to issue CERs for a CDM project activity, the registry administrator shall, in accordance with the transaction procedures set out in decision -/CMP.1 (*Modalities for the accounting of assigned amounts*):

(a) Issue the specified quantity of CERs into a pending account of the executive board;

(b) Forward the quantity of CERs corresponding to the share of proceeds to cover administrative expenses and to assist in meeting costs of adaptation, in accordance with Article 12, paragraph 8, to the appropriate accounts in the CDM registry for holding and transferring such CERs;

(c) Forward the remaining CERs to the registry accounts of project participants and Parties involved, in accordance with their request.

7. Each CER shall have a unique serial number comprising the following elements:

(a) Commitment period: the commitment period for which the CER is issued;

(b) Party of origin: the Party which hosted the CDM project activity, using the twoletter country code defined by ISO 3166;

(c) Type: this shall identify the unit as a CER;

(d) Unit: a number unique to the CER for the identified commitment period and Party of origin;

(e) Project identifier: a number unique to the CDM project activity for the Party of origin.

8. Where the accreditation of a designated operational entity has been withdrawn or suspended, ERUs, CERs, AAUs and/or RMUs equal to the excess CERs issued, as determined by the executive board, shall be transferred to a cancellation account in the CDM registry. Such ERUs, CERs, AAUs and RMUs may not be further transferred or used for the purpose of demonstrating the compliance of a Party with its commitment under Article 3, paragraph 1.

9. The CDM registry shall make non-confidential information publicly available and provide a publicly accessible user interface through the Internet that allows interested persons to query and view it.

10. The information referred to in paragraph 9 above shall include up-to-date information, for each account number in the registry, on the following:

(a) Account name: the holder of the account;

(b) Representative identifier: the representative of the account holder, using the Party/organization identifier (the two-letter country code defined by ISO 3166) and a number unique to that representative for that Party or organization;

(c) Representative name and contact information: the full name, mailing address, telephone number, facsimile number and e-mail address of the representative of the account holder.

11. The information referred to in paragraph 9 above shall include the following CDM project activity information, for each project identifier against which the CERs have been issued:

(a) Project name: a unique name for the CDM project activity;

(b) Project location: the Party and town or region in which the CDM project activity is located;

(c) Years of CER issuance: the years in which CERs have been issued as a result of the CDM project activity;

(d) Operational entities: the operational entities involved in the validation, verification and certification of the CDM project activity;

(e) Reports: downloadable electronic versions of documentation to be made publicly available in accordance with the provisions of the present annex.

12. The information referred to in paragraph 9 above shall include the following holding and transaction information relevant to the CDM registry, by serial number, for each calendar year (defined according to Greenwich Mean Time):

(a) The total quantity of CERs in each account at the beginning of the year;

(b) The total quantity of CERs issued;

(c) The total quantity of CERs transferred and the identity of the acquiring accounts and registries;

(d) The total quantity of ERUs, CERs, AAUs and RMUs cancelled in accordance with paragraph 8 above;

(e) Current holdings of CERs in each account.

DECISION 18/CP.7

MODALITIES, RULES AND GUIDELINES FOR EMISSIONS TRADING UNDER ARTICLE 17 OF THE KYOTO PROTOCOL

The Conference of the Parties,

Recalling its decision 5/CP.6 containing the Bonn Agreements on the implementation of the Buenos Aires Plan of Action,

Aware of its decisions 3/CP.7, 11/CP.7, 15/CP.7, 16/CP.7, 17/CP.7, 19/CP.7, 20/CP.7, 21/CP.7, 22/CP.7, 23/CP.7 and 24/CP.7,

1. *Decides* to adopt the modalities, rules and guidelines for emissions trading contained in the annex below;

2. *Decides further* that any future revision of the modalities, rules and guidelines shall be decided in accordance with the rules of procedures of the Conference of the Parties serving as the meeting of the Parties to the Kyoto Protocol as applied. The first review shall be carried out no later than one year after the end of the first commitment period, based on recommendations by the Subsidiary Body for Implementation drawing on technical advice of the Subsidiary Body for Scientific and Technological Advice, as needed. Further reviews shall be carried out periodically thereafter;

3. *Urges* the Parties included in Annex II to the Convention to facilitate the participation in emissions trading under Article 17 of the Kyoto Protocol of Parties included in

Annex I to the Convention with commitments inscribed in Annex B which are undergoing the process of transition to a market economy;

4. *Recommends* that the Conference of the Parties serving as the meeting of the Parties to the Kyoto Protocol, at its first session, adopt the draft decision below.

DRAFT DECISION -/CMP.1 (*ARTICLE 17*)

Modalities, rules and guidelines for emissions trading under Article 17 of the Kyoto Protocol

The Conference of the Parties serving as the meeting of the Parties to the Kyoto Protocol,

Aware of its decisions -/CMP.1 (*Mechanisms*), -/CMP.1 (*Article 6*), -/CMP.1 (*Article 12*), -/CMP.1 (*Land use, land-use change and forestry*), -/CMP.1 (*Modalities for the accounting of assigned amounts*), -/CMP.1 (*Article 5.1*), -/CMP.1 (*Article 5.2*), -/CMP.1 (*Article 7*) and -/CMP.1 (*Article 8*) and decisions 3/CP.7 and 24/CP.7,

1. *Decides* to confirm and give full effect to any actions taken pursuant to decision 18/CP.7 and to any other relevant decisions by the Conference of the Parties, as appropriate,

2. *Urges* the Parties included in Annex II to the Convention to facilitate the participation in emissions trading under Article 17 of the Kyoto Protocol of Parties included in Annex I to the Convention with commitments inscribed in Annex B which are undergoing the process of transition to a market economy.

ANNEX

Modalities, rules and guidelines for emissions trading under Article 17 of the Kyoto Protocol[1]

1. For the purpose of the present annex the definitions contained in Article 1[2] and the provisions of Article 14 shall apply. Furthermore:

(a) An 'emission reduction unit' or 'ERU' is a unit issued pursuant to the relevant provisions in the annex to decision -/CMP.1 (*Modalities for the accounting of assigned amounts*) and is equal to one metric tonne of carbon dioxide equivalent, calculated using global warming potentials defined by decision 2/CP.3 or as subsequently revised in accordance with Article 5;

(b) A 'certified emission reduction' or 'CER' is a unit issued pursuant to Article 12 and requirements thereunder, as well as the relevant provisions in the annex to decision -/CMP.1 (*Article 12*), and is equal to one metric tonne of carbon dioxide equivalent, calculated using global warming potentials defined by decision 2/CP.3 or as subsequently revised in accordance with Article 5;

(c) An 'assigned amount unit' or 'AAU' is a unit issued pursuant to the relevant provisions in the annex to decision -/CMP.1 (*Modalities for the accounting of assigned amounts*) and is equal to one metric tonne of carbon dioxide equivalent, calculated using global warming potentials defined by decision 2/CP.3 or as subsequently revised in accordance with Article 5;

(d) A 'removal unit' or 'RMU' is a unit issued pursuant to the relevant provisions in the annex to decision -/CMP.1 (*Modalities for the accounting of assigned amounts*) and is equal to one metric tonne of carbon dioxide equivalent, calculated using global warming potentials defined by decision 2/CP.3 or as subsequently revised in accordance with Article 5.

2. Subject to the provisions of paragraph 3 below, a Party[3] included in Annex I with a commitment inscribed in Annex B is eligible to transfer and/or acquire ERUs, CERs, AAUs, or RMUs issued in accordance with the relevant provisions, if it is in compliance with the following eligibility requirements:

(a) It is a Party to the Kyoto Protocol;

(b) Its assigned amount pursuant to Article 3, paragraphs 7 and 8, has been calculated and recorded in accordance with decision -/CMP.1 (*Modalities for the accounting of assigned amounts*);

(c) It has in place a national system for the estimation of anthropogenic emissions by sources and anthropogenic removals by sinks of all greenhouse gases not controlled by the Montreal Protocol, in accordance with Article 5, paragraph 1, and the requirements in the guidelines decided thereunder;

(d) It has in place a national registry in accordance with Article 7, paragraph 4, and the requirements in the guidelines decided thereunder;

[1] The annex to decision -/CMP.1 (*Modalities for the accounting of assigned amounts*) contains operational provisions and procedures relevant to this annex.

[2] In the context of this annex, 'Article' refers to an Article of the Kyoto Protocol, unless otherwise specified.

[3] In the context of this annex, 'Party' refers to a Party to the Kyoto Protocol, unless otherwise specified.

(e) It has submitted annually the most recent required inventory, in accordance with Article 5, paragraph 2, and Article 7, paragraph 1, and the requirements in the guidelines decided thereunder, including the national inventory report and the common reporting format. For the first commitment period, the quality assessment needed for the purpose of determining eligibility to use the mechanisms shall be limited to the parts of the inventory pertaining to emissions of greenhouse gases from sources/sector categories from Annex A to the Kyoto Protocol and the submission of the annual inventory on sinks;

(f) It submits the supplementary information on assigned amount in accordance with Article 7, paragraph 1, and the requirements in the guidelines decided thereunder and makes any additions to, and subtractions from, assigned amount pursuant to Article 3, paragraphs 7 and 8, including for the activities under Article 3, paragraphs 3 and 4, in accordance with Article 7, paragraph 4, and the requirements in the guidelines decided thereunder;

3. A Party included in Annex I with a commitment inscribed in Annex B shall be considered:

(a) To meet the eligibility requirements referred to in paragraph 2 above after 16 months have elapsed since the submission of its report to facilitate the calculation of its assigned amount pursuant to Article 3, paragraphs 7 and 8, and to demonstrate its capacity to account for its emissions and assigned amount, in accordance with the modalities adopted for the accounting of assigned amount under Article 7, paragraph 4, unless the enforcement branch of the compliance committee finds in accordance with decision 24/CP.7 that the Party does not meet these requirements, or, at an earlier date, if the enforcement branch of the compliance committee has decided that it is not proceeding with any questions of implementation relating to these requirements indicated in reports of the expert review teams under Article 8 of the Kyoto Protocol, and has transmitted this information to the secretariat;

(b) To continue to meet the eligibility requirements referred to in paragraph 2 above unless and until the enforcement branch of the compliance committee decides that the Party does not meet one or more of the eligibility requirements, has suspended the Party's eligibility and has transmitted this information to the secretariat.

4. The secretariat shall maintain a publicly accessible list of Parties that meet the eligibility requirements and of Parties that have been suspended.

5. Transfers and acquisitions between national registries shall be made under the responsibility of the Parties concerned in accordance with the provisions in decision -/CMP.1 (*Modalities for the accounting of assigned amounts*). A Party that authorizes legal entities to transfer and/or acquire under Article 17 shall remain responsible for the fulfilment of its obligations under the Kyoto Protocol and shall ensure that such participation is consistent with the present annex. The Party shall maintain an up-to-date list of such entities and make it available to the secretariat and the public through its national registry. Legal entities may not transfer and/or acquire under Article 17 during any period of time in which the authorizing Party does not meet the eligibility requirements or has been suspended.

6. Each Party included in Annex I shall maintain, in its national registry, a commitment period reserve which should not drop below 90 per cent of the Party's assigned amount calculated pursuant to Article 3, paragraphs 7 and 8, of the Kyoto Protocol, or 100 per cent of five times its most recently reviewed inventory, whichever is lowest.

7. The commitment period reserve shall consist of holdings of ERUs, CERs, AAUs and/or RMUs for the relevant commitment period which have not been cancelled in accordance with decision -/CMP.1 (*Modalities for the accounting of assigned amounts*).

8. Upon establishment of its assigned amount pursuant to Article 3, paragraphs 7 and 8, and until expiration of the additional period for fulfilling commitments, a Party shall not make a transfer which would result in these holdings being below the required level of the commitment period reserve.

9. If calculations under paragraph 6 above, or cancellations of ERUs, CERs, AAUs and/or RMUs raise the required level of the commitment period reserve above the Party's holdings of ERUs, CERs, AAUs and/or RMUs valid for the relevant commitment period, which have not been cancelled, the Party shall be notified by the secretariat and, within 30 days of this notification, shall bring its holdings to the required level.

10. Any provisions relating to the commitment period reserve or other limitations to transfers under Article 17 shall not apply to transfers by a Party of ERUs issued into its national registry which were verified in accordance with the verification procedure under the Article 6 supervisory committee.

11. The secretariat shall perform functions as requested.

DECISION 19/CP.7

MODALITIES FOR THE ACCOUNTING OF ASSIGNED AMOUNTS UNDER ARTICLE 7, PARAGRAPH 4, OF THE KYOTO PROTOCOL

The Conference of the Parties,

Recalling its decisions 1/CP.3, 1/CP.4, 8/CP.4, and 5/CP.6 containing the Bonn Agreements on the implementation of the Buenos Aires Plan of Action,

Noting the relevant provisions of the Kyoto Protocol to the United Nations Framework Convention on Climate Change, in particular its Articles 3, 4, 5, 6, 7, 8, 12, 17 and 18,

Being aware of its decisions 11/CP.7, 15/CP.7, 16/CP.7, 17/CP.7, 18/CP.7, 20/CP.7, 21/CP.7, 22/CP.7, 23/CP.7 and 24/CP.7,

1. *Requests* the Subsidiary Body for Scientific and Technical Advice to develop technical standards for the purpose of ensuring the accurate, transparent and efficient exchange of data between national registries, the clean development mechanism registry and the transaction log, based on the annex to the decision below, with a view to recommending to the Conference of the Parties, at its eighth session, a decision on this matter, for adoption by the Conference of the Parties serving as the meeting of the Parties to the Kyoto Protocol at its first session, to facilitate the early development and establishment of national registries, as well as of the clean development mechanism registry and transaction log;

2. *Requests* the secretariat to develop the transaction log referred to in the annex to the draft decision below, taking into account the technical standards referred to in paragraph 1 above, with a view to establishing it no later than the second session of the Conference of the Parties serving as the meeting of the Parties to the Kyoto Protocol;

3. *Requests* the Chairpman of the Subsidiary Body for Scientific and Technological Advice, with the assistance of the secretariat, to convene inter-sessional consultations with Parties and experts for the purposes of:

(a) Preparing draft technical standards, as referred to in paragraph 1 above, for consideration by the Subsidiary Body for Scientific and Technological Advice at its sixteenth and seventeenth sessions;

(b) Providing for the exchange of information and experience between Parties included in Annex I and Parties not included in Annex I, as well as the secretariat, in relation to the development and establishment of national registries, the clean development mechanism registry and the transaction log.

4. *Recommends* that the Conference of the Parties serving as the meeting of the Parties to the Kyoto Protocol, at its first session, adopt the draft decision below.

DRAFT DECISION -/CMP.1 (MODALITIES FOR THE ACCOUNTING OF ASSIGNED AMOUNTS)

Modalities for the accounting of assigned amounts under Article 7, paragraph 4, of the Kyoto Protocol

The Conference of the Parties serving as the meeting of the Parties to the Kyoto Protocol,

Recalling Article 7, paragraph 4, of the Kyoto Protocol, *Recalling decision* 19/CP.7,

Being aware of its decisions -/CMP.1 (*Mechanisms*), -/CMP.1 (*Article* 6), -/CMP.1 (*Article 12*), -/CMP.1 (*Article* 17), -/CMP.1 (*Land use, land-use change and forestry*), -/CMP.1 (*Article 5.1*), -/CMP.1 (*Article 5.2*), -/CMP.1 (*Article* 7) and -/CMP.1 (*Article 8*) and decision 24/CP.7,

1. *Adopts* the modalities for the accounting of assigned amounts under Article 7, paragraph 4, of the Kyoto Protocol, as contained in the annex to the present decision;

2. *Decides* that each Party included in Annex I with a commitment inscribed in Annex B shall submit to the secretariat, prior to 1 January 2007 or one year after the entry into force of the Kyoto Protocol for that Party, whichever is later, the report referred to in paragraph 6 of the annex to the present decision. After completion of the initial review under Article 8 and resolution of any question of implementation relating to adjustments under Article 5, paragraph 2, or its assigned amount pursuant to Article 3, paragraphs 7 and 8, the assigned amount pursuant to Article 3, paragraphs 7 and 8, of each Party shall be recorded in the database for the compilation and accounting of emissions and assigned amounts referred to in paragraph 50 of the annex to the present decision and shall remain fixed for the commitment period;

3. *Decides* that each Party included in Annex I with a commitment inscribed in Annex B shall submit to the secretariat, upon expiration of the additional period for fulfilling commitments, the report referred to in paragraph 49 of the annex to the present decision;

4. *Requests* the secretariat to begin publishing the annual compilation and accounting reports referred to in paragraph 61 of the annex to the present decision after completion of the initial review under Article 8 and resolution of any question of implementation relating to adjustments under Article 5, paragraph 2, or its assigned amount pursuant to Article 3, paragraphs 7 and 8, and to forward them to the Conference of the Parties serving as the meeting of the Parties to the Kyoto Protocol, the compliance committee and each Party concerned;

5. *Requests* the secretariat to publish, after the additional period for fulfilling commitments, the final compilation and accounting reports referred to in paragraph 62 of the annex to the present decision and forward them to the Conference of the Parties serving as the meeting of the Parties to the Kyoto Protocol, the compliance committee and each Party concerned.

ANNEX

MODALITIES FOR THE ACCOUNTING OF ASSIGNED AMOUNTS UNDER ARTICLE 7, PARAGRAPH 4, OF THE KYOTO PROTOCOL[1]

I. MODALITIES

A. Definitions

1. An 'emission reduction unit' or 'ERU' is a unit issued pursuant to the relevant provisions in these modalities for the accounting of assigned amounts and is equal to one metric tonne of carbon dioxide equivalent, calculated using global warming potentials defined by decision 2/CP.3 or as subsequently revised in accordance with Article 5.

2. A 'certified emission reduction' or 'CER' is a unit issued pursuant to Article 12 and requirements thereunder, as well as the relevant provisions in the annex to decision -/CMP.1 (*Article 12*), and is equal to one metric tonne of carbon dioxide equivalent, calculated using global warming potentials defined by decision 2/CP.3 or as subsequently revised in accordance with Article 5.

3. An 'assigned amount unit' or 'AAU' is a unit issued pursuant to the relevant provisions in these modalities for the accounting of assigned amounts and is equal to one metric tonne of carbon dioxide equivalent, calculated using global warming potentials defined by decision 2/CP.3 or as subsequently revised in accordance with Article 5.

4. A 'removal unit' or 'RMU' is a unit issued pursuant to the relevant provisions in these modalities for the accounting of assigned amounts and is equal to one metric tonne of carbon dioxide equivalent, calculated using global warming potentials defined by decision 2/CP.3 or as subsequently revised in accordance with Article 5.

B. Calculation of assigned amount pursuant to Article 3, paragraphs 7 and 8

5. The assigned amount pursuant to Article 3, paragraphs 7 and 8, for the first commitment period, from 2008 to 2012, for each Party included in Annex I with a commitment inscribed in Annex B to the Kyoto Protocol[2] shall be equal to the percentage inscribed for it in Annex B of its aggregate anthropogenic carbon dioxide equivalent emissions of the greenhouse gases, and from the sources, listed in Annex A to the Kyoto Protocol in the base year, multiplied by five, taking into account the following:

(a) The base year shall be 1990 except for those Parties undergoing the process of transition to a market economy that have selected a historical base year or period other than 1990, in accordance with Article 3, paragraph 5, and for those Parties that have selected 1995 as the base year for total emissions of hydrofluorocarbons, perfluorocarbons and sulphur hexafluoride, in accordance with Article 3, paragraph 8;

[1] 'Article' in these modalities refers to an article of the Kyoto Protocol, unless otherwise specified.

[2] Hereafter referred to as a 'Party included in Annex I'.

(b) Those Parties for which land-use change and forestry (all emissions by sources and removals by sinks under category 5 of the *Revised 1996 Intergovernmental Panel on Climate Change Guidelines for National Greenhouse Gas Inventories*) constituted a net source of greenhouse gas emissions in the base year or period shall include in their emissions during that year or period the aggregate anthropogenic carbon dioxide equivalent emissions by sources minus removals by sinks in that year or period from land-use change (all emissions by sources minus removals by sinks reported in relation to the conversion of forests (deforestation));

(c) Those Parties that have reached an agreement in accordance with Article 4 to fulfil their commitments under Article 3 jointly shall use the respective emission level allocated to each of the Parties in that agreement instead of the percentage inscribed for it in Annex B.

6. Each Party included in Annex I shall facilitate the calculation of its assigned amount pursuant to Article 3, paragraphs 7 and 8, for the commitment period and demonstrate its capacity to account for its emissions and assigned amount. To this end, each Party shall submit a report, in two parts, containing the information specified in paragraphs 7 and 8 below.

7. Part one of the report referred to in paragraph 6 above shall contain the following information, or references to such information where it has been previously submitted to the secretariat:

(a) Complete inventories of anthropogenic emissions by sources and removals by sinks of greenhouse gases not controlled by the Montreal Protocol for all years from 1990, or another approved base year or period under Article 3, paragraph 5, to the most recent year available, prepared in accordance with Article 5, paragraph 2, and relevant decisions of the Conference of the Parties serving as the meeting of the Parties to the Kyoto Protocol (COP/MOP), taking into account any relevant decisions of the Conference of the Parties (COP);

(b) Identification of its selected base year for hydrofluorocarbons, perfluorocarbons and sulphur hexafluoride in accordance with Article 3, paragraph 8;

(c) The agreement under Article 4, where the Party has reached such an agreement to fulfil its commitments under Article 3 jointly with other Parties;

(d) Calculation of its assigned amount pursuant to Article 3, paragraphs 7 and 8, on the basis of its inventory of anthropogenic emissions by sources and removals by sinks of greenhouse gases not controlled by the Montreal Protocol.

8. Part two of the report referred to in paragraph 6 above shall contain the following information, or references to such information where it has been previously submitted to the secretariat:

(a) Calculation of its commitment period reserve in accordance with decision -/CMP.1 (*Article 17*);

(b) Identification of its selection of single minimum values for tree crown cover, land area and tree height for use in accounting for its activities under Article 3, paragraphs 3 and 4, together with a justification of the consistency of those values with the information that has been historically reported to the Food and Agriculture Organization of the United Nations or other international bodies, and in the case of difference, an explanation of why and how such values were chosen, in accordance with decision -/CMP.1 (*Land use, land-use change and forestry*);

(c) Identification of its election of activities under Article 3, paragraph 4, for inclusion in its accounting for the first commitment period, together with information on how its national system under Article 5, paragraph 1, will identify land areas associated with the activities, in accordance with decision -/CMP.1 (*Land use, land-use change and forestry*);

(d) Identification of whether, for each activity under Article 3, paragraphs 3 and 4, it intends to account annually or for the entire commitment period;

(e) A description of its national system in accordance with Article 5, paragraph 1, reported in accordance with the guidelines for the preparation of the information required under Article 7 of the Kyoto Protocol;

(f) A description of its national registry, reported in accordance with the guidelines for the preparation of the information required under Article 7 of the Kyoto Protocol.

C. Recording of assigned amount pursuant to Article 3, paragraphs 7 and 8

9. After initial review under Article 8 and resolution of any questions of implementation relating to adjustments or the calculation of its assigned amount pursuant to Article 3, paragraphs 7 and 8, the assigned amount pursuant to Article 3, paragraphs 7 and 8, of each Party shall be recorded in the database for the compilation and accounting of emissions and assigned amounts referred to in paragraph 50 below.

10. Once recorded in the compilation and accounting database referred to in paragraph 50 below, the assigned amount pursuant to Article 3, paragraphs 7 and 8, of each Party shall remain fixed for the commitment period.

D. Additions to, and subtractions from, assigned amount pursuant to Article 3, paragraphs 7 and 8, for the accounting of the compliance assessment

11. At the end of the additional period for fulfilling commitments, the following additions to the assigned amount pursuant to Article 3, paragraphs 7 and 8, of a Party shall be made in accordance with Article 3, paragraphs 3, 4, 10, 12 and 13, for the accounting of the compliance assessment for the commitment period:

(a) Acquisitions by the Party of ERUs in accordance with Articles 6 and 17;

(b) Net acquisitions by the Party of CERs, where it acquires more CERs in accordance with Articles 12 and 17 than it transfers in accordance with Article 17;

(c) Acquisitions by the Party of AAUs in accordance with Article 17;

(d) Acquisitions by the Party of RMUs in accordance with Article 17;

(e) Issuance by the Party of RMUs on the basis of its activities under Article 3, paragraph 3, and its elected activities under Article 3, paragraph 4, where such activities result in a net removal of greenhouse gases, as reported in accordance with Article 7, reviewed in accordance with Article 8, taking into account any adjustments applied under Article 5, paragraph 2, accounted in accordance with the decision -/CMP.1 (*Land use, land-use change and forestry*) and subject to any question of implementation relating to those activities having been resolved;

(f) Carry-over by the Party of ERUs, CERs and/or AAUs from the previous commitment period, in accordance with paragraph 15 below.

12. At the end of the additional period for fulfilling commitments, the following subtractions from the assigned amount pursuant to Article 3, paragraphs 7 and 8, of a Party shall be made in accordance with Article 3, paragraphs 3, 4 and 11, for the accounting of the compliance assessment for the commitment period:

(a) Transfers by the Party of ERUs in accordance with Articles 6 and 17;

(b) Transfers by the Party of AAUs in accordance with Article 17;

(c) Transfers by the Party of RMUs in accordance with Article 17;

(d) Cancellation by the Party of ERUs, CERs, AAUs and/or RMUs on the basis of its activities under Article 3, paragraph 3, and its elected activities under Article 3, paragraph 4, where such activities result in a net source of greenhouse gas emissions, as reported in accordance with Article 7, reviewed in accordance with Article 8, taking into account any adjustments applied under Article 5, paragraph 2, and accounted in accordance with decision -/CMP.1 (*Land use, land-use change and forestry*);

(e) Cancellation by the Party of ERUs, CERs, AAUs and/or RMUs following determination by the compliance committee that the Party was not in compliance with its commitment under Article 3, paragraph 1, for the previous commitment period, in accordance with decision 24/CP.7;

(f) Other cancellations by the Party of ERUs, CERs, AAUs and/or RMUs.

E. Basis for the compliance assessment

13. Each Party included in Annex I shall retire ERUs, CERs, AAUs and/or RMUs for the purpose of demonstrating its compliance with its commitment under Article 3, paragraph 1.

14. The assessment, after the expiration of the additional period for fulfilling commitments, of the compliance of a Party included in Annex I with its commitment under Article 3, paragraph 1, shall be based on the comparison of the quantity of ERUs, CERs, AAUs and/or RMUs, valid for the commitment period in question, retired by the Party in accordance with paragraph 13 above, with its aggregate anthropogenic carbon dioxide equivalent emissions of the greenhouse gases, and from the sources, listed in Annex A to the Kyoto Protocol during the commitment period as reported in accordance with Article 7 and reviewed in accordance with Article 8, taking into account any adjustments in accordance with Article 5, paragraph 2, as recorded in the compilation and accounting database referred to in paragraph 50 below.

F. Carry-over

15. After expiration of the additional period for fulfilling commitments and where the final compilation and accounting report referred to in paragraph 62 below indicates that the quantity of ERUs, CERs, AAUs and/or RMUs retired by the Party in accordance with paragraph 13 above is at least equivalent to its anthropogenic carbon dioxide equivalent emissions of the greenhouse gases, and from the sources, listed in Annex A to the Kyoto Protocol for that commitment period, the Party may carry over to the subsequent commitment period:

(a) Any ERUs held in its national registry, which have not been converted from RMUs and have not been retired for that commitment period or cancelled, to a maximum of 2.5 per cent of the assigned amount pursuant to Article 3, paragraphs 7 and 8, of that Party;

(b) Any CERs held in its national registry, which have not been retired for that commitment period or cancelled, to a maximum of 2.5 per cent of the assigned amount pursuant to Article 3, paragraphs 7 and 8, of that Party;

(c) Any AAUs held in its national registry, which have not been retired for that commitment period or cancelled.

16. RMUs may not be carried over to the subsequent commitment period.

II. REGISTRY REQUIREMENTS

A. National registries

17. Each Party included in Annex I shall establish and maintain a national registry to ensure the accurate accounting of the issuance, holding, transfer, acquisition, cancellation and retirement of ERUs, CERs, AAUs and RMUs and the carry-over of ERUs, CERs and AAUs.

18. Each Party shall designate an organization as its registry administrator to maintain the national registry of that Party. Any two or more Parties may voluntarily maintain their respective national registries in a consolidated system, provided that each national registry remains distinct.

19. A national registry shall be in the form of a standardized electronic database which contains, *inter alia,* common data elements relevant to the issuance, holding, transfer, acquisition, cancellation and retirement of ERUs, CERs, AAUs and RMUs and the carry-over of ERUs, CERs and AAUs. The structure and data formats of national registries shall conform to technical standards to be adopted by the COP/MOP for the purpose of ensuring the accurate, transparent and efficient exchange of data between national registries, the clean development mechanism (CDM) registry and the independent transaction log.

20. Each ERU, CER, AAU and RMU shall be held in only one account in one registry at a given time.

21. Each national registry shall have the following accounts:

(a) At least one holding account for the Party;

(b) At least one holding account for each legal entity authorized by the Party to hold ERUs, CERs, AAUs and/or RMUs under its responsibility;

(c) At least one cancellation account for each commitment period for the purposes of cancelling ERUs, CERs, AAUs and/or RMUs under paragraph 12(d) above;

(d) One cancellation account for each commitment period for the purposes of cancelling ERUs, CERs, AAUs and/or RMUs under paragraph 12(e) above;

(e) At least one cancellation account for each commitment period for the purposes of cancelling ERUs, CERs, AAUs and/or RMUs under paragraph 12(f) above;

(f) One retirement account for each commitment period.

22. Each account within a national registry shall have a unique account number comprising the following elements:

(a) Party identifier: the Party in whose national registry the account is maintained, identified by means of the two-letter country code defined by the International Organization for Standardization (ISO 3166);

(b) A unique number: a number unique to that account for the Party in whose national registry the account is maintained.

B. Issuance of ERUs, AAUs and RMUs

23. Each Party included in Annex I shall, prior to any transactions taking place for that commitment period, issue a quantity of AAUs equivalent to its assigned amount pursuant to Article 3, paragraphs 7 and 8, calculated and recorded in accordance with paragraphs 5 to 10 above, in its national registry.

24. Each AAU shall have a unique serial number comprising the following elements:

(a) Commitment period: the commitment period for which the AAU is issued;

(b) Party of origin: the Party issuing the AAU, identified by means of the two-letter country code defined by ISO 3166;

(c) Type: an element identifying the unit as an AAU;

(d) Unit: a number unique to the AAU for the identified commitment period and Party of origin.

25. Each Party included in Annex I shall issue in its national registry RMUs equivalent to the net removals of anthropogenic greenhouse gases resulting from its activities under Article 3, paragraph 3, and its elected activities under Article 3, paragraph 4, accounted in accordance with decision -/CMP.1 (*Land use, land-use change and forestry*) as reported under Article 7, paragraph 1, following completion of the review in accordance with Article 8, taking into account any adjustments applied in accordance with Article 5, paragraph 2, and resolution of any questions of implementation related to the reported net removals of anthropogenic greenhouse gases. Each Party shall elect for each activity, prior to the start of the commitment period, to issue such RMUs annually or for the entire commitment period. The decision of a Party shall remain fixed for the first commitment period.

26. Where a question of implementation is identified by an expert review team under Article 8 in relation to the calculation of the net removals of greenhouse gases from the activities of a Party under Article 3, paragraph 3 or 4, or where adjustments exceed thresholds to be decided according to paragraph 2 of decision 22/CP.7, the Party shall not issue the RMUs relating to the reported net removals of anthropogenic greenhouse gases for each activity under Article 3, paragraph 3, and for each elected activity under Article 3, paragraph 4, until the question of implementation is resolved.

27. Each RMU shall have a unique serial number comprising the following elements:

(a) Commitment period: the commitment period for which the RMU is issued;

(b) Party of origin: the Party included in Annex I issuing the RMU, identified by means of the two-letter country code defined by ISO 3166;

(c) Type: an element identifying the unit as an RMU;

(d) Activity: the type of activity for which the RMU was issued;

(e) Unit: a number unique to the RMU for the identified commitment period and Party of origin.

28. Each Party included in Annex I shall ensure that the total quantity of RMUs issued into its registry pursuant to Article 3, paragraph 4, for the commitment period does not exceed the limits established for that Party as set out in decision -/CMP.1 (*Land use, land-use change and forestry*).

29. Prior to their transfer, each Party shall issue ERUs into its national registry by converting AAUs or RMUs previously issued by that Party and held in its national registry. An AAU or RMU shall be converted into an ERU by adding a project identifier to the serial number and changing the type indicator in the serial number to indicate an ERU. Other elements of the serial number of the AAU or RMU shall remain unchanged. The project identifier shall identify the specific Article 6 project for which the ERU is issued, using a number unique to the project for the Party of origin, including whether the relevant reductions in anthropogenic emissions by sources or enhancements of anthropogenic removals by sinks were verified under the Article 6 supervisory committee.

C. Transfer, acquisition, cancellation, retirement and carry-over

30. ERUs, CERs, AAUs and RMUs may be transferred between registries in accordance with decisions -/CMP.1 (*Article 6*), -/CMP.1 (*Article 12*), -/CMP.1 (*Article 17*) and -/CMP.1 (*Land use, land-use and forestry*), and may be transferred within registries.

31. Each Party included in Annex I shall ensure that its net acquisitions of CERs from afforestation and reforestation activities under Article 12 for the first commitment period do not exceed the limits established for that Party as set out in decision -/CMP.1 (*Land use, land-use and forestry*).

32. Each Party included in Annex I shall cancel CERs, ERUs, AAUs and/or RMUs equivalent to the net emissions of anthropogenic greenhouse gases resulting from its activities under Article 3, paragraph 3, and its elected activities under Article 3, paragraph 4, accounted in accordance with decision -/CMP.1 (*Land use, land-use change and forestry*) as reported under Article 7, paragraph 1, following completion of the review in accordance with Article 8, taking into account any adjustments applied in accordance with Article 5, paragraph 2, and resolution of any questions of implementation related to the reported net emissions of anthropogenic greenhouse gases, in accordance with paragraph 12(d) above, by transferring the ERUs, CERs, AAUs and/or RMUs to the appropriate cancellation account in its national registry. Each Party shall cancel ERUs, CERs, AAUs and/or RMUs for each activity for the same period for which it has elected to issue RMUs for that activity.

33. Each Party included in Annex I may cancel ERUs, CERs, AAUs and/or RMUs so they cannot be used in fulfilment of commitments under Article 3, paragraph 1, in accordance with paragraph 12 (f) above, by transferring ERUs, CERs, AAUs and/or RMUs to a cancellation account in its national registry. Legal entities, where authorized by the Party, may also transfer ERUs, CERs, AAUs and RMUs into a cancellation account.

34. Prior to the end of the additional period for fulfilling commitments, each Party included in Annex I shall retire ERUs, CERs, AAUs and/or RMUs valid for that commitment period for use towards meeting its commitments under Article 3, paragraph 1, in accordance with paragraph 13 above, by transferring ERUs, CERs, AAUs and/or RMUs to the retirement account for that commitment period in its national registry.

35. ERUs, CERs, AAUs and RMUs transferred to cancellation accounts or the retirement account for a commitment period may not be further transferred or carried over to the subsequent commitment period. ERUs, CERs, AAUs and RMUs transferred to cancellation accounts may not be used for the purpose of demonstrating the compliance of a Party with its commitment under Article 3, paragraph 1.

36. Each Party included in Annex I may carry over ERUs, CERs and/or AAUs held in its registry, that have not been cancelled or retired for a commitment period, to the subsequent commitment period in accordance with paragraph 15 above. Each ERU, CER and/or AAU carried over in this manner shall maintain its original serial number and shall be valid in the subsequent commitment period. ERUs, CERs, AAUs and RMUs of a previous commitment period held in the registry of a Party which have not been carried over in this manner shall be cancelled in accordance with paragraph 12(f) above once the additional period for fulfilling commitments has ended.

37. Where the compliance committee determines that the Party is not in compliance with its commitment under Article 3, paragraph 1, for a commitment period, the Party shall transfer the quantity of ERUs, CERs, AAUs and/or RMUs calculated in accordance with decision 24/CP.7 into the relevant cancellation account, in accordance with paragraph 12(e) above.

D. Transaction procedures

38. The secretariat shall establish and maintain an independent transaction log to verify the validity of transactions, including issuance, transfer and acquisition between registries, cancellation and retirement of ERUs, CERs, AAUs and RMUs and the carry-over of ERUs, CERs and AAUs.

39. A Party included in Annex I shall initiate issuance of AAUs or RMUs by directing its national registry to issue AAUs or RMUs into a specific account within that registry. The executive board of the CDM shall initiate issuance of CERs by directing the CDM registry to issue CERs into its pending account in accordance with the requirements in Article 12 and requirements thereunder, as well as the relevant provisions in the annex to decision -/CMP.1 (*Article 12*). A Party included in Annex I shall initiate issuance of ERUs by directing its national registry to convert specified AAUs or RMUs into ERUs within an account of that national registry. Subject to notification by the transaction log that there are no discrepancies pertaining to the issuance, the issuance shall be completed when specific ERUs, CERs, AAUs or RMUs are recorded in the specified account and, in the case of ERUs, the specified AAUs or RMUs are removed from the account.

40. A Party included in Annex I shall initiate any transfer of ERUs, CERs, AAUs or RMUs, including those to cancellation and retirement accounts, by directing its national registry to transfer specified ERUs, CERs, AAUs or RMUs to a specific account within that registry or another registry. The executive board of the CDM shall initiate any transfer of CERs held in the CDM registry by directing it to transfer specified CERs to a specific account within that registry or another registry. Subject to notification by the transaction log, where applicable, that there are no discrepancies pertaining to the transfer, the transfer shall be completed when the specified ERUs, CERs, AAUs or RMUs are removed from the transferring account and are recorded in the acquiring account.

41. Upon the initiation of any issuance, transfer between registries, cancellation or retirement of ERUs, CERs, AAUs or RMUs, and prior to the completion of those transactions:

(a) The initiating registry shall create a unique transaction number comprising: the commitment period for which the transaction is proposed; the Party identifier for the Party initiating the transaction (using the two-letter country code defined by ISO 3166); and a number unique to that transaction for the commitment period and initiating Party;

(b) The initiating registry shall send a record of the proposed transaction to the transaction log and, in the case of transfers to another registry, to the acquiring national registry. The record shall include: the transaction number; the transaction type (issuance, transfer, cancellation or retirement, further distinguished in accordance with the categories in paragraphs 11 and 12 above); the serial numbers of the relevant ERUs, CERs, AAUs or RMUs; and the relevant account numbers.

42. Upon receipt of the record, the transaction log shall conduct an automated check to verify that there is no discrepancy, with regard to:

(a) In all transactions: units previously retired or cancelled; units existing in more than one registry; units for which a previously identified discrepancy has not been resolved; units improperly carried over; units improperly issued, including those which infringe upon the limits contained in decision -/CMP.1 (*Land use, land-use change and forestry*); and the authorization of legal entities involved to participate in the transaction;

(b) In the case of transfers between registries: the eligibility of Parties involved in the transaction to participate in the mechanisms; and infringement upon the commitment period reserve of the transferring Party;

(c) In the case of acquisitions of CERs from land use, land-use change and forestry projects under Article 12: infringement of the limits contained in decision -/CMP.1 (*Land use, land-use change and forestry*);

(d) In the case of a retirement of CERs: the eligibility of the Party involved to use CERs to contribute to its compliance under Article 3, paragraph 1;

43. Upon completion of the automated check, the transaction log shall notify the initiating and, in the case of transfers to another registry, the acquiring registry of the results of the automated check. Depending on the outcome of the check, the following procedures shall apply:

(a) If a discrepancy is notified by the transaction log, the initiating registry shall terminate the transaction, notify the transaction log and, in the case of transfers to another registry, the acquiring registry of the termination. The transaction log shall forward a record of the discrepancy to the secretariat for consideration as part of the review process for the relevant Party or Parties under Article 8;

(b) In the event of a failure by the initiating registry to terminate the transaction, the ERUs, CERs, AAUs or RMUs involved in the transaction shall not be valid for use towards compliance with commitments under Article 3, paragraph 1, until the problem has been corrected and any questions of implementation pertaining to the transaction have been resolved. Upon resolution of a question of implementation pertaining to a Party's transactions, that Party shall perform any necessary corrective action within 30 days;

(c) If no discrepancy is notified by the transaction log, the initiating registry and, in the case of transfers to another registry, the acquiring registry, shall complete or terminate the transaction and send the record and a notification of completion or termination of the transaction to the transaction log. In the case of transfers to another registry, the initiating and acquiring registries shall also send their records and notifications to each other;

(d) The transaction log shall record, and make publicly available, all transaction records and the date and time of completion of each transaction, to facilitate its automated checks and the review under Article 8.

E. *Publicly accessible information*

44. Each national registry shall make non-confidential information publicly available and provide a publicly accessible user interface through the Internet that allows interested persons to query and view it.

45. The information referred to in paragraph 44 above shall include up-to-date information for each account number in that registry on the following:

(a) Account name: the holder of the account;

(b) Account type: the type of account (holding, cancellation or retirement);

(c) Commitment period: the commitment period with which a cancellation or retirement account is associated;

(d) Representative identifier: the representative of the account holder, using the Party identifier (the two-letter country code defined by ISO 3166) and a number unique to that representative within the Party's registry;

(e) Representative name and contact information: the full name, mailing address, telephone number, facsimile number and email address of the representative of the account holder.

46. The information referred to in paragraph 44 above shall include the following Article 6 project information, for each project identifier against which the Party has issued ERUs:

(a) Project name: a unique name for the project;

(b) Project location: the Party and town or region in which the project is located;

(c) Years of ERU issuance: the years in which ERUs have been issued as a result of the Article 6 project;

(d) Reports: downloadable electronic versions of all publicly available documentation relating to the project, including proposals, monitoring, verification and issuance of ERUs, where relevant, subject to the confidentiality provisions in decision -/CMP.1 (*Article 6*).

47. The information referred to in paragraph 44 above shall include the following holding and transaction information relevant to the national registry, by serial number, for each calendar year (defined according to Greenwich Mean Time):

(a) The total quantity of ERUs, CERs, AAUs and RMUs in each account at the beginning of the year;

(b) The total quantity of AAUs issued on the basis of the assigned amount pursuant to Article 3, paragraphs 7 and 8;

(c) The total quantity of ERUs issued on the basis of Article 6 projects;

(d) The total quantity of ERUs, CERs, AAUs and RMUs acquired from other registries and the identity of the transferring accounts and registries;

(e) The total quantity of RMUs issued on the basis of each activity under Article 3, paragraphs 3 and 4;

(f) The total quantity of ERUs, CERs, AAUs and RMUs transferred to other registries and the identity of the acquiring accounts and registries;

(g) The total quantity of ERUs, CERs, AAUs and RMUs cancelled on the basis of activities under Article 3, paragraphs 3 and 4;

(h) The total quantity of ERUs, CERs, AAUs and RMUs cancelled following determination by the compliance committee that the Party is not in compliance with its commitment under Article 3, paragraph 1;

(i) The total quantity of other ERUs, CERs, AAUs and RMUs cancelled;

(j) The total quantity of ERUs, CERs, AAUs and RMUs retired;

(k) The total quantity of ERUs, CERs, AAUs carried over from the previous commitment period;

(l) Current holdings of ERUs, CERs, AAUs and RMUs in each account.

48. The information referred to in paragraph 44 above shall include a list of legal entities authorized by the Party to hold ERUs, CERs, AAUs and/or RMUs under its responsibility.

III. COMPILATION AND ACCOUNTING OF EMISSION INVENTORIES AND ASSIGNED AMOUNTS

A. Report upon expiration of the additional period for fulfilling commitments

49. Upon expiration of an additional period for fulfilling commitments, each Party included in Annex I shall report to the secretariat and make available to the public, in a standard electronic format, the following information. This information shall only include ERUs, CERs, AAUs and RMUs valid for the commitment period in question:

(a) The total quantities of the categories of ERUs, CERs, AAUs and RMUs listed in paragraph 47(a) to (j) above, for the current calendar year until the end of the additional period for fulfilling commitments (defined according to Greenwich Mean Time);

(b) The total quantity and serial numbers of ERUs, CERs, AAUs and RMUs in its retirement account;

(c) The total quantity and serial numbers of ERUs, CERs and AAUs which the Party requests to be carried over to the subsequent commitment period.

B. Compilation and accounting database

50. The secretariat shall establish a database to compile and account for emissions and assigned amounts pursuant to Article 3, paragraphs 7 and 8, and additions to, and subtractions from assigned amounts pursuant to Article 3, paragraphs 7 and 8, for the accounting of the compliance assessment, in accordance with paragraphs 11 and 12 above. The purpose of this database is to facilitate the assessment of the compliance of each Party included in Annex I with its commitment under Article 3, paragraph 1.

51. A separate record shall be maintained in the database for each Party included in Annex I for each commitment period. Information on ERUs, CERs, AAUs and RMUs shall only include units valid for the commitment period in question and shall be recorded separately for each type of unit.

52. The secretariat shall record in the database for each Party included in Annex I the following information:

(a) The assigned amount pursuant to Article 3, paragraphs 7 and 8;

(b) For the first commitment period, the total allowable issuances of RMUs resulting from forest management activities under Article 3, paragraph 4, and limits on net acquisitions of CERs from afforestation and reforestation activities under Article 12 pursuant to decision -/CMP.1 (*Land use, land-use change and forestry*).

53. The secretariat shall record in the database, for each Party included in Annex I, whether it is eligible to transfer and/or acquire ERUs, CERs, AAUs and RMUs pursuant to decisions -/CMP.1 (*Article 6*) and -/CMP.1 (*Article 17*) and to use CERs to contribute to its compliance under Article 3, paragraph 1, pursuant to decision -/CMP.1 (*Article 12*).

54. The secretariat shall annually record the following information relating to emissions for each Party included in Annex I, following the annual review under Article 8, the application of any adjustment under Article 5, paragraph 2, and the resolution of any questions of implementation pertaining to emission estimates:

(a) Aggregate annual anthropogenic carbon dioxide equivalent emissions of the greenhouse gases, and from the sources, listed in Annex A to the Kyoto Protocol for each year of the commitment period that has been reported in accordance with Article 7;

(b) Any adjustments under Article 5, paragraph 2, recorded as the difference, in carbon dioxide equivalent terms, between the adjusted estimate and the inventory estimate reported under Article 7;

(c) Aggregate anthropogenic carbon dioxide equivalent emissions in the commitment period, calculated as the sum of the amounts in subparagraphs (a) and (b) above for all years of the commitment period to date.

55. The secretariat shall annually record in the database the following information for each Party included in Annex I related to accounting for net emissions and removals of

greenhouse gases resulting from its activities under Article 3, paragraph 3, and its elected activities under Article 3, paragraph 4, following the annual review under Article 8, the application of any adjustment under Article 5, paragraph 2, and the resolution of any relevant questions of implementation:

(a) The calculation of whether the activities under Article 3, paragraphs 3 and 4, that have been reported in accordance with Article 7 result in net anthropogenic emissions or net anthropogenic removals of greenhouse gases pursuant to decision -/CMP.1 (*Land use, land-use change and forestry*);

(b) For those activities for which the Party has elected to account annually, the net anthropogenic emissions and removals of greenhouse gases pursuant to decision -/CMP.1 (*Land use, land-use change and forestry*) for the calendar year;

(c) For those activities for which the Party has elected to account for the entire commitment period, the net anthropogenic emissions and removals of greenhouse gases pursuant to decision -/CMP.1 (*Land use, land-use change and forestry*) for the calendar year;

(d) Any adjustments under Article 5, paragraph 2, recorded as the difference in carbon dioxide equivalent terms between the adjusted estimate and the estimate reported under Article 7;

(e) The total net anthropogenic emissions and removals of greenhouse gases pursuant to decision -/CMP.1 (*Land use, land-use change and forestry*) for the commitment period, calculated as the sum for all years of the commitment period to date of the amounts referred to in subparagraphs (b), (c) and (d) above.

56. Where a Party submits recalculated estimates of emissions and removals of greenhouse gases for a year of the commitment period, subject to the review in accordance with Article 8, the secretariat shall make appropriate amendments to the information contained in the database including, where relevant, the removal of previously applied adjustments.

57. The secretariat shall record and update the required level of the commitment period reserve for each Party included in Annex I, in accordance with decision -/CMP.1 (*Article 17*).

58. The secretariat shall annually record in the database for each Party included in Annex I the following information related to transactions, for the previous calendar year and to date for the commitment period, following completion of the annual review under Article 8, including the application of any corrections, and resolution of any relevant questions of implementation:

(a) Total transfers of ERUs, CERs, AAUs and RMUs;

(b) Total acquisitions of ERUs, CERs, AAUs and RMUs;

(c) Net acquisitions of CERs resulting from afforestation and reforestation activities under Article 12;

(d) Total issuances of RMUs relating to each activity under Article 3, paragraphs 3 and 4;

(e) Total issuances of ERUs on the basis of Article 6 projects;

(f) Total of ERUs, CERs and AAUs carried over from the previous commitment period;

(g) Total cancellations of ERUs, CERs, AAUs and RMUs relating to each activity under Article 3, paragraphs 3 and 4;

(h) Total cancellations of ERUs, CERs, AAUs and RMUs following determination by the compliance committee that the Party is not in compliance with its commitment under Article 3, paragraph 1;

(i) Total of any other cancellations of ERUs, CERs, AAUs and RMUs;

(j) Total retirements of ERUs, CERs, AAUs and RMUs.

59. Upon expiration of the additional period for the fulfilment of commitments, and following review under Article 8 of the report submitted by the Party under paragraph 49 above, including the application of any corrections, and the resolution of any relevant questions of implementation, the secretariat shall record in the database the following information for each Party included in Annex I:

(a) The total additions to, or subtractions from, the assigned amount pursuant to Article 3, paragraphs 7 and 8, for the accounting of the compliance assessment, in accordance with paragraphs 11 and 12 above;

(b) The total quantity of ERUs, CERs, AAUs and RMUs in the retirement account of the Party for that commitment period.

60. Upon completion of the Article 8 review of the annual inventory for the last year of the commitment period, and the resolution of any related question of implementation, the secretariat shall record in the database the aggregate anthropogenic carbon dioxide equivalent emissions of the greenhouse gases, and from the sources, listed in Annex A to the Kyoto Protocol of the Party for the commitment period.

C. *Compilation and accounting reports*

61. The secretariat shall publish an annual compilation and accounting report for each Party included in Annex I and forward it to the COP/MOP, the compliance committee and the Party concerned.

62. After the commitment period and the additional period for fulfilling commitments, the secretariat shall publish a final compilation and accounting report for each Party included in Annex I and forward it to the COP/MOP, the compliance committee and the Party concerned, indicating:

(a) The aggregate anthropogenic carbon dioxide equivalent emissions of the Party for the commitment period as recorded under paragraph 60 above;

(b) The total quantity of ERUs, CERs, AAUs and RMUs in the retirement account of the Party for the commitment period, as recorded under paragraph 59 (b) above;

(c) Where applicable, the quantities of ERUs, CERs and AAUs in the registry available for carry-over to the subsequent commitment period;

(d) Where applicable, the quantity in tonnes by which the aggregate anthropogenic carbon dioxide equivalent emissions exceed the total quantity of ERUs, CERs, AAUs and RMUs in the retirement account of the Party for the commitment period.

Glossary

Annex I Countries: The countries listed in Annex I of the UNFCCC and the Kyoto Protocol.

Annex B Countries: Countries listed in Annex B to the Kyoto Protocol, which is based upon countries listed in Annex I to the UNFCCC and sets out the Quantified Emission Reduction and Limitation Commitment of each country.

Assigned Amount: The quantity of greenhouse gases that an Annex I country can release in accordance with Article 3 of the Kyoto Protocol, during the first commitment period of that protocol (2008–12).

Assigned Amount Unit: The unit quantifying Assigned Amounts, equivalent to one metric tonne of carbon dioxide emissions.

Baseline: The emissions of greenhouse gases that would occur without the contemplated policy intervention or project activity.

Carbon dioxide equivalent (tCO_2e): The universal unit of measurement used to indicate the global warming potential of each of greenhouse gases. Carbon dioxide—a naturally occurring gas which is a byproduct of burning fossil fuels and biomass, land-use changes, and other industrial processes—is the reference gas against which the other greenhouse gases are measured.

Certified Emission Reductions: A unit issued pursuant to Article 12 of the Kyoto Protocol for the purposes of the Clean Development Mechanism that is equal to one metric tonne of carbon dioxide equivalent.

Certification: is the written assurance by an independent third party that during the relevant period the Project has achieved the GHG Reductions as reported in the verification report.

Clean Development Mechanism: The mechanism provided by Article 12 of the Kyoto Protocol to assist developing countries in achieving sustainable development by permitting industrialized countries to finance emissions-avoiding projects in developing countries and receive credit for doing so.

Clean Development Mechanism Executive Board: The committee supervising and monitoring the implementation of Article 12 of the Kyoto Protocol.

Conference of the Parties (COP): The Conference of the Parties to the United Nations Framework Convention on Climate Change. The Kyoto Protocol was the outcome of COP3 in 1997, and the Marrakesh Accords were the outcome of COP7 in 2001.

Conference of the Parties serving as the Meeting of the Parties: The Conference of the Parties to the UNFCCC serving as the Meeting of the Parties to the Kyoto Protocol. The body serving as the decision-making body for the Kyoto Protocol before the Kyoto Protocol comes into force.

Designated National Authority: The national authority countries participating in the CDM are required to establish pursuant to the Marrakesh Accords.

Designated Operational Entity, or Operational Entity (OE): An independent third party accredited by the CDM Executive Board to validate CDM projects, and verify and certify emission reductions.

Emission Reduction Unit: A unit of ERs issued under Joint Implementation equal to one metric tonne of carbon dioxide equivalent.

European Emission Trading Scheme: The Scheme envisaged by Directive 2003/87/EC of the European Parliament and the Council of 13 October 2003 for imposing carbon dioxide emissions caps on installations in specified sectors and permitting the use and transfer of allowances for the purposes of achieving these caps as implemented in each EU Member State by national laws.

Greenhouse Gases: The six gases listed in Annex A of the Kyoto Protocol which are carbon dioxide (CO_2), methane (CH_4), and nitrous oxide (N_2O), as well as hydrofluorocarbons (HFCs), perfluorocarbons (PFCs), and sulphur hexafluoride (SF_6).

GHG Reduction: means the removal or mitigation of greenhouse gases by a project below the baseline emissions and measured in tCO_2e as monitored in accordance with a Monitoring Plan.

Host Country: The country where an emission reduction project is physically located.

International Emissions Trading: Mechanism provided by Article 17 of the Kyoto Protocol, which allows the acquiescence and transfer of AAUs.

Joint Implementation: Mechanism provided by Article 6 of the Kyoto Protocol, whereby an Annex I country may acquire emission reduction units from projects that reduce net emissions in another industrialized country (including countries with economies in transition (EITs)).

Kyoto Protocol: The Protocol to the UNFCCC setting out developed countries quantified emission limitation and reduction commitments, and establishing the three "Kyoto Mechanisms", Joint Implementation in Article 6, the Clean Development Mechanism in Article 12, and Emissions Trading in Article 17. Adopted at the third Conference of the Parties to the United Nations Convention on Climate Change held in Kyoto, Japan in December 1997, the Kyoto Protocol commits industrialized country signatories to reduce their greenhouse gas emissions compared with 1990 emissions, in the period 2008–12.

Marrakesh Accords: The set of rules agreed to by the Parties to the UNFCCC at the occasion of their seventh session, which provides additional implementation guidelines for the CDM.

Monitoring Plan: A set of requirements for monitoring and verification of GHG Reductions achieved by a project.

Project Design Document: A project-specific document required under CDM which will enable the OE to determine whether the project (i) has been approved by the parties involved in a project, (ii) would result in reductions of greenhouse gas emissions that are additional, (iii) has an appropriate Baseline and Monitoring Plan.

Quantified Emission Reduction and Limitation Commitment: Quantified Emissions Limitation Reduction Commitments. These are the target GHG emission reduction limitations that industrialized countries agreed to be bound to under the Kyoto Protocol.

Registration: The formal acceptance by the CDM Executive Board of a validated project as a CDM project activity.

Small-scale Projects: Projects which are compatible with the definition of "Small-scale CDM Project activities" set out in decision 17/CP.7. of the Conference of Parties to the UNFCCC.

United Nations Framework Convention on Climate Change: The multilateral Treaty adopted in June 1992 that commits Parties to stabilize anthropogenic (i.e., human-induced)

GHG emissions to levels that would prevent dangerous anthropogenic interference with the climate system. In December 1997, the Parties to the UNFCCC adopted the Kyoto Protocol.

Validation: The assessment of a project's design, including its baseline, before the implementation of the project against the requirements of the CDM, by an independent third party.

Verification: The periodic independent review and ex post determination by an independent third party of GHG Reductions that have occurred during the relevant period as a result of a project activity.

Verification Report: A report prepared by an independent third party, pursuant to a Verification, which reports the findings of the Verification process, including the amount of GHG Reductions that have been found to have been generated.

Index